WATER QUALITY INSTRUMENTATION

By Mark Spencer

Original Illustrations by Leah Spencer

2022

Water Environment Federation
601 Wythe Street
Alexandria, VA 22314-1994 USA
www.wef.org

ISBN: 978-1-57278-358-4

About WEF

The Water Environment Federation (WEF) is a not-for-profit technical and educational organization of 30,000 individual members and 75 affiliated Member Associations representing water quality professionals around the world. Since 1928, WEF and its members have protected public health and the environment. As a global water sector leader, our mission is to connect water professionals; enrich the expertise of water professionals; increase the awareness of the impact and value of water; and provide a platform for water sector innovation. To learn more, visit www.wef.org.

Contents

List of Figures

List of Tables

Preface

In late 2009 I purchased AquaMetrix. I had a PhD in physical chemistry and worked for years at NASA in molecular spectroscopy. Surely going from spectroscopy to electrochemistry should not be all that hard. So, what if these two fields are as close as a jet fighter is to a bicycle. They're still categories of physical chemistry, and with the help of the internet and a few trade rags, I would be up to speed in about a month.

Several years later I am still learning. My biggest challenge was the proliferation of explanations I've read in white papers, technical articles, and those marvelous trade rags. They offer an incredible wealth of information, and they are the best avenue for anyone to learn about our industry. But I repeatedly encountered statements in an otherwise informative article that didn't make sense.

For example, the Nernst equation gives the potential of an oxidation-reduction reaction occurring on an electrode's surface. This is totally appropriate for an ORP probe but not for a pH probe. There are no oxidation-reductions reactions on the surface of a pH electrode—just a bunch of hydrogen ions creating a positive charge. I wasn't the only one confused. One author of a paper in a respected academic journal declared that the Nernst was invalid for describing a pH probe and characterized it as a glorified capacitor (Cheng, 1989). What a coincidence that the Nernst equation actually predicts very accurately the output of the pH probe! Because I could not find a proper treatment of the Nernst equation applied to the workings of a pH electrode, I formulated my own. No one *needs* to read it, but, if you like to question "matters of fact" like I do, you might enjoy it.

Then there's the dissolved oxygen sensor. It measures the concentration of oxygen molecules in a medium that, at room temperature, is saturated to the tune of 8.4 parts per million (ppm). Our atmosphere is about 20% oxygen—or 200 000 ppm. So how is it that a dissolved oxygen probe in air gives the same reading as a dissolved oxygen probe in water? The answer is not intuitive, but it does come from an 18th century theory (Henry's law) that sounds patently obvious until you think about it. Then it sounds absurd. More thinking then makes it sound correct.

When I began writing this book, I wasn't clear on the type of book it would be. Would it be a textbook for engineers and academics? Or would it be a glorified manual for operators and technical support workers? I'm still not sure. Nor am I sure that it has to fit squarely into one category or the other. I wrote it because I was convinced that there must be others like me who want to have a firm understanding of how sensors work but also want some practical knowledge that only comes with working with them.

I firmly believe that, whether you sell, service, develop, or market a product, the more you know, the more effective you are. Reading the manual for one product model will help you understand that particular model. Understanding how that product works will enable you to understand it and every other product like it. If you operate equipment, then that understanding will enable you to switch from one manufacturer's instrumentation to another—without reading the manual. You become the company "expert." If you sell equipment, then you are better able to understand your customers' needs and suggest solutions—without calling an inside salesperson or field service representative. You close more sales faster.

Do I really believe that all salespeople or technical support people will flock to the Water Environment Federation bookstore to bone up on the dynamics of redox reactions on the surface of an electrode? Well, no. Maybe a few—maybe just those who wanted to be superstars. With that in mind I wrote the book in such a way that it would be a comprehensive explanation of how sensors work but that one could go right to the topic of interest without reading the tome cover-to-cover. Material that is more academic than practical is shaded in gray so that the intellectually curious can read this so-called sidebar while everyone else can just skip over it.

True understanding has a habit of making everything seem simple. Everyone who comes to that state of nirvana will admit to struggling with a set of disparate facts before coalescing them together into a cohesive whole. Time cements understanding but that time can be shortened with a well constructed narrative. My goal was to write that narrative.

While the book does some deep dives into chemistry, I've tried to write it so that it can be understood by anyone with a college level understanding of science. We all know the saying "Give a man a fish and he eats for a day. Teach a man to fish and he eats for a lifetime." Translated into our industry: "Give a man a manual for a probe and he can make the probe work. Teach him how the probe works, and he can figure out what went wrong when it breaks."

Let's take an example near and dear to our hearts. At the heart of wastewater treatment is the nitrification–denitrification cycle. One set of bacteria convert ammonia to nitrite. Another set convert nitrite to nitrate. And, in the absence of dissolved oxygen, a third set convert the nitrate into nitrogen, which floats away into the atmosphere. I could tell you that this process is a reduction reaction that requires a readily oxidized chemical, such as methanol, in an acidic medium to produce nitrogen and water. Or I can simply write:

$$6\ NO_3^- + 5\ CH_3OH \rightarrow 3\ N_2 + 5\ CO_2 + H_2O + 6\ OH^-$$

Basic knowledge of chemical reactions enables us to glean from this reaction, in short order, the following:

1. One needs a carbon source to "feed" the reaction. In this case it's methanol, but it could just as easily be corn syrup or vinegar.
2. Add 5 parts of methanol to 6 parts of nitrate or—with a little arithmetic—about 3 grams of methanol to every gram of nitrogen.
3. If there is any oxygen present, then it will oxidize the methanol, and the nitrate won't be reduced to nitrogen.
4. Denitrification will result in a rise in pH.
5. The release of carbon dioxide will build up alkalinity, which will limit the rise in pH.

All of this information came from writing out one chemical reaction. If this doesn't excite you, then close the book and give it to a coworker.

I'd like to acknowledge the help of several authorities in instrumentation in our field of water quality instrumentation. Though I consider myself well versed in this field, I would be guilty of chutzpah on a monumental scale if I fancied myself an expert in every corner of instrumentation. I am unusually fortunate that some of the foremost authorities in different sensor technologies gave their time. I am moreover pleased that many of them come from different manufacturers. They can all be considered to be my competitors, but I believe that they are my colleagues.

No one understands chlorine measurement better than Vadim Malkov of Hach Corporation. Vadim's suggestions were as thorough as his deep knowledge. I appreciate his frankness. Gerhard Reiss, the good doctor or Dr. Reiss, answered innumberable questions. Terry Engelhardt is another Hach guru that left no misstep hidden. CD Feng, of Rosemount, has an uncanny ability to break down the most arcane material and make it understandable. I can only hope that my attempts to explain technical matters is as lucidly as his. Mike Sadar is arguably the leading expert in turbidity. His experience traverses not just the technical underpinnings of turbidity but the regulatory aspects. If one company can be said to be identified with dissolved oxygen more than any other that company would be YSI. Laura St. Pierre and her colleagues at YSI not only opened the door to that company's excellent products but also to other instrumentation in the Xylem universe. I thank Patrick Higgins for wielding his sharp pen on the dissolved oxygen chapter. Professor Shigeru Amemiya has made his research focus at the University of Pittsburgh groundbreaking R&D in voltammetry.

Kerim Kollu of RealTech helped me wrap my head around chemometrics applied to spectroscopy and Christoph Wagner of S-CAN read through that final part of the book to make sure I didn't display any flashes of ignorance.

And, lastly, is my daughter, Leah, the most talented artist in a succession of artists, who gave up hundreds of hours of time to make her Dad's book look good.

1.0 REFERENCES

Cheng, K. L. (1989). pH glass electrode and its mechanism. In J. T. Stock & M. V. Orna (Eds.), *Electrochemistry, past and present*. American Chemical Society.

1

Introduction

According to the World Health Organization and UNICEF, 2.2 billion people—one third of the global population—do not have access to clean drinking water (UNICEF & World Health Organization, 2019). By 2030 half of the world's population will live without enough clean water.

Water is a $500 billion industry that is growing at a compound annual growth rate of 6% to 7% (Nahal & Lucas-Leclin, 2012). Water undergoes treatment, and instrumentation is key to every aspect of the process. Whether the process occurs in a municipal water treatment plant, a municipal wastewater treatment plant, industrial wastewater treatment, or part of a manufacturing process, instrumentation underlies every facet of operation. To control a process, one must measure it.

Instrumentation for water quality can be roughly broken down into two categories: real time measurements that are used for process control and laboratory instrumentation. As a rough distinction, laboratory equipment typically measures specific analytes or classes of analytes, such as organic compounds, heavy metals, and microorganisms. Measurements are made on a periodic basis to ensure that the water meets U.S. Environmental Protection Agency or state standards. The instrumentation is typically sophisticated and includes gas chromatography, mass spectrometry, optical emission spectroscopy, or atomic absorption.

Instrumentation for process control encompasses mostly the measurement of bulk parameters, and the measured parameters are fed into control mechanisms that maintain the treated water within boundaries; for example, dispensing caustic to maintain a pH above, say, 8.0. These bulk parameters include pH, oxidation-reduction potential (ORP), conductivity, turbidity, suspended solids, and biological oxygen demand. However, dissolved oxygen and even pH, which relates the concentration of H^+ ions, are specific analytes, and the various chlorine measurements—for example, free chlorine—include a small subset of chlorine species. Ionic compounds such as ammonium, nitrite, nitrate, fluoride, and orthophosphate are being incorporated into process monitoring and control to an ever-increasing degree. This book focuses on the bulk parameters mentioned above but includes a description of the more recent analyte specific instrumentation.

Process control means that either the probe is wired to an analyzer, which is armed with relays to control the process, or the probe is wired to a transmitter, which passes along the parameter value to a higher-level process control system such as a programmable logic controller (PLC) or supervisory control and data acquisition (SCADA) system.

Finally, instruments that measure physical parameters, including temperature, pressure, flow, and level, are a vital part of process control but are not included in this book.

To understand how a sensor works one must understand the science that drives it. The chemistry that goes on in water is very different from the chemistry that goes on in gas, such as the atmosphere, or in a solid, or for that matter, a liquid that is not water. Water's unique properties manifest themselves in chemical reactions that involve charged chemical species.

This chapter reviews the science—particularly chemistry—that underlies the workings of water quality instrumentation

1.0 DEFINITIONS

Precision in terminology is essential if we are all to be on the same page. In addition to acronyms and cryptic initials, even simple English terminology can mean two different things to different people. For instance, the term *transmitter* can mean any piece of electronics that connects directly to a probe and sends a signal representing the process parameter being measured to another piece of electronics. A transmitter might also mean an alternating current–powered unit that sends power to the probe, receives the probe output, converts it into a signal, activates relays to control the process, and retransmits the signal in either digital or analog form to a second device. Or transmitter might mean a piece of electronics that is powered by a

4-20 megaamperes (mA) loop, reads the probe signal, and retransmits the process signal.

This book adheres to several conventions:

- Every piece of analytical equipment measures something. That something is referred to as the *parameter*. The parameter may be physical, such as flow, or chemical, such as dissolved oxygen, or an aggregate measurement, such as conductivity. When the parameter is a specific chemical entity, such as dissolved oxygen or ferric ion, it is referred to as an *analyte*. An analyte may be a molecular compound (carrying no net charge), or it may be a charged ion.
- An electrode that has a negative charge sends electrodes to the rest of a circuit (as a result of oxidation) is called an *anode*. An electrode that has a positive charge (as a result of reduction) receives electrons from a circuit is called a *cathode*. A particularly useful mnemonic for remembering is the acronym ACID—"Anodic Current Into Device."
- The term *process* denotes any aqueous solution being measured. It can be a metal plating solution, or it can be the water coming out of a tap. One common synonym of "process" is "sample."
- In this book, the terms *sensor* and *probe* are used interchangeably. They refer to any transducer that measures a parameter and returns an electrical signal. The actual distinction is that a probe is a sensor that is inserted into the process.
- Most of the sensors discussed in this book are *electrochemical*. This means that the parameter being measured creates a direct electrical signal. That signal can be in the form of a voltage, current, or a resistance. A *potentiometric* sensor produces a change in voltage, an *amperometric* sensor produces a change in current at a fixed voltage, and a *conductometric* sensor produces a change in resistance. A *voltametric* sensor measures current as a function of a changing voltage. Voltage, current, and resistance are tied together via Ohm's law:

$$V = I \times R \tag{1.1}$$

- The term *potential* means the electric potential of any object that arises from a collection of charges. Any one point in space can be characterized by a potential. Its unit of measurement is *volts*. The term *voltage* denotes the difference between two electric potentials. One of those potentials could be earth ground. For instance, in our discussion of ORP, we discuss how to construct a complete electrochemical cell from two so-called half-cells. Each half-cell has a potential, but each does nothing until they electrically connect with each other. The sum of the two potentials—one positive and one negative—yields the voltage of the complete cell.

- An *analyzer* is loosely defined for any electronic device that powers the probe, reads the sensor input, and processes that input into an analog or digital signal that represents the magnitude of the parameter being measured. It may or may not contain relays for process controls, and it may or may not transmit the measured parameter.
- A *transmitter* is an analyzer that is either loop or direct current–powered and transmits either an analog or digital signal to another device. The other device performs process control by switching on and off a relay or a feedback signal (e.g., a 4-20 mA signal or a proportional, integral, derivative signal) to a pump or other device that actually controls the process. In the water industry, most transmitters output a 4-20 mA signal to a PLC.
- A *controller* is a transmitter that is separately powered and that contains relays or other process control elements. A sensor and a controller comprise a complete, self-contained process control system.
- For measurement, this book uses International Standard (SI) units. This includes the CGS system of centimeters, grams, and seconds or the alternate MKS system—meters, kilograms, and seconds. There are instances when the book uses British units such as gallons to denote the capacity of a treatment plant in gallons per day (gpd).
- Three units are used for measuring concentrations: Concentration by weight is expressed as *mg/L* (milligrams per liter) or *ppm* (parts per million). Although ppm appears to denote concentration in terms of numbers of molecules, it really is a weight ratio. The reason ppm is identical to mg/L arises from the fact that 1 cm^3 of water weighs 1 g. To denote concentration in terms of number of molecules, ions or atoms the term *moles/liter* or just *M* is used.
- An *analyte* is any chemical species that is measured by an analyzer. In an aqueous solution it is likely to be an ion. However, it can be a neutral molecule such as hypochlorous acid (HOCl) aka "free chorine." Conductivity and ORP are not analytes. They are properties of a water sample and arise from individual analytes, each at a different concentration.
- There is a big difference between *precision* and *accuracy*. To understand that difference, suppose you have two rulers. One has markings every quarter inch but the rulings that denote the distances are right on the mark. This ruler gives measurements that are accurate but, with

a full centimeter between markings, they are not precise. The other ruler has markings every millimeter, and the distances between these fine markings are also right on the mark but the zero of the ruler is indistinct. This ruler will give you measurements that are very precise but not accurate. Similarly, a pH analyzer that is out of calibration will still maintain the same precision that it had when it left the factory, but the measurement obtained is not accurate. (The term *reproducible* usually means the same thing as precise.)

2.0 THE ELEMENTAL TRUTH

To understand chemistry, one must understand elements and atoms. To that end the Periodic Table of the Elements, reproduced in Figure 1.1, is the repository of everything you need to know.

There are two numbers associated with each element of the Periodic Table. The number on the top of the element's box is the *atomic number*, and it is the number of protons or electrons for each element. I say "or" because the number of protons has to equal the number of electrons. If there

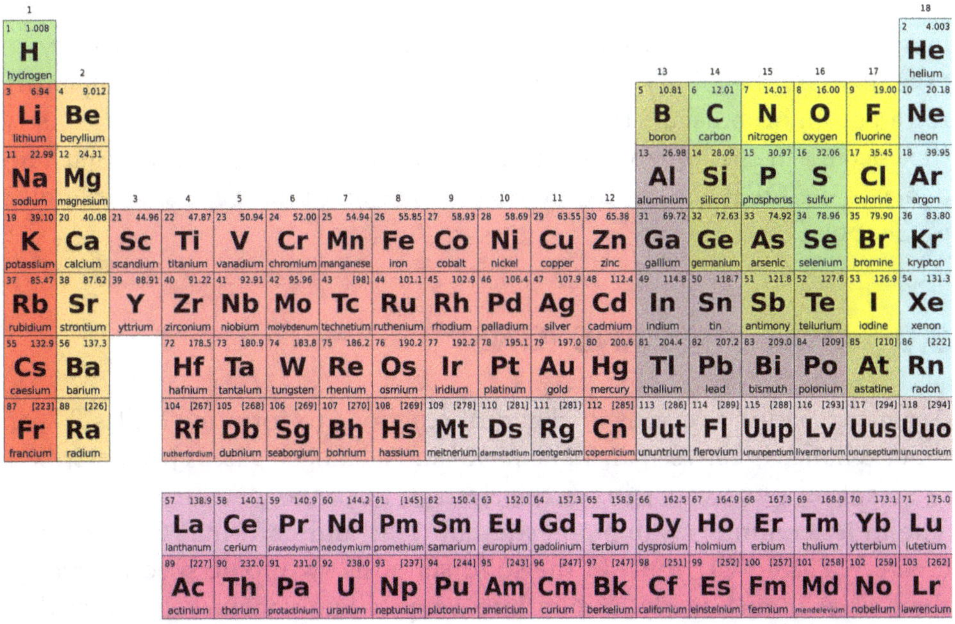

FIGURE 1.1 Periodic table of the elements (VectorStock). Put this on your wall, and you will be warmly welcomed into geekdom.

is one less electron, then the atom becomes a positive ion with a charge of +1. If the atom takes an additional electron, it becomes a negative ion with a charge of −1. The number on the bottom is the *atomic mass*, and it gives us the information we need to weigh out chemical compounds to just the right amount needed to do a job.

The placement of an element in the Periodic Table characterizes its tendency to react with other elements. The electrons in an atom are arranged in *shells*, much like the layers of an onion. The reactivity of an element is determined by the electrons in the outermost shell or shells. The left-most column consists of elements whose outermost shell contains only one electron. Sodium and potassium are the most common. Atoms prefer full shells, so the elements in this column—*alkali* metals—have a strong propensity to give up that lone element in a chemical reaction. The right-most column contains the *noble* gases, such as chlorine and fluorine. Their outer shell is completely filled and so "desire" neither to give nor take electrons. The column one over from the right contains the *halogen* elements. Their outer shells are one shy of a full shell. For them the perfect partner with which to react is an element from the alkali metals. The result is an ion pair, in which the alkali metal is now minus one electron and carries a single positive charge and the halogen ion has an additional electron and carries a single negative charge. This is a *salt*; it exists as a crystal in the solid state and breaks up into the constituent ions when dissolved in water.

While going from left to right across the Periodic Table, elements go from a strong propensity to give up electrons to a strong propensity to take them. In the middle that tendency is more ambiguous. The propensity for an atom or molecule to give or take electrons underlies most chemistry that occurs in the aqueous state. The parameter that quantifies that tendency is the ORP and forms the basis of the first sensor we shall examine in Chapter 2.

3.0 MOLECULES AND IONS

Atoms exchange electrons in chemical reactions to form molecules. When that exchange is between an alkali metal and a halogen gas, as just described, the result is a one-way trip for the electron from the metal to the halogen. The most common example is sodium (Na) and chlorine (Cl) atoms to create Na^+ and Cl^- ions. In the solid state the two ionic species arrange themselves in an orderly fashion into a crystalline structure. In water they simply go on their way as individual ions surrounded by water molecules.

Just one column over from the alkali metals are the alkaline earth metals, such as magnesium and calcium. They have two electrons in their outer shell and have a strong desire—though not as strong as the alkali metals—to give up both of them. They also form salts with the halogens though it takes two atoms of the latter to "balance the books," hence calcium chloride, $CaCl_2$. One column to the left of the highly reactive halogen gases—the most notable being oxygen—are the elements that are two electrons shy of a "full deck." Though these elements have strong tendencies to find two electrons from donors, there is not the reciprocity with the alkaline earth metals we might expect.

Between the alkaline earth and halogens lie the bulk of elements in the Periodic Table. The tendency to ship off an electron from one atom to another is replaced by a varying tendency of atoms to share electrons. That sharing of electrons gives rise to the very strong *covalent* bond, which keeps the resulting compound intact even when dissolved in water. The degree to which atoms share electrons determine the form of the resulting compound. For instance, one carbon atom (C) can partner with two oxygen (O) atoms to create carbon dioxide, CO_2, in which the electrons are shared but are pulled slightly toward the two oxygen atoms. CO_2 is a gas consisting of individual CO_2 molecules at room temperature. On the other hand, one silicon (Si) atom can bond with two oxygen atoms to create silicon dioxide (SiO_2), but the O atoms have so much greater pull on the electrons that the electrons are pulled much more strongly toward them, such that the resulting silicon dioxide molecules form a crystalline solid at room temperature—glass.

The reader will note that we use standard chemical notation to describe chemical species. The *molecular mass* or, as it is more commonly called, the *molecular weight*, is simply the sum of the masses of the individual atoms. A water molecule consists of one atom of oxygen and two of hydrogen. From the Periodic Table we know that the atomic mass of hydrogen is approximately 1 and that of oxygen is 16. That gives us a molecular weight of 18. (I say "approximately" because most elements consist of a mixture of atoms with differing numbers of neutrons, with each configuration of neutrons constituting an *isotope*. The atomic weight is a weighted average of all possible isotopes.

The dissolution of salts into positive and negative ions in water occurs thanks to the special properties of the ubiquitous but unique water molecule. Each water molecule consists of two hydrogen atoms and one oxygen atom. The hydrogen atom is eager to donate its one and only electron. An oxygen atom, on the other hand, occupies the column immediately adjacent to that of the halogen elements and, though not with the same voracious appetite as its halogen neighbors, is nonetheless eager to pull two electrons into its outer electron orbit. When two hydrogen atoms and one oxygen atom combine to form a water molecule, the two shared electrons spend most of their time in the vicinity of the oxygen atom. That leaves the two hydrogen atoms with a net positive charge and the oxygen atom with a net negative charge. The water molecule is neutral, but the H atoms carry a partial positive and the O atom has a partial negative charge. A molecule with charge separation is called *polar*.

When a sodium chloride (NaCl) crystal dissolves in water, the positively charged Na^+ ions lower their energy by coupling with the negatively charged oxygen atom while the negatively charged chloride ion lowers its energy by coupling with one of the two hydrogen atoms. The result is a solution of water with positively charged sodium ions—cations—and negatively charged chloride ions—anions. The two ions form complexes with the polar water molecules and are referred to as *solvated* ions (see Figure 1.2). Ions in aqueous solution under the influence of an electric field form electrical currents just as the electrons that flow through metallic conductors, such as copper.

A cation is a positively charged ion. We denote an ion that has lost one electron to become a singly charged cation via the superscript "+" (e.g., Na^+). An anion is a singly charged negative ion and is denoted by a superscript "−" (e.g., Cl^-). Both species are examples of *monovalent* ions. To denote an ion that carries more than one charge—for example, calcium that has lost electrons—some writers denote the resulting *divalent* ion as Ca^{+2} and some as Ca^{2+}. I shall use the latter.

Another convention that needs to be defined is that for writing the formula of a chemical compound that is made up of more than one of the same atoms, such as ethanol, which has two C atoms, five 6 H atoms, and one O atom. One could write this as C_2H_6O, but it is more common to write out the sequence of atoms in the actual molecule, or CH_3CH_2OH. Figure 1.3 shows

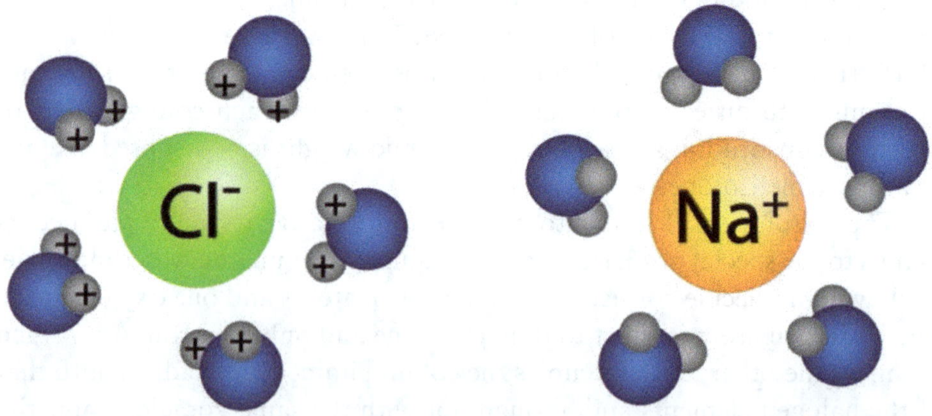

FIGURE 1.2 NaCl dissolving in water. Negatively charged Cl^- ions are attracted to the positively charged H atoms while the positively charged Na^+ ion settles next to the negatively charged O atom.

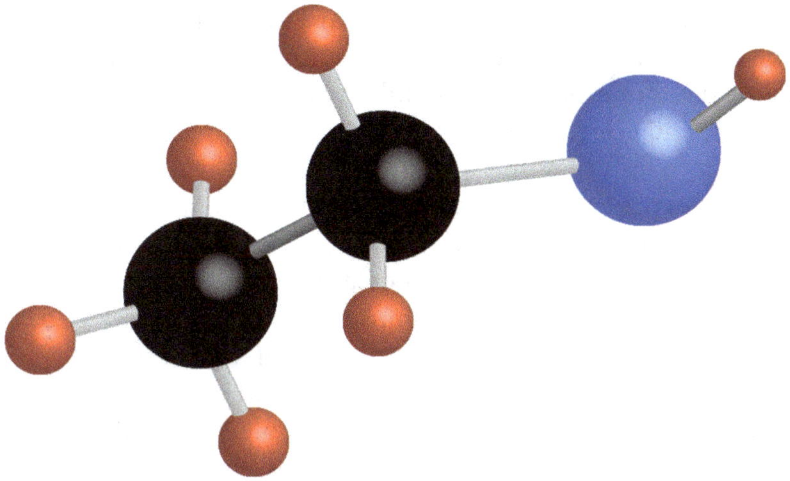

FIGURE 1.3 Atomic structure of an ethanol molecule. The carbon atoms (gray) are attached to 5 hydrogen atoms (red) and 1 oxygen atom (blue). The latter attaches to one H atom. One can think of this structure as a backbone of 2 carbon atoms and 1 oxygen atom. Hence the formula CH_3CH_2OH.

the structure. On one end is a carbon atom that connects to three hydrogen atoms. In the middle is a carbon atom that connects to two hydrogen atoms and one oxygen atom. And, on the end, is the oxygen atom that connects to a lone hydrogen atom. The longer version of notation gives the reader an understanding of molecular structure, so it is the one I prefer to use.

4.0 CHEMISTRY REACTIONS

Chemistry is the study of reactions that take one set of chemical species, rearrange some of the constituent atoms (*reactants*), and convert them into new chemical species (*products*). The notation for writing out a chemical reaction is as follows:

$$aA + bB + \ldots \rightarrow cC + dD + \ldots \tag{1.2}$$

Capital letters denote the chemical constituent—either an atom, compound, or ion—and small letters denote the number of that constituent. The state of that constituent is often notated as a solid (s), liquid (l), gas (g), or dissolved species (aq for "aqueous"). Take, for example, the reaction between gaseous hydrogen and solid magnesium oxide to form dissolved magnesium hydroxide:

$$\mathrm{MgO(s) + H_2O(l) \rightarrow Mg(OH)_2(aq)} \tag{1.3}$$

All chemical reactions can go backward and forward to some extent. Often the arrow symbol is replaced by a double arrow to denote a reaction that can go in either direction (↔) under typical conditions of temperature, pressure, pH, and concentration of reactants and products.

In the discussion of some topics, I invoke a chemical reaction whose chemical species are not compounds in the traditional sense. A particularly important example is the description of a H^+ ion that attaches to the silicon-oxygen unit of the glass electron to create a "quasi" species, S-O-H^+. By recasting a common phenomenon in this way we avail ourselves to the tools that chemistry offers to explain the workings of a pH probe.

In the world of water, the presence of charged species opens up a range of possibilities that are not open to neutral molecules in the gaseous state (unless the temperature is so high that the gases form a plasma). These possibilities involve the trading of electrons and protons.

Let's take the simple reaction of carbon dioxide dissolving in water to form carbonic acid (or carbonation). It can be broken down into two steps:

$$CO_2 + H_2O \leftrightarrow CO_2 + H^+ + OH^- \leftrightarrow HCO_3^- + H^+ \leftrightarrow H_2CO_3 \quad (1.4)$$

The first reaction is a decomposition one in which an H_2O molecule dissociates into an H^+ cation and an OH^- *hydroxyl* anon. The hydroxyl anion, then combines with the CO_2 molecule in a synthesis reaction to create the *bicarbonate* anion. Finally, the H^+ ion combines with the bicarbonate ion to form carbonic acid.

The double arrows, just introduced, denote that the reactions go in both directions. Equation 1.4 shows that carbon dioxide dissolves in water to create an acidic solution. In fact, when you dissolve carbon dioxide in water you get a mixture of all species included in Equation 1.4: carbonic acid, bicarbonate, and carbon dioxide all at the same time. How much of each depends on the pH of the starting solution. Chapter 3 on pH delves into this balancing act. Equation 1.4 is the basis for carbonation and the increasing acidity of the oceans from rising CO_2 levels in the atmosphere.

Whether a reaction goes from left to right, right to left, or in both directions depends mostly on the change in energy and the relative concentration of the chemicals on either side of the reaction. Consider what happens when a strong acid, hydrochloric acid (HCl), mixes with a strong base, sodium hydroxide:

$$HCl + NaOH \rightarrow H_2O + NaCl \quad (1.5)$$

In this reaction hydrochloric acid and sodium hydroxide (*caustic*) form table salt, NaCl, and water. This one only goes almost exclusively as written.

One cannot mix salt and water and expect to see HCl and NaOH. One can, however, take a solution of salt, insert two electrodes, impose a voltage across the electrodes, and drive the reaction backward. The addition of the voltage makes the backward reaction energetically feasible. The next section picks up the backward reaction.

5.0 CHEMICAL RECIPES

Equation 1.5 describes what happens when hydrochloric acid mixes with caustic. More importantly it is a recipe that tells one how much of each chemical to mix. It informs us that we need one molecule of HCl for every molecule of NaOH. The ratio of one chemical reactant to another is called the *stoichiometric ratio*. In this case the stoichiometric ratio is exactly one but, in other reactions, it can be any ratio of two whole numbers.

If we wish to mix NaOH and HCl in quantities that ensure they both react completely, we need to ensure that both chemicals have the same number of molecules. This is where the *mole* comes in. One mole is the number of molecules—for example, atoms or ions that yield that chemical's weight in grams. For instance, HCl is composed of one H atom, which is 1.008 atomic mass units (amu), and one Cl atom, which is 34.454 amu. So, one HCl molecule weighs 1.008 + 35.454 amu, or 36.462 amu. One mole of HCl weighs 36.462. As long as we work with moles, we can get the stoichiometry of any reaction right. In this example, we weigh out 36.462 g of HCl and 59.452 g of NaOH to get 18.015 g of water and 58.444 g of salt.

By the way, one mole contains 6.022×10^{23} molecules, atoms, or ions, and is known as *Avogadro's* number. This is why you should tape the Periodic Table to your wall.

In our world of aqueous chemistry, we don't weigh out dry chemicals. We mix solutions. We therefore define a 1 M solution as one that contains one mole of a chemical species in 1 liter of water.

6.0 WHAT MAKES CHEMISTRY "GO?"

Let's pick up the discussion, started in the last section, of why some reactions are one-way trips and others go both ways. The answer is energy. The reactants—everything on the left side of the arrow—have a specific amount of energy and the products—everything on the right side—also have a specific amount. To be exact, this energy is the *Gibbs free energy* and is denoted as *G*. It is the energy that is available to do work. The unit of energy in the SI system is the *joule* (J) or the *calorie* in the SI (metric) system and

BTU in the U.S. system. The difference in energy between the products and reactants, ΔG, is the energy change of the reaction. If ΔG is positive, then energy must be put into the reactants to make them go. Without that input nothing happens. A reaction that requires energy input is *endothermic*. If ΔG is negative, then the reaction can proceed by itself. A reaction that releases energy is *exothermic*. Adding energy to a reaction is similar to pushing a ball up a hill. One must add energy to the ball to push it against gravity up a hill. Once on the top the ball rolls back down and releases energy.

Let's take up the reverse of Equation 1.5. Water treatment facilities and quite a few backyard pools make their own disinfectant by passing a current through a saltwater mixture. The products of the reaction are, however, not the starting HCl and NaOH. They are sodium hypochlorite, or *bleach* and hydrogen gas:

$$NaCl + H_2O \rightarrow NaOCl + H_2 \qquad (1.6)$$

This is *electrolysis*, and it is the basis of on-site hypochlorite generation. Since we have not yet discussed electrochemical reactions, Equation 1.6 sidesteps the oxidation-reduction reactions that occur between the start and finish of hypochlorite generation. The important point, though, is that energy in the form of an electrical current is required to make the reactions go. Without it the salt and water will remain forever as a salt solution.

Before moving on, let's take this opportunity to note a standard shortcut that the notation of chemical reactions involving ions make. Some ions participate actively in reactions, and some can best be described as "innocent bystanders." Equation 1.7 is such an example. The Na^+ ion remains a Na^+ ion so we lose nothing by leaving it out:

$$Cl^- + H_2O \rightarrow OCl^- + H_2 \qquad (1.7)$$

Ensuring that ΔG is negative is necessary to drive a reaction in the direction as it is written but it is not sufficient. Mixing hydrogen and oxygen together has a very large negative ΔG value.

$$2H_2 + O_2 \rightarrow 2H_2O \qquad (1.8)$$

But, at room temperature, nothing happens. However, create a spark in the container containing the mixture, and the container goes "boom." At the bottom of the container will be a few drops of water.

What just happened is analogous to the ball rolling down a hill. Only this time, there is a little hill on top of the big hill that is preventing the ball from rolling down all on its own. One only has to give it a little kick to get up over the little hill, and it will roll down the hill that is now made taller by addition of the little hill. An energy barrier prevents the ball or a chemical reaction from proceeding as we expect.

Every chemical reaction has some energy barrier. Dynamite doesn't explode unless we light a fuse, and closer to home, reactions on electrodes that underlie all of electrochemistry have their own barrier before electrons are released and captured. Figure 1.4 shows this hill in context of the overall reaction between carbon dioxide and nitrogen dioxide to create carbon dioxide and nitric oxide:

$$CO + NO_2 \rightarrow CO_2 + NO \quad (1.9)$$

On the left side are the reactants—CO and NO_2—and on the right side, at the bottom of the hill, are the products—CO_2 and NO. A hill with a barrier of 134 kJ separates the two. The origin of that hill is beyond the scope of this book, but the short answer is that the two reactants form a

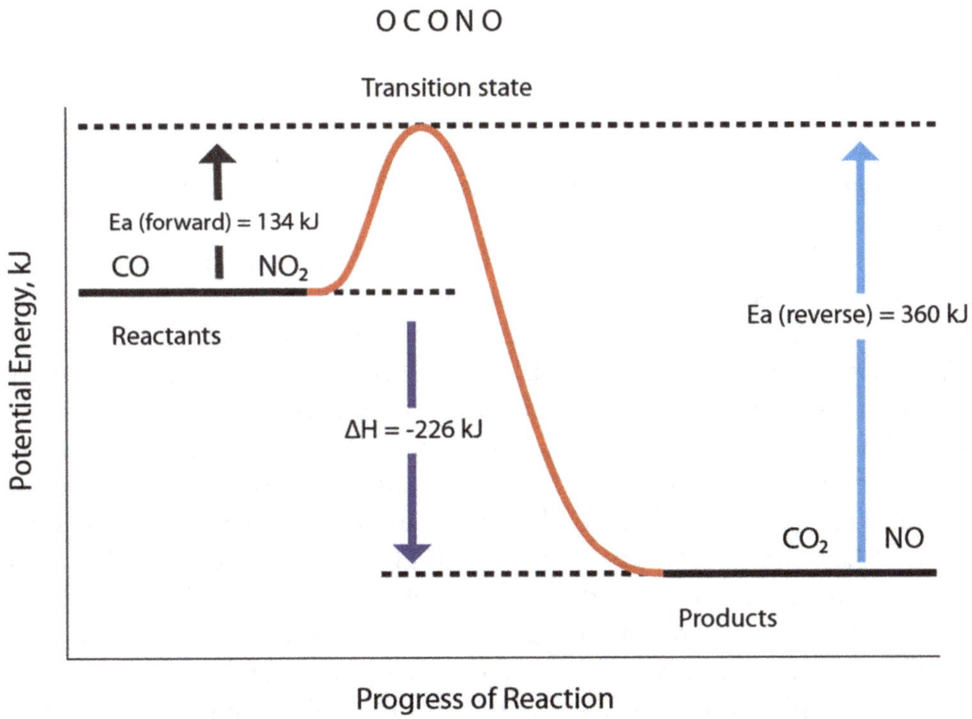

FIGURE 1.4 The reaction between CO and NO_2 to make CO_2 and NO. The reaction goes down hill but first it must go uphill before the big descent. Eact is the energy of activation and is the difference in energy between the top of the hill and the bottom on the product or reactant side.

chemical intermediary (aka *transition state*), OC-ONO, that can then break up into the final products.

The energy to go up the intermediate hill to create the transition state is called the *energy of activation*. Every reaction has one. We hear about catalysts in the context of automobile exhaust and our own bodies. A *catalyst* is a material, such as a platinum surface, that creates an easier-to-make intermediary and, thereby, lowers the energy of activation. It is important to remember that the energy of activation does not change the reaction's free energy change. For CO and NO_2, that energy release is 360 kJ/mole. What it does do is make the reaction go faster.

All reactions are fundamentally like the one portrayed in Figure 1.4. Some have such a small energy of activation that the energy of the reactants at room temperature is sufficient. Hence the reaction proceeds spontaneously—at room temperature. Others require significant energy, typically through heating, light absorption, or electricity.

Let's look at the reaction that occurs when carbon dioxide dissolves in water:

$$CO_2 + H_2O \leftrightarrow H_2CO_3 \leftrightarrow HCO_3^- + H^+ \leftrightarrow CO_3^{2-} + 2H^+ \quad (1.10)$$

The height difference between the left side and the right side, ΔG, is even less than the bunny hill in a New Jersey ski resort. So, the reaction goes both ways at room temperature. In fact, in any sample of water exposed to carbon dioxide, one will find every single species listed in Equation 1.10 dissolved in the water.

So far, the discussion of the direction in which chemical reactions proceed has assumed stoichiometric ratios of chemical reactants. For example, in Equation 1.10, there are exactly equal moles of CO_2 and H_2O. But what if this is not the case? Say we start this system at equilibrium with all the species listed in these coupled reactions remaining constant. Then we inject additional CO_2. The system restores itself back to a state of equilibrium by converting much of that CO_2 to HCO_3 and CO_3^{2-} ions. This driving of a chemical reaction in a particular direction by adding more of one of the species is the basis of *Le Chatelier's principle*. If one adds H^+ ions, then Equation 1.10 will restore equilibrium by proceeding from right to left and creating more CO_2.

What happens if one adds caustic, NaOH, to the carbonic acid mixture? At first glance NaOH, which dissolves to form Na^+ and OH^- ions, does not seem to play a role. However, OH^- anions will combine with H^+ ions to create H_2O molecules, thus leaving a shortage of H^+ ions. This will pull the reaction to the right and draw in CO_2 molecules to react and create bicarbonate and carbonate ions. Le Chatelier's principle is nature's push to restore systems of chemicals to a state of equilibrium.

We return to this topic in Chapter 6 when we use LeChatelier's principle to explain the important role of pH buffers like the carbonic acid mixture. Buffers seem to have a higher intelligence that allows them to resist outside forces trying to change their pH. In fact, it's just a simple law of nature.

With this information in hand, it's time to go back and look at the chemical equation introduced in the Preface and reproduced here:

$$6\ NO_3^- + 5\ CH_3OH \rightarrow 3\ N_2 + 5\ CO_2 + 7\ H_2O + 6\ OH^- \qquad (1.11)$$

Exchange reactions, of which this is one, account for the vast majority of chemical reactions in nature. They represent a chemical factory in which atoms enter the factory assembled as one set of molecules and leave packaged as new ones.

This reaction is something that every wastewater operator knows at some level. It is the denitrification of nitrate (NO_3^{2-}) to nitrogen (N_2). The methanol (CH_3OH) provides the necessary C atoms to make the reaction go forward. This chemical equation explains why an operator adds a carbon source, such as methanol, so that the denitrifying bacteria can do their job in the anoxic zone of a wastewater treatment plant.

The stoichiometry of this reaction—that is, the number of molecules and ions that balance out this reaction—is a lot more complex than any previous reaction in this chapter. Nevertheless, by knowing how to calculate the molar mass of each compound, we can use Equation 1.11 to weigh or measure out the right amount of methanol to mix into the denitrification chamber. Methanol has a molar mass of 32.042 g, and nitrate has a molar mass of 62.005 g. The stoichiometric ratio of methanol to nitrate is 5 to 6 means 5×34.042 g = 510.64 g methanol to $6 \times 62.005 = 372.03$ g nitrate. Equivalently, we add 1.372 g of methanol to every gram of nitrate. We only need to know the nitrate concentration of nitrate and the volume of the tank or chamber to calculate how much methanol to mix in.

This exercise should demonstrate that being able to read and understand a written chemical reaction yields more information than simply what goes in and what comes out of the reaction.

7.0 REFERENCES

Nahal, S., & Lucas-Leclin, V. (2012). ESG & sustainability: A blue revolution—global water. Bank of America Merrill Lynch.

UNICEF & World Health Organization. (2019). *Progress on drinking water, sanitation and hygiene: 2000–2017: Special focus on inequalities.*

2

Oxidation-Reduction Potential

1.0 WHY START WITH OXIDATION-REDUCTION POTENTIAL?

Section 3.0 in Chapter 1 introduced water's unique properties. The H_2O molecule is an example *par excellence* of a polar molecule; that is, one that has a high degree of charge separation—in this case between the two positively charged hydrogen atoms and the negatively charged oxygen atom. The resulting charge separation is higher than other compounds, such as polar molecules such as methanol (CH_3OH) or ethanol (CH_3CH_2OH). Water's high degree of polarity (its *dipole moment*) enables salts to easily dissolve. The ability of water to "host" ions opens up a world of chemistry that does not exist in air and in most other liquids. That aqueous environment is like a currency exchange consisting of only two currencies. Both are the two atomic particles that carry charge—the proton and the electron.

The proton is the same as the hydrogen ion, H^+, and forms the basis of pH measurement. The H^+ ion attaches itself to a water molecule to form the hydronium cation, H_3O^+, but as a matter of convenience, most writers use the shorter H^+ ion. We'll revisit this nomenclature shortcut again in the next chapter on pH.

In an aqueous environment, electrons attach to a molecule to make an anion. The tendencies of some atoms and molecules (e.g., sodium) to give away an electron and others, such as chlorine, to grab electrons result in exchanges of electrons and is the basis of electrochemical reactions. The potential of an electrochemical reaction to proceed is the basis of the oxidation-reduction potential (ORP) measurement. The potential of a dissolved compound or ion to release or take in protons is the basis of pH measurement.

Most books on electrochemical measurements start with the pH probe. After all pH measurements are everywhere and virtually all workers in the water industry are very much aware of pH. However, in order to understand pH, one has to understand oxidation-reduction reactions. For this reason, the first sensor that this book covers is the ORP sensor.

ORP is arguably the most mysterious measurement of all. pH measurements are everywhere. Conductivity is a measurement that anyone who studied basic electronics can readily understand. Dissolved oxygen and free chlorine refer to particular chemical species. Turbidity is a characteristic that anyone who peers into murky water readily understands. But there is little in the real world with which to connect one to ORP. Perhaps that is why pH sensors outsell ORP sensors in the water and wastewater industry by an order of magnitude. There are so many applications in the water industry in which an ORP sensor ought to be an indispensable tool and yet is overlooked. Nitrification, denitrification, disinfection, phosphorus removal, and sulfide reduction are just several examples in water treatment where an ORP probe is advantageous.

2.0 WHAT ORP MEANS

As discussed in Chapter 1, Section 3.0, some atoms, such as those belonging to the alkali elements, want to unload electrons, and some atoms, such as those belonging to the halogen elements, want to take on electrons. This concept of swapping electrons is common to all atoms and many chemical compounds.

In the chemical universe that exists in water, there are givers and there are takers, though a giver in one context can become a taker in another. The givers are atoms or compounds that give electrons to those in need, and the takers are those that take them. Electrons cannot just float in water. For every molecule or atom that donates an electron there must be another that can grab it.

The process of giving an electron is called *oxidation* and the process of taking one is called *reduction*. A chemical species that takes electrons is an *oxidizing agent* (or *oxidant*) while one that gives electrons is a *reducing agent* (or *reductant*). A species that is reduced is an oxidizing agent while one that is oxidized is a reducing agent. Though this nomenclature is logical, it can be confusing. In the example of sodium combining with chlorine, the Na atom is oxidized (i.e., is a reducing agent) while the Cl atom is reduced (i.e., is a oxidizing agent). Figure 2.1 shows this simple but celebrated redox reaction.

As previously mentioned, a molecule or atom can be oxidized when paired with one other species and yet be reduced by another. This is true of many metals.

A chemical reaction that involves an exchange of electrons is called an *oxidation-reduction* reaction. For brevity's sake this term is frequently shortened to *redox*. The propensity of an atom or molecule to give up one or more electrons is called the *oxidation potential*, and the propensity to gain an electron is called the *reduction potential*. If the reaction runs backward, the redox potential changes sign. Because one is the negative of the other, both terms are merged simply as *oxidation-reduction potential*, or ORP. At this point it is still unclear as to what "potential" really means.

One molecule is known more than all others for taking electrons—oxygen. The oxygen atom is two electrons shy of a full shell, so the oxygen molecule (O_2) wants four. Since the H atom has one electron to "give away," the hydrogen molecule (H_2) has two. Two H_2 molecules therefore have the

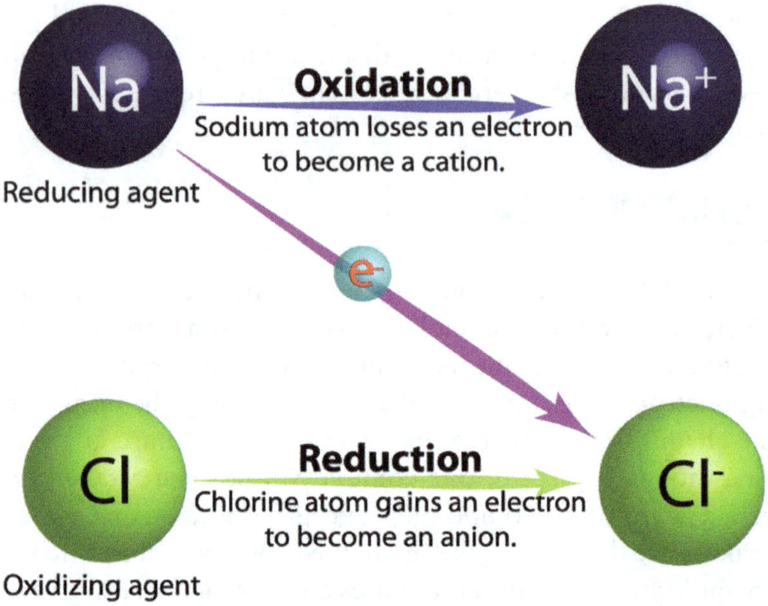

FIGURE 2.1 Schematic of an oxidation-reduction reaction. A Na atom gives up an electron to a Cl atom with the result that Na becomes the cation Na^+ and Cl becomes the anion Cl^- (chloride).

giving capacity to give four electrons to the O_2 molecule, and the chemical reaction for the swap is:

$$2\,H_2 + O_2 \rightarrow 2\,H_2O \tag{2.1}$$

Equation 2.1 hides the details of the electron swaps. It would be more illuminating if we were to break up this equation so that we can see this exchange take place. We break up Equation 2.1 into the sum of an oxidation reaction and a reduction reaction. Because two halves make a whole, they are called *half-reactions*. In this example the oxidation of hydrogen and reduction of oxygen is written as follows:

$$H_2 + 2\,H^+ \rightarrow 2\,e^- \tag{2.2}$$

$$O_2 + 4\,e^- \rightarrow 2\,O^{2-} \tag{2.3}$$

As Equation 2.1 shows, we need to double up the H_2 molecule in Equation 2.2:

$$2\,H_2 \rightarrow 4\,H^+ + 4\,e^- \tag{2.4}$$

The products on the right side of Equation 2.2 can then be added to the products on the right side of Equation 2.3 to formulate an alternate way of showing the formation of water from its constituent H^+ and O^{2-} ions.

$$4\ H^+ + 2\ O^{2-} \rightarrow 2\ H_2O \qquad (2.5)$$

For every redox reaction, there is an oxidized species on one side of the reaction and a reduced species on the other.

The convention for writing a half-reaction is to write it as a reduction, whether the reaction is, in fact, an oxidation or a reduction:

$$X^{n+} + ne^- \rightarrow X \text{ or } X + ne^- \rightarrow X^{n-} \qquad (2.6)$$

The oxidized and reduced species together make up one *redox couple*. For the oxidation of hydrogen, the redox pair consists of H_2 and H^+. For the reduction of oxygen, the redox pair consists of O_2 and O^{2-}. If the half-reaction proceeds as written—that is, from left to right—then the reaction is a reduction. If it goes from right to left it is an oxidation. Keep in mind that a redox couple may participate in an oxidation reaction when paired with a second redox couple, or it may engage in a reduction reaction when paired with another. Redox labels of "oxidation" and "reduction" are relative but, as the next section shows, they are predictable.

Because electrons do not float around all by themselves, there can be no oxidation reaction without an accompanying reduction reaction. It really does take two halves to make a whole. Whether it's love and marriage or oxidation and reduction, you can't have one without the other.

Let's look at another reaction involving oxygen—this time the oxidation of iron along with the reduction of oxygen. The half-reaction for the oxidation of iron (Fe) is:

$$Fe \rightarrow Fe^{3+} + 3\ e^- \qquad (2.7)$$

(Iron can give two electrons to form the *ferrous* Fe^{2+} cation or 3 electrons to form the *ferric* Fe^{3+} cation.) The reduction of oxygen is Equation 2.3.

For Equation 2.3 and its 4-electron withdrawal to match up with Equation 2.7 and its 3-electron release, we are going to have to combine four Fe atoms per three O_2 molecules so that a total of 12 electrons are exchanged. The stoichiometry of Fe to O_2 is 4:3. The overall reaction is:

$$4\ Fe + 3\ O_2 \rightarrow 2\ Fe_2O_3 \qquad (2.8)$$

This reaction is almost as ubiquitous as the formation of water. It is the formation of rust. Any similarity between the terminology *oxidation* and *oxygen* is not a coincidence. The former dates to the late 18th century when French scientist, Antoine Lavoisier, discovered oxygen and its propensity to rust iron. Lavoisier coined the term *oxidation* to denote the addition of oxygen.

Since the 18th century chemists have uncovered thousands of reactions in which oxidation is brought about by compounds other than just oxygen. In the formation of rust, iron atoms are oxidized. But, as previously cautioned, a chemical species can be oxidized when paired with one species and reduced when paired with another. Let's see what happens when iron meets zinc (Zn). Let's start with the ferrous ion, Fe^{3+}, as created in Equation 2.3. The Fe^{3+} atom will steal one electron from a neutral Zn atom to make one Fe^{2+} ion and one Zn^+ ion.

$$Fe^{3+} + Zn \rightarrow Fe^{2+} + Zn^+ \tag{2.9}$$

The ferrous ion and zinc atoms swap one electron. Notice that Equation 2.9 doesn't tell the whole story—there are no anions to balance the Fe^{3+} on the left and the Fe^{2+} and Zn^+ on the right. The Fe^{2+} cation is reduced—that is, is the oxidizing agent. In fact, zinc is so easily oxidized that manufacturers often coat metals such as iron, steel, copper, or aluminum with zinc. As long as the zinc is present it will oxidize and spare the underlying metal. When the zinc is depleted, workers recoat the protected metal with more zinc. This is the process known as *galvanizing*, and zinc is known as a *sacrificial anode*.

Before going further it's important to make an important distinction that is exemplified by the two common redox reactions described above. If one takes a gaseous mixture of hydrogen and oxygen and ignites the mixture with a spark, then Equation 2.1 results with a bang. The two half-reactions of Equation 2.2 and Equation 2.3 do not really occur as individual reactions. The H_2 molecules transfer their electrons directly to the O_2 molecules. The only way the two half-reactions can occur in the real world is to separate them so that each occurs in its own aqueous solution and the transfer of electrons happens through a conductor that joins the two. There is a common device that actually does this—a hydrogen fuel cell. Strictly speaking, a reaction is not an electrochemical one unless the transfer of electrons is a physically distinct process.

Separating redox reactions into separate oxidation and reduction reactions pays greater dividends than just illustrating what happens in an electrochemical reaction. The thousands of reactions that result from all the possible combinations of elements and compounds can be reduced to a manageable set of oxidation-reduction half-reactions. One can combine any two half-reactions to make one whole reaction. However, there is an ambiguity: A half-reaction can be an oxidation reaction when paired with another half-reaction but, when paired with another reaction, it can switch roles and become a reduction reaction. So, it is clearly impossible to label any reaction as unequivocally oxidation or reduction.

3.0 THE ELECTROCHEMICAL CELL—TWO HALVES MAKE A WHOLE

The two examples or redox reactions discussed above—the formation of water and the oxidation of iron—occurred in an environment in which the oxidant and the reductant were mixed together, and the transfer of electrons is invisible. To study the two half-reactions, we need to separate them. Let's imagine that the oxidation of sodium to Na^+ cations occurs in a chamber, and the liberated electrons travel through a conductor—say a wire—and enter a second chamber containing chlorine. There they reduce the chlorine gas to Cl^- anions. Of course, this all happens in an aqueous environment where the two ionic species can exist.

So far this is just a thought experiment so let's make it real. Instead of highly unstable sodium and deadly chlorine gas, let's use two common metals—zinc and copper. The zinc will be our choice for a metal that oxidizes to create Zn^{2+} ions and electrons.

$$Zn(s) \rightarrow Zn^{2+} + 2\ e^- \qquad (2.10)$$

All we need to do is to take a strip or rod of zinc and immerse it in a container of water into which the Zn^{2+} cations dissolve. Let's next take a strip or rod of copper and immerse it in a second container of a dissolved copper salt, say, copper sulfate, $CuSO_4$. The latter will dissolve into Cu^{2+} cations and inconsequential SO_4^{2-} anions. The Cu^{2+} cations can take up electrons and reduce to Cu, thereby adding to the copper electrode.

$$Cu^{2+} + 2\ e^- \rightarrow Cu(s) \qquad (2.11)$$

All we need to do is to connect the zinc anode and copper cathode with a wire so that the electrons liberated from the oxidation of zinc can make their way to the cathode where they can reduce the copper ions.

Each container hosts a half-reaction. The container is appropriately called a *half-cell*. The aqueous solution in either half-cell is called an *electrolyte*. This configuration of two half-cells is called a *galvanic cell*. Sometimes the term *Voltaic* is used, but some interpret a Voltaic cell as a single chamber containing both anode and cathode.

There's only one problem. As the zinc electrode oxidizes to produce Zn^{2+} ions, the electrolyte in this oxidation half-cell starts to accumulate a positive charge. In the other half-cell, electrons from the reduction of Cu^{2+} ions leave behind a net negative charge in the solution. The build-up of opposing charges in both cells creates an electromotive force, which opposes the two half-reactions and brings them to a halt very quickly.

The only way to restore electrical neutrality is to add a device that allows ions to freely roam between the two half-cells. That device is a *salt bridge*. In its simplest configuration it can be a U-shaped tube that connects the two half-cells. We might be tempted to simply fill the salt bridge with water and let ions from both chambers mix with each other to achieve charge neutralization but that would present a problem: the Cu^{2+} ions from the cathode side would migrate over to the anode side to balance out the increasing concentration of Zn^{2+} ions. That would deplete the pool of Cu^{2+} ions available for reduction.

A better solution is to fill U-tube with a high concentration of a salt-like potassium chloride, KCl, and plug both ends with a porous material, called the *porous junction*. Cotton works well for this benchtop setup. Why KCl? As we'll see in our discussion of conductivity in Chapter 4, the K^+ and Cl^- ions are roughly the same size and travel at about the same rate through water. As Zn atoms on the anode oxidize to Zn^{2+} cations, Cl anions diffuse through the porous junction into the anode chamber to neutralize the growing positive charge. As Cu^{2+} anions withdraw electrons from the copper cathode and become copper atoms, K^+ cations from the salt bridge flow into the cathode chamber to neutralize the growing negative charge.

Figure 2.2 illustrates the electrochemical cell. A light bulb is inserted between the anode and cathode to drive home the point that the flow of electrons from one to the other does electrical work. We could have just as easily inserted a voltmeter to show that the sum of the two half-reactions creates a voltage. It should be obvious by now that the electrochemical cell is a battery. The voltage of the battery is the sum of the oxidation and reduction reaction potentials. For this electrochemical cell in which the concentrations of Cu^{2+} and Zn^{2+} ions are both 1 M, the voltage is 1.1 V. More on this later in our discussion of standard state potentials.

To measure that potential in an electrochemical cell, one inserts a voltmeter between the two half-cells. An ideal voltmeter draws a negligibly small current—that is, has a nearly infinite resistance (or impedance). This is necessary because a current draw will result in a voltage drop across the voltmeter and an artificially low voltage reading. A *galvanometer* is a very accurate device that measures the potential difference by imposing its own

FIGURE 2.2 An electrochemical cell is composed of two half-cells. In the half-cell on the left zinc atoms in the anode oxidize to Zn^{2+} ions and release electrons, which travel over the wire to the half-cell on the right. In that cell Cu^{2+} ions withdraw electrons from the cathode to reduce to Cu and become part of the cathode. The salt bridge contains a high concentration of KCl, which diffuses into both half-cells to neutralize the resulting charge imbalance from the redox reactions.

voltage that nulls the voltage being measured. The technique of measuring the potential of a redox reaction is called *potentiometry* and is the basis of operation of ORP and pH probes. In Figure 2.4, a voltmeter replaces the light bulb in Figure 2.2. We'll pick up this description of the Galvanic cell-turned-potentiometric sensor in Section 7.1.

One deviation from the half-cell setup just described is in order. Sometimes we get lucky. Take the household alkaline battery. The anode material (reducing agent) is the familiar zinc, and the oxidizing agent is manganese dioxide, MnO_2. The redox reactions are a little more complicated, but they follow the same pattern:

Anode $$Zn(s) + 2\ OH^- \rightarrow ZnO + H_2O + 2\ e^- \quad (2.12)$$

Cathode $$2\ MnO_2\ (s) + H_2O + 2\ e^- \rightarrow Mn_2O_3 + 2\ OH^- \quad (2.13)$$

Note that hydroxide, OH^-, is consumed at the anode and produced at the cathode. How convenient! Just make up a solution (or paste) with a hydroxide, such as NaOH, and you take care of the electrolyte that also participates.

4.0 WHAT REALLY HAPPENS AT THE ELECTRODE

Electrochemistry is a complex science. A full description of the physics and chemistry that occurs on the surface of electrodes can be found in very advanced textbooks. For the purpose of this book let us address a fundamental question that any inquisitive reader will eventually pose: what, if anything, goes on at the surface of the cathode or anode? According to the previous section, an electrochemical cell with a high impedance voltmeter between the two electrodes allows no current to flow. If there is no current, then there can be no exchange of electrons that provides the basis of all oxidation-reduction reactions. If there are no oxidation-reduction reactions on the surface of the platinum band, then how can the ORP sensor measure anything? The same question applies to a battery. What goes on inside the battery when one connects it to a battery tester or voltmeter and measures its voltage?

The answer is that oxidation-reduction reactions really do continue to occur at the surface of an electrode—even though, paradoxically, no current results. Take the example of Fe^{+2} ions oxidizing to form Fe^{+3} ions. If there is an excess of Fe^{+2} over Fe^{+3} ions, then more of the latter will convert to the latter:

$$Fe^{2+} \rightarrow Fe^{3+} + e^- \tag{2.14}$$

The opposite reaction occurs at the same time—though not at the same rate. Only when the concentrations of both species are equal are the forward and reverse currents equal and opposite. That flow of electrons going in both directions is called the *exchange current*, and the magnitude of that current explains a property of ORP sensors that is brought up later in the chapter to explain why ORP sensors can be very slow. This is a crucial point to understand and that will resurface later.

5.0 THE HYDROGEN CONVENTION

The galvanic cell allows one to separate the two half-reactions from one redox reaction. This is a big step forward toward the understanding of oxidation and reduction reactions, but it still leaves us with an ambiguity: a chemical species, such as iron, oxidizes when in the presence of one partner, such as oxygen, but reduces when with another partner, such as zinc. If there are to be half-cell reactions, then there ought to be half-cell potentials. But there is no way of assigning an absolute potential because there is no way of studying one half-reaction without coupling it to another.

The solution is to arbitrarily choose one special half-reaction and reference every other half-reaction in the universe to that one. This very special reaction is the oxidation of dissolved hydrogen molecules to hydrogen ions:

$$H_2 \rightarrow 2\ H^+ + 2\ e^- \qquad (2.15)$$

We can use the setup in Figure 2.2 and make one of the half-cells the one in which Equation 2.15 occurs. By assigning the potential of this half-reaction, the value of zero volts the overall voltage measured between the two half-cells is always the potential of the other half-cell.

Assigning an absolute value to a physical quantity in a relative world is nothing new. A car's speedometer may show that the driver is driving at 60 mph but how is it possible to assign a speed to that car? After all the earth is rotating at about 1000 mph, so an observer in space might measure the car's speed at 1060 mph. The Earth rotates around the sun at 67 000 mph so an observer in deep space might observe that your car is in fact moving at 67 060 mph. As an earth-bound species we define velocity under the assumption that the Earth's velocity is zero. It's an arbitrary assignment, but it allows us to assign precise values to the speed of earthbound objects like cars, fastballs, and jets.

References exist everywhere, but they need to be defined very precisely. The reaction of Equation 2.15 must be carried out as follows: hydrogen gas at a pressure of 1 atm (15 psi) is bubbled through water at 25 °C. The reaction occurs on the surface of a platinum electrode. This reference half-cell is called the *standard hydrogen electrode*, or *SHE* (to differentiate it from a male electrode). The careful reader will notice that the term *electrode* is being used to describe two separate devices—the platinum metal, which provides the surface for the reaction to occur, and the entire half-cell. This is an unfortunate but common practice.

With this arbitrary appointment of the hydrogen electrode one can now assign any half-reaction as an oxidation or as a reduction and assign an absolute value of the potential for that half-reaction. This requires setting up a galvanic cell and making one of the half-cells a hydrogen cell. In the other half-cell is the reaction of interest. If that reaction gives up electrons to the SHE, so that Equation 2.6 runs in reverse, then it is an oxidation reaction. If it withdraws electrons from the SHE, so that Equation 2.6 runs as written, then it is a reduction reaction. The half-reaction to the hydrogen electrode is assigned a value of 0 V, the potential of the half-cell reaction in question is the voltage displayed on the voltmeter of the electrochemical cell. A potential can now be assigned to every half-reaction imaginable.

Let's now make the connection between a half-cell potential and the propensity of the half-cell reaction to proceed. As discussed in Chapter 1, Section 6.0 that propensity is given by a quantity called the reaction's change in free energy, ΔG. Recall that reactions go in one direction only if ΔG for that direction is negative. The only remaining question then is to connect ΔG with the voltage one reads on the galvanic cell.

That a change in redox potential generates a release of free energy is obvious to anyone who has used hand warmers. The active ingredient in this small heat engine is iron powder that is kept in an air-free environment provided by the plastic wrapper. When the user opens the wrapper and exposes the iron to air the oxidation reaction, Equation 2.14, starts. The release of heat from the reaction is proof positive that the reaction proceeds with a decrease in free energy.

The connection between ΔG and the voltage of the electrochemical cell is simple. They are proportional to each other, and that proportionality is given by:

$$\Delta G = -nFE \tag{2.16}$$

Here n is the number of moles of electrons per moles of product atom or molecule and F is the famous Faraday's constant, which is the charge (Coulombs or C) in a mole of electrons. Its value is 96 548 C/mole. E is the electric potential, which is in units of volts. A positive change in voltage gives a negative change in free energy and the reaction goes.

Table 2.1 lists some of the more common half-reactions. By convention, most authors write *all* half-reactions as a reduction—that is, with the loose electrons on the left side of the equation. A positive potential indicates that this reaction goes spontaneously when combined with a SHE. Because they are written as reduction half-reactions all oxidation half-reactions have negative signs. Remember that when a reaction switches direction the potential changes sign.

This table is enormously practical. It reduces the millions of possible redox reactions down to a number that is on the order of a hundred. One can simply take a redox reaction in question and break it down into its two half-reactions. By simply adding the two voltages—with the correct sign of course—one gets the potential of the full reaction.

As an example, consider what happens when one drops a piece of sodium metal in a beaker of water that has dissolved oxygen. This is a

TABLE 2.1 Standard redox potentials. Potential values are with respect to the standard hydrogen electrode and at 25 °C. Concentrations are 1 M (moles/liter) for the species in question.

Oxidation Half-Reactions	E^0	Reduction Half-Reactions	E^0
$2\,H^+ + 2\,e^- \rightleftharpoons H_2$	0	$2\,H^+ + 2\,e^- \rightleftharpoons H_2$	0
$Li^+ + e^- \rightleftharpoons Li$	−3.04	$Cu^{2+} + e^- \rightleftharpoons Cu^+$	0.158
$K^+ + e^- \rightleftharpoons K$	−2.924	$Cu^{2+} + 2\,e^- \rightleftharpoons Cu$	0.340
$Ba^{2+} + 2\,e^- \rightleftharpoons Ba$	−2.90	$Ce^{4+} + e^- \rightleftharpoons Ce^{3+}$	1.70
$Mg^+ + e^- \rightleftharpoons Mg$	−2.70	$Cl_2 + 2\,e^- \rightleftharpoons 2\,Cl^-$	1.36
$Ca^{2+} + 2\,e^- \rightleftharpoons Ca$	−2.76	$O_2 + 4\,H^+ + 4\,e^- \rightleftharpoons 2\,H_2O$	1.229
$Na^+ + e^- \rightleftharpoons Na$	−2.711	$O_2 + 2\,H_2O + 4\,e^- \rightleftharpoons 4\,OH^-$	0.401
$Mg^{2+} + 2\,e^- \rightleftharpoons Mg$	−2.375	$Cu^+ + e^- \rightleftharpoons Cu$	0.522
$H_2 + 2\,e^- \rightleftharpoons 2\,H^-$	−2.23	$I_3^- + 2\,e^- \rightleftharpoons 3\,I^-$	0.534
$Al^{3+} + 3\,e^- \rightleftharpoons Al$	−1.66	$O_2 + 2\,H^+ + 2\,e^- \rightleftharpoons H_2O_2$	0.70
$ZnO + H_2O + 2\,e^- \rightleftharpoons Zn + 2\,OH^-$	−1.25	$Hg^{2+} + 2\,e^- \rightleftharpoons Hg$	0.85
$Mn^{2+} + 2\,e^- \rightleftharpoons Mn$	−1.185	$Hg_2^{2+} + 2\,e^- \rightleftharpoons Hg$	0.796
$Fe(OH)_2 + 2\,e^- \rightleftharpoons Fe + 2\,OH^-$	−0.89	$HgO + H_2O + 2\,e^- \rightleftharpoons Hg + 2\,OH^-$	0.097
$Hg^{2+} + 2\,e^- \rightleftharpoons Hg$	−0.85	$Ag^+ + e^- \rightleftharpoons Ag$	0.800
$2H_2O^+ + 2\,e^- \rightleftharpoons H_2 + 2\,OH^-$	−0.83	$AgCl + e^- \rightleftharpoons Ag + Cl^-$	0.236
$Fe^{3+} + e^- \rightleftharpoons Fe^{+2}$	−0.768	$H_2O_2 + 2\,e^- \rightleftharpoons 2\,OH^-$	0.88
$Zn^{2+} + 2\,e^- \rightleftharpoons Zn$	−0.762	$HNO_3 + 3\,H^+ + 3\,e^- \rightleftharpoons NO + 2\,H_2O$	0.96
$Cr^{3+} + 3\,e^- \rightleftharpoons Cr$	−0.74	$Br_2 + 2\,e^- \rightleftharpoons 2\,Br^-$	1.087
$Ag_2S + 2\,e^- \rightleftharpoons 2\,Ag + S_2^-$	−0.69	$CrO_4^{2-} + 8\,H^+ + 3\,e^- \rightleftharpoons Cr^{3+} + 4\,H_2O$	1.195
$PnO + H_2O + 2\,e^- \rightleftharpoons Pb + 2\,OH^-$	−0.58	$Pt^{2+} + 2\,e^- \rightleftharpoons Pt$	1.2
$S + 2\,e^- \rightleftharpoons S^{2-}$	−0.51	$MnO_2 + 4\,H^+ + 2\,e^- \rightleftharpoons Mn^{2+} + 2\,H_2O$	1.208

(continued)

TABLE 2.1 *(Continued)*

Oxidation Half-Reactions	E^0	Reduction Half-Reactions	E^0
$2\ CO_2 + 2\ H^+ + 2\ e^- \rightleftharpoons H_2C_2O_4$	−0.49	$Cr_2O_7^{2-} + 14H^+ + 6\ e^- \rightleftharpoons 2Cr^{3+} + 7\ H_2O$	1.33
$Cr^{3+} + e^- \rightleftharpoons Cr^{2+}$	−0.41	$Cl_2 + 2\ e^- \rightleftharpoons 2\ Cl^-$	1.358
$Fe^{2+} + 2\ e^- \rightleftharpoons Fe$	−0.44	$HOCl + H^+ + 2\ e^- \rightleftharpoons Cl^- + H_2O$	1.49
$Cd^{2+} + 2\ e^- \rightleftharpoons Cd$	-0.40	$HOCl + 2H^+ + 2\ e^- \rightleftharpoons Cl_2 + H_2O$	1.63
$Co^{2+} + 2\ e^- \rightleftharpoons Co$	−0.28	$OCl^- + H_2O + 2\ e^- \rightleftharpoons Cl^- + 2\ OH^-$	0.81
$Ni^{2+} + 2\ e^- \rightleftharpoons Ni$	−0.23	$MnO_4^- + 8\ H^+ + 5\ e^- \rightleftharpoons Mn^{2+} + 4\ H_2O$	1.491
$AgI + e^- \rightleftharpoons Ag + I^-$	−0.152	$Au^+ + e^- \rightleftharpoons Au$	1.68
$Sn^{2+} + 2\ e^- \rightleftharpoons Sn$	−0.136	$H_2O_2 + 2\ H^+ + 2\ e^- \rightleftharpoons 2\ H_2O$	1.776
$Pb^{2+} + 2\ e^- \rightleftharpoons Pb$	−0.126	$Co^{3+} + e^- \rightleftharpoons Co^{2+}$	1.842
$CO_2 + 2H^+ \rightleftharpoons CO + H_2O$	−0.11	$O_3 + 2\ H^+ + 2\ e^- \rightleftharpoons O_2 + H_2O$	2.07
$Fe^{3+} + 3\ e^- \rightleftharpoons Fe$	−0.036	$F_2 + 2\ H^+ + 2\ e^- \rightleftharpoons 2\ HF$	3.03

perennial favorite science demonstration as the beaker erupts in flames. From the seventh row of the left column of Table 2.1:

$$Na \rightarrow Na^+ + e^-;\ E^0 = 2.711\ V \tag{2.17}$$

The reduction of oxygen, from the air and dissolved in the water, can be found in the sixth row of the reduction column:

$$O_2 + 2\ H_2O \rightarrow 4\ e^- + 4\ OH^-;\ E^0 = 0.401\ V \tag{2.18}$$

Add them and you get:

$$4\ Na + O_2 + 2\ H_2O \rightarrow 4\ Na^+ + 4\ OH^-;\ E^0 = 3.112\ V \tag{2.19}$$

Note that, to make the number of electrons donated by Equation 2.17 (one) equal to the number taken by Equation 2.18 (four), we must multiply Equation 2.17 by 4. The Na^+ and OH^- ions produced constitute dissolved sodium hydroxide, or caustic. The 3.11 V that we get from adding the potentials of both half-reactions is a large change. The explosion and fire that results is testimonial to the large free energy change that this reaction produces.

One might think that, if we multiply both sides Equation 2.17 by four to make the stoichiometry work, that we need to multiply ΔG by 4 as well. That this is incorrect becomes clear when we recall that a large D cell battery has the same 1.5 V as a much smaller AAA battery. The voltage doesn't increase, but the flow of electrons—that is, the current—does.

Table 2.1 also allows one to quantify an observation mentioned above—that is, that a chemical species can be an oxidizing agent when paired with one redox partner but, with another, can be a reducing agent. As long as their potentials are different any two half-reactions can be combined to make a full redox reaction that is energetically allowed. It isn't necessary to choose one species from the left column and one from the right. One can take both species from one column. If we combine two half-reactions with positive ΔG value to create a full reaction that proceeds spontaneously, then we must reverse the half-reaction with the lower potential. This ensures that the difference of the two potentials will be positive.

As an example, let's return to the oxidation of iron. From the left column of Table 2.1:

$$Fe \rightarrow Fe^{2+} + 2e^-; E^0 = 0.409 \text{ V} \quad (2.20)$$

We'll pair this with zinc:

$$Zn \rightarrow Zn^{2+} + 2e^-; E^0 = 0.762 \text{ V} \quad (2.21)$$

Because Equation 2.20 is the half-reaction that's the one to reverse (thus converting it to a reduction reaction):

$$Fe^{2+} + 2e^- \rightarrow Fe; E^0 = 0.409 \text{ V} \quad (2.22)$$

Adding Equation 2.21 and Equation 2.22 gives:

$$Fe^{2+} + Zn \rightarrow Zn^{2+} + Fe; E^0 = 0.762 - 0.409 \text{ V} = 0.353 \text{ V} \quad (2.23)$$

The voltage generated by a galvanic cell with ferrous (Fe^{2+}) ions on the cathode side and Zn atoms on the anode side will indeed make a battery that outputs a voltage of 0.353 V or less. (I say "or less" because of the overpotential that is required to overcome an energy barrier at the metals'surfaces. More on that in Chapter 5, Section 2.0.)

If Equation 2.23 runs backward, then E is now −0.353 V, and the reaction doesn't go. We can drive the reaction if we connect a direct current voltage source to the electrodes and provide at least 0.353 V. This setup is called an *electrolytic cell* and works like a galvanic cell running backward. Two very common examples of electrolytic cells are battery chargers and metal plating.

Up until now all redox reactions discussed have used a standard hydrogen electrode. That's easy to do on paper but much more inconvenient in practice. It requires a cylinder of hydrogen gas, a gas delivery system to dissolve the hydrogen in water and a means to measure the gas pressure in the electrode to ensure that it is at 1 atmosphere. On the left side of Figure 2.3 is a SHE. There must be a more convenient reference.

There is. It is a half-cell that contains a silver electrode that is coated with silver chloride and in equilibrium with a concentrated KCl solution. The silver wire has a large capacity to conduct current, and it is chemically stable. The half-reaction for the so-called silver/silver chloride (Ag-AgCl) electrode is:

$$AgCl\ (s) + e^{-} \rightarrow Ag(s) + Cl^{-};\ E^{0} = 0.197\ V\ (4\ M\ KCl) \quad (2.24)$$

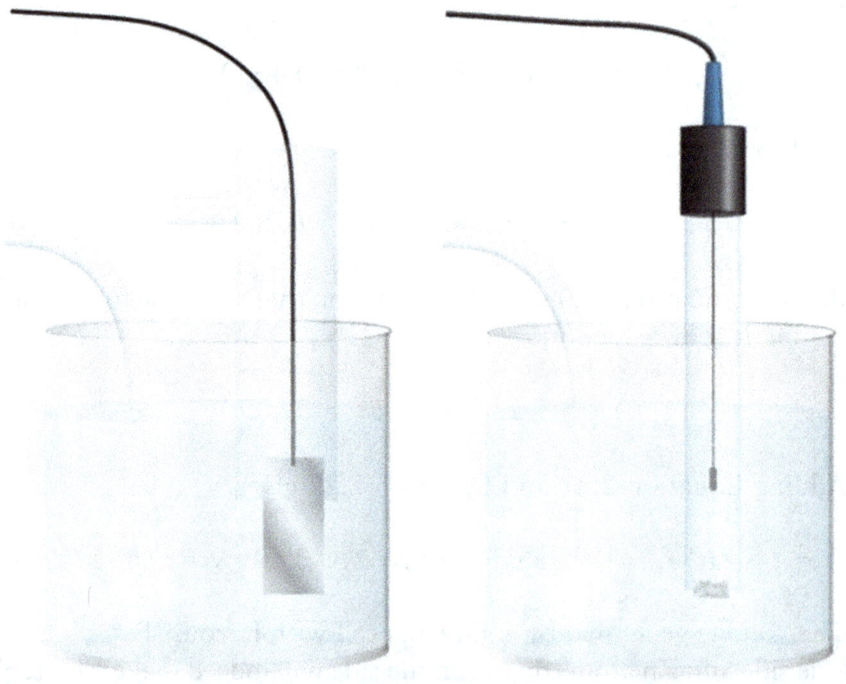

FIGURE 2.3 On the left is a standard hydrogen electrode. It requires a constant supply of hydrogen gas. On the right is a silver—silver chloride electrode. One adds 200 mV to the potential measured with a Ag-AgCl electrode to yield the ORP value one would have measured with a SHE.

The right side of Figure 2.3 shows an Ag-AgCl standard electrode.

The Ag-AgCl electrode has become the de facto standard reference electrode for nearly all potentiometric measurements, but the reference potential for reported ORP values is still that of the SHE. This means that any measurement done with an Ag-AgCl electrode must be corrected. According to Table 2.1, that suggests adding 236 mV to the measured ORP value (at 25 °C). However, the table is for species at the standard 1 M (molar) concentration. The Ag-AgCl electrode is usually immersed in 4 M or saturated KCl solution. The redox potential for such a half-cell is 197 mV at 25 °C and 4M KCl, as indicated in the parentheses of Equation 2.24. So, the correction is to add 197 mV to the actual reading.

It is very important to understand that the correction of potentials from the Ag-AgCl electrode is temperature dependent and that the corrections in the previous paragraph are only valid at 25 °C. The temperature dependence is approximately −0.48 mV per degree Celsius (Bard, et al., 1985).

There is one other reference electrode that has been used in the past as an alternate to the SHE. The calomel electrode is based on a mercury chloride (Hg_2Cl_2) electrode, and the redox reaction is:

$$Hg_2Cl_2 + 2\ 2e^- \rightarrow 2\ Hg + 2\ Cl^-;\ E^0 = 0.268\ V \qquad (2.25)$$

Because of its use of mercury this electrode has fallen out of favor.

6.0 THE NERNST EQUATION

It is now time to introduce the most important equation in all of electrochemistry. The ORP values listed in Table 2.1 are great for comparing different redox reactions. However, they only work if the reactants are at 1 M and the temperature is 25 °C. What if they aren't? The answer is the Nernst equation:

$$E = E^0 - \frac{RT}{nF} ln Q \qquad (2.26)$$

The equation is a prescription for determining the actual potential (E) if we know a few things about our system. E^0 is the redox potential under standard conditions—that is, the potential(s) listed in Table 2.1. T is the temperature in Kelvin. (The Kelvin scale is just like the Celsius scale except that it sets zero at true zero, which renders a Kelvin temperature 273 degrees higher than the Celsius value.) R is a constant known as the gas constant and is equal to 8.314 joules/mole-K. F is Faraday's constant, which was defined earlier as the charge of 1 mole of electrons—96 485 C (Coulombs)—and n is the number of electrons exchanged in the redox reaction. R is the universal gas constant—8.3145 J $mol^{-1}K^{-1}$.

The last variable left to define is Q, which is shorthand for "quotient." It is based on the ratio of products to reactants. For the reaction:

$$n_aA + n_bB + n_ee^- + \ldots \rightarrow n_cC + n_dD + \ldots \tag{2.27}$$

We define the ratio of the *activities* of the products and reactants:

$$Q = \frac{a_c^{\ nc} a_d^{\ nd} \ \ldots}{a_a^{\ na} a_b^{\ nb} \ \ldots} \tag{2.28}$$

The activity of a chemical species is a measure of its propensity to react with others. It is intuitive that, the more concentrated the species is, the faster the reaction. In fact, when a species is very dilute, concentration and activity are equal. However, as its concentration increases the species' activity becomes increasingly less than its concentration. That's because it increasingly "feels" the presence of other ions or molecules of its own kind and becomes less reactive. This is analogous to the difference between one person walking alone on a deserted street and another finding his way through midtown Manhattan in rush hour. It's convenient to relate the activity to the concentration through a simple proportional relationship:

$$a_X = \gamma_X[X] \tag{2.29}$$

$[X]$ is the concentration of species X and γ_X is the concentration of X. In most cases, writing Q in the Nernst equation in terms of concentrations—rather than the more precise activities—is an acceptable approximation:

$$Q \approx \frac{[C]^c [D]^d}{[A]^a [B]^b} \ldots \tag{2.30}$$

There will be cases when this approximation breaks down because the concentration of one or more species is too high. One particularly important example is the measurement of pH for concentrated acids.

In the Nernst equation as written in Equation 2.26, "*ln*" is the *natural logarithm*—that is, it is the logarithm that uses e (2.718) as the base. To convert a natural logarithm to a *common logarithm*—based on 10 as the base—simply multiply the natural logarithm by 2.303.

Let's finish the task, started in Equation 2.26, of evaluating the collection of constants in the front of this formula. We'll substitute the values for R and multiply the natural logarithm (base e) by 2.303 to change it to the common logarithm (base 10). At a temperature of 25 °C (298 K) this collection of constants equals 0.05916 V or 59.16 mV:

$$\frac{RT}{F} = \frac{8.314(\mathrm{J/mol-K}) \times 298\ \mathrm{K}}{96\ 485\ \mathrm{C}} = 0.05916\ \mathrm{V} = 59.16\ \mathrm{mV} \quad (2.31)$$

(As long as one uses the units specified by the SI system when calculating something new, that new result will always be in the correct units. In this case that means that E will be in volts.) This number is arguably the most often used number in all potentiometry.

Thus, at 25 °C, Equation 2.31 thus becomes:

$$E = E^0 - \frac{59.16\ \mathrm{mV}}{n} \log Q\ (\text{at } 25\ °\mathrm{C}) \quad (2.32)$$

Recall that n is the number of electrons exchanged in the redox reaction. For a redox reaction in which one electron is exchanged then, for every tenfold increase in the product concentrations or tenfold decrease in the reactant concentrations, the voltage increases by 59.16 mV (at 25 °C). And that's why 59.16 (59 for government work) is a number you should never forget.

Let's now apply the Nernst equation to a real example—the reaction that occurs in galvanized steel in which Zn reduces the ferrous ion (Fe^{2+}) to neutral Fe.

$$Fe^{2+} + Zn(s) \rightarrow Zn^{2+} + Fe(s);\ E^0 = 0.762 - 0.409\ \mathrm{V} = 0.353\ \mathrm{V} \quad (2.33)$$

Note that the neutral zinc and iron atoms are denoted by "*s*" to signify that they exist as solids. This means that they are not dissolved in solution and do not enter the calculation of Q. According to Equation 2.33 the potential of this reaction (E^0) is 0.353 V. This means that an electrochemical cell containing 1 mole/liter (1 M) of Fe^{2+} in one half-cell and 1 M Zn^{2+} in the other outputs a voltage of 0.353 V. Now suppose the ferrous iron concentration stays at 1 M, but the Zn^{2+} concentration drops to 0.1 M. The Nernst equation predicts that the voltage is:

$$Q = \frac{[Zn^{2+}]}{[Fe^{2+}]} = \frac{0.1}{1} = 0.1 \quad (2.34)$$

The common logarithm of 0.1 is +1 ($10^{-1} = 0.1$). Equation 2.33 shows that 2 electrons change partners so Equation 2.32 gives a voltage of

$$E = 353\ \mathrm{mV} - \frac{59.16\ \mathrm{mV}}{2} X(-1) = 353\ \mathrm{mV} + 29.58\ \mathrm{mV} = 382.58\ \mathrm{mV}$$

This makes sense. There are now more Fe^{2+} reactants than Zn^{2+} products so that Equation 2.33 is driven to the right, which increases the potential. The iron rusts and zinc comes to its rescue and changes the Fe^{2+} back to its native metal.

One important point to remember is that the Nernst equation works for any oxidation-reduction reaction, whether it is a half-reaction that occurs at an anode or cathode, or it is the full reaction that combines both oxidation and reduction half-reactions. In this case the half-cell potentials simply add to create a total redox potential. Keep in mind, as well, that, for half-reactions, the Nernst equation and the potential it predicts for a half-reaction is always written for the reduction reaction. Therefore, the quotient, Q, is always the ratio of reductants to oxidants. Remember that an oxidation reaction has a negative voltage associated with it.

7.0 CONSTRUCTION OF AN ORP SENSOR

7.1 The ORP Sensor on Paper

With an understanding of what ORP, or redox potential, is and how to calculate it, it's time to make a sensor for measuring it. Before we build a real analyzer let's build a conceptual one. We'll start with the electrochemical cell pictured in Figure 2.2. From the discussion in Section 3.0 we already know that, to measure the potential of the two half-cell reactions, we need to substitute a voltmeter for the light bulb. The left side of Figure 2.4 shows

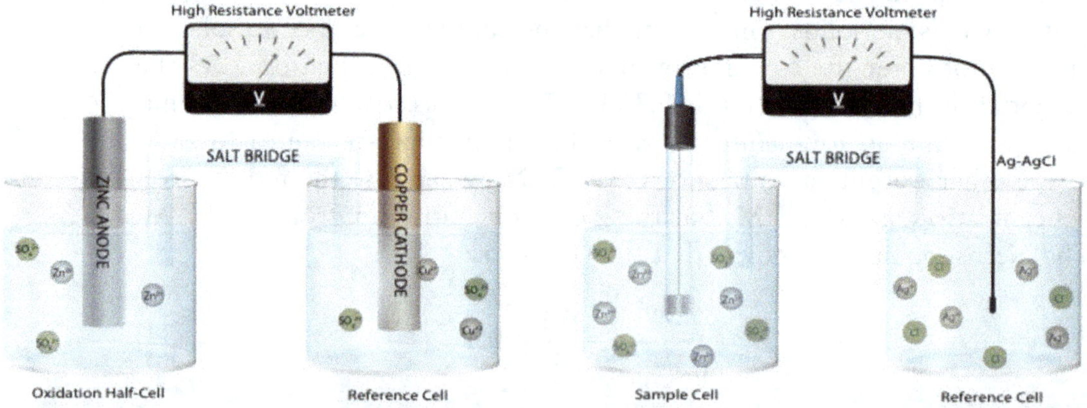

FIGURE 2.4 Left: Two half-cells containing an oxidation reaction at a zinc anode and a reduction reaction at a copper cathode. The voltmeter measures the potential between the two half-cells. Right: An Ag-AgCl reference electrode replaces the copper cathode and a platinum electrode upon which redox reactions can occur replaces the zinc anode.

the modification. The ideal voltmeter, or galvanometer, ensures that no electrons flow from the anode to cathode. (Because current is considered the flow of positive charge, we can say that no current flows from the cathode to the anode.) As explained in Section 4.0, this condition of zero net current means that each electrode is at equilibrium with respect to the oxidation or reduction reaction occurring at each.

If this setup is to become an ORP analyzer, then one of these cells—the reference cell—must contain an Ag-AgCl electrode in a saturated or 4M solution of KCl and the other—the sample cell—needs to be include some kind of surface upon which redox reactions for all dissolved compounds and ions can occur. If a dissolved species participates in an oxidation reaction it will give up electrons at the surface and, if another is involved in a reduction reaction it will take up electrons at the surface. From where do the electrons come or go? The Ag-AgCl electrode.

The best option for such an electrode is one made of platinum or gold. These materials do not react readily with most chemicals—that is, they are chemically *inert* or *noble*. Platinum is the material found in a car's catalytic converter, where it accelerates the oxidation of carbon monoxide, CO, from the car's exhaust to CO_2. When immersed in water most oxidation and reduction reactions occur on its surface. The half-cell with the platinum (or gold) electrode becomes the sample cell on the left of the right-hand figure in Figure 2.4.

Whether the reference half-cell is an anode or cathode depends on the sum of all ORP values for all the electrochemically active species dissolved that react in the sample half-cell. If the overall ORP is positive, then reduction occurs in the sample half-cell and oxidation occurs in the reference half-cell. The former is the cathode, and the latter is the anode. An overall negative ORP turns the reference electrode into a cathode and the process electrode into an anode.

This ORP analyzer works but there is no such device on the market. It is simply too cumbersome to be practical. To measure the ORP of a sample of water one would need to pour the process solution into the sample half-cell. We need a design for a one-piece probe that attaches to a piece of electronics containing the voltmeter. Probes are convenient because they can go into a pipe, a beaker, a tank, or a flowing stream.

7.2 Constructing an Actual ORP Sensor

To make an ORP analyzer that is compact and easy-to-use, we need to first focus our attention on modifying the awkward salt bridge. This is the U-tube usually containing concentrated KCl (typically in soaked cotton wool) that ensures electrical neutrality in each half-cell and electrical continuity between the two. As this salt diffuses out of the bridge into both half-cells the KCl becomes increasingly diluted with ions from the cells and needs to be recharged.

The solution is to enclose the reference electrode—along with its surrounding KCl solution—in a glass envelope and cap it off with a piece of material that allows ions and water to pass through. We can now immerse the reference electrode assembly into the sample. That piece of porous material is typically a frit made of ceramic, polymer material such as PVDF, or even a tiny hole. It may also be a capsule of concentrated KCl with porous frits on both ends. Whatever the actual configuration, this structure is called the *liquid junction.*

To this major design innovation, we add the process electrode, reference electrode, and voltmeter to measure the potential difference between the two. The following is a summary of these four components.

1. **Reference electrode.** The reference electrode is a silver wire coated with AgCl and immersed in a solution of KCl that is saturated, 4 M or 3.5 M. To prevent too much silver from leaching out of the wire to make the silver chloride coating, the KCl solution also contains a saturated solution of AgCl. (AgCl has such a low solubility that very little is in solution.)
2. **Process (or working or measuring) electrode.** This electrode is an inert platinum (Pt) or gold (Au) surface upon which oxidation and reduction reactions can take place. The Pt electrode is usually in the shape of a band that encircles an empty glass ampoule. A wire intercepts the glass wall to connect to the Pt band and provides a circuit to the voltmeter. Oxidation and reduction reactions occur on the metallic surface, and a potential builds up that is equal to the sum of the individual reactions.
3. **Liquid Junction** or **Salt Bridge.** In most *combination* probes the liquid junction is a small porous disk as described above. The flow of KCl (or KNO_3) through the frit maintains circuit continuity. *Differential probes* use a capsule packed with KCl (or KNO_3) and having porous frits on both ends. The choice of KCl or KNO_3 as an electrolyte in the reference chamber or in the salt bridge is based on the fact that K^+ and Cl^- (or NO_3^-) ions migrate at roughly the same rate. If one ion migrated faster than the other there would arise a potential difference along the path of the junction that would change the voltage read on the voltmeter. Though this so-called potential junction is always present the use of this salt minimizes it. Chapter 3, Section 6.5 goes into the diffusion potential at greater depth.

Some combination probes contain just one liquid junction between the reference electrode and sample. The single junction represents an economical design choice. More expensive combination probes use a double junction. It consists of an inner glass compartment containing the Ag-AgCl wire and the frit inside a larger glass compartment with a frit—a junction-within-a-junction. The double junction slows the migration of salt out of the reference chamber.

Figure 2.5 shows the reference electrode and process electrodes side-by-side.

4. **Voltmeter.** Just as in the conceptual design first discussed, a voltmeter reads the potential difference, or voltage, between the reference and process electrode. Any reading of a voltage requires that zero current passes through the voltmeter. Any non-zero current constitutes a voltage drop and reduces the apparent redox potential. Just for fun, try measuring the voltage from an ORP probe using an inexpensive voltmeter. You will notice that the voltage on the voltmeter is lower than the ORP reading on the analyzer to which the probe connects.

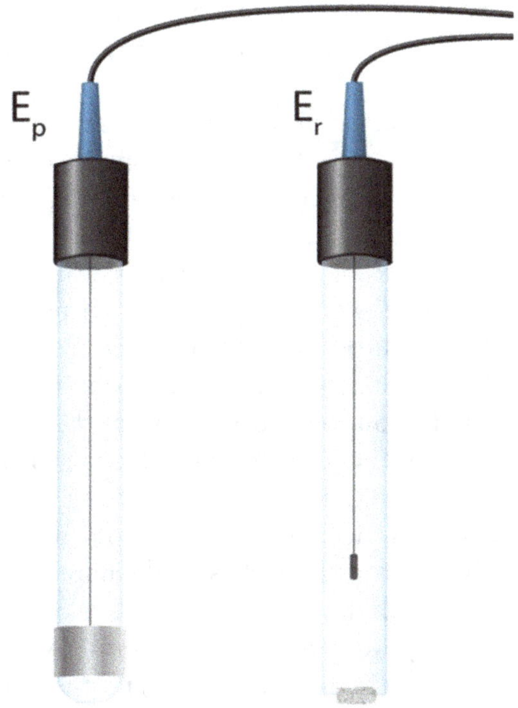

FIGURE 2.5 ORP sensor constructed from separate reference and process electrodes. The reference electrode is on the right. The porous frit is on the bottom of the reference electrode.

If you then replace the cheap voltmeter with an expensive one, like a Fluke or Agilent meter, you read a higher voltage closer to the proper ORP value. A good voltmeter has an input impedance on the order of a 10^9 Ω.

Let's now look at two evolutionary pathways that the ORP sensor has taken to meet two different goals.

7.3 The Combination ORP Probe

The combination probe is a tube-inside-a-tube. The outer tube is filled with KCl and houses the usual Ag-AgCl wire that passes through the top of the glass envelope so that it can connect to one terminal of the analyzer. It is the reference electrode. A smaller diameter tube passes right through the reference tube so that it sticks out the bottom of the tube. This tube is totally isolated from the larger one. A platinum or gold band wraps around the exposed bottom of the tube and connects to a wire that passes through the glass enclosure. The wire connects to the other terminal of the voltmeter. The liquid junction, which is either a porous frit, a tiny hole, or glass sleeve, is at the bottom or side of the larger tube where it can provide the interface between the redox products in the sample and the Ag-AgCl electrode inside the reference chamber.

Because the process electrode is combined with the reference electrode, this probe is called a *combination probe* and is shown on the left side of Figure 2.6. On the right is an actual combination probe. The term *combination* is often misunderstood as it conjures in the mind a probe that combines two or more probes together into one device, such as a pH and an ORP sensor.

The combination probe is a marvel of engineering. Its compact design enables it to be sold for as little as $100. Most ORP, pH, and other potentiometric probes sold on the market today are combination probes.

Most of the discussion on the construction of the ORP probe is analogous to that of the pH probe and the ion-selective probe—in short, any potentiometric sensor. The major differences among configurations of any given probe are the choices of the process electrode and of the liquid junction. This is especially true for pH probes, and Chapter 3, Section 4.2 summarizes the predominant choices. For an ORP sensor, the choices for the process electrode are very limited. Platinum is the overwhelming favorite because of its low-cost relative to gold and its inertness in highly oxidizing environments. Gold is used in highly reducing environments, such as solutions containing cyanide.

As briefly mentioned above, the liquid junction comes in a variety of designs, including a ceramic frit, polymer fit, small hole, and glass sleeve.

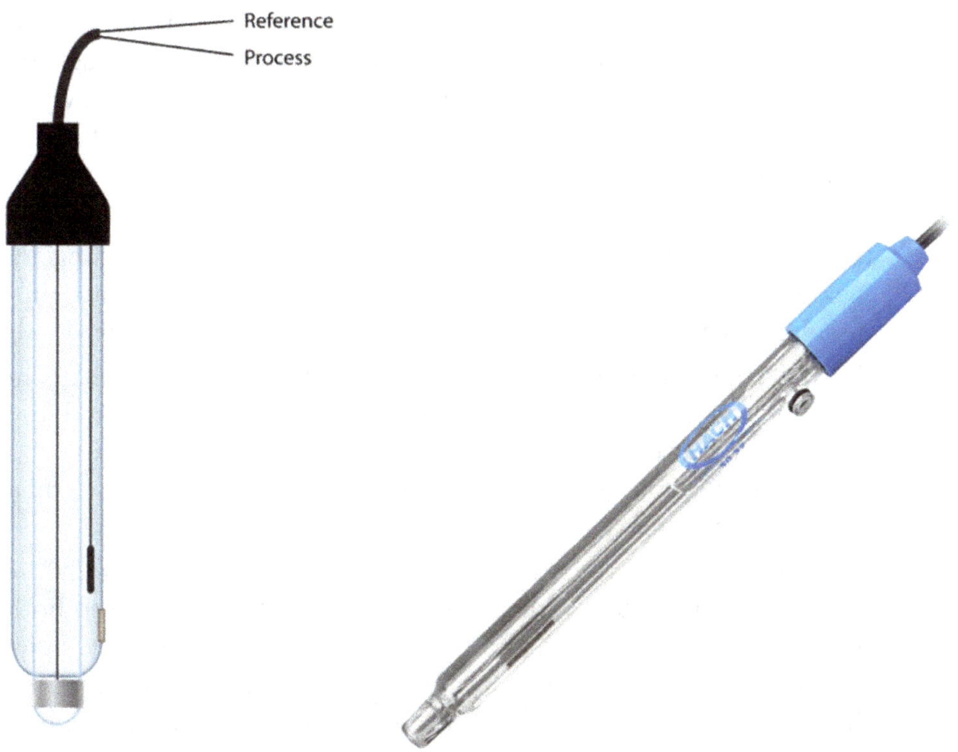

FIGURE 2.6 Left: Combination probe schematic. The process electrode is the inner tube with the platinum band. The reference electrode is the wire on the right side inside the larger tube. The liquid junction is the tiny porous frit on the large glass envelope. Right: A combination probe made by Hach. The small port at the top of the probe allows the user to replenish the KCl reference solution (Courtesy Hach).

These different designs strike a particular balance between the response time and longevity of the reference salt solution. A faster response comes at the cost of a shorter-lived solution. Because the liquid junction is more of an issue with pH probes, the next chapter delves into them in more detail.

The development of the combination probe "rewrote the book" on pH and ORP measurement, but it does have its limitations.

1. In its role of completing the electrical circuit the KCl solution slowly leaks into the water in which it is immersed. As the ions leak out, the sample solution leaks in to replace it. The reference half-cell becomes contaminated and gradually changes its half-cell potential and, thus, diminishes the probe's efficiency. The use of a double junction greatly retards the rate of electrolyte leakage. Some combination probes feature a hole in the body with a removable rubber cap that allows the user to refill the chamber.

2. The reference electrode is an Ag-AgCl wire immersed in KCl electrolyte and saturated AgCl. It connects to the process through the liquid junction. Impurities, such as sulfides, can coat the wire, which decrease its efficiency. Dissolved AgCl can react with H_2S to form Ag_2S, which precipitates out and plugs the liquid junction.
3. The liquid junction is subject to fouling. The source can be either from the outside in the form of suspended particulates or on the inside by precipitates of silver. The dissolved AgCl will react with dissolved hydrogen sulfide (H_2S) to form insoluble Ag_2S compounds.
4. For the sake of completeness, we include ground loops. Perhaps no issue will frustrate users of pH probes more than ground loops. However, their effect on ORP probes is much less than that of pH probes so we introduce the concept of ground loops here and return to them in the next chapter.

   A ground loop exists whenever one electrical device "finds" an electrical pathway to another, the signal ground is common to both, both devices are grounded and there is continuity between the signal ground and device grounds. If a current finds its way into the loop then, in accordance with Ohm's law, the small resistances in the loop create a different voltage, with respect to earth ground, at each instrument ground.

   As long as the probe is in a process with no other electrical device there can be no ground loop. But immerse a device, like a mixer or pump, that is grounded through the submerged device and ground connection can materialize.

   The ORP probe is unique because its platinum (or gold) electrode provides a highly conductive path to any ground loop that is established as a result of a submerged device. Figure 2.7 shows this scenario. Because of the low resistance the differences in voltages at the probe and analyzer are minor. The difference is totally different in pH probes, and we'll dive into the topic of ground loops in much greater detail then.

7.4 The Differential Probe

The differential design addresses the drawbacks just listed. *Before proceeding I must disclose that my company derives much of its revenue from the sale of differential probes. I have made every attempt to present a balanced presentation of all probe technologies and to portray instrumentation fairly from all manufacturers. I can only hope that this discussion is not colored by any biases I may harbor.*

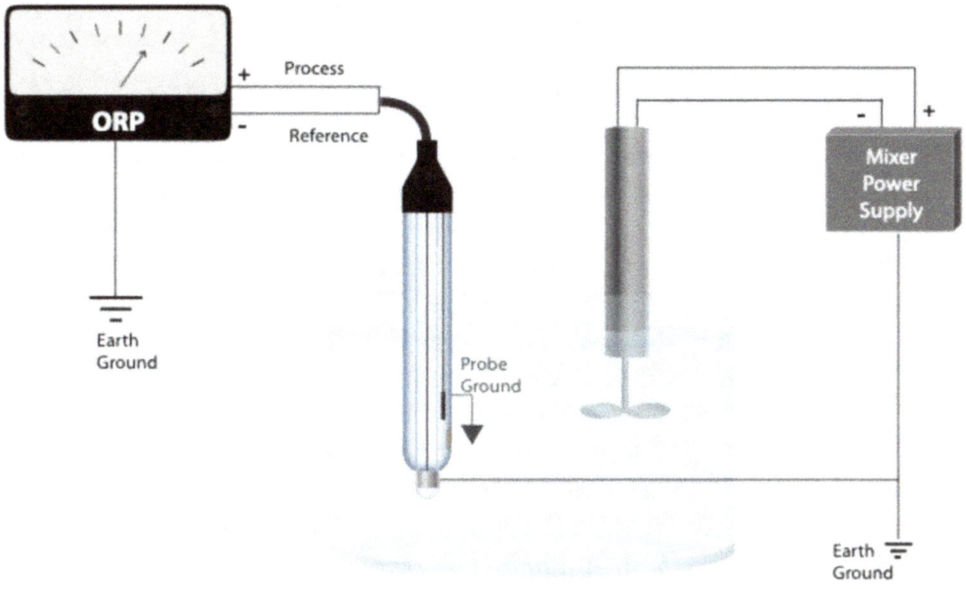

FIGURE 2.7 Establishment of a ground loop by virtue of the additional electronics (a mixer). The probe's pathway to ground is through the metallic process electrode.

The differential pH probe was developed in 1975 by Karl King of Great Lakes Instruments (now Hach). It uses two identical pH electrodes for the process and reference electrodes and a ground rod common to both. Adopting the pH differential probe design to an ORP one is done simply by replacing the process pH electrode to an ORP one. However, the reference pH electrode remains intact for the simple reason that it is the Ag-AgCl wire immersed in neutral KCl solution inside this pH electrode that provides the same reference potential as an Ag-AgCl wire immersed in the reference chamber of a combination probe. This is a convenient shortcut, but the resulting ORP probe is not a "pure" differential probe.

We defer a description of the differential probe to Chapter 3, Section 4.3 in the chapter on pH. To summarize, the differential probe splits apart the process and reference electrode and pairs each with one common ground rod. Chapter 3, Section 4.3 lists the consequences and advantages of the differential design change. One consequence that differs for the differential ORP probe is the subtraction of ground loops by the ground rod that is common to both the ORP electrode and the reference pH electrode. As discussed in the last section ground loops and other stray voltages are much less of a nuisance than they are for pH probes. The ORP metal electrode has an impedance of close to zero while the reference electrode has an impedance on the order of several kilohms. The impedance of the pH electrode is hundreds of megaohms. Therefore, a pH probe is more susceptible to stray grounds. We'll go into more detail on ground loops in Chapter 3, Section 7.

8.0 CALIBRATION AND MECHANICS

8.1 Why Calibration Is a One-Point Exercise

Most analytical instruments are calibrated using two points. Sometimes calibration is only done with one point, but that's because the analyzer output for zero is assumed to be zero. This may be the case with dissolved oxygen or conductivity measurements. Even though one performs what appears to be a one-point calibration the analyzer is actually performing a two-point calibration by adding the assumed zero as the second point.

Look at a generalized two-point calibration as exemplified in Figure 2.8. The numbers on both axes are arbitrary. The numbers on the x-axis represent the process variable, such as pH or dissolved oxygen concentration. The numbers on the y-axis represent the response of the analyzer to the process, such as mV or μA. The line that connects the points can be characterized by a *slope*, or *span*, and *intercept*, or *offset*. The slope is the change in response of the analyzer to the change in the process variable. The offset is the response of the analyzer when the process variable is zero. For most sensors the offset in a perfect world is zero. In the real world the offset is always some non-zero value. In Figure 2.8 the line through the two points has a slope of 2.0375 and an offset of 1.2. (The variables were not defined so these numbers are meaningless.) The one anomaly to the expectation of a (0,0) point is pH calibration. The zero output of a pH probe occurs at the finite pH value of 7. The next chapter discusses this point.

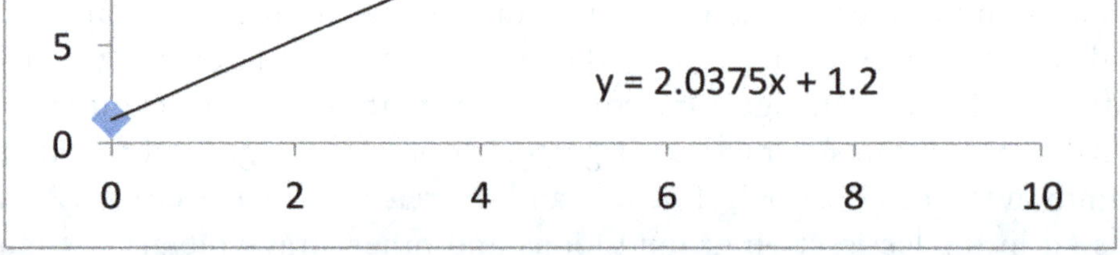

FIGURE 2.8 Generalized two-point calibration. The values on the x-axis represent the process variable (e.g., dissolved oxygen concentration) while the values on the y-axis are the response to the variable that we measure. In this example the numbers are meaningless.

ORP sensors are calibrated at only one point, but in this case, calibration is a true one-point exercise. That's because the measured response of an ORP probe—its output voltage—is the ORP parameter. The voltage and the process variable are one and the same. To put it another way: calibration is an exercise in which the measurement units of the analyzer (volts in this case) are *mapped* to the units of the process parameter—for example, pH for a pH probe or parts per million (ppm) for a chlorine sensor. But, for an ORP sensor, the voltage is the final answer. There is no mapping, which means there is no calibration.

An ORP analyzer measures a redox potential in the same way that a voltmeter reads the voltage of a battery. If the voltmeter is inaccurate one can adjust the zero adjustment, aka the offset, of the voltmeter to adjust its output but that's it. There is no span to adjust.

We can't calibrate an ORP sensor, but we can adjust the offset so that the analyzer's reading matches the known ORP value of a standard solution. This is *standardization*. An analyzer's voltage might differ from the true value because of variability in the reference electrode, poisoning of the platinum or gold electrode, or a resistive load along the cable that connects the probe to the analyzer. These errors are typically just a few millivolts, and the standardization typically adds, or subtracts, just the offset to the raw reading.

Unlike pH calibration solutions, ORP calibration solutions come in a wide range of ORP values. Moreover, unlike pH standards, which are buffered and resist changes in pH, ORP calibration solutions are not buffered and will drift in time as actual redox reactions take place. Most calibration solutions are either Light's solution—476 mV—or Zobell's solution—229 mV.

Both use a familiar redox pair—Fe^{2+} and Fe^{3+}. The actual iron salt compounds that comprise these are listed in Table 2.2. To these two common standards are added a third—pH 4 or pH 7 buffer saturated with quinhydrone. This third option allows anyone with pH buffer to create a do-it-yourself ORP standard by ordering an additional, inexpensive ingredient.

In addition to the lack of buffering ORP, standard solutions change their ORP values with temperature. The ORP of Zobell's solution decreases with temperature by approximately −1.5 mV/°C (Nordstrom, 1977). The full temperature correction potential, with respect to a Ag-AgCl electrode is:

$$E(mV) = 231.45 - 1.5220(T - 25\ °\mathrm{C}) - 2.2449\ (T - 25\ °\mathrm{C})^2 \quad (2.35)$$

Our own measurements of Light's solution give a change of the ORP value with temperature as −1.7 mV/°C. Keep in mind that these ORP values of both standards, as well as their temperature corrections, are with respect to the Ag-AgCl half-cell potential. Not only must one add approximately

TABLE 2.2 Three commonly available ORP standard solutions.

	Oxidant/ Reductant	Recipe	ORP (vs. Ag/ AgCl)
Light's Solution	Fe^{+2}/Fe^{+3}	2.64 g K4[Fe(CN)$_6$] × 3 H_2O and 2.06 g $K_3[Fe(CN)_6]$ × H_2O in 500 mL buffer pH 7.00 at 25 °C	476 mV
Zobell's Solution	Fe^{+2}/Fe^{+3}	1.861 g $Fe(NH_4)_2$ $(SO_4)_2$ × 6 H_2O and 2.411 g Fe $NH_4(SO_4)_2$ × 12 H_2O in 500 mL 1 M H_2SO_4 at 25 °C	229 mV
Quinhydrone	Quinone/ Quinhydrone	pH 4 or pH 7 buffer saturated with Quinhydrone	pH 4 – 263 mV pH 7 – 86 mV

200 mV to convert the ORP values to that for the SHE electrode, but one must also take into account the temperature variation of the Ag-AgCl electrode itself, as first mentioned in Section 5 (Bard et al., 1985).

8.2 A Question of Accuracy

Check the datasheet for a pH probe, and you will see a precision of 0.01 pH units. Many users mistakenly think that this means that their readings will be accurate to within 0.01 units of the correct pH value. As was pointed out in Chapter 1, Section 1.0, precision and accuracy are two different things. Accuracy assesses how close to the real value a measured parameter is, and precision assesses how fine that measurement is. In actuality, pH measurements are only accurate to within about 0.1 pH unit. The same is true for ORP measurements. A difference of 0.1 pH units is a difference of 5.9 mV. To the degree they are both potentiometric sensors, then one might expect the precision for an ORP sensor to be approximately 5 mV.

Copeland and Lytle (2014) made measurements of ORP and pH of common oxidizing agents at three different pH values (7, 8, and 9) and at varying concentrations. The oxidants include oxygen (O_2), free chlorine (HOCl and OCl^-), chlorine dioxide (ClO_2), monochloramine (NH_2Cl), dichloramine ($NHCl_2$), and permanganate (MnO_4^-). They compared their ORP measurements to the theoretical values predicted by the Nernst equation. Discrepancies ranged from less than 1% to 18% (HOCl). All measurements used two separate probes to enable the observation of variations of ORP measurements made with real probes. The first value in each cell of Table 2.3 is the average discrepancy between the theoretical and actual measurements. The mV readings are divided by the measured standard state potential to give a percentage discrepancy. These values are the second listed in each cell of Table 2.3.

An interesting observation was that the two probes agreed to within 0.1 mV of each other when placed in Zobell's solution (ORP value of 220 mV) but displayed the wide variations shown in Table 2.3 when placed in solutions of the oxidizing agents. The proposed reason for this contradictory behavior lies in the actual redox reactions that occur in the calibration solution versus those that occur with the oxidizing agents. As the next section explains, the redox reaction in Zobell's solution at the surface of the ORP electrode is a fast, one-electron exchange between Fe^{+2} and Fe^{+3}, whereas typical redox reactions of the oxidizing agents are slow two-electron exchanges. Fast one-electron exchanges dominate the electrode kinetics and the measured potential, whereas the slower two-electrode exchanges are more influenced by other redox reactions caused by fouling of the source electrode.

Oxydation-reduction potential measurements are also more susceptible to changes in the calibration standard than pH ones. The reason is that ORP standards are not buffered, whereas pH calibration standards are. A buffered standard is one that resists changes in its value. One can add a few drops of acid in a pH 7 buffer and the solution pH will change very little. As the next chapter shows, the chemistry of weak acids underlies this stability. ORP standard solutions, on the other hand, changes their value upon addition of other chemicals. For instance, any recipe for making up ORP standards calls for allowing the solution to sit for several hours before taking final measurements of its value. As oxygen from the air dissolves into the solution its ORP rises. As the calibration solution sits, the chemistry between iron compounds changes the actual ORP reading. Fortunately, most applications

TABLE 2.3 Average variation of standard state potential made by two ORP electrodes (Copeland and Lytle, 2014).

Oxidant	pH 7	pH 8	pH 9
Oxygen	59 mV (10.2%)	38 mV (6.9%)	23 mV (4.6%)
Monochloramine	11 (1.2)	48 (6.7)	47 (7.1)
Permanganate	68 (8.4)	50 (6.3)	73 (10.8)
Chlorine Dioxide	50 (5.1)	78 (8.2)	112 (12.3)
Chlorine	25 (2.5)	24 (2.8)	55 (7.1)

do not require exacting ORP values. For example, effective disinfection requires that the ORP of the water be between 650 and 800 mV. Accuracy to within 10 mV is good enough.

One important point to always remember when making ORP measurements is that the ORP value you measure is with respect to Ag-AgCl reference electrode—not the SHE electrode, which constitute most tabulations of standard ORP values. Just remember that, to go from the ORP reading on your analyzer to the ORP reading relative to the SHE, add 199 mV.

8.3 Why ORP Measurements Are Usually Very Slow

One point bears repeating: an ORP measurement is an aggregate of every possible redox reaction in the water. In a standard solution such as Zobell's or Light's, the redox reactions consist mostly of conversion between Fe^{+2} and Fe^{+3} in which the iron redox partners are in concentration on the order of parts per thousand. The redox reactions are fast, and it takes less than a minute for an ORP sensor to equilibrate with the solution.

However, an ORP sensor reacts very differently when immersed in a glass of tap water. It might take more than an hour for the sensor reading to reach its final value. Figure 2.9 shows the output response of a combination ORP probe to immersion in a calibration solution and then in tap water. At time close to zero, two ORP probes are immersed in 600 mV calibration solution, and at 14 minutes, both probes are then immersed in tap water. The red line is from a probe with a Pt wire electrode whose tip represents the

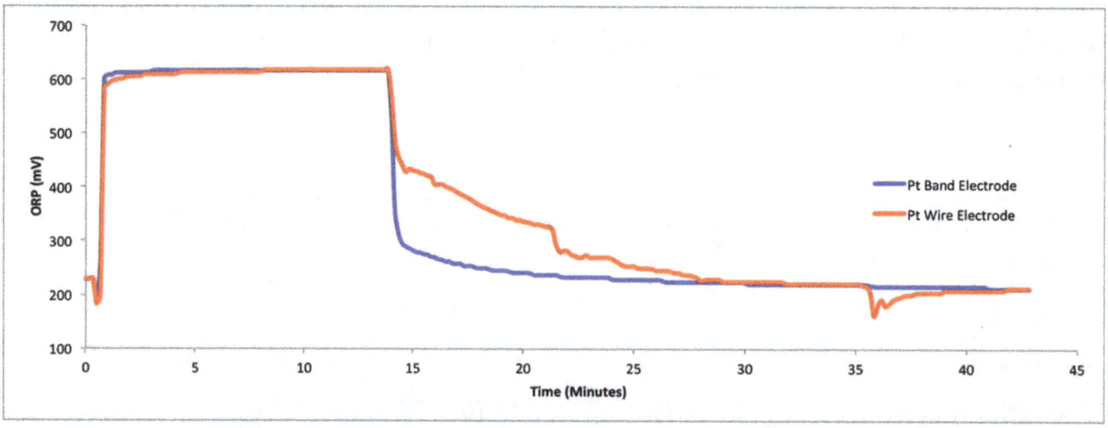

FIGURE 2.9 Time response of ORP sensor in chlorinated water. The red line is for a probe with a wire Pt electrode with a surface area of about 1 mm^2. The blue line is for a probe with a much larger Pt band electrode of about 50 mm^2. First the probes are immersed in 600 mV calibration solution and, at 14 minutes, both are immersed in tap water (Water Analytics).

entire surface area—about 1 mm^2. The blue line is from a probe consisting of a Pt band of approximately 50 mm^2. Its faster response is because of the larger surface area of the electrode. Even for this probe the response time is 10 minutes. In fact, both agree to less than 1%.

The reason behind this slow response is because of the *exchange current density*. This term was introduced back in Section 4.0 to explain the microscopic dynamics of redox reactions at the surface of an electrode. The exchange of electrons between partners of a redox couple occurs at a rate that varies widely. The conversion between ferrous (Fe^{+2}) and ferric (Fe^{+3}) ions is very fast so the redox reaction in either direction is characterized by a large current exchange. In general, redox reactions that involve the exchange of only one electron are fast. Those that involve swapping multiple electrons are slow. These reactions typically involve breaking bonds among reactant molecules and forming them to make product molecules or ions. The predominant oxidizing species in tap water—free chlorine (HOCl) and oxygen—swap two electrons to form their reduced partners:

$$O_2 + 4\,H^+ + 4\,e^- \rightarrow 2\,H_2O \qquad (2.36)$$

$$HOCl + H^+ + 2\,e^- \rightarrow Cl^- + H_2O \qquad (2.37)$$

Not surprisingly these reactions occur more slowly than an iron atom simply losing or adding an electron. Chapter 5, Section 2.0 delves once more into the dynamics of electrodes in its discussion of potentiometric and amperometric sensors.

The presence of dissolved oxygen and free chlorine in ordinary tap water give it a positive ORP value. The H_2O molecule itself is very stable and will not oxidize or reduce without the application of an external voltage. Minerals dissolved in tap water will only react in the presence of reducing agents, and these are rarely present. The only chemical species that can participate in redox reactions with the Ag-AgCl electrode is dissolved oxygen and residual chlorine (or, in some municipalities, monochloramine).

8.4 Why pH Measurements Are Temperature Compensated but ORP Measurements Are Not

As the next chapter shows, pH measurements must be temperature compensated to give an accurate pH measurement. This is because a pH probe doesn't measure the hydrogen ion concentration (or, to be exact, its activity) directly. It measures the difference in potential between the reference and process electrodes. The analyzer inserts this voltage into the Nernst equation to determine the pH. As the Nernst equation shows, the measured voltage varies linearly with temperature.

The ORP value is the probe output voltage—the *E* in the Nernst equation. Therefore, changing the temperature changes the potential of the chemical mixture being studied as well as of the Ag-AgCl electrode itself. Figure 2.10 dramatically illustrates the difference between ORP values in a boiler feedwater between room temperature and 200 °C. Of course, this is not a surprise—we understand that reactions increase with temperature, and redox reactions are no exception. The increase in ORP is real. No correction for temperature needs to be made.

Does this mean that there is no need to attach a temperature element to an ORP probe? The answer is a resounding "no." One still needs to understand how effective a chemical is at oxidizing or reducing substances in the water. If, for example, a treatment facility tries a new disinfectant to treat water, the operator needs to know how effective is that disinfectant. That's difficult to do if it turns out that a high ORP reading was the result of a very hot sample instead of an adequate concentration of the oxidizing agent. When we report ORP values, we also report the temperature. To compare one redox chemical to another we report ORP values at 25 °C. How do we report the ORP value at 25 °C when we measured the process at, say, 37 °C? Easy—just use the Nernst equation. Are you getting the hint that, even if you hate math, you really should know the Nernst equation?

9.0 THE CONNECTION BETWEEN pH AND ORP?

Though they look almost identical pH and ORP probes are very different. But is there any correlation between pH and ORP?

The answer is "usually but not always." If the redox reaction has hydrogen ions participating, then their concentration and, hence, the pH of the system, will affect the redox reactions.

First let's look at a redox reaction involving electron swapping between copper and iron:

$$2\ Fe^{3+} + Cu \rightarrow 2\ Fe^{2+} + Cu^{2+} \quad (2.38)$$

Nowhere are hydrogen ions to be found, so there is no pH dependence on the oxidation of copper by iron. The example of a chemical titration of Fe^{2+} by Ce^{4+} in Chapter 5, Section 4.5 is another example of a redox reaction with no H^+ participation.

Now let's look at another reaction—the reduction of oxygen to make water:

$$O_2 + 4\ e^- + 4\ H^+ \rightarrow 2\ H_2O \quad (2.39)$$

This reaction involves hydrogen ions, so pH will affect the direction of the reaction. Simply looking at this chemical equation tells us that this reduction reaction consumes H^+ ions—that is, it increases the pH. In other words, decreasing the pH increases the ORP value. One can think of the positively charged H^+ ions as balancing out the negatively charged electrons, so the pattern embodied in this simple reaction is one that holds in general.

The Nernst equation allows us to quantify this relationship:

$$\begin{aligned} E &= E^0 - \frac{0.05916\text{ V}}{4}\log\frac{1}{[O_2][H^+]^4} \qquad (2.40) \\ &= E^0 + 0.01479\text{ V}\log[O_2] + 0.05916\text{ V pH (at 25 °C)} \end{aligned}$$

A couple of mathematical tricks with logarithms led to the final expression in Equation 2.40. The salient point of this Nernst equation calculation is that, for every unit decrease in pH, the ORP will increase by 59.16 mV (at 25 °C).

The principle of decreasing pH to increase ORP plays out in the disinfection of water by free chlorine, aka hypochlorous acid (HOCl). The standard potential for the reduction of HOCl is 1.49 V. However, as Equation 2.41 shows, hydrogen ions do alter the reaction. Adding acid drives the reaction to the right, thus increasing the ORP. Because two electrons are involved, the Nernst equation predicts that every unit decrease in pH generates an increase in ORP 0.05916/2 V, or 29.08 mV.

$$HOCl + H^+ + 2\,e^- \rightarrow Cl^- + H_2O \qquad (2.41)$$

This is why the pH of the disinfection process is so important. As the pH rises chlorination is less effective. Figure 2.10 (Journal AWWA, 2014) shows actual ORP values for laboratory samples of free chlorine at different pH values.

(The change in the ORP value of HOCl with pH is only half the story. As Chapter 3 on pH and Chapter 6 explain, the most important effect of pH on disinfection is the relative concentration of HOCl compared to its companion anion, OCl^-. HOCl is a more effective disinfectant than OCl^-).

Both nitrification and denitrification depend on pH. This dependence is:

Nitrification: $2\,NH_3 + 3\,O_2 \rightarrow 2\,NO_2^- + 2\,H_2O + 2\,H^+$ (2.42)

Denitrification: $6\,NO_3^- + 5\,CH_3OH \rightarrow 3\,N_2 + 5\,CO_2 + 7\,H_2O + 6\,OH^-$ (2.43)

Le Chatelier's principle predicts that, as $[H^+]$ increases, the pH decreases, nitrification decreases, and Equation 2.42 shifts to the left. The pH dependence may not be so obvious in the denitrification of Equation 2.43 but, as

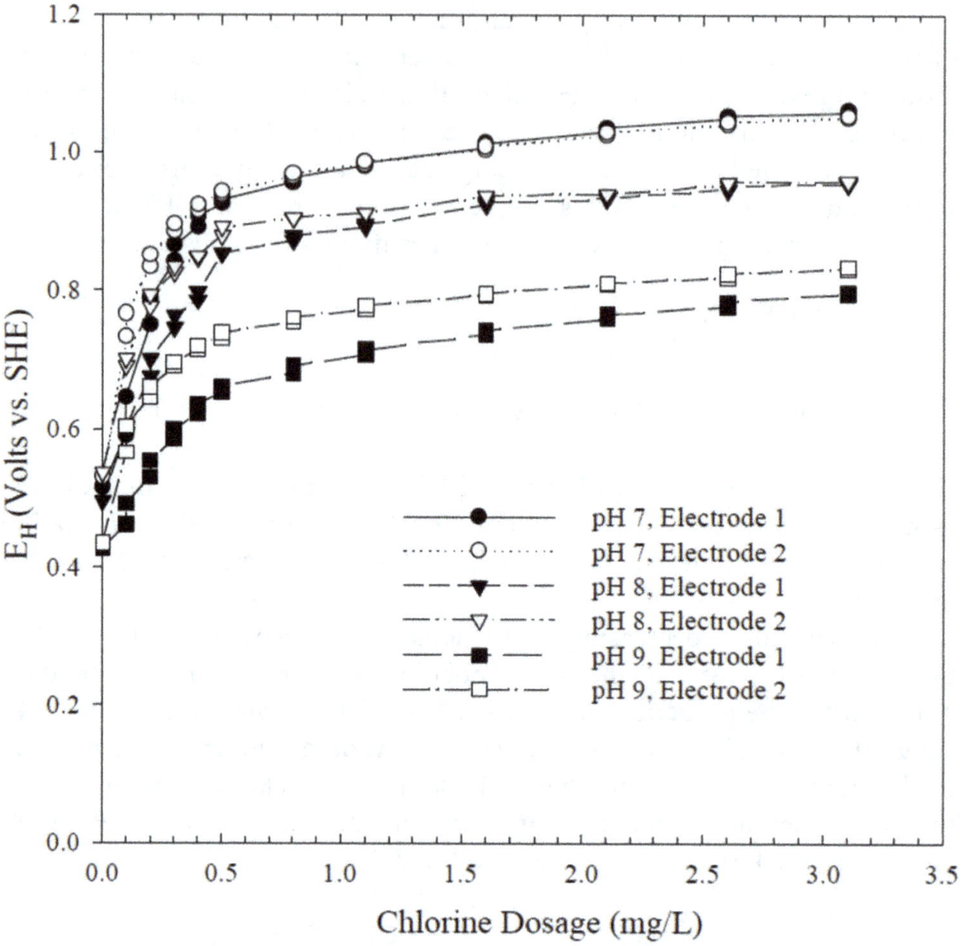

FIGURE 2.10 Effect of pH and chlorine concentration on ORP of water. The 0.4 V potential at zero added chlorine reflects the oxidation potential of the carbonate buffer solution (Reproduced with permission, *Journal AWWA*, 106:01:E10. Copyright © AWWA 2014. All rights reserved.).

will be clear in the next chapter, the hydroxyl ion concentration, $[OH^-]$, is the flip side of $[H^+]$. An increase in the former corresponds to an increase in pH. Therefore, as the pH decreases during denitrification, the rate increases.

To quantify the change in ORP for these two reactions, we recast these two chemical reactions as oxidation half reactions involving hydrogen ions. They are:

$$NH_3 + O_2 \rightarrow NO_2^- + 3\,H^+ + 3\,e^- \tag{2.44}$$

$$NO_2^- + H_2O \rightarrow NO_3^- + 2\,H^+ + 2\,e^- \tag{2.45}$$

Both are oxidation reactions that generate an equal number of electrons and H^+ ions. Therefore, both oxidation reactions increase in ORP by 59.16 mV (at 25 °C) for every unit decrease in pH. The general rule of thumb for redox reactions that are pH sensitive is: the ORP change is equal to the ratio of H^+ ions to electrons multiplied by 59.16 mV.

10.0 ORP APPLICATIONS

As stated in the beginning of the chapter, ORP does not refer to any specific chemical species, like pH, dissolved oxygen, or free chlorine. Neither is it a familiar physical parameter, such as pressure, temperature, or flow. To even experienced faclity operators it's a parameter shrouded in mystery.

The fact is, however, that ORP is a very powerful parameter because it gives a "bottom line" answer to the question: is the process an oxidizing or reducing environment? Can it oxidize ammonia to nitrite or reduce nitrate to nitrogen? Can it break down organic matter or scavenge oxygen? The contents of the water may be a "black box." An ORP sensor doesn't care. It simply measures the redox potential of everything in the water—whether it's ozone, chlorine, sulfite, or dissolved oxygen

The following sections review some common applications for ORP measurement.

10.1 ORP to Monitor Disinfection

The most common use of ORP is to measure the disinfection of treated water.

Disinfection by chemicals is usually via chlorine. What is commonly called chlorine is really hypochlorous acid, or HOCl. Dissolving chlorine gas (Cl_2) in water creates HOCl:

$$Cl_2 + H_2O \leftrightarrow HOCl + HCl \tag{2.46}$$

All acids contain a hydrogen atom that dissociates from its parent molecule as the hydrogen cation, H^+. It is the hydrogen ion that is reactive; the anion that is left does very little. Acids lose their hydrogen ions to varying degrees. A strong acid, like hydrochloric acid (HCl) or sulfuric acid (H_2SO_4), readily dispense with its H^+ ion. HOCl is a weak acid, which means it does not completely dissociate to form H^+ and OCl^-. It forms a mixture of the dissociated form and the neutral precursor. The double arrow denotes the equilibrium that exists between the neutral HOCl molecule and the dissociated H^+ and OCl^-.

$$HOCl \leftrightarrow H^+ + OCl^- \tag{2.47}$$

The next chapter delves more into the topic of weak acids as the foundation of pH buffers. The point is that the undissociated HOCl is a far more effective disinfectant than the OCl^- anion. Equation 2.47 explains why adding H^+—thereby lowering the pH—drives the dissociation reaction to the left and increases the disinfecting power of hypochlorous acid.

There are two relatives of hypochlorous acid: sodium hypochlorite, NaOCl, and calcium hypochlorite, $Ca(OCl)_2$. Both dissolve in water to form OCl^- and the corresponding cation—Na^+ or Ca^{2+}. If one lowers the pH of an OCl^- solution, it will form HOCl. A solution of NaOCl, of concentration as high as 18%, is commonly called *bleach*.

HOCl kills bacteria through a wide range of reactions affecting cell processes such as glucose oxidation, DNA replication, and ATP (energy storage molecule) synthesis (Virto et al., 2005). All of these oxidation reactions involve the reduction of HOCl as summarized in the two HOCl reactions that are listed in Table 2.1:

$$HOCl + H^+ + e^- \rightarrow Cl^- + H_2O \; ; \; E^0 = 1.49 \text{ V} \tag{2.48}$$

$$2\ HOCl + 2\ H^+ + 2\ e^- \rightarrow Cl_2 + 2\ H_2O \; ; \; E^0 = 1.63 \text{ V} \tag{2.49}$$

Which of the preceding reactions take place depends on pH. At lower pH values (higher H+ concentrations). Equation 2.49 is the dominant one.

The hypochlorite ion, OCl^-, undergoes one of the following reduction reactions:

$$OCl^- + H_2O + e^- \rightarrow \frac{1}{2}Cl_2 + OH^- \; ; \; E^0 = 0.42 \text{ V} \tag{2.50}$$

$$OCl^- + H_2O + 2\ e^- \rightarrow Cl^- + 2\ OH^- \; ; \; E^0 = 0.89 \text{ V} \tag{2.51}$$

The much larger potentials of Equation 2.48 or 2.49 versus that of Equation 2.50 or 2.51 explain, partly, why HOCl is a more powerful oxidizing agent than OCl^-. (The other part is that, as a neutral species, HOCl is able to penetrate the bacterial cell wall more easily.)

The term *free chlorine* refers to the sum of HOCl and OCl^-. In addition to disinfecting at a water treatment facility, free chlorine leaves the facility as *residual chlorine* to persist throughout the distribution system. However, a chorine analyzer only measures HOCl concentration (unless acid is added to the aqueous sample to convert OCl^- to HOCl). For a pH greater than 7.5 the HOCl concentration is less than 50% of total chlorine and drops to less than 15% at a pH of 8.5. To obtain the missing OCl^- concentration the analyzer uses a separate pH probe or manual input of the pH value and Equation 6.10 to obtain the total free chlorine concentration. But there is a problem. Two identical concentrations of free chlorine but at two different pH values will have different fractions of HOCl and, thus, different disinfecting powers. This is typically not a big problem in water treatment facilities because the pH of water undergoing treatment does not vary much over time. But it does add uncertainty to the interpretation of free chlorine.

Table 2.4 lists the survival times of several pathogens of great concern to any water or wastewater facility. The table clearly shows that if there are any bacteria in water and chlorine is added to bring the ORP to 665 mV, the microbes will die.

Spa and pool vendors will often sell an ORP analyzer for measuring the disinfection of the water. In fact, some so-called chlorine analyzers are actually ORP analyzers that have been calibrated to free chlorine concentrations.

Another way of expressing disinfection efficacy is to plot the efficacy removal at a logarithm of fractional removal of *E. coli* bacteria as a function of ORP. Figure 2.11 shows data taken from Victorin et al (1972). The kill fraction is expressed as a logarithm of 10—that is, a log-10 removal value of 2 means 99% deactivation, or 10^{-2} survival. Survival rates were measured after 3 minutes of exposure to free chlorine. As Figure 2.11 shows, a log-10 removal of 2 is achieved at an ORP value of approximately 610 mV, which is consistent with Table 2.4.

The point of this section is not to suggest that ORP probes are preferable to chlorine analyzers for measuring disinfecting efficacy. In fact, treatment facilities dose chlorine, so it's only natural that they measure it. As the saying goes, "Don't dose what you can't measure." Another caveat is that the ORP values listed in Table 2.4 apply only to chlorine. Different oxidizing agents kill pathogens through different oxidation reactions, so their optimum working ORP values are not going to be the same as that for chlorine. For instance, peracetic acid (PAA) is increasingly used as a disinfectant for food because it does not oxidize organic matter to create disinfection by-products. Its germicidal actions are complex and include

TABLE 2.4 Survival times of pathogens at three ORP values (Suslow, 2004).

	Survival (seconds) at ORP (mV)		
ORP Value (mV)	<485	550 to 620	665
E. Coli 0157:H7	>300	<60	<10
Salmonella spp.	>300	>300	<20
Listeria	>300	>300 S	<30
Thermotolerant Coliform	>48 hours	>48 hours	<30

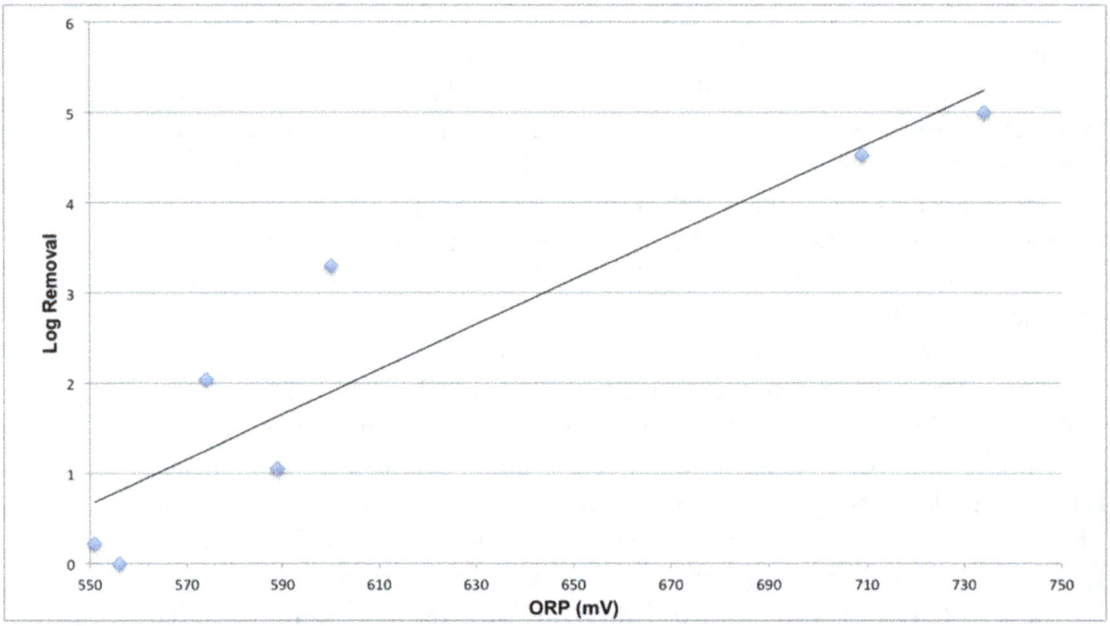

FIGURE 2.11 Log-10 removal of *E. coli* after 3 minutes as a function of ORP value. A log-10 removal means 99% removal. The graph is the author's display of data reported by Victorin et al. (1972).

denaturing proteins, disrupting cell walls, and oxidizing chemical bonds on proteins and enzymes (Buchanan, 2015). Studies by Luukonen et al. (2014) showed that ORP values corresponding to disinfecting levels of PAA—greater than 1 ppm—leveled off at 600 mV. (The authors hypothesized that, as the PAA concentration increase beyond 2 ppm, it decomposes to hydrogen peroxide, H_2O_2.)

Last, ORP sensors have limitations that are discussed in Chapter 6. The point is that, if you know that chlorine is the predominant oxidizing species and you want to know how effectively it will disinfect a sample of water, then measuring ORP gives the answer.

10.2 Monitoring Nitrification and Denitrification

The two primary goals of wastewater treatment are to break down organic matter and convert toxic ammonia, a by-product of organic decay, into harmless nitrogen. The first step of ammonia breakdown is the oxidation of ammonia (NH_3) to nitrite (NO_2) by a group of ammonia-oxidizing bacteria:

$$2\ NH_3 + 3\ O_2 + \rightarrow 2\ NO_2^- + 2\ H_2O + 2\ H^+ \tag{2.52}$$

A second set of bacteria further oxidize nitrite to nitrate:

$$2\ NO_2 + O_2 \rightarrow 2\ NO_3^- \tag{2.53}$$

Because both ammonia and nitrite are oxidized by oxygen, it makes sense that a dissolved oxygen analyzer be used to ensure that there is just enough dissolved oxygen to do the job. Aeration of wastewater is one of the largest uses of electricity in a wastewater treatment facility so excess dissolved oxygen means wasted money. In fact, many treatment facilities set the dissolved oxygen concentration in the aerobic zone to between 2 and 4 ppm. Studies show that levels of dissolved oxygen as low as 0.5 ppm can be sufficient to ensure complete nitrification (Bellucci et al., 2011).

Measuring such low concentrations of dissolved oxygen is subject to error when the dissolved oxygen level is less than 2 ppm. But measuring the ORP corresponding to low levels of dissolved oxygen is straightforward. For the first 80 hours of treatment, the dissolved oxygen concentration is so low that it is immeasurable. Yet measurements of ORP (and pH) easily track the progress of nitrification. Typical ORP values for nitrification range from 100 to 350 mV.

That the ORP measurement can provide complementary, and sometimes more accurate, information of dissolved oxygen than the measurement of dissolved oxygen itself should come as no surprise. We only need to take another look at the Nernst equation to remember that the voltage from an ORP sensor varies linearly with the logarithm of the concentration of oxygen. Let's suppose that the dissolved oxygen concentration changes from 10 ppm to 1 ppm the ORP will decrease by 59.16 mV (at 25 °C). This is easily measurable but not preferable to the direct dissolved oxygen measurement. On the other hand, let's suppose that the dissolved oxygen concentration now drops from 0.1 ppm to 0.01 ppm. In this case a dissolved oxygen analyzer is effectively stuck at zero yet an ORP analyzer records the same −59.16 mV. Of course, the dependence of the ORP signal on the logarithm of the analyte concentration is a double-edged sword. The change from, say, 8 ppm to 9 ppm registers a 12.5% increase in the reading of a dissolved oxygen analyzer but a 5.1% in an ORP analyzer. For changes of low levels of dissolved oxygen, the ORP analyzer works great but, for changes in moderate levels, the dissolved oxygen probe is superior. The ORP probe is also subject to misleading ORP readings caused by other oxidizing agents.

After nitrification converts ammonia to nitrate, a third set of bacteria reverses the direction of the ORP and denitrifies nitrate to nitrogen. This was the reaction introduced in the beginning of the book:

$$6\ NO_3^- + 5\ CH_3OH \rightarrow 3\ N_2 + 5\ CO_2 + 7\ H_2O + 6\ OH^- \quad (2.54)$$

The bacteria that reduce nitrate to nitrogen are *facultative*, which means that they can use oxygen for metabolism but, in the absence of oxygen, can resort to using nitrate to oxidize a carbon source, the latter of which, in

2.54, is methanol (CH_3OH). To force these "ambidextrous" bacteria to use a carbon source over oxygen means that a wastewater treatment facility must maintain a dissolved oxygen as close to zero as possible. Note that, in the first two nitrification reactions, the nitrogen species (ammonia and nitrite) are oxidized while oxygen is reduced. In this third reaction the nitrogen species (nitrate) is reduced while methanol is oxidized. Because wastewater treatment is nitrogen-centric we say that the first two reactions are "oxidations" and the third is a "reduction." In truth all three reactions consist of the necessary oxidation and reduction half-reactions.

An environment in which dissolved oxygen is absent, but nitrate is present is an *anoxic* environment. The term *anaerobic* denotes an environment devoid of oxygen or any form of combined oxygen. A convenient mnemonic device for connecting terminology with function is to rewrite "anoxic" as aNO_xic. Here NOx refers to any nitrogen oxide, mainly NO_2^- or NO_3. During denitrification the ORP value is maintained between −50 and +50 mV. This neutral ORP value is the consequence of the oxidizing agent, NO_3^-, and the reducing agent, CH_3OH, being in approximately equal concentration.

It might seem odd that, if denitrification is the third stage in nitrogen removal, this process occurs in the front end of the nitrogen removal train rather than the back. The reason is that incoming wastewater is naturally low in dissolved oxygen and high in organic content, which is just what Equation 2.54 requires. To enable denitrification to work in the proper time sequence the effluent from the aerated section of the oxidation ditch is returned to the front of the line to bring fresh nitrate into the mix.

Figure 2.12 is a graphical display of the ORP values for common biochemical reactions found in wastewater treatment. These ORP values are listed with respect to the standard hydrogen electrode. As just mentioned, ORP values in the aerobic regime, where nitrification dominates, is at values greater than 100 mV. ORP values under anaerobic conditions are less than −50 mV and facilitate the generation of fatty acids, hydrogen sulfide, and phosphate release. Denitrification occurs between −50 and +50 mV.

Figure 2.13 shows ORP readings measured during actual nitrification and denitrification through wastewater processing. Keep in mind that these values are relative to an Ag-AgCl electrode, which means they are 200 mV lower than values in Figure 2.12. The so-called *Nitrate Knee* shows the point at which nitrification stops and denitrification begins.

10.3 Phosphorus Removal

There are two major nutrients that must be treated and minimized in wastewater treatment: nitrogen and phosphorus. The removal of both by bacteria is lumped together as *biological nutrient removal*. The previous section

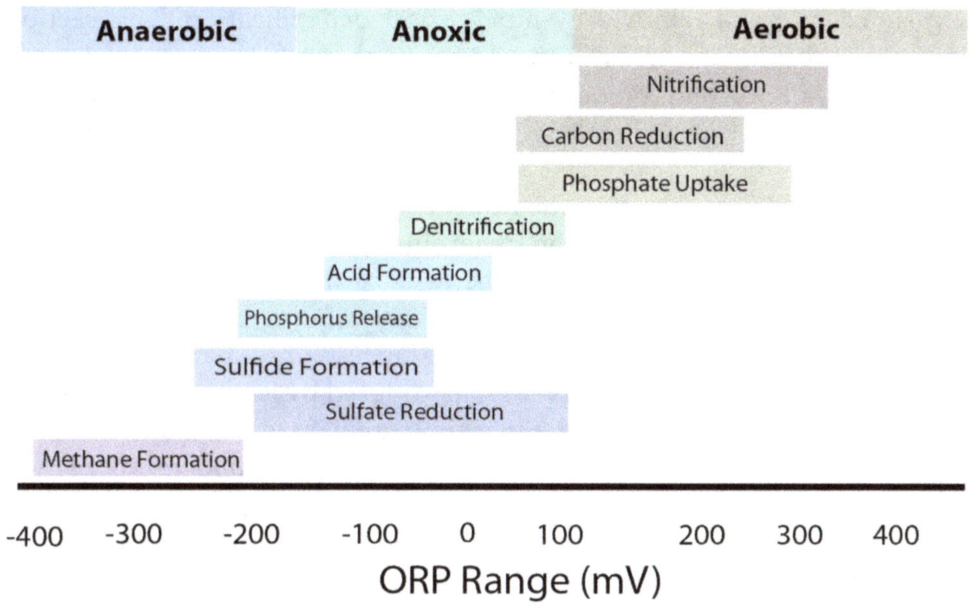

FIGURE 2.12 Graphical display of ORP values for biochemical reactions in wastewater treatment (Adapted from Kelly, 2020).

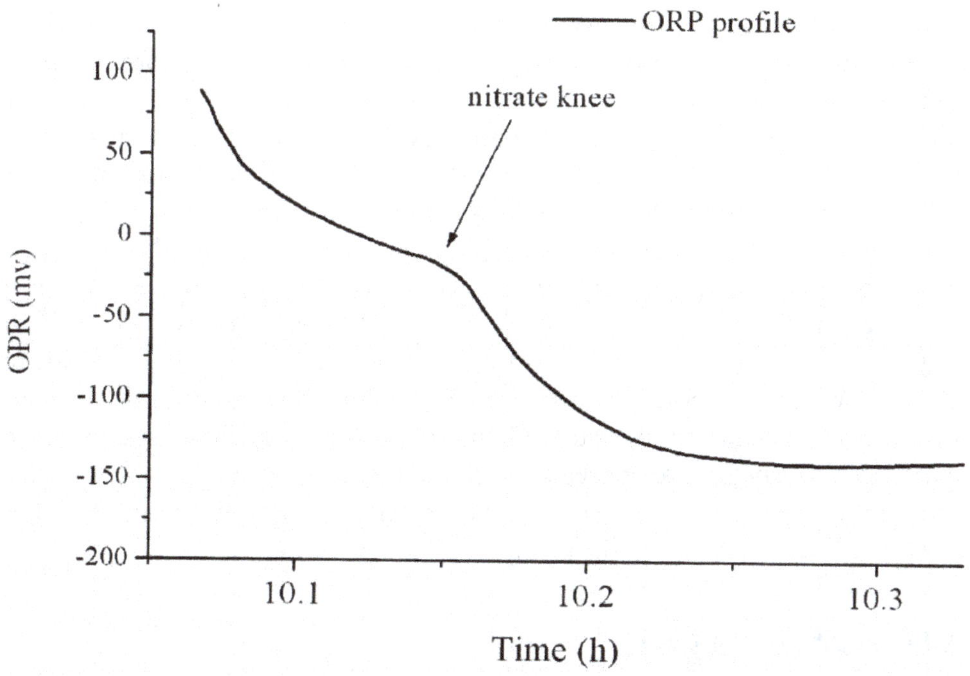

FIGURE 2.13 Monitoring of denitrification through ORP monitoring. These ORP values are reported relative to an Ag-AgCl reference electrode (Filali-Meknassi et al., 2005).

explained nitrogen removal by nitrification–denitrification. This section explains phosphorus removal. In one sense, this is a tougher nutrient to address because it only takes on the order of 0.1 mg/L of phosphorus to cause the eutrophication in lakes, rivers, and streams that manifest itself as algae blooms.

Biological phosphorus removal, like biological nitrogen removal, requires an aerobic stage and an anaerobic one. Both biochemical pathways are very different, yet, unlike the case for nitrogen, both involve the same set of bacteria called *phosphorus-accumulating organisms* (PAOs). The first phase is anaerobic in which PAOs ingest a group of organic compounds called *volatile fatty acids* (VFAs) produced by fermentative bacteria. (The simplest VFA is acetic acid, the main ingredient in vinegar). The PAOs ingest VFAs and convert them to fuel (*polyhydroxyalkanoate*s or PHAs). The energy to do this comes from breaking up chains of phosphate (PO_4^{2-}), called *polyphosphate*, into their individual phosphate units. Phosphate is also referred to as *orthophosphate* to distinguish it from polyphosphate. The PAOs release the orthophosphate as waste products. This stage actually releases phosphate into the wastewater liquor, so it appears to be contrary to the goal of phosphorous removal. However, it is actually the prelude to something much more promising. The anaerobic condition of this stage—known as *fermentation*—also kills bacteria that would otherwise interfere with the subsequent aerobic stage.

In the aerobic stage that follows, the PAOs burn their stores of fuel, PHA, to grow and reproduce. To stock up on food that they will use in their anaerobic stage, they ingest large quantities of PO_4^{2-} and store it as energy-rich polyphosphate. This uptake in phosphorus more than compensates for the phosphorus that was released in the anaerobic phase.

Figure 2.14 (Samstag, 2000) displays the ORP and dissolved oxygen levels in a wastewater treatment facility in Washington state during and after aeration. As aeration is active the ORP rises along with dissolved oxyen the ORP. Ammonia levels drop as it is oxidized to nitrite and nitrate (NOx). When aeration is turned off, the dissolved oxygen levels drop to zero (an anoxic condition) and ORP values become negative. The ammonia concentration begins to increase, and the NOx concentrations decrease. Note that, when the dissolved oxygen concentration values read zero, the ORP readings continue to decrease.

10.4 Sulfide Removal

The rotten egg smell coming out of sewage and volcanoes is hydrogen sulfide—H_2S. It's a good thing this chemical is so pungent because, molecule-for-molecule, it is more lethal than hydrogen cyanide. Hydrogen sulfide

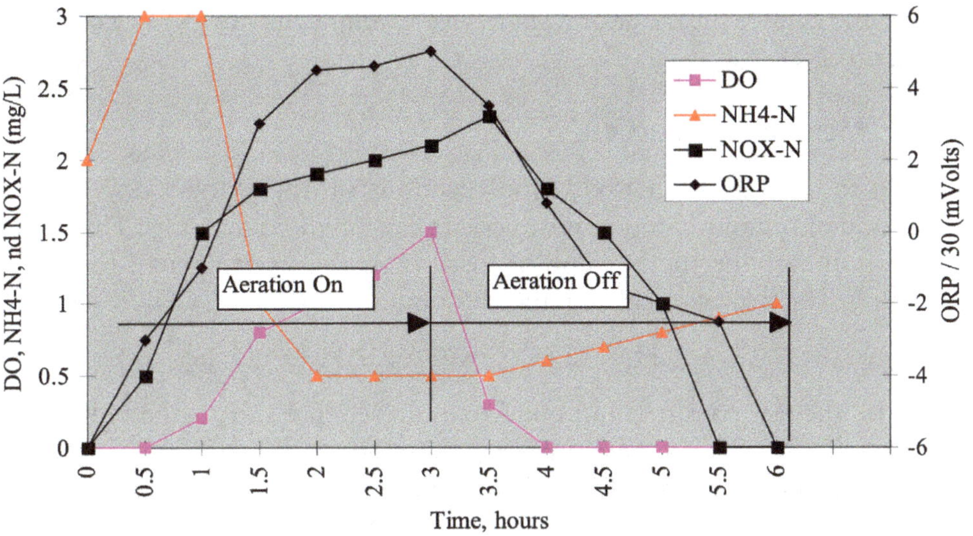

FIGURE 2.14 ORP, dissolved oxygen, ammonia, and nitrite/nitrate (NOx) values in a wastewater treatment during and after aeration. Note that the ORP readings continue to display real (negative) values even when dissolved oxygen values are zero (Samstag, 2000).

forms in an anaerobic environment that is common in wastewater and septic systems. In this environment several species of bacteria reduce sulfate (SO_4^{2-}) to sulfide (HS^{2-}). Figure 2.12 shows that this happens over an ORP range of −50 to −250 mV. Hydrogen sulfide is a strong acid that dissociates into sulfide, HS^-, and H^{+-}:

$$H_2S \rightarrow HS^- + H^+ \tag{2.55}$$

The oxidation of sulfide to sulfate is just the reverse of the formation of HS^-:

$$HS^- + 4\ O_2 \rightarrow 2\ SO_4^{2-} + 2\ H^+ \tag{2.56}$$

(A pilot demonstration [Duangmanee et al., 2007] showed that the ORP of an anaerobic digester started at −260 mV and increased to 0 after aeration.)

You may have noticed that nitrate (NO_3^-) and sulfate (SO_4^-) are compounds of nitrogen and sulfur, respectively, that have been oxidized by oxygen. The reduced forms of these two elements are ammonia (NH_3) and hydrogen sulfide (H_2S). As a general rule of thumb, an element with one or more H atoms is in a reduced form. Conversely, an element attached to one or more O atoms is likely to be in an oxidized form. The more O atoms, the greater the oxidation state—for example, nitrite (NO_2^-) oxidizes further to NO_3^{2-}.

10.5 Methane Production

One of the most famous oxidation reactions on the globe today are the oxidation of organic compounds to create carbon dioxide (CO_2). This is the reaction underlying the burning of fossil fuels that is attributed to global warming. The prototypical organic compound is methane, CH_4.

$$CH_4 + 2\ O_2 \leftrightarrow CO_2 + 2\ H_2O \tag{2.57}$$

With the last paragraph in mind one can see that CH_4 is the reduced C compound, and CO_2 is the oxidized compound.

The reverse of Equation 2.57 is the reduction of CO_2. While it is disadvantageous in a sewer, it is increasingly valuable in a wastewater treatment facility or landfill. *Methanogenesis*, also known as biogas generation, transforms wastewater facilities from consumers of electricity to generators of it.

An alternate reduction of CO_2 is reaction reduction by hydrogen gas, the latter of which yields both methane and water:

$$CO_2 + 4\ H_2 \rightarrow CH_4 + 2\ H_2O \tag{2.58}$$

Dissolved organic compounds found in wastewater, such as VFAs, are a main source for methane. Acetic acid is a common VFA that reduces *and* oxidizes to CO_2 and CH_4. Its methanogenesis reaction is:

$$CH_3COOH \rightarrow CH_4 + CO_2 \tag{2.59}$$

As Figure 2.12 shows, methanogenesis occurs in an ORP environment of −180 to −320 mV. By ensuring that there is just enough oxygen to keep the ORP between −180 and −230 mV, Khanal and Huang discovered that they could maximize the methane production and minimize the sulfide generation (Khanal & Huang, 2003).

10.6 Redox Titration for Precise Measurement

Titration is a common technique for determining the concentration of a dissolved substance using another one—the reagent—of known concentration to react with the former. One adds an increasing measured volume of the reagent until it reacts with all of the chemical of interest. In most titrations, a colored reagent is used and, when that point is reached, the color changes dramatically. Colorimetric measurements use reagent-based measurements. The user adds a fixed amount of reagent and the color of the mixture changes gradually or suddenly. For the former, one measures the tint of the reagent color and compares it to a lookup chart that matches tint to concentration. For the latter one determines the volume of reagent required to induce the sudden change of color. A common example is the N,N-diethyl-p-phenylenediamine [DPD]-based measurement of free chlorine—explained in Section 3 of the chapter on chlorine. In a *redox titration* one measures the concentration of one member of a redox pair by adding known concentrations of the other member and watching the ORP reading swing from one redox state to the other. The substance of interest is the *titrant* or *analyte*, and the titrating agent is the *titrant*.

Take the simple example of Ce (IV) (Ce^{4+}) reduced as Fe (II) (or *ferrous* or Fe^{2+}) is oxidized. From Table 2.1 we get the redox potentials:

$$Ce^{4+} + e^{-} \rightarrow Ce^{3+} \; ; \Delta E = 1.70 \text{ V} \tag{2.60}$$

$$Fe^{2+} \rightarrow Fe^{3+} + e^{-} \; ; \Delta E = +0.768 \text{ V} \tag{2.61}$$

$$Fe^{3+} + Ce^{4+} \rightarrow Fe^{2+} + Ce^{4+} \; ; \Delta E = 0.932 \text{ V} \tag{2.62}$$

The stoichiometry is a simple 1:1 of oxidant to reductant. Suppose we start off with a solution containing one mole of Fe^{2+} ion. (If this were a real titration, we would not know we had one mole solution of ferrous ion. All we would know is the volume of the sample.) Until we oxidize some of the ferrous ions to Fe^{3+} the ORP value is ill-defined because the denominator in the Nernst equation is zero. We add a known volume and concentration of Ce^{4+} titrant. As we do, Fe^{+2} oxidizes to Fe^{+3} and that same amount (in moles) of Ce^{4+} reduces to Ce^{3+}. The right-hand side of Figure 2.15 shows a generalized schematic of the setup. As the first drops (*aliquot*) of titrant react, we read an ORP slightly higher than 0.77 V, the exact value determined by the Nernst equation. Rather than calculate the ORP as we add progressively more Ce^{4+}, let's jump all the way to the end of the titration, at which point we have added a considerable (e.g., tenfold) excess of the Ce (IV) titrant. At that point we expect to read a value close to 1.70 V. All along the way we note the ORP value as we add one aliquot of Ce (IV) after another.

The point that is interesting is in the middle—the point at which just enough Ce^{4+} is added to oxidize all the Fe^{2+} in solution (1 mole in this case). At this point $[Ce^{4+}] = [Fe^{2+}]$ and $[Ce^{3+}] = [Fe^{3+}]$ and is called the *equivalence point*. If we plug these into the Nernst, the ratio of product concentrations to reactant concentrations—$[Ce^{3+}][Fe^{3+}] / [Ce^{4+}] [Fe^{2+}]$ is exactly one. The logarithm of one is zero, and the ORP value is the average of both standard state ORP values, 0.768 V and 1.70 V, or 1.23 V. (Note that if the stoichiometry of the redox Fe^{2+} to Ce^{4+} were not one, then the ORP value of the equivalence point would be a weighted average.) If we work through the Nernst equations for several steps along the way, we get

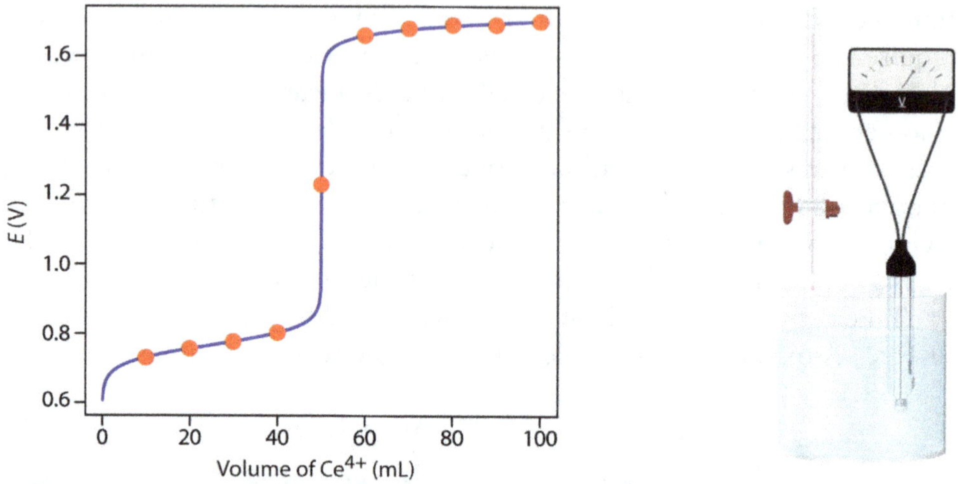

FIGURE 2.15 Left: Titration of Fe^{+2} by Ce^{+4} to yield Fe^{3+} and Ce^{4+}. The end points reflect the standard state potentials Fe^{2+} at the beginning and Ce^{3+} at the end. The sharp inflection at the equivalent point occurs at the average of the two standard state potentials (Harvey, 2016). Right: A schematic of the titration setup. A burette dispenses measured amounts of titrant into the sample while the ORP analyzer measures the changing ORP value.

a curve like Figure 2.15 (Harvey, 2016). The sharp inflection at the equivalence point is a consequence of the denominator ($[Ce^{4+}]$ $[Fe^{2+}]$) becoming very small and the ratio therefore becoming very large. This behavior makes the redox titration very accurate at determining concentrations of a redox-active species.

Redox titrations can also be done through color change, hence eliminating the need for a ORP analyzer. Some oxidizing agents change color as they convert from their oxidized state to a reduced state. For instance, permanganate (MnO_4^-) changes from bright purple to pink when it reduces to Mn^{2+}. We could just have easily used a solution of potassium permanganate ($KMnO_4$) to determine the concentration of Fe^{2+}.

Let's look at another example—one that involves pH changes. Although the heyday of chrome bumpers and appliances is past, chrome and other metal plating is still a big industry. One oxidation state of chromium, the highly oxidized Cr^{6+}or Cr (VI), is a serious health threat in our drinking water. In nature Cr (VI) actually exists as a mixture of two oxides.

$$Cr_2O_7^{-2} + H_2O \leftrightarrow 2\ CrO_4^{-2} + 2\ H^+ \qquad (2.63)$$

As Equation 2.63 shows, the ratio of one species to the other depends on pH. The more reduced species, Cr^{3+}, or Cr (III), is relatively innocuous.

It also forms insoluble salts, such as $CrCl_3$, that can be easily removed. All that is needed is a reducing agent to donate the three electrons needed to reduce Cr^{6+} to Cr^{3+}. Sulfur and iron compounds are both commonly used reducing agents. For example, the ferrous (Fe^{+2}) salt, $FeSO_4$, reduces the Cr(VI) anion chromate (CrO_4^{2-}). The two half-reactions can be found in Table 2.1. For convenience they are reproduced here:

$$CrO_4^{-2} + 8\ H^+ + 3\ e^- \rightarrow Cr^{3+} + 4\ H_2O\ ;\ E^0 = 1.195\ V \quad (2.64)$$

$$Fe^{2+} \rightarrow Fe^{3+} + e^-;\ E^0 = -0.768\ V \quad (2.65)$$

The sulfate anion was left out because it does not participate in the oxidation of Fe (II). We call it a *spectator* ion. However, it does enter into the process as the anion of both insoluble salts, Cr(III) and Fe(III). The overall reaction is:

$$CrO_4^{-2} + 3\ Fe^{2+} + 8\ H^+ \rightarrow Cr^{3+} + 3\ Fe^{3+} + 4\ H_2O\ ;\ \Delta E = 0.427\ V \quad (2.66)$$

Chromate, CrO_4^{-2}, oxidizes the ferrous ion to a ferric—Fe(III) or Fe^{+3}—to yield the insoluble Cr^{+3} salt, $Cr_2(SO_4)_3$ and the Fe (III) salt, $Fe_2(SO_4)_3$. In sum, the Cr^{6+} cation reduces to Cr^{3+}, and the Fe^{2+} cation oxidizes to Fe^{3+}.

Prior to adding the titrant, Fe (II), the ORP is that of the analyte, which, in the case of CrO_4^{-2}, is 1.195 V under standard conditions. As we add the reducing Fe (II) titrant, the ORP drops. When the molar concentration of the titrant approaches the stoichiometric ratio of the redox reaction—three moles of $FeSO_4$ to one mole of H_2CrO_4—the ORP drops precipitously. At the point at which the concentration ratio exactly equals the stoichiometric ratio, the change in ORP is at its greatest—that is, it reaches the *equivalence point* of the ORP curve. This middle of the inflection determines the concentration of the analyte. If we continue to add titrant, the ORP decrease more slowly and levels off. The ORP value will approach that of the titrant—in this case $FeSO_4$, or −0.768 V at standard conditions.

Keep in mind that this final ORP value does not have to be negative. In this example it is entirely possible that both half-reactions can be reductions with respect to the standard hydrogen electrode. As long as one half-reaction has a smaller potential than the other, it will proceed as the oxidation half-reaction when paired with its "stronger" oxidizing partner.

As long as one knows the stoichiometry of the complete redox reactions and the correct standard state potentials, one can calculate the potential for the equivalence point. For the simplest case, in which there are only two half-reactions and the number of moles of both species of each redox couple is the same—that is,

$$aA_{red} + bB_{ox} \rightarrow aA_{ox} + bB_{red} \tag{2.67}$$

the potential at the equivalence point is given by:

$$E_{eq} = \frac{aE_b^0 + bE_a^0}{a+b} \tag{2.68}$$

In Equation 2.67 the oxidation half-reaction is A_{red} to A_{ox}, and the reduction half-reaction is B_{ox} to B_{red}. For the case of Cr(VI) being reduced by Fe(II) there are actually three half-reactions—the reduction of H^+ to H_2O is the third. For this case the calculation of E_{equiv} is straightforward but not as simple as Equation 2.68. The additional reduction half-reaction is folded into the chromate reduction half-reaction. With this additional pH dependent variable, the calculation of E_{eq} is:

$$E_{eq} = \frac{3\ E_{Fe}^0 + E_{Cr}^0}{4} = \frac{2RT}{F} pH = 0.875\ V - 0.1183\ pH\ (at\ 25\ °C) \tag{2.69}$$

$E^0{}_{Cr}$ is the standard state potential for the reduction of Cr(VI) and $E^0{}_{Fe}$ is the standard state potential for the reduction of Fe(III). The second term can be thought of as a pH correction.

Loosely speaking, E_{eq} is a weighted average of the two E^0's. To see the complete formulation for the equivalent point potential for any redox reaction, see the classic paper by Bard and Simonsen (1960).

As in all ORP titrations, this example of Cr(VI) reduced by Fe(II) can be used for neutralizing Cr(VI) or for determining its concentration.

10.7 Cyanide Destruction

Another common redox reaction in industrial wastewater treatment is the destruction of cyanide (CN^-). This is done in two stages, both using sodium hypochlorite (NaOCl).

$$CN^- + OCl^- \rightarrow CNO^- + Cl^- \tag{2.70}$$

$$2CNO^- + OCl^- + H_2O \rightarrow 3\ Cl^- + N_2 + 2\ HCO_3^- \tag{2.71}$$

As in the previous example, these two redox reactions are written without the spectator Na^+ ions—for example, CN^- is written instead of NaCN. In the first reaction, cyanide (CN^-) oxidizes to cyanate (CNO^-), and in the second, cyanate oxidizes to harmless Cl^- and bicarbonate, HCO_3^-. An idealized redox titration curve for both reactions is shown Figure 2.16 (based partly on Liptak, 2005). Since the starting analyte, CN^-, is a reducing

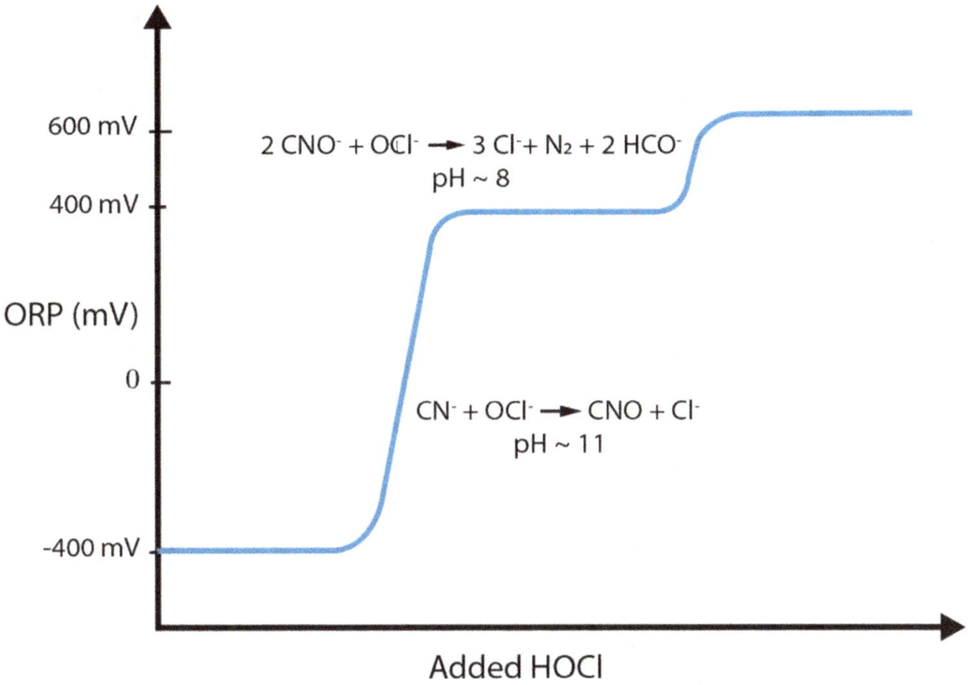

FIGURE 2.16 ORP of cyanide destruction by hypochlorite in two reactions. The first oxidizes lethal cyanide to nasty cyanate, and the second oxidizes nasty cyanate to harmless sodium chloride and sodium bicarbonate.

agent, the ORP is negative, and it increases toward positive ORP values as one adds oxidizing agents. You can clearly see that the titration curve is a combination of both oxidation reactions. As in the Cr(VI) example above, the purpose of the redox titration is not to measure the concentration of cyanide but to use just enough hypochlorite to destroy it. Once again, an inexpensive ORP analyzer ensures that one adds just the right amount of chemical to do the job.

10.8 Protection Against Corrosion

The many applications of ORP measurement can fill many pages. I'll end this section with just one other application that can be monitored and controlled with an ORP sensor.

The corrosion of steel, iron, and copper at normal temperatures is a very familiar sight. At the high temperatures that predominate in boilers, corrosion can corrode stainless steel pipe in a matter of days. This is because of the oxidation of steel that accelerates with temperature. For this reason, the water that is fed into boilers is de-aerated to less than 10 parts per billion (ppb) of dissolved oxygen. In addition to the chemical degradation of steel, flow can remove the protective oxide coat on steel piping in a process known as flow accelerated corrosion (FAC). In 1986 FAC in Virginia Power's Surry Nuclear Power Plant resulted in a high-pressure explosion that killed four workers.

One might assume that a dissolved oxygen analyzer that measures very low concentrations of oxygen would be the obvious choice for this process. However, as already discussed in Section 10.2, measuring minute amounts of dissolved oxygen can sometimes be done more accurately via ORP than by dissolved oxygen itself. In the case of boiler feedwater, the case for ORP measurement is even stronger. That's because the ultimate goal for the plant operator is not to minimize dissolved oxygen in the boiler feedwater but to maintain a reducing environment. A positive ORP means that oxidation of the iron in the steel pipe will occur. Operators add a reducing chemical—usually hydrazine—to ensure that the ORP is negative. Hydrazine reacts with dissolved oxygen—that is, a *scavenger*—and reduces the concentration of dissolved oxygen below 10 ppb.

In a study of power plant steam generators, Sampson and Hicks measured the ORP at the *in-situ* temperature of the boiler feedwater (app. 200 °C), ORP at room temperature, and dissolved oxygen (Sampson & Hicks, 2009). Figure 2.17 shows the results of introducing trace amounts (<10 ppb) of dissolved oxygen. The ORP readings at the feedwater temperature show a stronger response to the oxidizing environment generated by the oxygen than the dissolved oxygen measurement itself. The ORP sensor can easily track changes in ORP resulting from trace amounts of dissolved oxygen and hydrazine.

In 2004, the Electric Power Research Institute (2000) recommended ORP measurement at the boiler feedwater temperature as the primary means to monitor corrosion potential in copper pipes.

A very analogous application is the neutralization of chlorine—*dichlorination*—in the effluent of wastewater treatment. Sodium bisulfite ($NaHSO_3$) is added to eliminate residual free chlorine (HOCl) to ensure that the latter does not enter.

$$HOCl + Na\ HSO_3 \rightarrow H_2SO_4 + 2\ HCl + Na_2SO_4 \qquad (2.72)$$

As the bisulfite reduces the free chlorine, the ORP value drops from approximately 600 mV to approximately 300 mV.

10.9 The Final Word

The ORP sensor is fundamentally unique in water quality instrumentation. It measures a property of water rather than a chemical species. That property

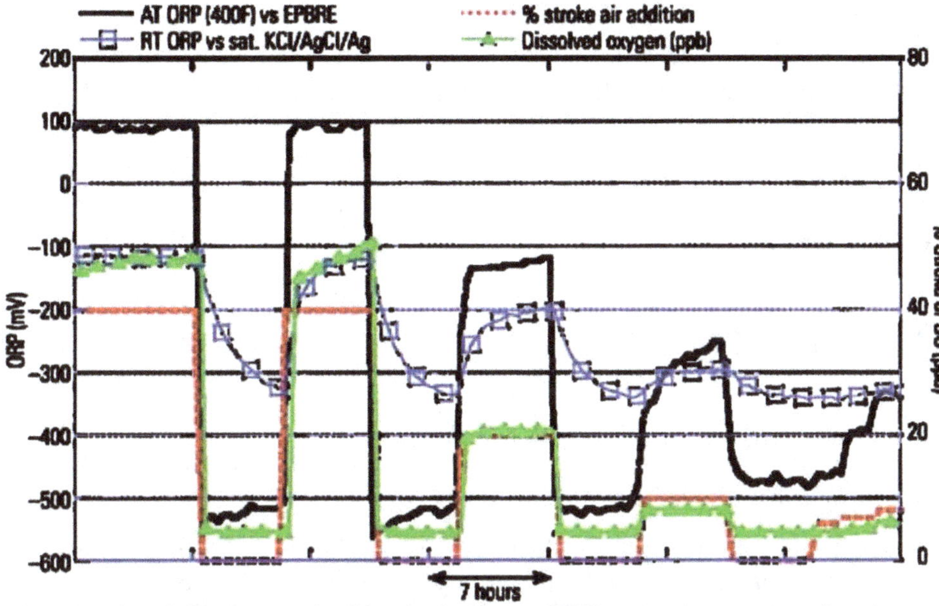

Notes: Ag = silver, AgCl = silver chloride, DO = dissolved oxygen, EPBRE = external pressure-balancing reference electrode, KCl = potassium chloride, ORP = oxidation-reduction potential, RT = room-temperature

FIGURE 2.17 Oxygen is introduced at ppb levels into a boiler (red). ORP at boiler feedwater temperature (AT ORP) in black, room temperature (RT-ORP) is in blue, and dissolved oxygen concentration is in green. The ORP reading provides an even more sensitive indicator of the presence of an oxidizing environment than direct measurements of dissolved oxygen (Sampson & Hicks, 2009).

is oxidation-reduction, or redox, potential. Like the conductivity sensor the redox potential that the ORP sensor measures is an aggregate property of all chemical species in water—in this case of the potentials of all dissolved species. The world of water chemistry is overwhelmingly dominated by reactions in which the chemical constituents swap electrons among themselves. Those reactions drive such vital processes such as disinfection, nitrification, denitrification, phosphate removal, methane production, hydrogen sulfide removal, precipitation of heavy metals, and corrosion. This wide range of chemical reactions that an ORP sensor monitors underscores just how powerful it is.

The lack of understanding of ORP is the main reason why many users shy away from using it in their installation. A second reason is that, like any sensor technology, there are limitations that users do not properly understand. One limitation is that the measurement can be very slow when done in relatively clean water. If, for example, disinfectant concentration changes on the time scale of minutes, then there is no way an ORP analyzer will be able to keep up with the change and control the release of disinfectant to maintain a constant level. However, if the slow response is not an issue—which is usually the case in water treatment—the ORP sensor is a valuable tool deserving of a second look. A second limitation is that an ORP sensor measures the aggregte redox potential of all species in the sample; it has no ability to distinguish one from the other. For applications in which the species are known or the need to know their identities is not important, ORP measurement is a critical one.

11.0 REFERENCES

Bard, A. J., & Simonsen, S. H. (1960). The general equation for the equivalence point potential in oxidation-reduction titrations. *Journal of Chemical Education, 3*(7), 364–366.

Bard, A. J., Parson, R., & Jordan, J. (1985). Standard potentials in aqueous solution. *Monographs in Electroanalytical Chemistry and Electrochemistry*. Marcel Dekker, Inc.

Bellucci, M., Ofiteru, I. D., Graham, D. W., Head, I. M., & Curtis, T. P. (2011). Low-dissolved-oxygen nitrifying systems exploit ammonia-oxidizing bacteria with unusually high yields. *Journal of Applied and Environmental Microbiology, 77*(21), 7787–7796.

Buchanan, J. R. (2015). Peroxyacetic acid as an alternative disinfectant for onsite and decentralized wastewater systems. *NOWRA Proceedings*.

Copeland, A., & Lytle, D. A. (2014, January). Measuring the oxidation-reduction potential of important oxidants in drinking water. *Journal AWWA*, 2014.106.0002.

Duangmanee, T., Khanal, S. K., & Sung, S. K. (2007). Micro-aeration for sulfide removal in anaerobic treatment of high solid wastewater: A pilot scale study. *Proceedings of the Annual Water Environment Federation Technical Exposition and Conference*.

Electric Power Research Institute. (2000). *Cycle chemistry guidelines for fossil plants: Phosphate continuum and caustic treatment* (Report 1000457).

Filali-Mekanassi, Y., Auriol, M., Tyagi, R. D. Comeau, Y., & Surampalli, R. Y. (2005). Design strategy for a simultaneous nitrification/denitrification of a slaughterhouse wastewater in a sequencing batch reactor: ASM2d modeling and verification. *Environmental Technology*, *26*(10), 1081–1100.

Harvey, D. T. (2016). *Analytical Chemistry, 2.1*. http://dpuadweb.depauw.edu/harvey_web/eTextProject/AC2.1Files/Chapter9.pdf

Kelly, J. T. (2020). The future of biological selectors. *Water Environment & Technology, 32*(6), 39–42.

Khanal, S. K., & Huang, J. (2003). ORP-based oxygenation for sulfide control in anaerobic treatment of high-sulfate wastewater. *Water Research, 37*(9), 2053–2062.

Liptak, B. G. (Ed.). (2005). *Instrument engineers' handbook (4th ed., Vol. 2). Process control and optimization.* CRC Press, 1325.

Luukkonen, T., Teeriniemi, J., Prokkota, H., Ramo, J., & Lassi, U. (2014). Chemical aspects of peracetic acid-based wastewater disinfection. *Water SA, 40*(1), 73–80.

Ndegwa, P. M., Wang, L., & Vadella, V. K. (2007). Potential strategies for process control and monitoring of stabilization of dairy wastewaters in batch aerobic treatment systems. *Process Biochemistry, 42*(9), 1272–1278.

Nordstrom, D. K. (1977). Thermochemical redox equilibria of Zobell's solution. *Geochimica et Cosmochimica Acta, 41*(12), 1835–1841.

Sampson, D. C., & Hicks, P. D. (2009). How to measure corrosion processes faster and more accurately. *Power,* May, 2009. https://www.powermag.com/how-to-measure-corrosion-processes-faster-and-more-accurately/

Samstag, R. W. (2000). Simulation and verfication of a cyclic aeration control system. *Proceedings of the Annual Water Environment Federation Technical Exposition and Conference.*

Suslow, T. V. (2004). Oxidation-reduction potential for water disinfection monitoring, control and documentation. *ANR Publication 8149, University. of California at Davis.*

Victorin, K., Hellstrom, K. G., & Rylander, R. (1972). Redox potential measurements for determining the disinfecting power of chlorinated water. *Journal of Hygiene, Cambridge University, 70*(313).

Virto, R., Manas, P., Alvarez, I., Condon, S. J., & Raso, S. J. (2005). Membrane damage and microbial inactivation by chlorine in the absence and presence of a chlorine-demanding substrate. *Applied and Environmental Microbiology, 71*(9), 5022–5028.

3

pH

1.0 THE KING OF SENSORS THAT IS MISUNDERSTOOD

It's the most common, the oldest sensor for electrochemical measurements. It measures the concentration of a subatomic particle. And yet it determines the fate of the chemistry of more reactions that occur on our planet and living things than any other species in water. Still, few users really understand how a pH probe works.

Some users believe that the H^+ ions that comprise pH penetrate the glass envelope in which they react with the Ag-AgCl electrode. In fact, they never make it deeper than a micron. Many believe that the voltage from a pH probe emanates totally from the glass electrode as H^+ ions accumulate on it. In fact, the Ag-AgCl wire is the key voltage source; the pH glass just modifies the potential of the wire. Some probe manuals strongly caution the user not to clean the probe too vigorously to avoid irreversibly removing the gel layer or they warn that, if the probe dries out, the gel layer is forever gone. In fact, the gel forms spontaneously as a microscopically thin layer when the glass is hydrated for several hours.

The truth is that the pH probe is unique in the world of sensors. It is different from the oxydation-reduction potential (ORP) sensor even though it is identical except for one component—the process electrode. Hydrogen ions never get past the glass envelope, and they don't react. Contrast this with a typical electrochemical analyzer such as a dissolved oxygen or free chlorine sensor. Chlorine really does diffuse through the membrane of a chlorine sensor and reacts at the cathode to generate a current. Even an ORP sensor, which is also a potentiometric device (i.e., outputs a voltage at near zero current) hosts a range of redox reactions on the process electrode. But hydrogen ions appear to just "sit there." They don't seem to participate in any reactions. So how can a probe output a voltage in accordance with the Nernst equation when there are no redox reactions that the Nernst equation describes?

The first mystery that confronts most users is just what the term "pH" means. Before delving into the answer let's back up and go through a short primer on water. Almost everyone knows its chemical formula—H_2O. This molecule bears some resemblance to table salt, NaCl. Just as sodium chloride dissolves in water to form Na^+ cations and Cl^- anions, water breaks (*dissociates*) up into H^+ cations and OH^- anions. The difference between water and salt is that NaCl completely dissociates into Na^+ and Cl^- ions, whereas only a tiny percentage of water molecules do—about one molecule in ten million. However, this miniscule fraction of ions makes a world of difference in the chemistry that goes on in water.

Before going further it's important to recall a nomenclature shortcut introduced in the previous chapter. Protons really don't exist by themselves in water. Each one attaches to a neutral water molecule to create a *hydronium* ion, H_3O^+. The actual dissociation of water takes two water molecules, not one:

$$2\,H_2O \Leftrightarrow H_3O^+ + OH^- \tag{3.1}$$

It would be more accurate to replace all instances of H^+ in this text with H_3O^+ but use of the former better illustrates the chemistry that underlies pH-dependent reactions. Simply be aware that H_3O^+ is the real cation and that H^+ is a "stand-in."

There are 10^{-7} moles/liter of H^+ ions in pure water. Because every water molecule that dissociates makes one H^+ cation and one OH^- anion, there must also be 10^{-7} moles/liter of OH^- ions. Remove the minus sign from "-7" and you have 7. That's the pH of pure water. Another name for exponent is "power"—that is, the concentration of hydrogen ions is "ten to the minus seven power." This is where the term pH comes from—the power of hydrogen. Some sources claim that the "p" stands for "potential." In either case, the pH value is shorthand for 10^{-pH}.

Suppose we take a container of pure water and add acid, say, hydrochloric acid (HCl). This is a strong acid that dissociates completely, so if we add a 1 molar (1 M) solution of HCl, then there will be 1 mol/L (M) of H^+. Does that mean there will be 1 mole per liter of H^+ ions in addition to the existing 10^{-7} M of H^+ ions and 10^{-7} M of OH^-?

The answer is "no." That's because of the following universal law:

$$[H^+][OH^-] = 10^{-14}\,M^2 \text{ (at 25 °C)} \tag{3.2}$$

For pure water, both $[H^+]$ and $[OH^-]$ equal 10^{-7} M. Equation 3.2 means that if the H^+ ion concentration increases, the OH^- concentration decreases by the same proportional amount. So, if we add a 1 M of HCl, then the pH is 0 because $10^{-0} = 1$ and $[OH^-] = 10^{-14}$. If we add 0.1 mol/L HCl, then the pH $= 1$ because $10^{-1} = 0.1$ and $[OH^-] = 10^{-13}$. In both cases the two exponents add up to 14 as Equation 3.2 demands (but only at 25 °C).

Now suppose that, instead of adding HCl, we add sodium hydroxide, NaOH, or *caustic*, or *lye*. Caustic is a very strong base, which means that the NaOH completely dissociates into Na^+ and OH^- ions. If we dissolve, say, 1 M of NaOH then, by Equation 3.2, [H+] = 10^{-14} and the pH is 14.

One interesting finding that should be obvious is that a solution can easily have a pH that is less than 0 or greater than 14. For instance, to reach a pH of −1 dissolve 10 moles of a strong acid like HCl in a liter of water. That's because $10^{-(-pH)} = 10^{-(-1)} = 10^1 = 10$. The pH of concentrated HCl is 1.1.

Just as pH is defined as a $[H^+]$ concentration expressed as 10^{-pH}, a parallel definition for pOH can be defined to quantity $[OH^-]$—that is, 10^{-pOH}. Using this nomenclature Equation 3.2 can be recast as:

$$10^{-pH} \times 10^{-pOH} = 10^{-14} \text{ (at 25 °C)} \tag{3.3}$$

When two exponentials with the same base are multiplied together, the following equality holds:

$$\text{If } A^x \times B^y = C^z \text{ then } x + y = z \tag{3.4}$$

This means Equation 3.3 is equivalent to:

$$pH + pOH = 14 \text{ (at 25 °C)} \tag{3.5}$$

The only constraint on pH values is the solubility of an acid or base. If the pH of a solution is −1, then pOH is 15. If the pH is 15, then pOH is −1, or there is 10 mol/L of a strong base, like NaOH, in solution. As pH goes up, pOH goes down by the same amount. If the pH is less than 7, it is *acidic* and, if it is greater than 7, it is *basic* or *alkaline*. Unfortunately, the term *alkalinity* sometimes refers to the buffering capacity of water. The use of the term pOH in the literature is virtually nonexistent. If you know the pH of a solution, then you also know its pOH. Just subtract the pH value from 14.

The definition of pH explains "neutral" for a pH measurement is not 0 but 7. No other measurement in water quality instrumentation uses a positive number to reflect a zero condition. This also explains why a broken pH probe usually reads 7, which fools some users into thinking their probe is working.

2.0 WHY A PROTON CHANGES EVERYTHING

How can a proton, a subatomic particle, change the chemistry that occurs in an aqueous environment? Why should a tank of water that has 10^{-6} mol/L of H^+ ions (pH 6) be significantly different that one with 10^{-8} mol/L?

Just as swapping negatively charged electrons underlies most chemical reactions that occur in the water, swapping positively charged protons does too. Take, for example, the largest water treatment system on earth—the ocean. This sample of water that covers 70% of the earth is a complex mixture of salts—mostly sodium and chloride. One minor species that plays an outsized role is carbon dioxide. We first met the dissolution of CO_2 in water in Equation 3.4. The reaction governs the equilibrium between CO_2 and the carbonate species that result when the gas dissolves:

$$CO_2 + H_2O \leftrightarrow HCO_3^- + H^+ \leftrightarrow H_2CO_3 \quad (3.6)$$

The HCO_3^- anion is *bicarbonate* and H_2CO_3 is *carbonic* acid. As usual, the double arrows reinforce the fact that the two coupled reactions can go from left to right or right to left—depending on the relative proportions of each of the species. This mixture of species is the consequence of carbonic acid being a weak acid. Unlike a strong acid like hydrochloric acid, it does not dissociate completely into bicarbonate ion and H^+. Adding CO_2 drives Equation 3.6 to the right. That means more HCO_3^- and H^+—that is, more acid and lower pH.

The lowering of pH by adding CO_2 to the ocean has been going on since the onset of the Industrial Revolution in the late 18th century and the accompanying burning of fossil fuels. This so-called *acidification* of the ocean has resulted in a decrease in the pH of the ocean from an average of 8.25 to 8.14. Because pH is a logarithmic scale that 0.1 pH drop is a 29% increase in the H^+ concentration. The acidification of the ocean manifests itself in very tangible ways. Much of ocean life consists of invertebrates that build calcium carbonate ($CaCO_3$) skeletons. These include corals, mollusks, starfish, and crustaceans. Unfortunately, another reaction similar to Equation 3.6 involves the decomposition of $CaCO_3$:

$$CaCO_3 + 2\,H^+ \leftrightarrow Ca^{2+} + CO_2 + H_2O \quad (3.7)$$

As the pH of the ocean drops, calcium carbonate dissolves. Equation 3.7 is the underlying chemistry that creates limestone caverns and destroys coral reefs. Models predict that the pH of the ocean will drop from the current 8.3 to 7.8 by 2100 (Woods Hole Oceanographic Institute, 2021). At this level the skeletons of coral polyps will have totally dissolved.

The same carbonic acid buffer system takes place in a more personal ecosystem—the human bloodstream. Blood maintains its pH at 7.4 though it does do this through other physiological mechanisms than just Equation 3.6. If the pH of a person's blood were to fall lower than 7.35 or rise above

7.45, he or she will die. Later sections in this chapter cover more examples of chemical systems that maintain a range of pH.

The point is that one proton indeed has a major influence on the chemistry of water. Another way to appreciate the importance of the proton is to appreciate its role as the counterweight to the electron. As explained in Section 9 of the previous chapter, an exchange of electrons in an aqueous sample, as measured by the solution's ORP, is usually balanced by the exchange of protons, as measured by the solution's pH.

3.0 HOW A PH PROBE WORKS

3.1 The Secret Ingredient of a pH Probe

The previous chapter described the construction of an ORP sensor from concept to the actual probe. The chemistry introduced in that discussion—oxidation-reduction reactions—is fundamental to all electrochemical sensors, so in this chapter, we build upon this foundation to construct a pH probe. In fact, we only need to make one change to an ORP sensor to convert it into one for pH.

Take a look at the ORP probe shown on the left side of Figure 2.5. The process electrode consists of a platinum (or gold) band encircling a glass envelope. A wire runs from the band through the inside of the envelope and to the signal input of the analyzer. The Pt band provides an inert surface

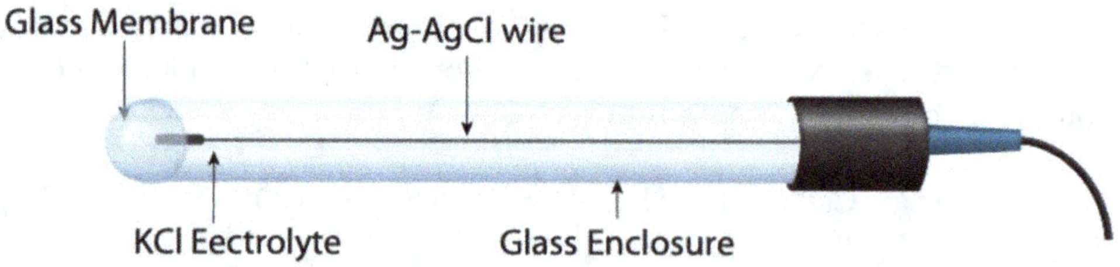

FIGURE 3.1. Schematic of a pH electrode. The specificity of the electrode toward H^+ ions lies in the thin glass membrane that forms the front of the electrode. This is only the process electrode and must be paired with a reference electrode to make a working pH probe.

Every electrochemical sensor that measures a specific analyte, regardless of its operating principle, contains a component in the sensor that is selective toward that analyte. For instance, a chlorine sensor uses a Teflon membrane that is selective toward free chlorine. It does allow other species to permeate but it highly favors the passage of the HOCl molecule. But a pH probe has nothing more than a glass bulb at the end. This prompts the question: "Does glass have a natural affinity for H^+ ions?"

upon which redox reactions take place. The pH electrode also uses a simple glass enclosure, but in this case, the front of the glass envelope has a unique affinity for H^+ ions. This specificity is what makes the pH electrode the ion-selective device that it is.

The answer is more or less "yes." Glass is a very loose (*amorphous*) structure of mostly silica—silicon (Si) atoms that bond to four oxygen (O) atoms. The structure of one Si bonding to four O atoms is tetrahedral though, and the crystalline unit is SiO_4—quartz. The quartz crystalline structure is perfect but in amorphous glass, the tetrahedra are distorted, as shown in Figure 3.2. Each O atom in the structure bonds to two Si atoms, and because of this sharing, the stoichiometry of silica glass is not SiO_4 but SiO_2.

It gets more complicated. In the manufacture of soda lime glass, the most common glass variety, one adds sodium carbonate, Na_2CO_3. The molten

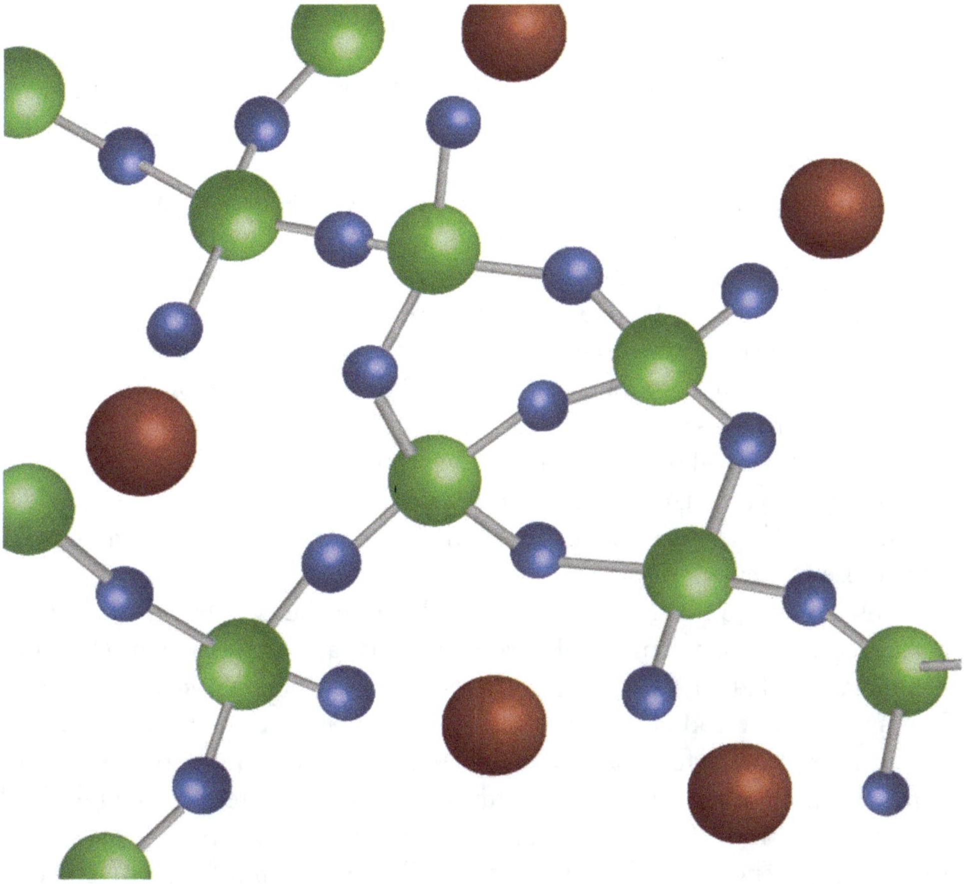

FIGURE 3.2 Silica framework of glass in which sodium (Na^+) ions replace Si^{4+} ions. Green = Si^{4+}, Blue = O^{2-}, Red = Na^+. Most pH-sensitive glass today uses lithium in place of sodium.

glass strips CO_2 from the carbonate and leaves behind Na_2O, which is incorporated into the glass structure as Na^+ cations and (more) O^{2-} anions. Soda lime glass contains about 15% Na_2O and 10% of another oxide—usually CaO. Likewise, borosilicate glass contains about 15% B_2O_3. Both glasses also contain trace amounts of oxides of aluminum (Al), calcium (Ca), and rare earths such as lanthanum (La) and praseodymium (Pr). We'll ignore these additional metals and focus on the role of the Na^+ ions.

In the absence of Na^+ ions, each O^{2-} ion bridges two Si^{4+} ions—thus forming a tetrahedral network. A Na^+ ion can only bond to one O^{2-} ion, so the latter no longer can bond to another Si^{4+}, and the network ends up with a break. Many schematic drawings show a complete tetrahedral Si-O network intact and intersperse the Na^+ ions nestled in the crystalline structure. This would result in highly positively charged glass! The sodium cations replace silicon cations into the Si-O network such that overall charge remains neutral.

Older pH probes used soda lime glass for pH sensing but most of today's probes incorporate Li_2O (from the addition of Li_2CO_3) into silica glass. The Li^+ ions integrate into the silica network just like Na_2O. Why lithium? It's the smallest of the alkali metals—just below hydrogen—and can easily integrate into the silica network.

When pH glass is immersed in water, the outer 10–100 nm of the lithium "adulterated" silica glass adsorbs water and becomes hydrated. The bond between an O^{2-} anion and H^+ cation is much stronger than that between O^{2-} and Li^+ (or Na^+). Hydrogen ions displace the lithium ions. This outer layer is also known as the *gel* layer, and it forms over a period of hours when the glass is first immersed in water. This is the basis of the working end of a pH probe—a round or flat piece of H^+ sensitive glass grafted onto a stem of ordinary glass. There are two surfaces on every piece of glass, so the same gel layer forms on the interior of the pH glass as on the outside. This interior of the electrode is sealed and contains a KCl solution at a pH value close to neutral (7).

This description of hydration of a pH probe sheds light on a common misconception—that letting a pH probe dry out can ruin it forever. The truth is that glass will always form the gel layer after a sufficient hydration period. That sufficient period is several hours for a probe that has been left to dry out for more than a few days. The real reason most manufacturers recommend that users keep a pH probe hydrated is to minimize the time required for the formation of the gel layer—the so-called *conditioning period.* A probe that has not been fully hydrated will show both a large offset—due to the difference in gel layers on both sides of the glass electrode—and a lower efficiency. Some users report a shortened lifetime of a probe that has been left to dry, but I am not aware of any data that confirms this.

It makes sense that one ought to store the probe in the same KCl solution that fills the interior of the process electrode. As a practical matter, storage in any ionic solution will preserve the elemental composition of the glass. Many users use pH 4 buffer as it hydrates and, by virtue of its mild acidity, keeps the probe clean.

3.2 The Connection Between the Nernst Equation and the pH Probe

On the surface (no pun intended), the pH probe seems like an easy sensor to understand. A potential difference between the two surfaces of the pH probe generates a voltage proportional to the pH of the solution. One might conclude—and many texts state—that the pH probe works by building up a charge difference between H^+ ions on both sides of the pH glass. There are two problems with this interpretation. First, if H^+ ions just replaced Li^+ or Na^+ ions, then there would be no change in the net charge. Second, if there were a net change of charge due to the adsorption of H^+ ions onto the outer glass surface, the result would be a capacitor. In fact, the mathematical description of a pH sensor based on a double layer capacitor has been proposed (Cheng, 1989) as a replacement of a Nernst formulation. The 59.16 mV/pH potential output by a pH probe is too much of a coincidence not to be the result of a Nernstian foundation. The following explains how this is so.

To start, let's consider the pH probe electrode as a pair of surfaces upon which redox reactions occur and from which a potential difference is generated. The Nernst equation quantifies a voltage that arises from a redox reaction or, more generally, from a reaction involving the transfer of charge:

$$E = E^0 - \frac{RT}{nF} ln\ Q \tag{3.8}$$

But where is the reaction when we're just talking about a buildup of H^+ ions on glass? Without a redox pair of chemical species, there can be no E^0, no Q, and no E.

We might, as just about every explanation on pH does, simply equate Q to the ratio of potentials of the outside phase boundary potential to the inside phase boundary potential, ignore E^0 and get the right answer:

$$E = -59.16 \text{ mV} \log \frac{H^+_{\text{proc}}}{H^+_{\text{ref}}} = 59.16 \text{ mV } (7 - \text{pH}) \text{ (at 25 °C)} \tag{3.9}$$

It works and few question the equation's validity but, to the curious, it does not explain why the Nernst equation is even applicable.

The following explanation shows how a pH probe or, for that matter, any potentiometric ion selective electrode, works within the framework of a redox reaction—a generalized one at that transfer reaction—and the ensuing Nernst equation. It is an exercise that is helpful for understanding the electrochemical principle of operation at work. I have not found it in any other text. That's why I present it as the following section. But it can be skipped without the reader suffering the loss of any practical knowledge.

3.3 Making the Nernst Equation Fit a Proton

Let's generalize the Nernst equation to extend beyond the redox reactions that involve an exchange of electrons. In fact, the equation is valid for the exchange of any charge between two chemical species.

As the gel layers form on the surface of pH-sensitive glass, H^+ ions replace Li^+ (or Na^+) ions that have bonded to O^{2-} ions. We can view this as an exchange of two positively charged species:

$$H^+ + SiO^-Li^+ \leftrightarrow SiO^- H^+ + Li^+ \tag{3.10}$$

The displaced Li^+ ions enter the process and have no effect. The Nernst equation for the absorption of H^+ ions into the gel layer described by Equation 3.10 is accordingly:

$$E_1 = E_1^0 - \frac{RT}{nF} ln \frac{[SiO^- H^+][Li^+]}{[H^+][SiO^- Li^+]} \tag{3.11}$$

When the probe's electrode is first immersed in water Equation 3.10 proceeds from left to right. The ratio of [Si-O^--H^+] to [Si-O^--Li^+] is much less than one. As the glass surface becomes increasingly saturated with Si-O^--H groups, the ratio increases and the logarithm in the right side of Equation 3.11 swings toward increasingly positive values and E_1 swings toward larger negative values. This is the process that all pH probe users observe when placing a dry probe into service. This is the mechanism behind the formation of the gel layer, also called the conditioning of the probe or the rejuvenation of the glass surface.

The same process occurs on the inside of the glass membrane facing the constant pH KCl solution. The difference, of course, is that the inside glass surface was conditioned the day the probe left the factory and stays that way for the life of the probe. The change in potential from the process electrode reflects the difference in the potentials between the inner and outer glass surface and, hence, the change in just the outer surface.

We can simplify Equation 3.11 be recalling that $\ln(xy) = \ln(x) + \ln(y)$. Let's break up the logarithm into three separate terms as follows:

$$E_1 = E_1^0 - \frac{RT}{nF}\ ln[Li^+] + ln\frac{[SiO^-H^+]}{[SiO^-Li^+]} + ln\frac{1}{[H^+]} \qquad (3.12)$$

As the glass becomes hydrated H^+ ions replace Li^+ ions. When the glass is fully hydrated the concentration of Li^+ ions in solution becomes negligible and that fist term in the bracket become negligible. The second term becomes constant and it can be pulled out to join the mysterious E_1^0 term. The third term defines pH:

$$ln\frac{1}{[H^+]} = -\ln[H^+] = pH \qquad (3.13)$$

We now have a simplified form of Equation 3.12:

$$E_1 = E_1{}^0 - \frac{RT}{nF} ln\frac{[SiO^-H^+]}{[SiO^-Li^+]} - \frac{RT}{nF} pH \qquad (3.14)$$

When the pH electrode is fully hyrdrated the second terms becomes a constant and can be combined with E_1^0 to make a new constant.

$$E_1 = Const + 2.303\ \frac{RT}{nF} pH_1 \qquad (3.15)$$

Now we see that the fully hydrated pH-sensitive glass responds to changes in the pH of the process just like we expect it to.

Let's re-examine the step in which the seond term in $SiO\text{-}H^+/SiO\text{-}Li^+$ becomes constant.

It is tempting—and many texts state—that the concentration H^+ ions adsorbed by the gel layer (as $SiO^-\text{-}H^+$ groups) rises and falls as the pH of the sample changes. That's wrong. If the concentration of H^+ ions did vary, then the ratio between H^+ ions in the gel layer to H^+ ions in the sample would be constant and the probe would produce a constant voltage. Another way of saying this is to note that the ratio $SiO^-\text{-}H^+/SiO^-\text{-}Li^+$ would vary and counter the change in pH. In actuality, the formation of the gel layer is the process of attaining a steady population of $SiO^-\text{-}H^+$ groups that remains relatively constant as long as the pH glass remains wetted.

In Chapter 8's discussion of ion selective electrodes (ISEs), of which the pH probe is but one example, the stable concentration of the analyte within the sensing membrane is a universal and necessary feature. The molecular entity that attracts and stabilizes the analyte ion is the *ionophore*. In the case of the pH probe the ionophore is Si-O^-. A necessary condition that enables ionophores to work is that it is energetically favorable for the analyte ion to attach to the ionophore—that is, that the free energy change, ΔG, is negative. It's also necessary for the decrease in free energy of complexing the ion to the ionophore to be larger than the decrease of the ionophore combining with other ions. That's what gives the pH probe its *selectivity*. For pH-sensitive glass, the silica network has a greater affinity for H^+ ions than the Li^+ ions they displace. That's why Equation 3.10 proceeds from left to right. This, again, is the essence of the formation of the gel layer. It is also the principle of operation behind all potentiometric sensors. We return to this often-misunderstood issue in the discussion of ISEs in Chapter 8, Section 1.0.

The same chemistry and mathematical treatment hold for the inner glass surface of the electrode. We'll denote the potential as E_3, and its value is determined by one equivalent to Equation 3.15.

$$E_3 = Const + 2.303 \frac{RT}{nF} pH_3 \tag{3.16}$$

In fact, the only difference between E_1 and E_3 is the term involving pH. (The value for "Const" is ideally the same though, in reality, a very small difference will arise due to differences in the glass surface.) The pH probe is an example of a symmetrical electrode in which both sides of the electrode are composed of the same material. In all other ISEs, only the face of the electrode facing the sample has a structure containing the ionophore.

3.4 Channeling the Inner Potential

If one could only measure the potential across the glass membrane, a pH probe would be so simple. Unfortunately, no one has ever managed to measure this potential. The only practical way to do so is to add the Nernstian potentials to potentials that can be measured—the potentials at the interface between the Ag-AgCl wire and KCl (typically 4 M) and saturated AgCl solution. That's the *raison d'être* for the Ag-AgCl electrode. One might say that the AgCl electrode "channels" the ethereal pH glass potential difference.

Figure 3.3 illustrates the chain of potentials that add up to create the voltage in the process electrode that is read by one side of a high impedance voltmeter.

The process electrode sports the potentials E_1, E_2, E_3, and E_4. E_1 and E_3 are the pH-dependent potentials on the outer and inner surfaces of the glass,

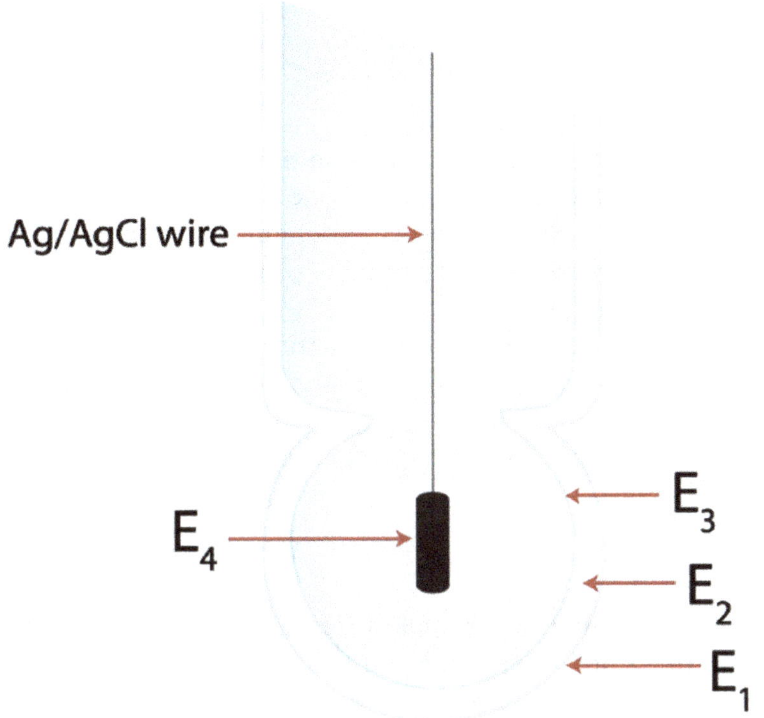

FIGURE 3.3 The actual potential in a combination probe is the sum of a series of six potentials that extend from the outer glass surface of the process electrode all the way to the Ag-AgCl reference electrode.

respectively. The potential E_2 makes a minor contribution to the potential of the process electrode. Between the two glass surfaces of the pH-sensitive membrane is the interior of the membrane. This bulk medium consists of the same silicate lattice in which Li^+ (or Na^+) ions and other metal cations replace Si^+ ions. These cations are essential as they provide a conductive pathway between the two surfaces. Without the tiny but finite conductivity provided by these ions the glass interior would provide an open in the circuit, and the probe would not work. The potential E_2 reflects the potential in the interior, dry glass layer. If it were composed of a uniform structure, then E_2 would be zero. A non-zero E_2—the *asymmetry potential*—reflects a non-uniform structure that grows as the electrode ages.

The potential E_4 is the familiar potential for the AgCl wire immersed in a solution of 4 M KCl or saturated AgCl. The redox reaction was already defined in Equation 2.24. I repeat it for convenience:

$$AgCl\ (s) + e^- \rightarrow Ag\ (s) + Cl^-;\ E^0 = 0.197\ V\ (4\ M\ KCl)$$

The Nernst equation for this reaction in the glass electrode is:

$$E_4 = 0.197\ V - \frac{RT}{nF} ln[Cl^-]_{proc,inner} \tag{3.17}$$

The reason that only the concentration of Cl^- enters the Nernst equation is because the other two chemical species, AgCl and Ag, are solids and have activities of one. The subscript "proc, inner" reminds us that this is the chloride concentration inside the glass process electrode. Because it is sealed it will not change.

The sum of all the potentials from Equations 3.15, 3.16, and 3.17 combine to create the potential coming from the process electrode:

$$\begin{aligned} E_{process} = E_1 - E_3 + E_2 + E_4 = 2.303\ \frac{RT}{nF}(7 - pH) + \\ E_2 + 0.197\ V - \frac{RT}{nF} ln[Cl^-]_{proc,inner} \end{aligned} \tag{3.18}$$

Note that we assume that the pH of the electrolyte inside the electrode is 7. Even if the pH is not 7 the final result will not change as the electrolyte in the reference solution will have a pH very nearly the same. Any discrepancy will simply add to the contant voltage that we call the offset.

Let's now look at the reference electrode. There are only two potentials to consider. E_5, the potential of the Ag-AgCl wire, also in a KCl solution, saturated with AgCl.

$$E_5 = 0.197\ V - \frac{RT}{nF} ln[Cl^-]_{ref,inner} \tag{3.19}$$

The liquid junction completes the circuit by diffusing KCl electrolyte out of the reference chamber and ions from the process into the chamber. The diffusion rate differs for different ions and that difference creates a charge separation. The reason KCl and KNO_3 are usually chosen as the electrolyte is that the K^+ cation and Cl^- or NO_3^- cation diffuse at roughly the same speed so that the charge separation is minimal. More on this in Chapter 3, Section 6.5. We denote the diffusion potential at the liquid junction as E_6.

The voltmeter reads the difference in potential between the process and reference electrode. That is:

$$\begin{aligned} E_{probe} &= E_{proc-}E_{ref} = (E_1 - E_3 + E_2 + E_4) - (E_5 + E_6) \\ &= 2.303\ \frac{RT}{nF}(7 - pH) - \frac{RT}{nF} ln\frac{[Cl^-]_{proc,inner}}{[Cl^-]_{ref,inner}} + E_2 + E_6 \end{aligned} \tag{3.20}$$

When the probe is new, the concentration of KCl in both the reference and process chambers is the same (usually 4 M). So $\frac{[Cl^-]_{proc,inner}}{[Cl^-]_{ref,inner}}$ is 1.0, the logarithm of 1 is zero and the third term in Equation 3.18 disappears. As the probe ages K^+ and Cl^- ions leak out and are replaced by ions in the sample; $\frac{[Cl^-]_{proc,inner}}{[Cl^-]_{ref,inner}}$ decreases from 1.0, the logarithm of a number less than 1.0 is negative, and the third term increases from zero. We'll call the third term E_{elec}. E_2 reflects the asymmetry of the pH glass so we'll relabel it E_{asym}. E_4 is the diffusion potential across the liquid junction so we'll relabel it E_{diff}. The final expression of Equation 3.18 is thus:

$$E_{probe} = E_{proc} - E_{ref} = 59.16\ mV \frac{T}{298\ K}(7 - pH) - E_{elec} + E_{asym} - E_{dif} \quad (3.21)$$

This very methodical treatment of the pH probe yields the same result from the simplistic treatment that led to Equation 3.9—with the addition of the asymmetry potential, offset and diffusion potential. The 59.16 mV (at 25 °C) constant is the *slope* of the voltage versus pH curve: every unit change in pH gives a change of 59.16 mV (at 25 °C) to the output of the probe. At temperatures other than 25 °C, we need only apply the correction factor. (Recall that the Kelvin scale equals the Celsius scale with an offset of 273 K. Recall also that the Kelvin scale dispenses with the superscript "*o*".) This information is invaluable during calibration. One calibrates a probe by measuring its voltage output at two or more solutions of known pH. The difference between the two outputs (at 25 °C) should be:

$$\frac{Pr\,obe\ Output\ (Buffer\ 2) - Pr\,obe\ Output\ (Buffer\ 1)}{pH\ (Buffer\ 2) - pH\ (Buffer\ 1)} = 59.16\ \frac{mV}{pH}(\text{at } 25\ °C) \quad (3.22)$$

In the real world the actual slope is likely to be close, but not exactly equal, to 59.16 mV/pH. Calibration of a pH probe—as well as any instrument that produces an output proportional to a process variable—is the measurement of the slope (and offset) of the actual voltage output versus pH response.

Figure 3.3 graphically displays the potentials in a combination probe and Table 3.1 summarizes them.

TABLE 3.1 Summary of potentials (voltages) generated through a combination pH probe. The voltage measured by the probe is the sum of all 6 potentials.

Electrode	Potential	Origin
Process Electrode	E_1	Potential across the outer gel layer, given by the Nernst Eqn. for $[H^+]$ for the sample pH.
	E_2	Asymmetry potential between the inner and outer gel layers.
	E_3	Potential across the inner gel layer, given by the Nernst Eqn. for $[H^+]$ in the sealed electrolyte solution in the process electrode.
	E_4	Potential between the Ag-AgCl wire and reference KCl solution. Given by the Nernst Eqn. for $[Cl^-]$.
Reference Electrode	E_5	Potential between the Ag-AgCl wire and process KCl solution. Given by the Nernst Eqn. for $[Cl^-]$.
	E_6	Diffusion potential from flow of ions though the liquid junction. (K^+, Cl^- from the ref. chamber and ions from the sample.)
Total	**Sum of Above**	$E_1 - E_3 + E_2 + E_4 - E_5 - E_6$. E_1 is the potential that determines the pH. **Actual probe voltage measured by a voltmeter.**

3.5 Adapting to a Differential Probe

The analysis of the voltage output of the differential probe draws on the same potentiometric principles as that of the combination probe. There are two major differences in the design:

1. In a combination probe the reference electrode is an Ag-AgCl wire in a KCl solution that interfaces with the sample through the liquid junction. In a differential probe the KCl solution is enclosed in a pH-sensitive glass envelope that is identical to the process electrode. This so-called *stem electrode* is immersed in a neutral solution of either pH 7 buffer or salt. The liquid junction borders the outer compartment that contains the neutral solution.
2. In a combination probe the leads from the two electrodes connect directly to the voltmeter. The output is a voltage whose high impedance is due to the high impedance of the glass electrode (E_2). In a differential probe the electrodes connect to a buffer amplifier along

with a ground rod. The buffer amplifier subtracts the reference-to-ground voltage from the process-to-ground voltage and converts a high impedance voltage to a low impedance one plus a ground signal.

The potentials across the reference electrode now have the same origins as those for the process electrode with one exception—the addition of the diffusion potential across the salt bridge. To modify Equation 3.20 for the differential probe requires two changes:

1. Replace the potential at the reference electrode, E_5, with a sum of terms equivalent to those for the process electrode.
2. Subtract the potential of the ground rod, E_G, from both electrodes.

Since there are now two parallel sets of potentials, we'll use the subscripts, "*P*" and "*R*," to distinguish between those for the process electrode from those for the reference electrode.

$$E_{Proc} = E_{1,P} - E_{3,P} + E_{2,P} + E_{4,P} - E_G \tag{3.23}$$

$$E_{Ref} = E_{1,R} - E_{2,R} + E_{4,R} + E_{dif} - E_G \tag{3.24}$$

$E_{1,R}$ and $E_{3,R}$ are the potentials on both sides of the pH glass membrane in the reference chamber. Since both sides face neutral (pH 7) solution, these terms cancel each other out. In the real world they don't cancel out exactly, but any finite difference can be rolled into the offset potential. E_{2R} and E_{2P} are the asymmetry potentials. Every piece of glass is unique so the difference between the two is finite. We'll set $E_{2P} - E_{27} = \Delta E_{asym}$. E_{4R} and E_{4P} are the potentials at the Ag-AgCl wire in KCl solution. Now both Ag-AgCl wires are both in identical sealed stem electrodes and ought to be identical. Like the asymmetry potentials, the difference, called the offset potential, will be small but non-zero. We'll call the difference in potential between the two wires ΔE_{AgCl}. Because both electrodes use the same ground rod to complete their circuits, the E_G components really do cancel each other. Only the diffusion potential, E_{dif}, is left by itself.

Subtracting E_R from E_P now gives us the following voltage of the differential probe:

$$E_{Diff\ probe} = 59.16\ mV \frac{T}{298\ K}(7.0 - pH) + \Delta E_{asym} + \Delta E_{AgCl} - \Delta E_{dif} \tag{3.25}$$

The form of Equation 3.25 is identical to that of Equation 3.21. The part of the probe voltage that results from the sum of the asymmetry potential,

Ag-AgCl potential and diffusions potential can be rolled into one relatively constant term, E_{offset}. (I say "relatively" because, as the next section shows, the diffusion potential changes on the order of a few millivolts depending on the ionic makeup of the sample.) E_{offset} is the voltage we measure when we insert the probe into pH 7 solution.

$$E_{Diff\ probe} = 59.16\ mV \frac{T}{298\ K}(7.0 - pH) + E_{offset}(T) \qquad (3.26)$$

This non-zero offset must be accounted for during the calibration procedure. The "T" in parentheses reminds us that the offset is, itself, a function of temperature. The point of this exercise in arithmetic is not to steer this book into academic waters but to give insight into why pH probes behave the way they do. Armed with this information we are now able to diagnose problems.

4.0 REVISITING SENSOR TYPES

Chapter 2 covered the three different designs of an ORP sensor: a benchtop system consisting of separate process and reference electrodes, the combination probe, and the differential probe. The previous section discussed the workings of a pH probe—for both a combination and differential probe. Until now, the only discussion of the consequences of the differential probe design were discussed in the previous chapter. There are also design improvements of the combination probe that have not yet been discussed. Let's delve into those practical details now.

4.1 Separate Electrodes

Let's start with a pH sensor that consists of separate reference and process electrodes. Figure 3.4 showed one for an ORP sensor. Figure 3.4 shows it for a pH probe. The process electrode uses the same design as that for the combination and differential probe. In most cases it is just larger, and in some cases, a foil shield envelops the glass section for reducing radiofrequency (RF) noise. The reference electrode is of a similar size to the process electrode but comes in a variety of configurations. In place of pH-sensitive glass, the liquid junction terminates the bottom of the reference electrode. The flow of K^+ and Cl^- ions out of the body of the electrode through the liquid junction provides the electrical continuity to the process.

Because the reference electrode is separate it can be refilled with fresh electrolyte. This eliminates the problematic contamination of the electrolyte

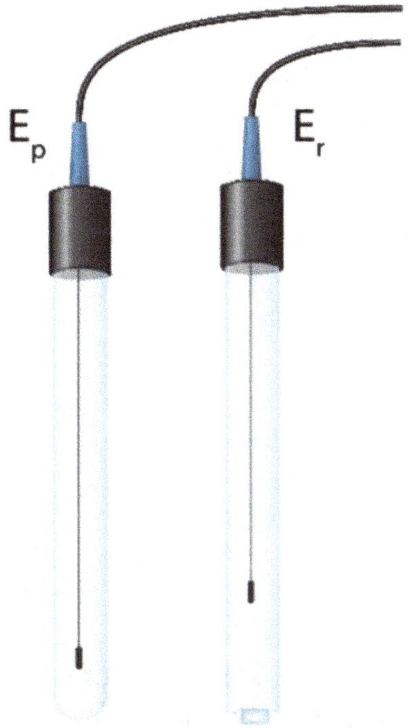

FIGURE 3.4 pH measurement using separate reference and process electrodes. The liquid junction completes the circuit by providing a conductive pathway for ions.

that limits the lifetime of most combination probes. The different possible liquid junctions give rise to different configurations. The main consideration is the balance between low impedance and long-term stability:

1. The most common liquid junction is a *porous frit* commonly constructed of ceramic. Other choices for the junction material include polymer—usually PVDF (Kynar) or PTFE (Teflon)—and natural fiber—typically linen.
2. Solutions with very low ionic concentrations generate a very slow response by virtue of their low conductivity. An electrode with a *pinhole*—also referred to as an *open pore* or *porous pin junction*—allows free flow of electrolyte into the solution. The rapid loss of electrolyte is an obvious drawback to this electrode, though this type of electrode usually contains a hole for refilling.
3. Similarly, a *sleeve electrode* replaces the junction with a ground glass sleeve that slides over the bottom of the electrode that is similarly ground. The sleeve covers a hole in the side of the electrode, and a

thin film of solution diffuses through the ground glass interface and through the hole. This electrode is ideal for low ionic strength (i.e., pure) water samples and—on the opposite end of the scale—solutions that have contaminants that can rapidly contaminate the reference solution—such as suspensions, emulsions, and oil. Like the open pore electrode its electrolyte must be replaced on a frequent basis. Electrolyte (KCl or KNO_3) is inexpensive and refilling an electrode is quick. Some probes (e.g., Yokogawa's Bellomatic reference electrode) use a pressurized reservoir to ensure that fresh electrolyte flows into the chamber.

4. One configuration of reference electrode is tailored toward minimizing the contamination of the reference solution and thereby extending its lifetime. The *double junction* electrode takes the conventional single junction electrode and inserts it into an outer chamber with its own junction. This additional chamber typically contains an electrolyte of lower ionic content so that the sample is compromised more slowly by the electrolyte. For solutions containing chemicals that can react with silver (such sulfides), the outer electrolyte can be free of AgCl that is always present in the other electrodes.

Figure 3.5 illustrates the configurations.

Separate electrodes are at home in a laboratory where its accuracy and flexibility are highly valued. However, it is not very convenient in a treatment facility or factory when the sensor must be immersed in the process or inserted into a plumbing fitting.

4.2 The pH Combination Probe

Just as with the ORP combination probe we investigated, the pH combination probe "combines" the reference electrode and process electrode into one glass or plastic envelope. For the sake of completeness, we'll review its construction. Figure 3.6 shows a schematic of the combination probe. It is a tube inside a tube. The inner tube is the process electrode with the pH-sensitive glass bulb at the bottom. The reference electrode surrounds the process electrode though the two are totally isolated. In the illustration the porous junction is now on the side of the outer glass envelope though it can also be on the bottom. As the drawing shows it really is very small—on the order of a couple millimeters.

Just as the reference electrode comes in several configurations so does the process electrode. In general, each strikes a different balance between utility and signal-to-noise.

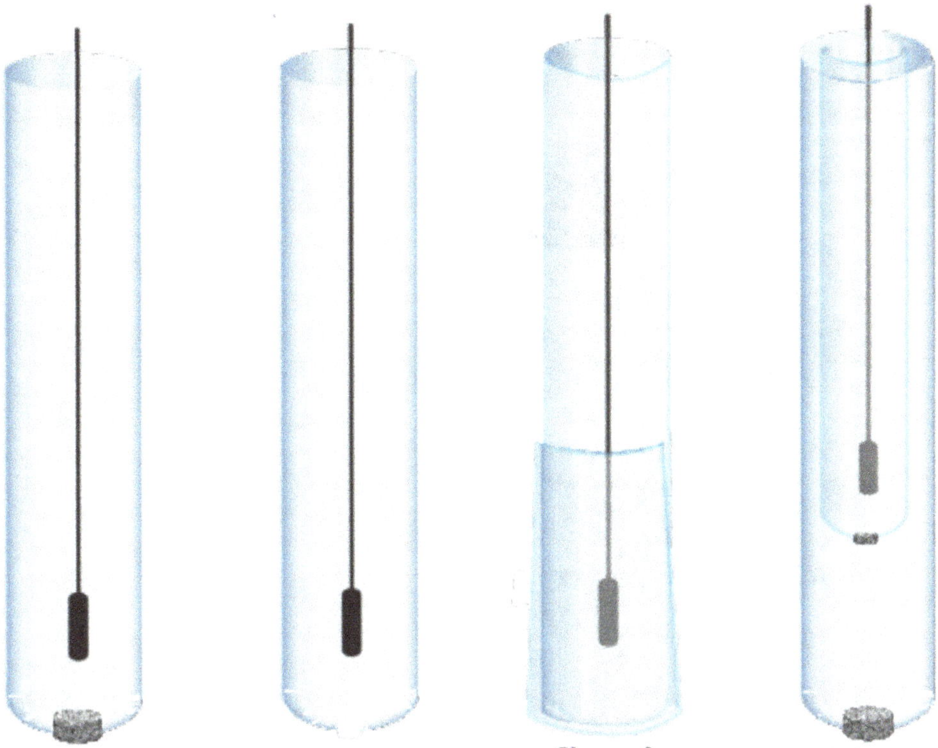

FIGURE 3.5 Reference electrodes. The most common are the porous junction, double junction, and reversed sleeve.

The spherical electrode has the largest surface area and thinnest glass, which translates into a high signal-to-noise ratio and fast response. However, its shape and thinness make it more prone to breakage. A hardened glass electrode (not shown) also has a spherical shape, but its thicker glass makes it less prone to breakage. A thicker glass means a higher impedance and slower response. The flat electrode is typically flush with the front of the probe, so that it does not protrude beyond the front. It is less prone to breakage, and its flat surface is sometimes characterized as "self-cleaning." However, its minimal surface area means a lower signal-to-noise. For most applications, this is not significant. The hemispherical electrode represents a compromise between the flat and spherical electrode. The microelectrode is miniaturized to fit into biological samples. The spear electrode is designed to insert into solid objects—mostly food.

Many combination probes contain a temperature element—a resistance temperature detector (RTD) or thermistor—that allows the analyzer to compensate for temperature. The combination probe represents a huge

FIGURE 3.6 Combination pH probe. The probe is the usual process electrode inside a concentric reference electrode. The junction is the little black rectangle.

step forward in convenience, especially when used in a process control environment. It's compact and inexpensive. The vast majority of all pH probes in operation are combination style probes. The combination probe does have its limitations that were listed in a discussion of the ORP combination probe in Chapter 2, Section 7.3. The next section recaps those limitations and reintroduces the differential probe.

For most users these constraints are a small price to pay for a probe that, without a protective housing, pre-amplifier, or temperature element, typically costs between $80 and $300. Probes for process control are encased in a protective housing. They are either inserted into a pipe or container via a compression fitting or submerged in an open tank using a threaded connection to a submersion arm (usually just a pipe with a female thread on the end). Figure 3.7 shows a typical industrial pH probe. The threads near the front end allow it to be mounted in a compression fitting while the threads at the rear end allow it to be mounted on a submersion arm. The protruding tabs that surround the electrode protect the latter from being broken

FIGURE 3.7 Combination probe for process control. It may be mounted into a pipe or side of a tank using the front threads. The rear threads are for mounting on a submersion arm (Courtesy Rosemount).

by a technician in a hurry. For this particular probe the white annulus that surrounds the process electrode is the porous junction.

4.3 The Differential pH Probe

As we discussed in Chapter 2, Section 7.4, the differential probe addresses the limitations of the combination probe—that is, contamination of the reference solution, poisoning of the Ag-AgCl reference electrode, stray ground potentials, and transmission distance. Figure 3.8 is a schematic that shows how the differential probe works. This design splits the two half-cell circuits—process and reference—into two individual circuits.

Whereas the reference electrode of a combination probe is an Ag-AgCl wire in a KCl-AgCl electrolyte solution and open to the process via the liquid junction, the reference electrode of a differential probe is identical to the process electrode—a sealed glass tubes containing an Ag-AgCl wire in an electrolyte solution. (The ORP probe uses the same sealed pH electrode.) Both identical electrodes—usually called *stem electrodes*—are electrically tied to one *ground rod* that sits in the process—that is, they share the same ground.

The voltage produced by a combination probe is the difference between the process and reference electrodes:

$$E = E_{process} - E_{reference} \tag{3.27}$$

When a stray ground potential, often in the form of a ground loop, is inadvertently introduced into the process, the potential can no longer find its

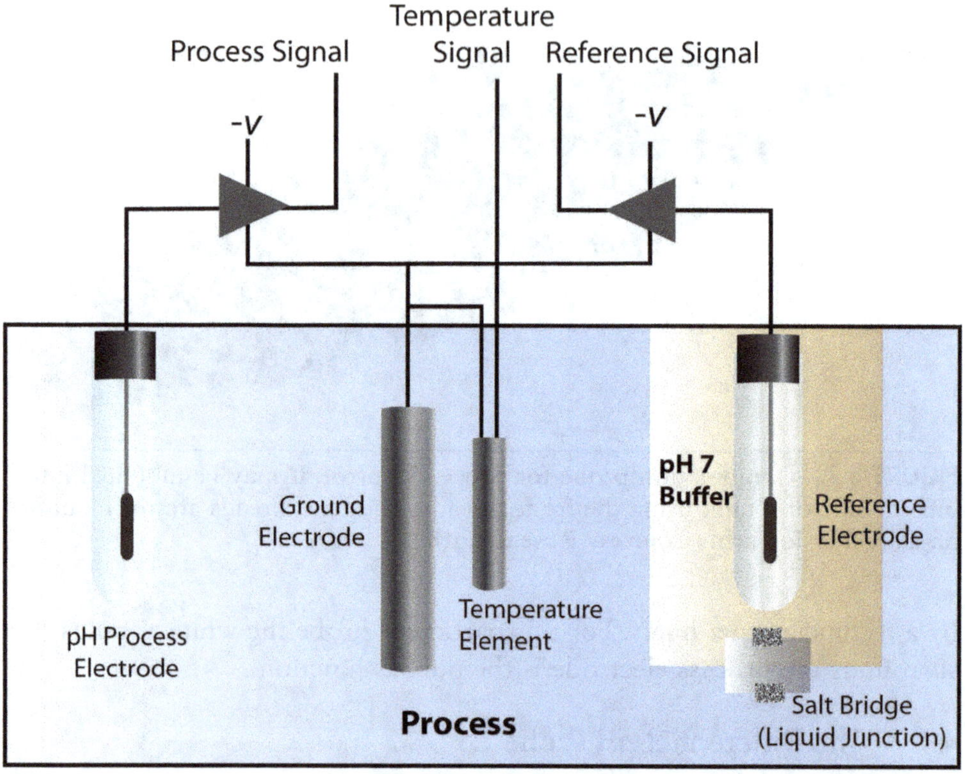

FIGURE 3.8 Schematic of a differential probe. The process and reference electrodes are split, and their half-cell circuits are completed by the same ground rod. Any stray voltages that enter through the process are subtracted out automatically.

way along the low impedance pathway up to the reference electrode, which is now safely enclosed in a glass exterior. Instead, the unwanted potential creates a conductive path to the ground rod. Because the ground rod is common to both reference and process electrode half circuits, the addition of the stray potential (E_{gnd}) changes the potential equally on both so that Equation 3.27 becomes:

$$E = (E_{proc,signal} + E_{gnd}) - (E_{ref,signal} + E_{gnd}) = E_{process} - E_{reference} \tag{3.28}$$

The stray voltage is therefore eliminated. The probe output is the difference between the process half-circuit and the reference half-circuit—hence the term "differential."

The differential probe confers another advantage. As long as there is a circuit to subtract two coupled individual circuits, a buffer amplifier can be easily added to the circuit—either in the form of two amplifiers to each half circuit before the differential voltage is generated or in the form of one amplifier after the differential voltage is generated. A buffer amplifier does not change the voltage of an input signal. It changes the impedance from high input impedance to low output impedance. Low input impedance means that the current coming out of the buffer amplifier is multiplied by the ratio of the input impedance to the output impedance. Again, this is due to Ohm's law, $I = V/R$:

$$I_{out} = \frac{V/R_{in}}{V/R_{out}} = \frac{R_{in}}{R_{out}} \tag{3.29}$$

Impedance and resistance are not really the same. The former is a complex quantity and is denoted by Z. For this book—especially since the imaginary component of the impedance is near zero—the terms impedance and resistance are used interchangeably. A typical value for R_{in} is 100 MΩ or greater and, for R_{out}, is on the order of 1 MΩ so the current increase is a factor of 100 or greater. That's a strong signal that can now travel hundreds of feet without degradation, thus enabling very long cable lengths.

Figure 3.9 shows a typical differential probe by Hach showing the salt bridge, process electrode, and ground rod (partially obscured).

The stem electrode on the reference side is immersed in the *reference solution*—typically pH 7 buffer or a KCl or KNO_3 solution. By sealing off the Ag-AgCl electrode from the sample it is protected from contaminants that can coat it. The well in which the reference solution is contained interfaces to the sample via a salt bridge. As stated above the salt bridge of a

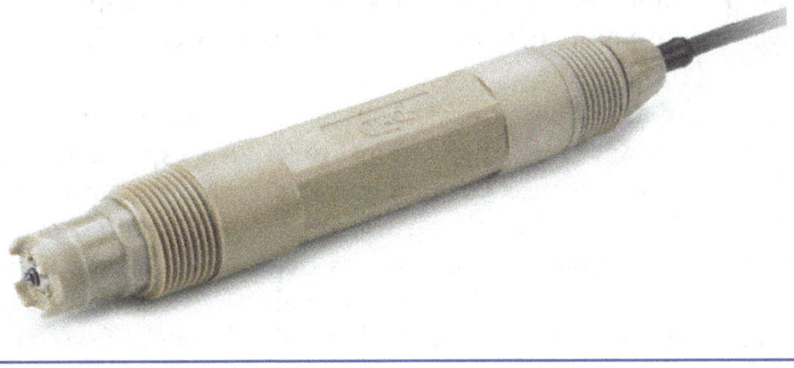

FIGURE 3.9 Differential pH probe. The ground rod is a non-corrosive (usually titanium) on the far side of the probe front. The salt bridge is the hexagonal piece on the bottom (Courtesy Hach).

differential probe is a capsule packed with KCl gel with porous frits on both ends. One end faces the reference solution and the other faces the process. By making the salt bridge a self-contained capsule the user can simply throw it away when it becomes contaminated by the process. The reference solution inside the reference chamber can also be easily replaced when it, too, becomes contaminated.

The benefits of the differential probe design are

1. The reference solution can be recharged when it becomes contaminated with ions from process water. However, combination probes and laboratory reference electrodes are available with reference chambers that can be refilled.
2. The Ag-AgCl wire electrode is sealed from the environment by virtue of being part of a stem electrode that is identical to the process electrode.
3. A salt bridge replaces the liquid junction. This is essentially a double junction containing KCl, and it is replaced when it becomes fouled. Because of its large size, the salt bridge enables both low impedance and fast response.
4. The encapsulated pre-amp increases the possible cable length from 5 m to 1000 m.
5. The ground rod subtracts out ground loops or other stray voltages that enter the probe. The additional ground shifts the difference between the process and reference electrode and changes the pH value. In our discussion of the differential ORP probe, we treaded lightly on the issue of ground loops because it's much less of a problem than with pH probes. The reason is that the Pt electrode of the ORP probe provides a low impedance pathway to voltmeter, whereas the glass bulb of the pH probe provides a very high impedance route. We'll go into more detail on ground loops in Section 7.

The differential probe is considerably more expensive than a combination probe. For clean water applications, the added expense is not likely worth the additional expense. Ground loops and other stray potentials can be remediated in ways other than resorting to a differential probe. For long transmission distances, pre-amplifiers that convert a high impedance voltage to a low impedance voltage are available for combination probes. But, for very aggressive processes, the longevity of the differential probe resulting from a sealed reference electrode, changeable reference solution, and changeable salt bridge, may enable the differential probe to generate a lower total cost of ownership.

5.0 CALIBRATION

5.1 Two Points (or More)

In a perfect world every pH probe would produce a voltage that looks exactly like the left side of Figure 3.10. Calibration would be unnecessary. A probe in pH 7 solution at 25 °C would always read 0 mV and one in pH 4 solution would always read 177 mV. But, in the real world, the offset is almost never zero, and the slope is almost never 59.16 mV/pH at 25 °C. (To be rigorous, the diffusion potential would prevent the offset from ever reading zero even in the most ideal probe.)

Calibration is the exercise of determining just what the true offset and the true slope are for a real pH probe. The Nernst equation predicts that the voltage output is linear with pH. That means that only two points are needed to construct the voltage output versus pH calibration curve. (The term "curve" is actually a misnomer when applied to a straight line.) Because every measurement has an uncertainty associated with it, the larger the number of data points used to construct the line, the more precise the calibration slope and offset. In practice a two-point calibration is more than sufficient if the two points cover the range experienced by the process. Some analyzers allow multi-point calibration.

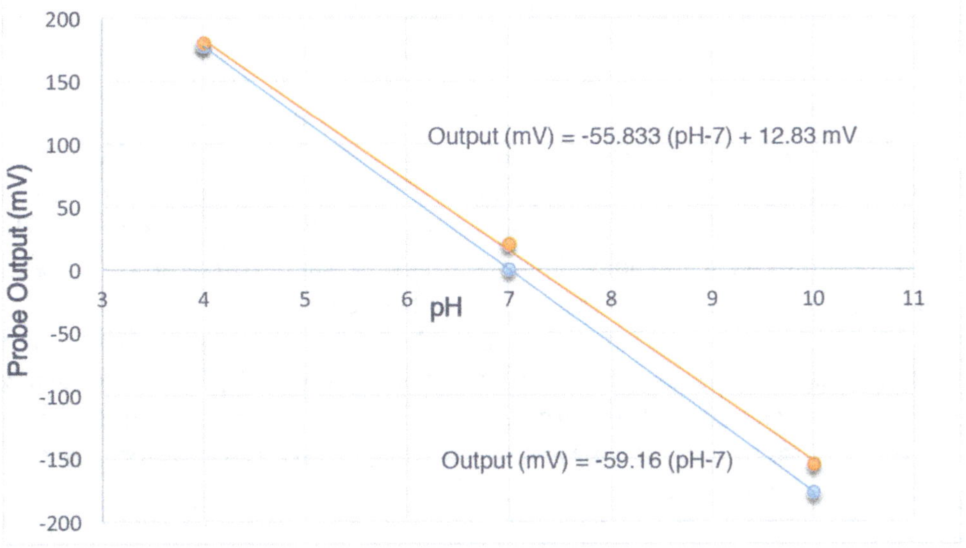

FIGURE 3.10 Calibration curve for a pH probe using 3 calibration standards (4, 7, and 10). The blue points represent an ideal probe with zero offset and 59.16 mV/pH slope predicted by the Nernst equation. The red points represent the actual probe voltages. The efficiency is 55.833 mV/pH. The equations show the results of linear least squares fit of the three points.

For the sake of brevity, we round up the values of the calibration standards to 4, 7, and 10. Figure 3.10 shows a calibration using the three pH standards at 25 °C. The blue line represents the ideal probe in which the offset at pH 7 is zero and the slope is 59.16 mV/pH. The red line represents an actual probe output at the three buffers. The probe in pH 7 solution outputs 15 mV. A linear-least-squares fit of the three voltages yields an offset of 13 mV and a slope of 55.83 mV/pH. The offset differs from the 15 mV measured in pH 7 because the line through the 3 points shifts to best match the points at pH 4 and 10.

The least squares fit on the graph actually yields an offset of several hundred mV. That's because pH 7 is our zero but not the graph's zero. If we were to deduce the actual offset from the graph we would need to subtract (7 × calculated slope) from the calculated intercept to get the offset for pH 7. This is just another anomaly of zero being seven.

With the slope and offset of the actual probe, it is easy to calculate the expected probe output at any pH value.

$$Probe\ Output\ (mV) = 55.833\frac{mV}{pH} \times (Actual\ pH - 7) + 12.83\ mV \tag{3.30}$$

Or, by rearranging this:

$$Actual\ pH = 7 - \frac{Probe\ Output\ (mV) - 12.83\ mV}{55.833} \tag{3.31}$$

For example, if our probe measures 231 mV the pH is 3.87.

The efficiency of the probe is defined as the measured slope divided by ideal slope. Remember that, only for 25 °C, is the ideal slope 59.16 mV. The efficiency for this particular probe is 55.833/59.16 = 94.4%. Nearly every pH analyzer sold today displays either the slope or the efficiency and the offset immediately following a calibration exercise.

The frequency of calibration is highly dependent on the accuracy desired and the process being measured. For an accuracy of ±0.1 pH, the frequency can vary between every three months for a clean water application, in which the pH is close to neutral, to every week for wastewater or water in which the water is highly acidic (<4) or highly caustic (>10).

5.2 Why Calibration Should Be Done at the Same Temperature as the Process

Of the six potentials summarized in Table 3.1, E_1 is the one that responds to the pH of the solution. The remaining five potentials that make up the combination probe (or nine for a differential probe) add or subtract to E_1 to create the probe voltage that the analyzer measures and converts to a pH value. All of them have some temperature dependence. The potentials on both sides of the pH glass electrode exhibit a temperature dependence that is already accounted for by the analyzer when it converts the voltage to pH. There are two potentials that are not. They are:

1. The potentials at the Ag-AgCl wire in KCl are governed by Equation 2.24. As with all redox reactions with a positive E, this one increases with temperature. Presumably both electrodes are at the same temperature, so any temperature variations should mostly cancel out.
2. The asymmetry potential has a temperature dependence due to the amorphous, non-uniform structure of the silicate-based glass.
3. The diffusion potential, E_6, creates an offset that varies with the chemical makeup of the process and with temperature.

These potentials add up to create a temperature dependent offset. Calibration measures the offset as the voltage reading when the probe is in pH 7 buffer. Some refer to the calibration curve as an *isotherm* because the measurements of pH output at two or more pH points are done at the same temperature. Two isotherms for two temperatures are straight lines with differing slopes, each given by Equation 3.21; with any two lines of differing slope they, must intersect somewhere. That point of intersection is called the *isothermal intersection point.*

In the ideal world, the probe offset would be zero for any calibration at any temperature—that is, all isotherms would intersect at the point of pH 7, 0 mV, the so-called *zero point.* The closer to the isothermal intersection the measurement is, the less the error of the analyzer's pH value due to an uncertainty in temperature. Because most pH measurements occur near pH 7, having the isothermal intersection point at the zero point is advantageous. The left side of Figure 3.11 shows this case for two isothermals at 25 °C and 60 °C. Every probe manufacturer wants its probes to have isothermal intersection points as close to the zero point as possible.

Unfortunately, the real world is not as neat and tidy as the one exemplified in the left side of Figure 3.11. The right side of Figure 3.11 shows the same two isothermals but with the 60 °C one containing an offset of +10 mV. The isothermal intersection point is at approximately (pH 8.5, −90 mV)—far from the zero point. Some texts that describe the isothermal intersection point disregard the temperature dependence of the offset.

The wandering nature of the isothermal intersection point underlies one of the most important prescriptions for accurate pH measurements: to minimize the temperature dependence of the isothermal point, one should

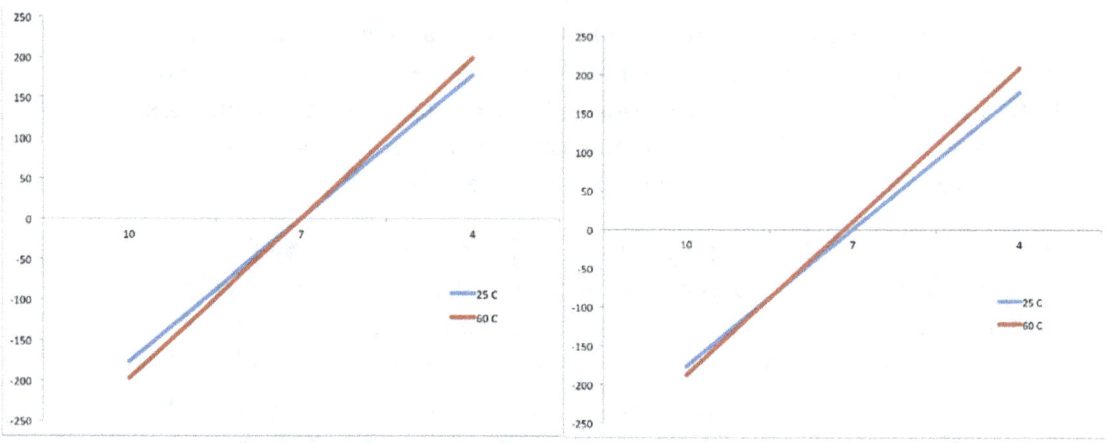

FIGURE 3.11 Two calibration curves (isothermals)—at 25 °C and 60 °C—that ideally intersect at the zero point, (pH 7, 0 mV). (Two calibration curves at the same temperature but with the 60 °C line containing an offset.)

calibrate the pH probe at the same temperature of the process in which it will be used. The non-zero offset will be the same for the calibration as it is for the measurement and, therefore, will not affect the accuracy of the measurement. Keep in mind that the pH of calibration solutions changes with temperature. Table 3.3 is good enough for most measurements, and Table 3.4 gives very precise pH values over a very wide temperature range.

6.0 THINGS THAT AFFECT PH MEASUREMENT

No other sensor in the instrumentation toolbox can match the enormous measurement range of 14 orders of magnitude. Still, the pH probe is not without limitations. Let's discuss some common issues that affect real pH measurement.

6.1 It's the Activity, Not the Concentration

When we introduced the Nernst equation in Chapter 2, Section 6.0 we noted that the voltage of a galvanic cell varies with the activity of the species—not the concentration—involved in the oxidation and reduction reactions. The activity for a species is proportional to the concentration but the so-called activity coefficient, γ, is not constant. It varies with the concentration.

$$Q = \frac{a_c^{\,n_c}\, a_d^{\,n_d} \ldots}{a_c^{\,n_a}\, a_b^{\,b} \ldots};\ a_X = \gamma_X[X] \qquad (3.32)$$

Chapter 2, Section 6.0 discussed the connection between activity and concentration. For very dilute the activity and concentration are nearly equal—that is, the activity coefficient is close to one. As the concentration of a dissolved species increases, its activity decreases and its rate of reaction with other species decreases. For most potentiometric devices in most common applications, the deviation between activity and concentration is small. For pH probes at low pH the deviation can be significant.

According to the Nernst equation the pH varies logarithmically with the H^+ activity, not concentration. The Nernst equation for a pH probe is therefore:

$$E = -\frac{RT}{nF}\log a_{H^+}\ ;\ a_{H^+} = \gamma_{H^+}[H^+] \qquad (3.33)$$

The point of this section is that a pH probe measures H^+ activity, not its concentration. To extract the concentration from the pH value out of the Nernst equation will give a value that is increasingly incorrect as the pH drops. For instance, a 0.01 M HCl solution ought to have a pH of 2. The activity coefficient at that concentration is 0.915. If we use the rigorous formulation of the Nernst equation, Equation 3.33, we find that the pH is actually 2.04. It's very important to recognize that this is not an error. The pH of 2.04 that we measure is real. It's also important to understand that the activity coefficient depends not just on the hydrogen ion concentration but also to the other ions in solution (in this case Cl^-). It is possible that two different solutions can have the same $[H^+]$ but different pH values. *Ionic strength* refers to the total concentration of ions, and it's important to know the concentration of not just H^+ ions, but all ions, to interpret the pH.

6.2 Temperature Compensation

The output voltage of a pH probe is proportional to temperature with the caveat that the temperature must be expressed on an absolute scale—that is, in units of Kelvin. One adds 273 (to be exact, 273.15) to the temperature in Celsius to obtain the temperature in Kelvin. Since 25 °C (298 K) is the standard temperature for electrochemical measurements, it is convenient to express the Nernst equation at that temperature and use a scaling factor to correct the probe voltage for other temperatures. From Equation 3.21 or 3.26, and ignoring the temperature-dependent offset:

$$E(T) = 59.16\ mV \times \frac{T}{298\ K} \qquad (3.34)$$

For example, suppose we start with a solution at 25 °C, or 298 K. If we heat the sample to 40 °C, or 313 K, the probe will output a voltage that rises by the ratio 313/298 = 1.05. Now the slope of the probe output is actually 62.14 mV/pH—not the 59.16 mV/pH often assumed.

Most pH probes contain a temperature element that is wired to the analyzer and calculates the pH reading based on the temperature. That element is either a thermistor or an RTD. RTDs are resistors made from platinum whose resistance varies linearly with temperature. They are commonly available in 100 Ω and 1000 Ω configurations, which is the nominal resistance at 25 °C. Thermistors are resistors made of polymer or ceramic and exhibit a negative temperature correlation (i.e., the resistance decreases with temperature). They come in a wide range of nominal resistance from 100 to 3000 Ω. They are less expensive and faster than RTDs, but they are less linear and have lower long-term stability.

Compensation by the simple linear correction of Equation 3.32 at the analyzer is called *automatic temperature compensation*. In fact, this so-called "correction" is really just a calculation of the pH of a sample using the Nernst equation and the values of the probe voltage and actual temperature.

6.3 A Temperature Effect That Is Real

Temperature compensation is necessary for getting the correct pH value from the pH probe's voltage output. There is, however, a temperature effect, which is very real and does not need correction. It does need understanding.

Take a look back at the chemical equilibrium equation that defines pH:

$$[H^+][OH^-] = 10^{-14}\,M^2 \text{ (at 25 °C)} \tag{3.2}$$

The equation contains a "gotcha" implied by the text inside the parentheses. The dissociation constant of 10^{-14} is only valid at 25 °C.

Imagine what happens if we heat up a container of water. Most of the additional energy transferred into the water goes into making the water molecules move faster, which we perceive as heat. But some goes into tearing apart neutral water molecules to create more H^+ and OH^- ions. The dissociation constant would therefore increase above 10^{-14}. In fact, at 40 °C, the new dissociation constant triples to 2.916×10^{-14}. If we replace 10^{-14} with this new dissociation constant of 2.916×10^{-14} in Equation 3.2, then a simple calculation tells us that, at 40 °C, $[H^+] = [OH^-] = 10^{-6.77}$. Therefore, the pH of pure water at 40 °C is 6.77. This is a significant correction. It is important to understand that water at 40 °C with a pH of 6.77 is not acidic. It is, in fact, neutral. The magic number 7 is so ingrained that it is very easy to forget that it's only true at 25 °C.

Every chemical compound that dissolves in water will have its own dissociation constant with a unique temperature dependence. For instance,

0.001 M NaOH at 25 °C has a pH of 11. (If [OH] of 10^{-3}, then $[H^+] = 10^{-11}$.) However, at 0 °C the pH of this mixture is 11.94 and, at 50 °C, the pH is 10.26. It is reassuring to know that, if the analyzer does temperature compensation correctly, then the pH reading it displays is correct. It is important to know that, even after temperature compensation, the pH value is not going to be the same as one at 25 °C.

6.4 Buffer Solutions

While on the topic of dissociation constants, let's dive into the science of buffer solutions. They have a seemingly magical property of resisting changes in pH when an acid or base is added. Here is how this remarkable property works.

At the heart of every buffer solution is a weak acid. An acid dissociates into hydrogen ions (H^+) and the remainder of the chemical species—what is called the *conjugate base* (A^-), or salt:

$$HA \rightarrow H^+ + A^- \tag{3.35}$$

In a strong acid Equation 3.35 proceeds all the way to the right, and HA dissociates completely. Examples of strong acids are hydrochloric acid (HCl), sulfuric acid (H_2SO_4), and nitric acid (HNO_3). For a weak acid dissociation is only partial, and there is a mixture of dissociated and undissociated molecules—that is, the acid (HA) and its conjugate base (A^-). We characterize the degree of dissociation by the dissociation equilibrium constant, K_s, defined as the ratio of the product of reaction product concentrations ($[H^+]$ and $[A^-]$) to the reactant concentration ([HA]):

$$K_a = \frac{[H^+][A^-]}{[HA]} \tag{3.36}$$

We'll ignore the temperature dependence. To see how the pH depends on the dissociation constant, let's take the logarithm of both sides of Equation 3.36. In doing so we make use of the rule that a logarithm of a product of two quantities is equal to the sum of the logs of each quantity:

$$\log K_a = -pH + \log\frac{[A^-]}{[HA]} \tag{3.37}$$

Recall that the purpose of the *p* in pH is to denote the negative logarithm of the H^+ concentration. Following this convention, we can express $\log K_s$ as $-pK_s$. With one final step of switching the two *p* terms, we get:

$$pH = -pK_a + \log\frac{[A^-]}{[HA]} \tag{3.38}$$

This is the famous *Henderson-Hasselbach equation*, and it explains how buffers do their magic. It happens when the pH of the buffer is close to the pK_a value, which occurs when $[A^-] = [HA]$ and the ratio in Equation 3.38 is 1. The logarithm of 1 is zero.

Let's see what happens at this special point, denoted in Figure 3.12. First let's explain with words. When one adds a strong acid the excess H^+ ions latch onto A^- anions to form HA molecules:

$$A^- + H^+ \rightarrow HA \tag{3.39}$$

When one adds a strong base the excess OH^- ions combine with HA molecules to form the conjugate base:

$$HA + OH^- \rightarrow A^- + H_2O \tag{3.40}$$

Either way, A^- (conjugate base) ions and HA (acid) (sometimes called the *conjugate acid*) "mop up" any external acid or base added.

Now let's explain buffers using the Henderson-Hasselbach equation. Acetic acid, the major component of vinegar, is a common weak acid. It partially dissociates to the conjugate base, *acetate,* and H^+ ions:

$$CH_3CH_2COOH \leftrightarrow CH_3CH_2COO^- + H^+ \tag{3.41}$$

The dissociation constant, K_a, is 1.8×10^{-5}. Take the example of a solution containing 0.25 M CH_3CH_2COOH and 0.50 M $CH_3CH_2COO^-$.

$$pH = -\log 1.8\times 10^{-5} + \log \frac{0.50}{0.25} = 4.74 + 0.30 = 5.04 \tag{3.42}$$

According to the Henderson-Hasselbach equation, the pH of a buffer system varies with the logarithm of the ratio $[A^-]/[HA]$ rather than just the ratio. This explains why the addition of acid or base produces a small change in the pH. Logarithms of numbers change much more slowly than the numbers themselves. Figure 3.12 is a generalized depiction of Equation 3.36. The graph is a plot of the pH of a buffer versus the ratio of the conjugate base to acid. It is easy to show, using Equation 3.38, that the change in pH is minimal over the range of $pK_a \pm 1$. When the pH is less than $pK_a - 1$, the added acid exhausts the conjugate base. When the pH exceeds $pK_a + 1$, the added base exhausts the conjugate acid. From a mathematical standpoint, it is the logarithm of the conjugate base to acid.

Perhaps the most famous of all buffer reactions is that of carbonic acid. We encountered this reaction involving three carbonate species in Chapter 1,

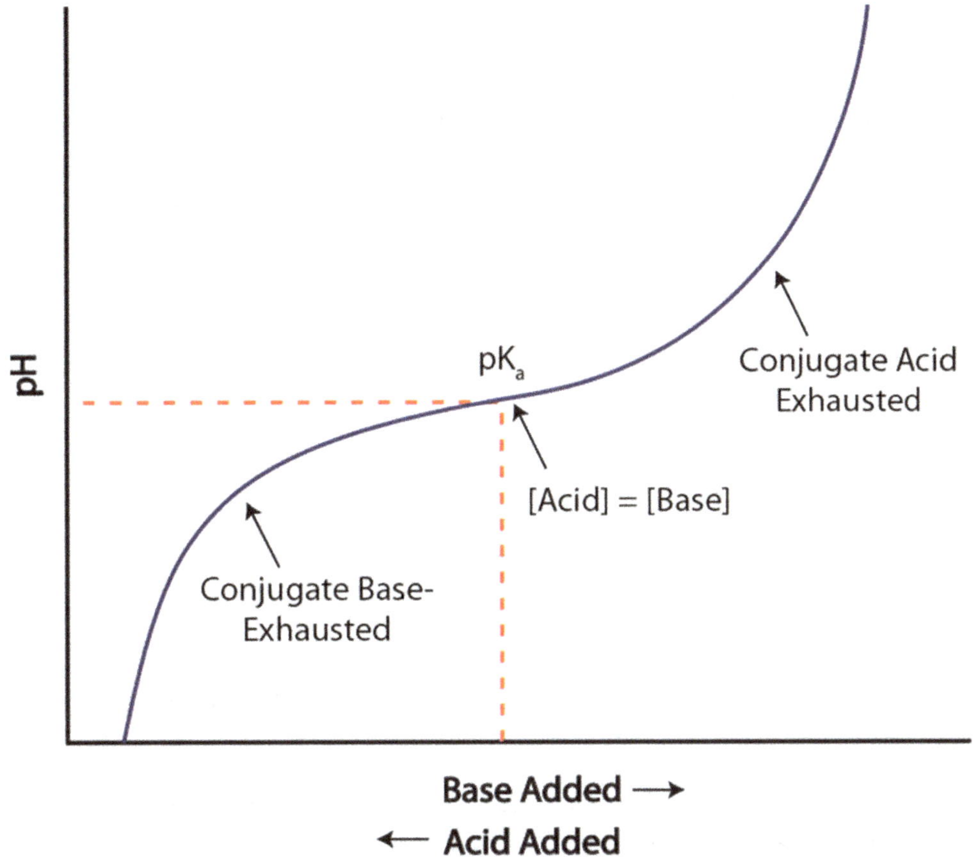

FIGURE 3.12 Henderson-Hasselbach equation. When the pH is close to pKa, any addition of acid or base to change the pH produces a minimal change.

Section 6.0 when we discussed the reversibility of chemical reactions. The carbonic acid system consists of three species whose concentrations are linked through Equation 1.6 and which I repeat here:

$$CO_2 + H_2O \leftrightarrow H_2CO_3 \leftrightarrow HCO_3^- + H^+ \leftrightarrow CO_3^{2-} + 2\,H^+ \quad (3.43)$$

One can think of this system as consisting of one neutral species (CO_2), one acid (H_2CO_3), one salt or conjugate base (CO_3^{2-}), and one species that is both an acid and salt (HCO_3^-). It is not possible to describe the behavior of this three-way buffer system with one *pK*a value. However, one can use the three equilibrium constants in Equation 3.41 to calculate the relative concentrations of carbonic acid (H_2CO_3), bicarbonate (HCO_3^-), and carbonate (CO_3^{2-}). This is shown in Figure 3.13. Between pH 6.5 and 10.5, at least half of the dissolved CO_2 is tied up as bicarbonate.

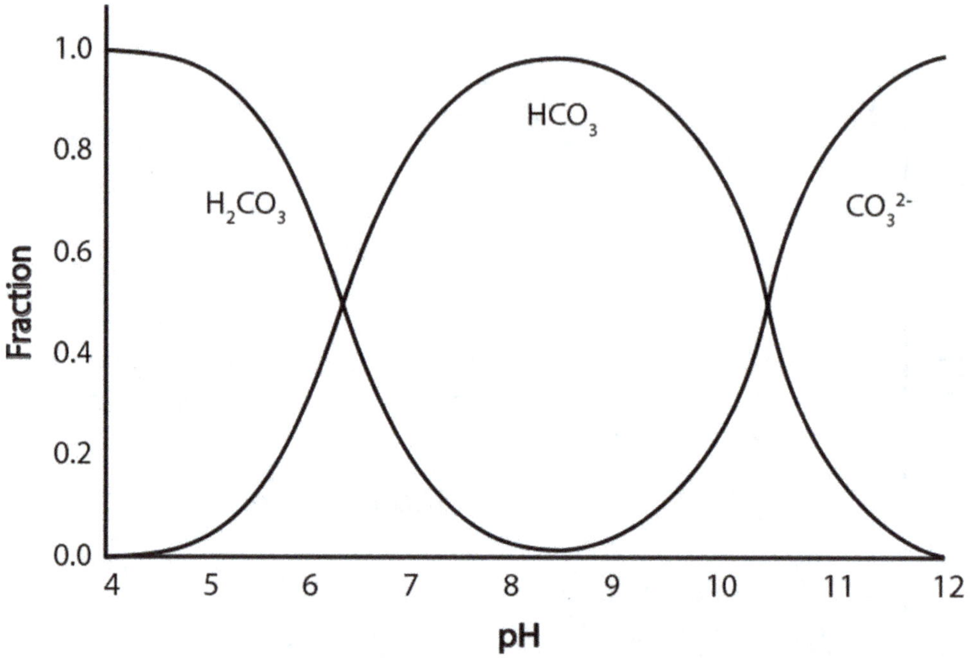

FIGURE 3.13 Equilibrium concentrations of carbonic acid, bicarbonate, and carbonate as a function of pH.

Now that we know how buffers work, we can appreciate how useful they are as calibration standards. We use standards for every analytical instrument, but most are susceptible to changes either through degradation or change by outside contaminants. For instance, a conductivity reference solution will change its conductivity in proportion to the amount of added salt. However, one can add acid or base to a pH buffer solution and the change will be small. As long as the pH of our calibration solution is close to that of its pK_a value, it will resist additions of acid or base.

There are three buffers of particular importance to pH measurement. These are the three buffers that are used for calibration. They are the buffers for pH 4, 7, and 10. However, these nice round numbers are a matter of convenience; they are not exactly at the *pK*a of the weak acid that comprise them. After all, that would be an amazing coincidence if they were. For instance, pH 7 buffer is made up of potassium dihydrogen phosphate (KH_2PO_4). This highly soluble powder dissolves almost completely into K^+ and $H_2PO_4^-$ ions. The $H_2PO_4^-$, in turn, dissociates into HPO_4^{2-}.

$$KH_2PO_4 \leftrightarrow H_2PO_4^- + K^+ \leftrightarrow K^+ + H^+ + HPO_4^{2} \quad (3.44)$$

For maximum buffering capacity one would make a 1:1 mixture of the acid $H_2PO_4^-$ and the conjugate base, HPO_4^{2-}. The acid is provided by

KH_2PO_4, and the conjugate base is provided by $NaHPO_4$. The pH would therefore equal the *pK*a for this dissociation reaction, which is 7.20. But there's a problem. The NIST primary buffer, which consists of 0.05 M KH_2PO_4 and 0.05 M Na_2HPO_4, lists 6.86 as the pH of that mixture. What's the catch?

The explanation for the discrepancy lies in recasting the Henderson-Hasselbach equation, Equation 3.36 in a more accurate form—that is, one in which the activities of the conjugate base and acid replace their concentrations. Ions, by virtue of their electrostatic attraction, have activity coefficients significantly less than one. Section 7.3 invokes this correction to explain acid error. Multiply charged ions, like HPO_4^{2-}, have even lower coefficients. The result is that the actual *pK*a value at 0.05 M of both acid and conjugate base is actually 6.83. The weak acid used to make pH 4 buffer is made up of 0.05 M potassium hydrogen phthalate, $KHC_8H_4O_4$, aka "KHP", has a *pK*a value of 4.0008. For pH 10 buffer, the weak base is 0.01 M $NaB_4O_7{\bullet}10H_20$, aka "Borax," has a *pK*a of 9.180.

Table 3.2 lists most, if not all, of the common buffers in use today. Note that $Ca(OH)_2$ is actually not a buffer but a strong base.

The equilibrium between the acid and the anion of a weak acid is temperature dependent. That means that the actual pH of a buffer is dependent on temperature, with the nominal value being the pH value at 25 °C. The right-hand column of Table 3.2 lists the temperature coefficient of the standard. Table 3.3 lists the pH values for several temperatures for the three common calibration standards.

For greater accuracy over a wide temperature range, the following empirical relationship can be used to calculate the temperature dependence of calibration standards:

$$pH(T) = pH(25\ °C) + \frac{A}{T} + B + CT + DT^2 \qquad (3.45)$$

Table 3.4 Temperature dependence of calibration standards. The formula is pH 5 pH(25 °C) 1 A/T 1 B 1 CT 1 DT2. lists the four coefficients for the three standard buffers plus a strong acid and strong base.

6.5 The Diffusion Potential at the Liquid Junction

Section 3 described the diffusion, or junction, potential as a major contributor to the offset of a pH probe. Let's dig a little deeper into the cause and effect of the junction potential.

There are several names for the permeable gateway that electrically connects the reference electrode to the sample. They include the liquid junction, porous junction, and salt bridge. All of them strive to do one thing well:

TABLE 3.2 Recipes for common calibration standards. Included is the temperature coefficient, which relates the change in pH to change in temperature.

Standard Solution	pH (25 °C)	Temp (°C)	Formula (g/L)	ΔpH/ΔT
Potassium Hydrogen Tartrate (saturated at 25 °C)	3.557	25–95	> 7 g $KHCH_4O_6$ [a]	n/a
0.05 M Potassium Dihydrogen Citrate	3.776	0–50	11.41 g $KH_2C_6H_5O_7$	−0.0022
0.05 M Potassium Hydrogen (Phthalate)	4.004	0–95	10.12 g $KHC_8H_4O_4$	+0.0012
0.025 M Potassium Dihydrogen Phosphate + 0.025 M Disodium Hydrogen Phosphate	6.863	0–95	3.387 g KH_2PO_4 + 3.533 g Na_2HPO_4 [b]	−0.0028
0.008695 M potassium Dihydrogen Phosphate + 0.03043 Disodium Hydrogen Phosphate	7.415	0–50	1.179 g KH_2PO4 + 4.303 g Na_2HPO_4	−0.0028
0.01 Sodium Borate Decahydrate (Borax)	9.183	0–95	3.80 g $Na_2B_4O_710H_2O$	−0.0082
0.025 Sodium Bicarbonate + 0.025 Sodium Carbonate	10.014	0–50	2.092 g $NaHCO_3$ + 2.64 g Na_2CO_3 [4]	−0.0096
0.05 Potassium Tetroxalate	1.679	0–95	12.61 g $KH_3C_4O_82H_2O$	n/a
Calcium Hydroxide (saturated at 25 °C)	12.454	0–60	> 2 g $Ca(OH)_2$	−0.0033

[a]Approximate solubility.

[b]Prepare with boiled and cooled distilled water to remove carbon dioxide.

they ensure that the cations and anions of the salt move to complete the electrical circuit and do so at speeds that are as close as possible to each other. We call the latter condition *equitransferent*.

The parameter that describes how fast an ion moves in solution is its *ionic mobility*. The electrical *conductivity* of a solution is proportional to the sum of the ionic mobility values of the constituent ions. The subject of conductivity constitutes the next chapter. For now, let's just discuss the ionic mobilities of the ions that make up the reference solution. A convenient measure of an ion's mobility is the conductivity of that ion divided by its concentration—the *molar conductivity*. To be precise, the molar conductivity

TABLE 3.3 Dependence of pH on temperature for the three commonly used pH buffers.

Temp (°C)	pH 4	pH 7	pH 10
10	4.00	7.07	10.18
15	4.00	7.04	10.14
20	4.00	7.02	10.06
25	4.00	7.00	10.00
30	4.01	6.99	9.95
35	4.02	6.98	9.91
40	4.03	6.97	9.85
50	4.05	6.96	9.78
60	4.08	6.96	9.75

decreases as the concentration increases—much like the decreasing activity of a species as its concentration increases—so it is calculated at very dilute concentrations. Chapter 4, Section 2 discusses in greater detail the subject of molar conductivity and ionic molar conductivity.

As one might expect, small ions, such as hydrogen and lithium ions, move faster than big ones, though this relationship is subject to other factors. Suppose we make a liquid junction with table salt—NaCl. The sodium ion is larger than the chloride ion, so its mobility is lower. As the ions diffuse from the higher concentration reference to the lower concentration process, the faster chloride ions diffuse faster than the sodium ions and generate a separation of charge. This creates a *junction potential*—also known as a *diffusion potential*—that is on the order of several millivolts. That's enough

TABLE 3.4 Temperature dependence of calibration standards. The formula is pH = pH(25 °C) + A/T + B + CT + DT^2.

x	pH Calibration Standard (pH at 25 °C)				
Coefficients	M HCl 1.095	Phthalate 4.005	Phosphate 7.000	Carbonate 10.012	$Ca(OH)^2$ 12.454
A	0	0	1722.78	2557.10	7613.65
B	1.0148	6.6146	−3.6787	−4.2846	−38.5892
C (×102)	0.0062	−1.8509	1.6436	1.9185	11.9217
D (×105)	0.0678	3.2721	0	0	−11.2918

to change a pH measurement by several percent. For that reason, we want the salt in the electrolyte to be composed of cations and anions that are as close to each other in mobility, or molar conductivity, as possible. Ion mobilities correlate with ion size, expressed as their atomic or molecular weights. Potassium has an atomic weight of 39 amu, and chlorine has a weight of 35.5 amu. These are relatively close, which explains why KCl is ubiquitous as an electrolyte in the reference cell and in the salt bridge. This little rule of thumb only works when salts are composed of a one-electron cation and a one-electron anion. To be sure one must look up the molar conductivity of a potential ion. Table 3.5 is a compilation of molar conductivities of common ions. You will notice that, as good a match as K^+ and Cl^- are, that between K^+ and NO_3^- is even better. That's why KNO_3 is also frequently found in electrolytes and salt bridges. Lithium acetate ($LiCH_3COO^-$) is widely used in ion selective electrodes where K^+, Cl^-, or NO_3^- ions might interfere with the analyte. The most common example is the ammonium probe, in which the ionophore has a cross selectivity to K^+ ions, thereby invalidating the use of KCl.

The liquid junction potential is not just dependent on what's on the inside of the junction. It also depends on what's on the outside. This is why the junction potential is not strictly a constant. As K^+ and Cl^- ions migrate out of the reference compartment, ions from the process take their place. These external ions add to the liquid junction potential. This effect is most apparent when measuring concentrated hydrochloric acid in which the tiny H^+ ion is about five times more mobile than Cl^-. Table 3.6 lists junction potentials for some common solutions. It shows that a 1 M solution of HCl, which we

TABLE 3.5 Molar conductivity values of common cations and anions. A more thorough discussion of mobility is in Chapter 4, Section 2.3.

Cation	Molar Conductivity ($S\ cm^2\ mol^{-1}$)	Anion	Molar Conductivity ($S\ cm^2\ mol^{-1}$)
Li^+	38.7	CH_3COO^-	40.9
Na^+	50.1	ClO_4^-	67.4
K^+	73.5	NO_3^-	71.5
NH_4^+	73.6	Cl^-	76.4
Mg^{2+}	106.2	B^{r-}	78.1
Ca^{2+}	119.0	SO_4^{2-}	80.0
H^+	349.8	OH^-	198.3

TABLE 3.6 Liquid junction potentials for several samples (Radiometer Analytical, 2007).

Sample	Liquid Junction Potential (mV)
1 M HCl	14.1 mV
0.1 M HCl	4.6 mV
0.01 M HCl	3.0 mV
0.1 KCl	1.8 mV
pH 1.68 Buffer	3.3 mV
pH 4.01 Buffer	2.6 mV
pH 4.65 Buffer	3.1 mV
pH 7.00 Buffer	1.9 mV
pH 10.01	1.8 mV
0.01 M NaOH	2.3 mV
0.1 M NaOH	−0.4 mV
1 M NaOH	−8.6 mV

know to be pH 0 ($10^0 = 1$), will read 0.2 pH units high. It is interesting to note that even our standard pH buffers of 4, 7, and 10 give non-negligible errors but calibration using these buffers will take into account this error.

The liquid junction in a combination probe is usually a porous material such as Kynar (PVDF), ceramic and Teflon (PTFE). The design of the liquid junction strikes a balance between longevity of the reference solution and minimization of the diffusion potential. As the junction becomes more porous, the diffusion rate increases and the diffusion potential decreases. Pure water measurements require a high rate of leakage to provide the necessary conductivity essential for stable readings. However, a faster diffusion rate also results in faster depletion or contamination of the reference solution. Some manufacturers incorporate a pressurized reservoir of electrolyte to continuously replenish the rapidly depleting solution.

The liquid junction in the differential probe is a salt bridge—a plastic capsule containing KCl gel with ingress on both ends through a porous plug of the materials just mentioned. The KCl gel is a buffer between the reference chamber and the external sample. With porous junctions on both ends it allows a fast diffusion through the salt bridge while minimizing contamination of the reference solution. The reference solution can be equitransferent

KCl or KNO_3 solution to minimize the diffusion potential or pH 7 buffer to maximize the pH stability over time.

The liquid junction—whether it be a porous frit or salt bridge—is susceptible to fouling. The result is a slower migration of ions through the junction, which, in turn, generates a slower response and a changing potential. The change in probe output is a few millivolts, so it is unlikely to be the cause of a significantly lowered efficiency.

6.6 The Ever-Changing Reference Electrode

From the moment a pH probe enters a solution the electrolyte in the reference compartment leaks into the sample at a rate that varies widely, depending on the type of liquid junction. As the KCl electrolyte leaks out the process leaks in to take its place. If the probe is in relatively clean water the reference solution will lose its ionic strength and the potential of the Ag-AgCl electrode will decrease, resulting in an increasingly negative offset and lower pH reading. Whichever direction the offset drifts, eventually it becomes so great—approximately larger than 50 mV—that the probe will not calibrate.

The reference chamber of most combination probes is sealed but some have an inlet port that allows the user to replace the reference solution. An alternate strategy to extend the life of the reference solution is the double junction electrode as described in Section .

Differential probes consist of a sealed electrode immersed in KCl solution. Changing the solution is simple. Some manufacturers recommend off-the-shelf pH 7 calibration buffer while others recommend KCl or KNO_3 solution. Still others use a combination of the two. The benefit of using pH 7 buffer is that, as a true buffer, it maintains its pH longer than a simple KCl solution. The benefit of KCl or KNO_3 is that KCl and KNO_3 are equitransferent, which minimizes the diffusion potential across the junction.

An aging reference solution will also cause the probe to lose efficiency if the contaminants react with the AgCl coating on the Ag-AgCl wire. This is especially a problem with processes containing sulfides, which react with silver to form a black precipitate, silver sulfide (Ag_2S). The precipitate coats the Ag-AgCl wire, thereby poisoning it, and creates precipitates that clog the junction. Other compounds that react with silver or the KCl electrolyte include mercury, cyanide, and ammonia.

6.7 The Process Electrode in Harm's Way

The most common cause of low efficiency is a fouled glass bulb of the process electrode. Fats, oils, and grease, paper pulp, bacterial film ("biofilm"), scaling, and plating chemicals are all capable of coating the electrode and reducing its efficiency. Calcium (or magnesium) carbonate precipitates out at pH values greater than 7. Dilute HCl (2%) or household white vinegar is sufficiently acidic to dissolve scaling. Biofouling is the buildup of a bacterial film. In this case an oxidizing agent like hypochlorite will clear the electrode. An alternative treatment is soaking in a stronger organic acid such as acetic, oxalic, or glycolic acid.

Whether the process electrode is separate, combined, or differential, it has a finite lifetime. Various explanations have been proposed on chemical or mechanical grounds. The most prevalent one invokes the extended leaching of the alkali (Li^+ or Na^+) ions (Morf, 1995) into solution and the subsequent breakdown (hydrolysis) of the silica network. The exchange of alkali ions with H^+ ions is necessary for the formation of the so-called gel layer and functioning of the electrode. Over time, more and more hydrogen ions replace alkali atoms, and the depth of the gel layer increases, thus decreasing the conductivity of the glass electrode. The result is a longer response time, which may look like a decreased efficiency if the probe is not left in buffer solution long enough during calibration. In processes of high temperature and/or low pH the lifetime of the electrode can be as short as 1 year. Figure 3.14 shows data obtained by Yokogawa that show decrease in lifetime for a range of temperatures. For a combination probe with a sealed reference chamber the inevitable contamination of the reference

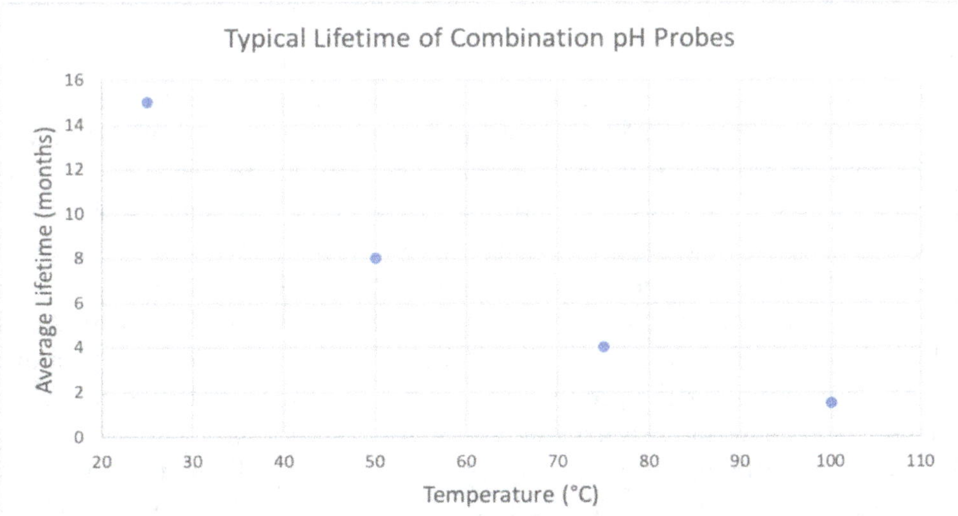

FIGURE 3.14 Lifetimes of typical combination pH probes at different temperatures (Data courtesy of Yokogawa).

solution or poisoning of the Ag-AgCl electrode will be the limiting factor for its lifetime.

There are several remediation steps that might rejuvenate a probe with a depleted electrode:

1. Immerse the probe in successive cycles of baths of acid and caustic. One cycle consists of a bath in 0.1 M HCl for about 20 seconds followed by a bath in 0.1 M NaOH for another 20 seconds. Repeat this procedure for 3 cycles.
2. Etch the glass with a strong acid to expose fresh glass. A dip in a 20% solution of ammonium hydrogen fluoride (NH_4HF_2) for no more than 30 seconds followed by 15 seconds in 6M HCl will do the job. (I can tell you from experience that you have as good a chance of destroying the probe as you do of resuscitating it.)
3. Immerse the probe in 5% HF for no more than 15 seconds. Follow with a 20 second rinse in 0.1 M HCl. This procedure will remove a layer of glass and expose a fresh one. There is perhaps even more dangerous step than the previous tactic as it involves one of the most dangerous chemicals (HF) to be found in a lab.

6.8 Diagnosing a Probe

A potentiometric probe is a simple device. It generates a voltage in response to the property or analyte being measured. That voltage is determined by the Nernst equation. Therefore, diagnosing the health of a probe is straightforward. Most analyzers include an option for reporting the output voltage of the probe. In a perfect world, and at 25 °C, a pH probe outputs a voltage of 0 mV at pH 7, 177 mV in pH 4 buffer at 25 °C (3 × 59 mV/pH), and −177 mV at pH 10. In the real world, there will be an offset at pH 7 as explained in Section 2, so the important metric is the difference between the probe voltage at pH 4 and pH 7 or 10. So, if a pH 7 buffer gives a probe output of 20 mV, then we expect to see a reading of 197 mV in pH 4 buffer for a probe operating at 100% efficiency. If the output is 180 mV, then the efficiency is (180−20)/177 = 90%. If you're wondering how one obtains millivolt readings for the probe, rest assured that most analyzers contain a diagnostics menu that reports the raw probe voltage.

Remember from the first chapter that there is a distinction between accuracy and precision. A calibrated probe with 50% efficiency will deliver pH readings that are just as accurate as a probe with 100% efficiency, but its precision will be much lower. The precision of the probe scales linearly with efficiency, so if a probe is giving 0.1 pH units of precision when 100% efficient, it will only give 0.2 pH units when 50% efficient.

7.0 THE CURSE OF THE DREADED GROUND LOOP

7.1 The Key Word Is Loop

There is arguably nothing more frustrating than taking a probe out of its process, calibrating it, and placing it back in the process—only to find that it doesn't read correctly. This is the curse of the ground loop. The previous chapter on ORP introduced ground loops but did not dive too deeply into the subject. The platinum or gold band that comprises the ORP process electrode provides a low resistance pathway to a stray voltage. So does the Ag-AgCl reference electrode, which is separated from the process by the liquid junction. For these reasons ground loops do not contribute to significant voltage changes to the signal. A pH probe is a very different story. The following discussion is pertinent to a combination probe.

To properly understand a ground loop, we need to understand electrical grounds:

All buildings (at least those that meet building codes) are wired with one ground being an actual earth ground. Current electrical codes mandate an 8-foot rod made of stainless steel, copper-coated steel, or zinc-coated steel. The main electrical panel is grounded to this earth ground, and the individual circuits originating in the panel tie their grounds to the main ground. In the United States the neutral of a 3-conductor circuit is also tied to ground. As we venture further downstream from the electrical panel, we find instrumentation and controls in clusters. Each electrical device has its own ground and, if the enclosure is metallic, it's a safe bet that the enclosure is grounded. All grounds of electrical devices that are powered by the same building circuit are grounded through that circuit's ground. In the ideal world the circuit ground is the same as the earth ground, but, in the real world, the pathways from the various circuits to earth ground are all different and have a finite impedance. Current through an impedance creates a voltage so currents that flow between grounds generate different potentials.

If a device on one circuit sends an electrical signal to another on a different circuit and that signal ground connects to the grounds of both devices, then the result is a continuous circuit completed by a pathway between the two grounds—a ground loop. The current that finds its way into the loop generates a voltage at the signal ground by virtue of the impedance and Ohm's law. Typical impedances are on the order of a gigaohm so even a small current can generate a large voltage.

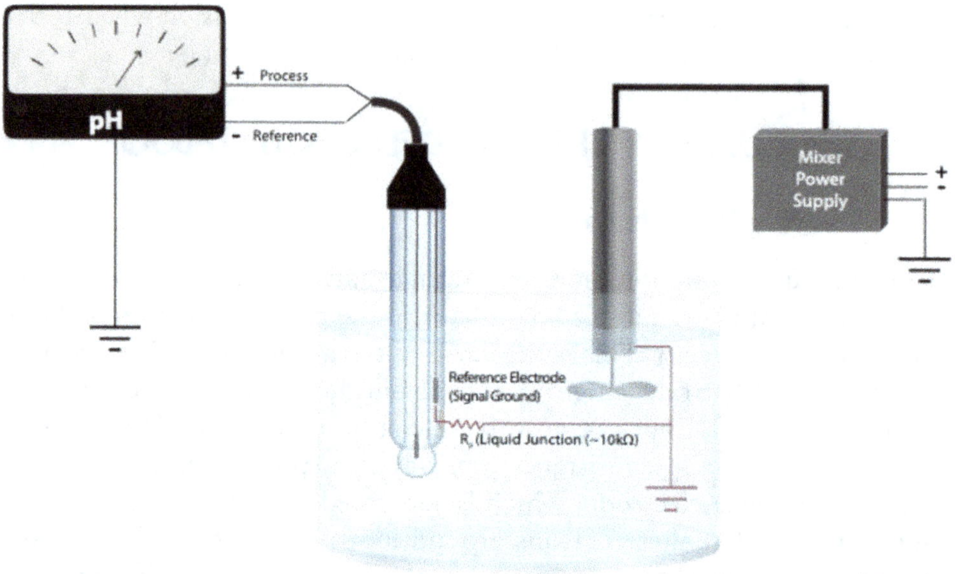

FIGURE 3.15 Electrical schematic of a ground loop in a pH probe. The ground of the probe is tied to the ground of the mixer through the reference electrode. The ground of the meter is at a different potential, so the current loop is from the probe signal to the analyzer to the analyzer ground to the probe ground, the latter of which is tied to the mixer ground.

Let's now show how a pH probe in a process can be part of a ground loop. Figure 3.15 shows a pH probe in a container of water. If the probe is in the container by itself, there is no ground loop. But immerse a grounded device, like the mixer shown in the figure, and the water in the container is now at the same potential as the ground of the mixer power supply (shown separately in the figure for illustration purposes).

The process electrode of a pH probe is glass. Even with the alkaline ions (lithium or sodium) of pH glass, the latter is still a good insulator with a resistance on the order of hundreds of megaohms. The path of least resistance is now through the liquid junction and to the Ag-AgCl reference electrode. The impedance of the junction is on the order of kiloohms. Figure 3.15 shows that a ground loop configuration now consists of the reference electrode in a pathway through the liquid junction and into the grounded solution.

There is one more condition that must be met in this setup: the signal ground—that is, the Ag-AgCl wire of the reference electrode—must be electrically connected to the ground of the pH meter. If, as is often the case, the probe is electrically isolated from the analyzer, then there can be no ground loop.

When the above conditions are met there is a ground loop. Now all we need is current injected into the loop to impart a voltage, courtesy of Ohm's law. The source of the current can be of two types:

7.1.1 Induction

A continuous conductive loop can act just like the secondary winding of a transformer. Michael Faraday showed more than two centuries ago that pulling a magnet through a coil induces a voltage in the coil and, hence, a current. That's how a transformer works. Alternating current flows in the windings of the primary and generates an alternating current magnetic field. The magnetic field, in turn, loops around a second set of windings—the secondary—and induces a current, sometimes referred to as an electromotive force (EMF). The ratio of windings between the primary and secondary determines the change in voltage between the two.

The ground loop is the secondary. So, what's the primary? It's usually a power cable somewhere near the probe that carries a fair amount of 60 Hz alternating current—the power line frequency in the United States—on the order of 100 mA or greater. Other sources include pumps, mixers, fluorescent lights, heaters, microwave ovens, and communication equipment. The result is an alternating current magnetic field—at 60 Hz for a power cord but at other frequencies for other sources of electrical noise. The alternating current magnetic field generates a current in the ground loop, which, in turn, generates a voltage. The result is that the grounding point at each electrical device is at a slightly different potential. This is the source of the hum from audio equipment.

The induced AC voltage is frequency dependent. The preceding discussion assumed a 60 Hz alternating current signal. In general, interfering signals of higher frequency induce a larger induced voltage. That's because, according to Faraday's law, the induced EMF is proportional to the *change* in the magnetic field. A direct current flowing through a conductor creates a direct current magnetic field and, therefore, zero-induced EMF or current. The higher the frequency of the alternating current the larger the EMF. Because the induced voltage is an alternating current one, induction manifests itself as noise imposed upon the signal. It does not result in a direct current voltage offset.

7.1.2 Common Mode Impedance Coupling

The second scenario in which current makes its way through the ground loop is for a second circuit having the same ground in common into which current can "leak." Figure 3.16 illustrates one possible route that a stray

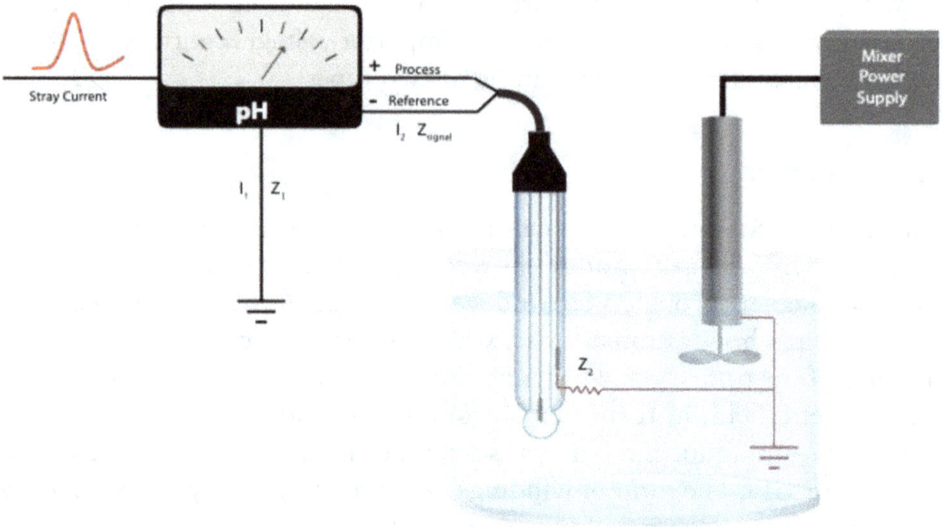

FIGURE 3.16 Common mode impedance coupling. Stray current enters the analyzer and splits into two paths to ground. The potential at the reference electrode can shift by hundreds of *millivolts*.

current or voltage can intrude into a ground loop. A stray current finds its way into the analyzer, where it divides into two paths:

1. I_1 travels from the analyzer to the ground. The impedance of this path is Z_1.
2. I_2 travels along the signal ground from the analyzer to the probe and into the ground through the mixer's ground. The impedance of this path is Z_{signal} along the signal ground and Z_2 along the reference electrode, liquid junction, and sample.

Because both pathways ultimately end at the same ground potential, we can set up two Ohm's law equations that are coupled:

$$I_1 Z_1 = I_2 (Z_{signal} + Z_2) \text{ or } \frac{I_2}{I_1} = \frac{Z_1}{Z_{signal} + Z_2} \tag{3.46}$$

The result is that the voltage at the reference electrode is now shifted by the amount $I_2 Z_2$. Since Z_2 is hundreds of kΩs, even a fraction of a μA of stray current can induce a voltage on the order of 100 mV—more than 1 pH unit.

Note that device 1 cannot be a handheld meter, because it is battery powered and has no connection to ground. Note also that the only way a ground loop is possible is if there is a grounded device, like the mixer pictured in the water. This is an important point to understand. It explains why the oversold prescription of always grounding your sample—even if you don't have a problem—can be a big mistake.

A variation on the theme described above is the injection of current from the ground of another circuit into the ground of the circuit containing the analyzer. A typical example is a high current drawing device, like a pump, in which current leaks from the hot to the neutral side of the circuit. In 120 V alternating current circuits in the United States the neutral conductor is tied to ground. The ground current splits between the ground wire to the analyzer and the ground wire leading from the analyzer's outlet to earth ground. The result is the same voltage divider along the ground loop characterized above.

7.1.3 What to Do About Ground Loops

Confirming a ground loop is easy:

1. Take a grab sample of the process and measure its pH. If the value changes from the *in-situ* measurement, then you have a ground loop.
2. With the probe in a grab sample electrically connect the grab sample to the actual process via a wire. If the reading jumps the moment the wire connects the two, then you have a confirmed ground difference between the isolated buffer and process.

So how does one minimize noise from an alternating current source? To start, an ounce of prevention is worth a pound of cure. If there are no other electrical devices in the sample that are grounded through their enclosure, then there is no path to ground from the sample. At the other end of the loop, a well-engineered controller (or transmitter) will isolate the probe input from ground. This breaks the continuity of ground between the probe's ground and the controller's ground.

For a controller that doesn't isolate the probe input, an isolation barrier (or simply "isolator") will break the loop. A second solution is to isolate the probe input to galvanic or optical isolators. The former inserts a transformer into the circuit. The latter converts an electrical signal to an optical one and then back to an electrical signal.

The fix for a ground loop is simple in theory and, in many cases, impossible in practice. Simply remove the interfering device and its accompanying pathway to ground. If the process is in a metal pipe or a pump, mixer, and so forth, the process train removing the pathway to ground is impractical.

To minimize the interference from inductive coupling:

1. Maintain a healthy distance between signal cables and power cables. Some installations run all power lines and cables through the same

channel and that's a recipe for serious noise issues. Remember that the magnetic field drops off as the square of the distance, so a little separation goes a long way.

2. Induction can only occur between two conductors that are parallel. So, run power cable perpendicular to signal cables.
3. Twist the signal and ground probe wires. Twisted pairs are a staple of noise reduction techniques. Remember that the magnetic field from a disrupting source (like a power cord) circulates in one direction. So does the induced current in the affected conductor. If the signal and ground wires twist around each other, then the induced current will flip directions at every twist, and the induced currents will cancel each other. Most combination probes use cable in which the ground conductor is a coaxial shield surrounding the inner signal conductor. Though it cannot become a twister pair, coaxial cable shields the signal conductor from an alternating current field.

To address a ground loop from common mode impedance, first try to find the offending grounding source. Disconnect each piece of equipment and see if the pH value suddenly changes. Even if you are successful, it may be impractical to remove the offending device. The only corrective action is to minimize the impedance between the probe and earth ground. One simple way to equalize the ground of the process to that of the analyzer is to simply run a grounding wire from the process to the instrument ground. That can be done simply by running a wire from earth ground straight into the process, as shown in Figure 3.17. This is the prescription provided by nearly every discussion of ground loops, but it is important to realize that it does not solve ground loops. It only minimizes it.

A differential probe is mostly immune from ground loops. Whereas a combination probe measures the difference directly between the process and reference electrodes, a differential probe decouples the two. Instead, it pairs each electrode with a ground rod that is common to both. A ground loop will develop between the ground rod and the instrument ground. Because the ground rod is common to both electrodes it will add or subtract the same voltage to both. Hence the difference between the reference and process electrodes will remain constant.

7.2 Further Abuse by Static Charge

The pH probe is a potentiometric device, so it responds to voltage. We are all familiar with the consequences of rubbing our feet over a carpet in the low humidity environment of a building or home in winter. The human body has a high dielectric constant that allows it to build up a static charge

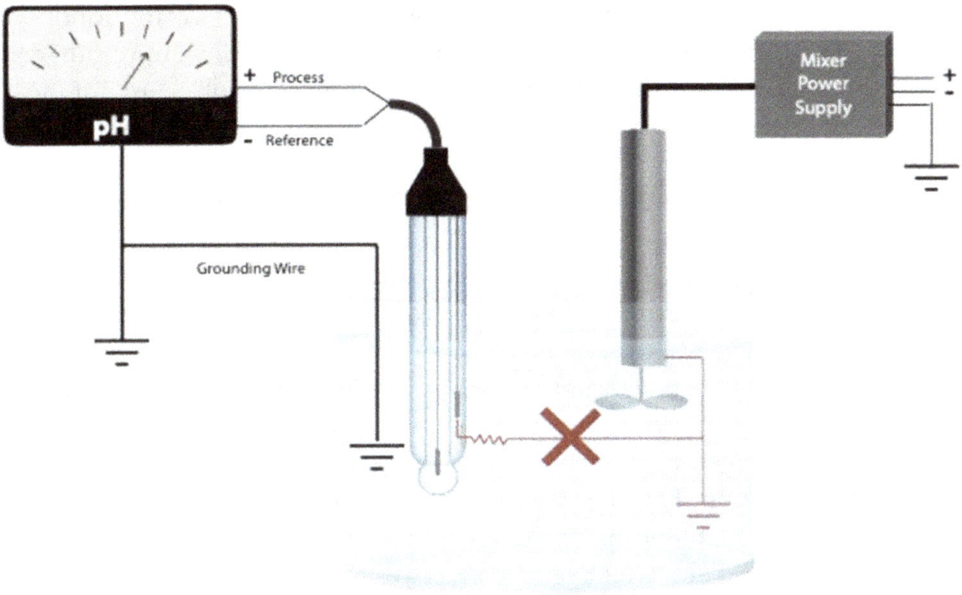

FIGURE 3.17 Short circuiting a ground loop by running a wire from the analyzer ground to the process.

just as in a capacitor. When we touch a conductor, such as a doorknob, the static charge releases as a spark. In an analogous process, relatively pure water flowing through a pipe made of an insulating material such as polyvinyl chloride will deposit charge onto the pipe. The deposition of a static charge on the pipe yields a new potential called a *streaming potential*. The streaming potential adds to the diffusion potential but is unique in that its magnitude increases with flow.

The easiest way to diagnose a probe suffering from static charge buildup is to watch the pH reading as the flowrate is reduced. Alternately, one can remove the probe from the pipe and measure its response in a separate grab sample of the process. The remediation of a static charge buildup is to insert a ground into the pipe that leads to earth ground. Keep in mind that a conductive pipe, such as stainless steel, will not sustain a static charge. Neither will water with conductivity greater than approximately 50 µS/cm. As with ground loops static charge is a more serious issue with a combination probe than a differential probe.

7.3 Acid and Alkaline Error

The pH sensor is an analytical marvel for its ability to measure the concentration of a chemical species (H^+ ions) over a whopping 14 orders of magnitude. This range is analogous to a tape measure that can measure the distance between Manhattan and Los Angeles to within two millionths of an inch. The ideal pH probe measures pH linearly over this entire range as dictated by the (logarithmic) Nernst equation. But real probes are not ideal, and the output of the probe wanders from its ideal value at both ends of the pH scale.

However, at both ends of the pH scale—less than 1 or greater than 12—the straight-line response rolls off. Nearly every measuring device that has a linear response, such as a photodetector or microphone, shows this type of behavior. After all every device has a range of operation beyond which it has a diminished response.

Figure 3.18 is a conceptual rendering of the response of a pH sensor. The green part of the response curve is a straight line and denotes the linear operating range of a typical probe. The yellow and red regions are the end portions of the pH scale where the voltage output rolls off as the pH increases on the high end or decreases on the low end.

On the low pH end we have *acid error*. Remember that, at very low pH values, the concentration of H^+ ions is very high. Recall, as well, that pH is proportional to activity—not concentration. At most commonly found pH levels, the two are nearly equal. However, at low pH—that is, high $[H^+]$ concentrations—the activity is substantially less than the concentration thus leading to a higher pH reading. This effect is real—the observed pH is the true pH. It's just higher than one would expect from the known acid concentration. There is, however, an additional effect that does lead to error because it is caused by the limitations of the pH electrode: there are only

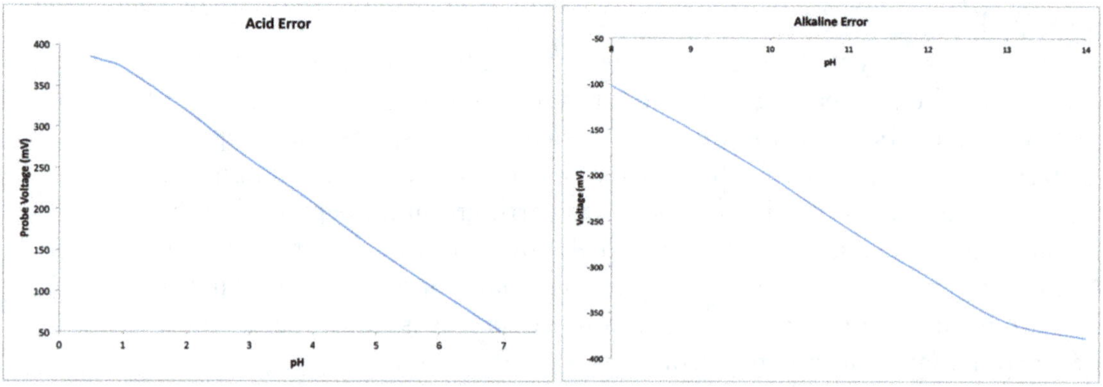

FIGURE 3.18 Left: Acid error. The difference of 60 mV at pH 0 results in the pH reading 1 full unit higher (pH 1) than the real pH 0 (data obtained at water analytics). Right: Alkaline error. At pH values greater than 10, Na^+ ions replace H^+ ions in the glass membrane. Both plots used a spherical, general purpose, glass electrode (Water Analytics).

so much H^+ ions the gel layer in our glass electrode can absorb before it starts to saturate. For this reason, acid error contains an effect that is real and one that is artificial. Figure 3.18 shows actual values of a pH probe in concentrated acid.

On the other end of the pH scale is *alkaline error*, also known as *sodium error*. The latter term is more appropriate because it is a problem that arises from the use of caustic (NaOH) to raise pH. Recall that pH glass has a higher affinity for H^+ ions than Na^+ ions. That high affinity is required to form a stable bound population of H^+ ions in the gel layer to generate a Nernstian response. However, at pH levels greater than 12 the activity of Na^+ ions greatly exceeds that of H^+ ions and displaces the latter from the gel layer. Without a stable population of H^+ ions in the gel layer necessary for a Nernstian response, the slope of the probe voltage falls. (One might ask that, since the pH glass has some affinity for Na^+ ions, why there isn't a Nernstian response to these ions that adds to that for H^+ ions. Because the affinity of glass for Na^+ is less than that for H^+ there is no stable concentration of Na^+ ions in the gel layer that is necessary for a Nernstian response. There will be a potential addition from Na^+ but it will be reduced.) Some manufacturers make a process electrode that is less susceptible to alkaline error. It is important to note that alkaline error really is an error—that is, it results from an imperfect measuring device.

7.4 Hydrofluoric Acid

Hydrofluoric acid (HF) etches glass (and bones). A glass electrode will not last long in a solution containing HF. For this application an antimony (Sb) electrode is the only solution. The electrode works through the reaction:

$$2\ Sb + 3\ H_2O \leftrightarrow Sb_2O_3 + 6\ H^+ + 6\ e^-;\ E^0 = 144\ mV \qquad (3.47)$$

Any acid will shift the balance of Equation 3.45 through the H^+ concentration. This is an oxidation reaction, so an antimony probe is more akin to an ORP probe than a pH one. It will work in any medium, but it is especially suited for HF because antimony is relatively inert to it.

The antimony electrode has several restrictions, which render it suitable for use in HF solutions where a glass-based pH probe would disintegrate:

1. The standard redox potential for Equation 3.47 is 144 mV. This means that the analyzer will display a pH of 4.6 prior to calibration. This large offset will likely incapacitate most analyzers' automatic pH buffer recognition feature and require manual calibration.

2. Any oxidizing agent, including oxygen and chlorine, will affect the reading.
3. The flow rate affects the rate of Equation 3.47 and, hence, the pH reading.
4. Common metals, such as copper, lead, and tin, will react with antimony and coat the surface with the reaction products that "poison" the electrode and prevent it from responding to pH.
5. The measurement range is restricted to a pH range of 3 to 9. At lower pH the removal of Sb shortens the life of the probe. At pH levels greater than 6, HF dissociates into H^+ and F^- ions. The F^- ion is not destructive to glass so an ordinary pH probe can be used.

Note that some manufacturers brand a probe model as "HF resistant." In most cases this means that the glass is of extra thickness and will last longer than a typical probe.

7.5 The pH of Mixed Solvents

This is a book on water quality instrumentation, so a discussion of nonaqueous solutions is not very pertinent. As general rule of thumb, one can measure the pH of water if it constitutes at least 5% of the volume of a solution. However, a nonaqueous solvent can indeed be characterized by a pH value. Nearly all organic solvents have hydrogen atoms attached to carbon atoms. For most the C atoms bond to their H atoms too tightly to allow them to dissociate as H^+ ions. Solvents that can dissociate into H^+ and OH^- ions include alcohols, such as methanol, ethanol, propanol and phenol, and carboxylic acids, such as acetic acid, and nitrogen containing solvents such as acetonitrile, ammonia, and aniline. The pH values in these solvents span a different range from that of water. For instance, ammonia dissolved in water dissociates as:

$$NH_3 + H_2O \rightarrow NH_4^+ + OH^- \leftrightarrow NH_3 + H^+ + OH^-; K_{diss} = 10^{-24} \quad (3.48)$$

The chemical equilibrium rule, for water at 25 °C, that is $[H^+][OH^-] = 10^{-14}$, no longer applies so pH 7 is no longer neutral. For ammonia neutral is pH 12 ($10^{-12} \times 10^{-12} = 10^{-24}$). The only rigorous way to calibrate a probe for a mixed solvent is to do so in a calibration standard containing a known fraction of the nonaqueous solvent. The difference in miscibility (caused mostly by the difference in polarity) between the KCl electrolyte and nonaqueous sample can result in a phase separation, which stops the flow of ions through the liquid junction. In these cases, a nonaqueous reference solution, such as LiCl in ethanol or acetic acid, is often used.

7.6 Summary of Probe Issues

To summarize, there are three components of a pH probe that change over time: the liquid junction, the reference electrode, and the process electrode. As the above sections have shown, any of them can become compromised. Table 3.7 summarizes the several issues users are likely to encounter.

TABLE 3.7 Troubleshooting a "sick" probe.

Problem	Cause	Solution
Slow response	Clogged liquid junction	Soak probe in weak acid. Differential probe: Change salt bridge.
Decreasing efficiency	Fouled process electrode	Scaling: Soak in weak acid. Biofilm: Soak in organic acid or bleach.
Low efficiency and slow response	Process electrode dried out	Soak in 4 M KCl or pH 4 buffer for at least 4 hours.
Drifting offset	Contaminated reference solution	Combination probe that is refillable: Replace reference electrolyte. Differential probe: Change pH 7 buffer.
Glass eroded	HF in solution	Switch to antimony (Sb) probe. Works for HF concentration $< 4\%$
Precipitate in reference solution	Sulfide in solution reacting with Ag	Soak electrode in thiourea/HCl.
pH reading stuck at app. 7	Process electrode cracked	Buy a new probe.
Efficiency is low and won't improve	Aged electrode. Na^+ or Li^+ ions leached out of glass.	Probe is likely dead. Try soaking in 20% ammonium bifluoride (NH_4HF_2) for 15 seconds to expose fresh glass.
Reading shifts from the calibration level	Ground loop	Either isolate process container or pipe from ground or tie its ground to the analyzer's ground.
Reading is erratic	RF interference	Isolate probe cable from alternating current power cables and high voltage sources (e.g., motors and pumps).

8.0 REAL WORLD APPLICATIONS FOR PH MEASUREMENT

8.1 Water Treatment

For water treatment facilities that draw on surface water, the pH of the raw water typically ranges from neutral to slightly acidic (about pH 6). After the initial screening of raw water, coagulants are added to clump together suspended solids. The suspended solids are typically negatively charged colloidal particles and are too small to settle out on their own. Most coagulants are chloride (Cl^-) or sulfate (SO_4^-) salts of either the ferric cation, Fe^{3+}, or aluminum cation, Al^{3+}. Ferric chloride, ferric sulfate, aluminum chloride, and aluminum sulfate ("alum") are the most common.

When dissolved in water the metal forms a hydroxide. In the case of ferric chloride:

$$Fe^{3+} + H_2O \rightarrow Fe(OH)_3 + 3\ H^+ \tag{3.49}$$

The positive ferric ion combines with the negatively charged suspended particles to form neutral particles. The latter then agglomerate to form larger particles, flocs, which then settle out of solution. Equation 3.49 shows that the concentration of ferric ions is pH dependent. Clearly the pH of the water must be acidic to favor dissolved Fe^{3+} ions over solid $Fe(OH)_3$. The same chemistry applies to aluminum-based coagulants.

The exact pH range depends on the choice of coagulant. Figure 3.19 shows results from jar testing of the coagulation by two alum compounds—poly-aluminum chloride (PACl) and poly-aluminum chloride sulfate (PACS)—as a function of pH (Lotfi et al., 2014). For PACl, the optimum pH is 6.5, whereas, for PACS, the optimum pH is 9.5. Turbidity is measured as the indicator of removal of suspended matter and will return as such in Chapter 7.

Following coagulation and flocculation, the facility operator raises the pH toward a more neutral value (6 to 8) using either caustic or lime [CaO or $Ca(OH)_2$]. The neutral pH is required for optimum filtering and biological breakdown of organic matter.

The final stage in water treatment is disinfection. The optimum pH is typically between 7 and 8 and represents a compromise between keeping the value low enough to favor the formation of the more effective HOCl (to be discussed in Section 8.9) but high enough to prevent corrosion of plumbing.

8.2 Municipal Wastewater Treatment

The previous chapter on ORP described the process of nitrification–denitrification sequence that is the basis of nitrogen removal in wastewater treatment. Chapter 12 , Section 2 of that chapter discussed the role of ORP

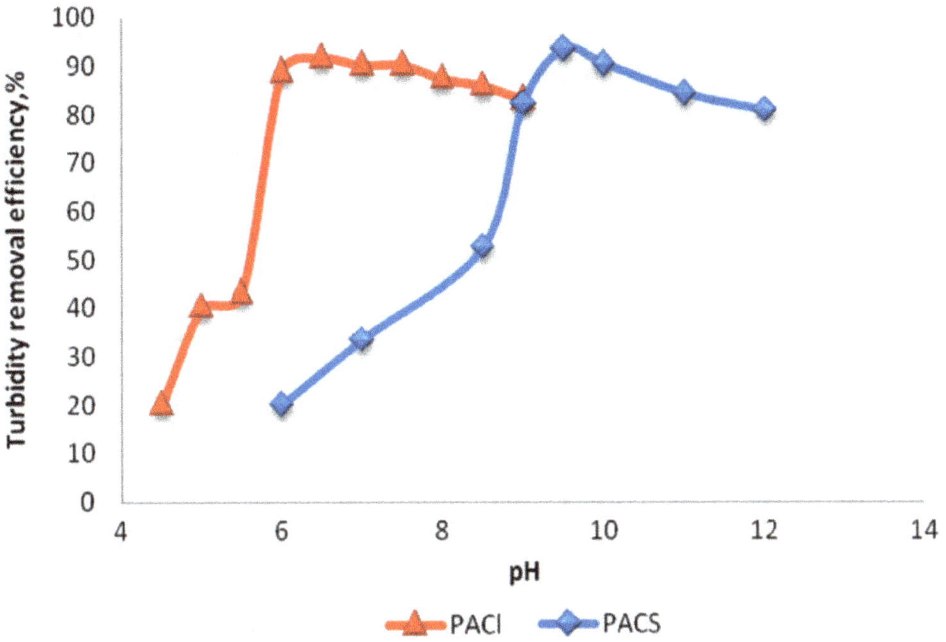

FIGURE 3.19 Effect of coagulation by poly-aluminum chloride and poly-aluminum chloride sulfate, measured by turbidity, as a function of pH (Lotfi et al., 2014). Optimum pH values are 6.5 for PACl and 9.5 for PACS.

as a robust way to measure the efficacy of ammonia and nitrite oxidation during nitrification and reduction of nitrate during denitrification. These fundamental processes are also dependent on pH.

Ammonia dissolves in water to create a mixture of ammonia and ammonium (NH_4^+):

$$NH_3 + H_2O \rightarrow NH_4OH \leftrightarrow NH_4^+ + OH^- \tag{3.50}$$

Adding acid protonates ammonia to form ammonium:

$$NH_3 + H^+ \rightarrow NH_4^+ \tag{3.51}$$

Figure 3.20 shows the ratio of ammonium to ammonia as a function of pH and at 3 temperatures (Culp & Culp, 1971). Typical pH values of influent wastewater range from 6.6 to 8.6 with an average of 7.4 (Lowe et al., 2010). It's clear that, even at this high end of this range, the mixture of the two species is predominantly NH_4^+.

During nitrification *Nitrosomonas* bacteria oxidize ammonium to nitrite as follows:

$$2\ NH_4^+ + 3\ O_2 \rightarrow 2\ NO_2^- + 2\ H_2O + 4\ H \tag{3.52}$$

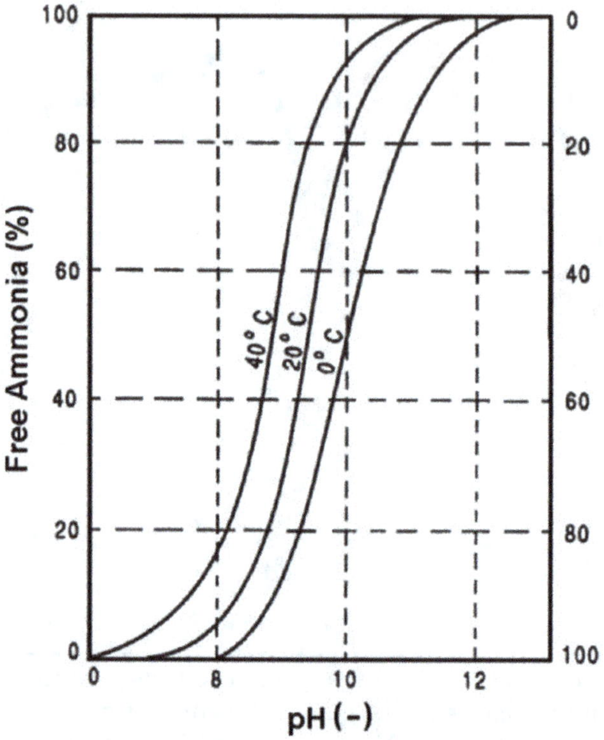

FIGURE 3.20 Proportions of NH_3 to NH_4^+ as a function of pH and for three temperatures (Culp & Culp, 1971).

As reaction Equation 3.52 proceeds the pH drops. The *Nitrosomonas* bacteria that oxidize ammonium suffer if the pH drops below 6.6. We know from our discussion of buffers that we can minimize pH swings by adding a buffer solution. The carbonic acid buffer also discussed in the last section mitigates pH swings. To introduce this buffer system into the wastewater, operators add carbonate (CO_3^{2-}), which converts partially to bicarbonate (HCO_3^-) in accordance with Equation 3.41. The term *alkalinity*, in this case, refers to the added carbonate. The bicarbonate is not just essential to prevent the pH from dipping below 6.6. It is also a carbon source for bacterial growth. As the bacteria convert HCO_3^- to CO_2, the decreasing bicarbonate forces Equation 3.41 to shift toward the right, which causes the pH to drop. Hence operators must add caustic to keep the pH stable.

The second step of nitrification is the oxidation of nitrite (NO_2^-) to nitrate (NO_3^{2-}): by the *Nitrobacter* species of bacteria:

$$2\ NO_2^- + O_2 \rightarrow 2\ NO_3^- + 3\ H + 2\ e^- \qquad (3.53)$$

This step results in the lowering of pH. The *Nitrobacter* bacteria need an environment in which the pH is between 7 and 8.5 to thrive. In particular, as the pH increases, the ratio of toxic NH_3 to NH_4^+ increases. At a pH of 8, that fraction is about 10%, and at higher pH values, bacteria die off.

The denitrification stage converts nitrate into nitrogen gas. The process has fewer variable inputs that influence the chemistry, and it can be summarized with one reaction first stated in the very beginning of the book.

$$6\ NO_3^- + 5\ CH_3OH \rightarrow 3\ N_2 + 5\ CO_2 + 7\ H_2O + 6\ OH \quad (3.54)$$

This chemical expression uses methanol (CH_3OH), but it could have used other sources of organic carbon as food source for the denitrifying bacteria that drive this reaction. A substantial amount of carbon comes in as the influent sewage, which is one reason why the denitrification stage is placed ahead of the nitrification stage even though the former follows the latter. Another reason is the naturally low concentration of dissolved oxygen in the influent. Phosphorous removal can occur first in an anoxic or anaerobic environment. To accommodate this reversal of reaction order sludge from the aerobic chamber is returned to the front of the process (return activated sludge). Still, carbon in the form of methanol will drive denitrification at a rate three times faster than carbon only in the form of sewage.

As Equation 3.54 shows, denitrification raises the pH. Therefore, the goal during this phase is to keep the pH level between 6.5 and 8. In fact one study in Austria (Wett et al., 1998) showed that automated process control of a sequencing batch reactor could vary the aeration and anoxic periods that produce nitrification and denitrification, respectively, based only on pH measurement.

The final stage of wastewater treatment is the dewatering of sludge. One set of experiments showed that sludge, initially at a pH of 7.62, can be dewatered by about 5% by lowering the pH to 6.78 (Texier, 2008). The dependence of this process on pH is not strong. The precipitation of phosphorus in the form of struvite takes place, chemically, at pH levels above 8.3 or, biologically, at pH levels between 7.3 and 7.8 (Simoes et al., 2018).

8.3 Acid Mine Remediation

One industry that produces wastewater that can devastate the environment is mining. The damage starts as a side effect of the mining operation. Crushing sulfide containing rocks exposes them to oxygen in the atmosphere. One major target is the mineral pyrite (aka "fool's gold"), FeS_2. When it is exposed to air and water the oxidation reaction produces sulfuric acid and a mixture of the two iron hydroxides, $Fe(OH)_2$.

$$2\ FeS_2 + 7\ O_2 + 2\ H_2O \rightarrow 2\ Fe^{2+} + 4\ SO_4^{2-} + 4\ H^+ \quad (3.55)$$

The ferrous (Fe^{2+}) ion reacts with water to give $Fe(OH)_2$, which precipitates out. Fe^{2+} can oxidize further to Fe^{3+} and form $Fe(OH)_3$, which also precipitates out. The latter is colloquially known as "yellow boy" and gives mine effluent streams their yellow color—so-called *acid mine drainage*. The oxidation of pyrite accelerates by several orders of magnitude in the presence of the bacteria, *Thiobacillus ferrooxidans*. Figure 3.21 is an Eh-pH diagram (Lenntech or Monhemius, 2017) that shows oxide states of iron that predominate over a range of pH and ORP values. The y-axis shows an increasing oxidation potential of the aqueous environment. The diagram shows that $Fe(OH)_2$ is predominant over $Fe(OH)_3$ at pH values less than 6 and in an oxidizing environment

The major consequence of Equation 3.55 is the production of sulfuric acid (H_2SO_4) and results in pH that can be as low as 2.5. At these low pH

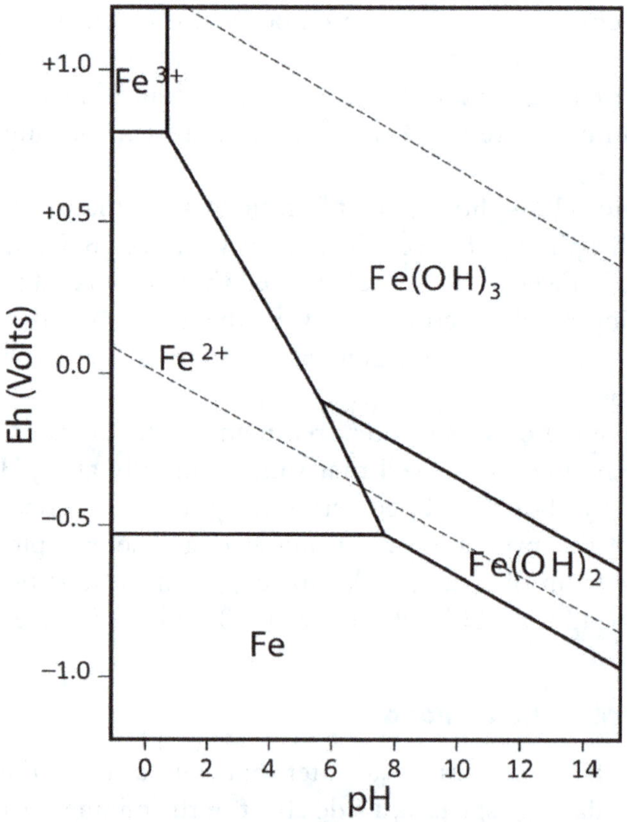

FIGURE 3.21 Eh-pH diagrams of oxidation states of iron, showing that $Fe(OH)_3$ is the preferred oxide over $Fe(OH)_2$ at pH <7 and in an oxidizing environment. (Figure 1 from "The iron elephant: A brief history of hydrometallurgists' struggles with element no. 26," published in CIM Journal, Vol. 8, No. 4, 2017. Reprinted with permission of the Canadian Institute of Mining, Metallurgy and Petroleum.).

values other metals dissolve and create an environmental nightmare (Stumm & Morgan, 1981).

The remediation of acid mine damage is lime softening. Quicklime, CaO, is added to the mine effluent and converts H_2SO_4 to the precipitate, $CaSO_4$.

$$SO_4^{2-} + 2\ H^+ + CaO \rightarrow CaSO_4 + H_2O \qquad (3.56)$$

8.4 Corrosion

Tap water that comes from a water treatment facility (versus a well) has a pH that is typically slightly basic—somewhere around 8. The reason is simple: low pH accelerates the oxidation of metals. Take the example of oxidation of iron—accompanied by the reduction of oxygen:

$$Fe \rightarrow Fe^{2+} + 2\ e^- \qquad (3.57)$$

$$O_2 + 4\ H^+ + 4\ e^- \rightarrow 2\ H_2O\ (\text{or } O_2 + 2\ H_2O + 4\ e^- \rightarrow 4\ OH^-) \qquad (3.58)$$

The combined reaction is:

$$2\ Fe + 4\ H^+ + O_2 \rightarrow 2\ Fe^{2+} + 2\ H_2O \qquad (3.59)$$

This combined reaction yields iron in its +2 oxidation (ferrous) state. Further oxidation results in the higher oxidation +3 (ferric) state:

$$4\ Fe^{2+} + O_2 \rightarrow 4\ Fe^{3+} + 2\ O_2 \qquad (3.60)$$

In the presence of water the ferrous—Fe(II)—ion and the ferric—Fe(III)—ion form hydroxides:

$$Fe^{2+} + 2\ H_2O \rightarrow Fe(OH)_2 + 2\ H^+ \qquad (3.61)$$

$$Fe^{3+} + 3\ H_2O \rightarrow Fe(OH)_3 + 3\ H^+ \qquad (3.62)$$

As these two equations show, lowering the pH will drive these two reactions backward and keep the ferrous and ferric ions in solution. Raising the pH facilitates the formation of the hydroxides. The ferrous and ferric hydroxides can shed water molecules to form oxides (that we call "rust"):

$$Fe(OH)_2 \leftrightarrow FeO + H_2O \qquad (3.63)$$

$$2\ Fe(OH)_3 \leftrightarrow Fe_2O_3 + 3\ H_2O \qquad (3.64)$$

FeO consists of Fe(II) whose +2 oxidation state matches oxygen's −2 state. Fe_2O_3 consists of Fe(III) whose +3 oxidation state matches the two O atoms. (For Fe^{+3} $2 \times +3 = +6$ electrons and for O^{-2} $3 \times -2 = -6$ electrons.) But there is a third iron oxide, Fe_3O_4 for which the arithmetic doesn't seem to work ($3 \times 3 \neq 4 \times 2$). That's because two of the three Fe atoms are in the +3 state, and the third is in the +2 state for a total of +8. This mixed oxidation state is denoted as Fe(II,III). This oxide is the first discovered magnetic compound, magnetite, or lodestone.

In the presence of oxygen and absence of water FeO is the predominant oxide and adheres to the underlying metal. This *passivation layer* protects the iron from further oxidation. However, in the presence of water, Equation 3.63 goes from right to left, converting FeO to $Fe(OH)_2$, the latter of which flakes off to expose more iron to rust. (Aluminum oxide, Al_2O_3, forms a very strong passivation layer that effectively prevents the formation of the hydroxide, thereby limiting the corrosion of aluminum).

Figure 3.22 shows the dependence of the corrosion rate on pH (Prawoto et al., 2009). Between pH 4 and 10, corrosion is relatively constant. The rapid increase of corrosion below pH 4 is the result of hydrogen evolution:

$$Fe + 2\,H^+ \rightarrow Fe^{2+} + H_2 \qquad (3.65)$$

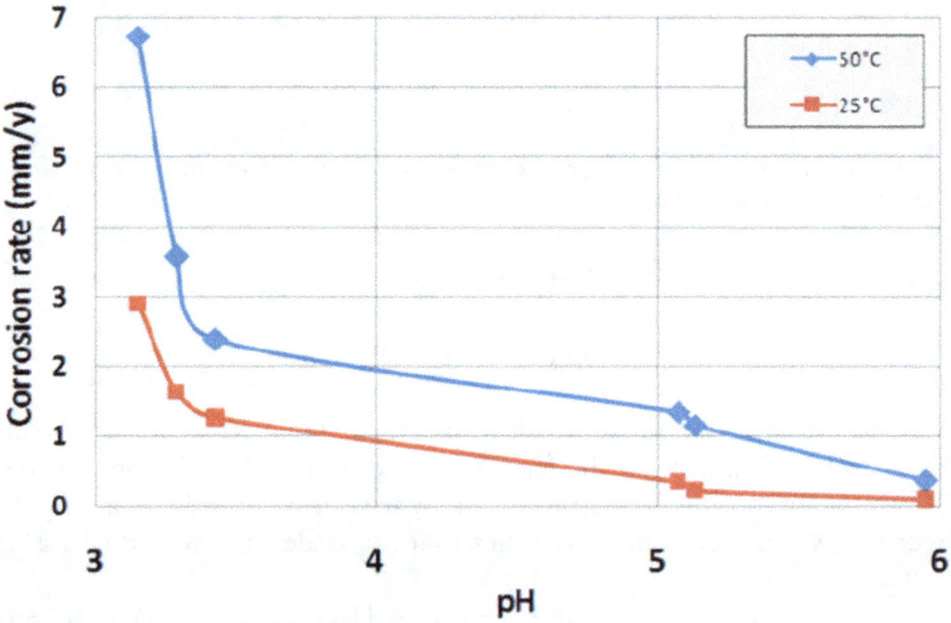

FIGURE 3.22 Effect of pH on corrosion. The rapid increase at pH <3.5 is from evolution of H_2 (Prawoto et al., 2009).

Iron is just one example of corrosion. The control of lead leaching from pipes that use lead-solder is a major problem recently addressed by U.S. Environmental Protection Agency (U.S. EPA)'s Lead and Copper Rule. The oxidation of lead atoms to lead ions is influenced by pH. *White rust* results from the unwanted chemistry of zinc on a steel substrate. Zinc is plated onto steel to make galvanized steel. The zinc oxidizes to form a passivation layer of ZnO but may also oxidize to form a mixture of oxides, including $Zn(OH)_3$. Maintaining the pH between 7 and 8 during the galvanizing process maximizes the formation of the ZnO layer. Many factors affect corrosion, such as sulfates, flow, dissolved oxygen, chlorides, cations, and bacteria. The intent of this section is to show just one example—oxidation of iron—and to show how pH plays a role in the chemistry.

8.5 Metal Precipitation

The previous section covered the oxidation of metals due to environmental factors—that is, oxygen and water. The intentional reaction of dissolved metal ions to create suspended solids that can be removed is a simple method for removing these metals. Dissolved metals in drinking water, industrial effluent or acid mine drainage are serious problems for our health and the environment. In accordance with the Clean Water Act, the U.S. EPA sets effluent limits for inorganic chemicals, which includes mostly metals. Metals of concern include arsenic, cadmium, copper, lead, nickel, manganese, mercury, and zinc.

The most widespread method for lowering the concentrations of dissolved metal ions is to precipitate them out as the hydroxide. In the previous section we looked at the case of iron hydroxides. Other metal ions follow the same chemistry. For example Mg^{2+} ions can be precipitated as $Mg(OH)_2$:

$$Mg^{2+} + 2\ OH^- \rightarrow Mg(OH)_2 \tag{3.66}$$

The generalized reaction for precipitating metals is:

$$M^{n+} + n\ OH^- \rightarrow M(OH)_n \tag{3.67}$$

The minimum of each curve represents the optimum pH for formation of the metal hydroxide.

NaOH and $Ca(OH)_2$ (lime) or CaO (quicklime) are used to raise the pH. Lime is less expensive and also serves as a coagulant to help with the settling of the metal hydroxide.

As an alternative to precipitating metal ions out as hydroxides, treatment facilities can add sulfide salts (e.g., Na_2S) at a very high pH. Sulfide salts of metals, such as silver, lead, cadmium, and zinc are very insoluble and have concentrations less than 10^{-6} M. However, this process requires the use of coagulants and has the potential to release poisonous H_2S gas. Carbonate precipitation is often used for metals such as lead, cadmium, and nickel. It is an attractive option when there already is a concentration of carbonate ions. Because this process is done at high pH, the metal carbonate precipitates out along with hydroxides. After metals have precipitated out, the next stage is to bring the pH back toward neutral. This is done simply by the addition of acid—typically sulfuric acid.

8.6 Industrial Wastewater Treatment

The most prevalent use of pH control in wastewater treatment is to simply ensure that water used in processing is discharged to the environment and meets effluent standards. The Clean Water Act requires that the pH of all wastewater effluent be between 6.5 and 8.5. This means that pH control of wastewater occurs of the final stages before discharge. Operators typically use caustic (NaOH) to raise pH and either hydrochloric (HCl) or sulfuric (H_2SO_4) to lower it. Carbon dioxide injection is increasingly being used as a safer alternative to these two strong acids. It has the added benefit that CO_2 injection forms alkalinity—that is, carbonate, which buffers the water against pH swings.

Paper manufacturing from the initial pulping stage to bleaching to final wastewater treatment requires pH control. The pulping process of converting wood to pulp is done in at very high pH solution of NaOH and the reducing agent, sodium sulfide, Na_2S. Bleaching—the process of purifying and whitening the pulp—is typically carried out using chlorine dioxide, ClO_2, at pH levels of 3.5 to 4.0. The extraction phase is done at pH levels between 10 and 11. For the final manufacturing step, the paper stock is treated with a variety of dyes, sizing, colorants, and finishes. Sulfuric or hydrochloric acid is typically used for pH adjustment. The wastewater contains high amounts of suspended solids, which are removed by flocculation as in a water treatment facility. Biological nutrient removal is dependent on pH just as in the municipal process. Optimum pH is between 7.0 and 7.8 for aerobic nitrification and between 6.8 and 7.2 for anaerobic denitrification (Kanget et al., 2009). Final disinfection by chlorine also follows the same guidelines as that for water treatment.

8.7 Scaling—The Problem With High pH

Metal plumbing and fixtures exposed to an acidic environment accelerates corrosion. On the high side of the pH scale, scaling leads to calcium carbonate deposits that can clog piping. The scaling is the progression of a reaction we've already discussed—the formation of carbonic acid when carbon dioxide dissolves in water. The pertinent reaction is repeated here along with the reaction that follows whenever calcium ions are present:

$$CO_2 + H_2O \leftrightarrow H_2CO_3 \leftrightarrow HCO_3^{-} + H^{+} \leftrightarrow CO_3^{2-} + 2\,H^{+} \quad (3.68)$$

$$Ca^{2+} + CO_3^{2-} \leftrightarrow CaCO_3\,(s) \quad (3.69)$$

The resulting calcium carbonate ($CaCO_3$) forms the scaling that builds up inside plumbing. As we can see from Equation 3.69, if we add acid then the reaction goes from right to left and carbonate (CO_3^{2-}) converts to CO_2. If we remove acid—that is, add base—then we make more carbonate, which, in turn, makes mores calcium carbonate. Thus, raising the pH generates more carbonate.

A pH that is too low not only accelerates corrosion but also leaches lead out of plumbing and creates a health problem (as in Flint, Michigan). A pH that is too high deposits calcium carbonate. There ought to be a "Goldilocks" range for pH to minimize both. Before revealing what that is, it's worth nothing that there is an additional parameter that affects scaling, thus making the minimization of scaling and corrosion a three-dimensional problem. The other parameter is alkalinity, which is the carbonate and bicarbonate ion concentration, CO_3^{2-} and HCO_3^{-} ions in solution. As they precipitate out as $CaCO_3$, at higher pH values, they form a coating that protects against corrosion. Alkalinity also serves as a buffer to moderate swings in pH.

To determine the optimum pH given a known alkalinity value—or vice versa—one can consult a graph called the *Bayliss Curve*. This is shown in Figure 3.23. The blue area in the lower part of the curve shows pH and alkalinity values that promote corrosion, and the tan area in the upper part shows values that promote scaling. The optimum values of pH and alkalinity are enclosed in the white middle section. In this area $CaCO_3$ is at a stable equilibrium, and there is just enough to form a protective layer of calcium carbonate on an iron surface.

There are several ways to add alkalinity to water. They include injecting carbon dioxide or adding sodium bicarbonate, ($NaHCO_3$), soda ash (Na_2CO_3), and lime ($CaOH$).

8.8 Water Softening

The last chapter discussed minimizing the deposition of calcium carbonate on the inside of pipes by optimizing the pH and alkalinity. But wouldn't it

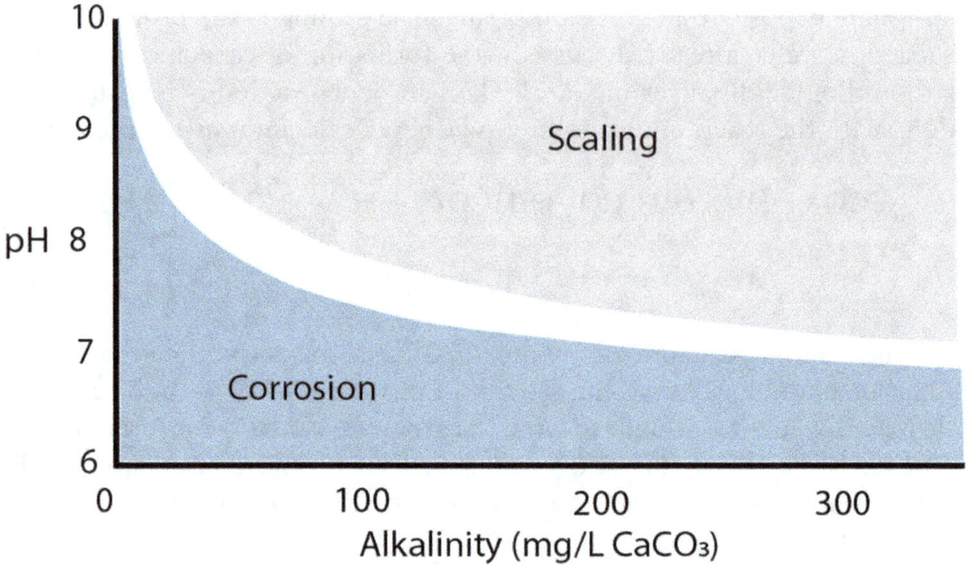

FIGURE 3.23 Bayliss Curve showing pH and alkalinity values optimized to prevent excessive scaling and corrosion. The pink areas represent values of pH and alkalinity conducive to the deposition of calcium carbonate (scaling) while the blue gray area represents values of pH and alkalinity conducive to corrosion. The white area in between the two represents the optimal zone.

be better to remove as much calcium as possible so that it is not necessary to optimize the pH and alkalinity? Because metals can precipitate out as insoluble hydroxides, it would also be advantageous to remove the metal cations that can form these precipitates. These include metals such as manganese, iron, barium, zinc, aluminum, and strontium. The metal with the highest concentration is magnesium (Mg^{2+}). Together, the sum total of Ca^{2+} and Mg^{2+} concentration in water is referred to as *general hardness*. These ions affect the chemistry of aquatic organisms. Their major effect on humans is through their propensity to precipitate out on the inside of pipes, our cookware, wine glasses, clothing, showers, and anything else that gets wet.

When we discuss calcium and magnesium ions in solution, we must remember that they entered solution as a result of a salt or compound dissolving in solution. For the sake of clarity, discussion will focus mostly on calcium. Keep in mind that, in the reactions that follow, Mg ions can replace Ca ions.

The most common calcium compound is calcium carbonate, $CaCO_3$—often dissolved from limestone. As stated in the previous section, CO_2 dissolves in water to form the weak acid, carbonic acid, H_2CO_3. This weak acid reacts with calcium carbonate to form calcium bicarbonate:

$$H_2CO_3\ (s) + CaCO_3\ (s) \leftrightarrow Ca(HCO_3)_2 \qquad (3.70)$$

When Ca^{2+} enters solution through this channel, it is referred to as *carbonate hardness*. Carbonate hardness is reported as if all calcium were converted to $CaCO_3$—that is, the equilibrium in Equation 3.68 is shifted all the way to the left. Units are in mass of $CaCO_3$: ppm or mg/l.

Calcium can also enter into solution through calcium sulfate, $CaSO_4$, or calcium chloride, $CaCl_2$. This calcium is referred to as *noncarbonate hardness*. Even though there is no carbonate in solution, noncarbonated hardness is measured as if all calcium were tied up as $CaCO_3$. This enables carbonate hardness and noncarbonated hardness to reflect the total concentration of Ca^{2+} in solution.

Just as increasing hardness leads to scaling decreases it removes it. Appropriately enough the process of removing Ca^{2+} and Mg^{2+} ions is called *water softening*.

There are several ways to soften water, including distillation, nanofiltration, reverse osmosis, and ion exchange, the latter in which a resin exchanges Ca^{2+} ions for more benign Na^+ ions. The most popular method used in a treatment facility is *lime softening*, first mentioned in Section 8.3. This approach involves adding calcium hydroxide, $Ca(OH)_2$, aka *lime* or *slaked lime*. Lime dissociates into Ca^{2+} and OH^- and reacts with calcium bicarbonate to make calcium carbonate, the latter of which precipitates out:

$$Ca^{2+} + 2\ OH^- + Ca(HCO_3)_2 \rightarrow 2\ CaCO_3^-\ (s) \rightarrow 2\ H_2O \qquad (3.71)$$

Keep in mind that precipitating $CaCO_3$ inside plumbing is a sin, whereas doing so in a treatment facility before making it into the distribution system is a virtue. Figure 3.24 shows the lowering of calcium and total hardness as lime is added (Dey et al., 2007). The lime dosages are expressed as the stoichiometric ratio between lime and bicarbonate. According to Equation 3.71 a ratio of 100% means 1 mole of $Ca(OH)_2$ to 1 mole of bicarbonate. As Figure 3.24 shows, the optimum pH for lime softening occurs between pH 10 and 11. Note that, for lime addition greater than 100%, the extraneous lime just adds Ca^{2+} to the water and increases the hardness.

Following softening to precipitate out calcium carbonate treatment, facilities lower the pH to around 8 for coagulation—flocculation or filtration to remove the suspended precipitate from the water. The former produces a sludge that must be dewatered and disposed.

Some treatment facilities make lime by dissolving calcium oxide, CaO—*quicklime*—in water:

$$CaO + H_2O \rightarrow Ca(OH)_2 \qquad (3.72)$$

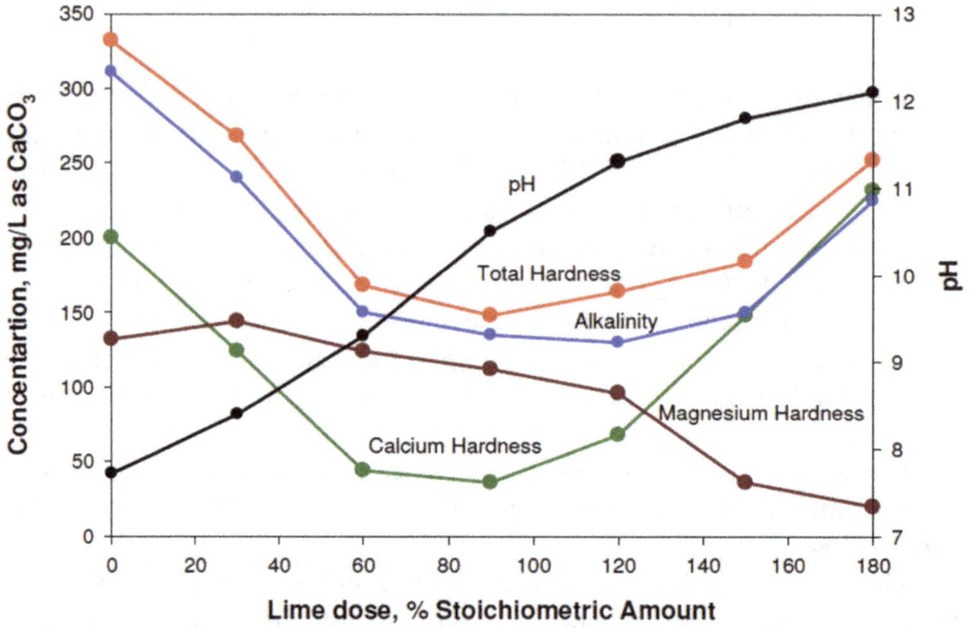

FIGURE 3.24 Dependence of hardness on lime dosing. Note that increasing lime beyond the 100% stoichiometric ratio deposits the excess calcium (Dey et al., 2007).

The tremendous heat released and the speed at which this highly exothermic reaction occurs is the reason why most plants prefer to start with just lime.

Lime also reacts with dissolved CO_2 to create calcium carbonate:

$$Ca(OH)_2 + CO_2 \rightarrow CaCO_3 + H_2O \tag{3.73}$$

For this reason, water with excessive CO_2 will undergo aeration to purge it.

The equivalent reaction for reacting magnesium bicarbonate is similar but with an important difference:

$$2\ Ca(OH)_2 + Mg(HCO_3)_2 \rightarrow Mg(OH)_2\ (s) + 2\ CaCO_3\ (s) + 2\ H_2O \tag{3.74}$$

The difference is that two precipitates result—the second being magnesium hydroxide.

We finish our discussion with noncarbonated hardness. The sulfate and chloride salts of calcium are removed through precipitation with sodium carbonate, Na_2CO_3 *(soda ash)*:

$$Na_2CO_3 + CaSO_4 \rightarrow CaCO_3(s) + Na_2SO_4 \tag{3.75}$$

$$Na_2CO_3 + CaCl_2 \rightarrow CaCO_3(s) + 2\ NaCl \tag{3.76}$$

There are analogous reactions for the chloride and sulfate of magnesium. The absence of H^+ ions in Equation 3.75 and Equation 3.76 explains why the removal of noncarbonated hardness has no dependence on pH.

8.9 Disinfection

Disinfection by hypochlorous acid, HOCl—known colloquially as chlorination—is an oxidation reaction. Chapter 6, on chlorine sensors, covers this topic in greater detail.

Equation 2.41 in Chapter 2, Section 9 showed that every unit of pH increase lowers the ORP by 29 mV. But that tells only half the story. The equation does not take into account HOCl's anion, OCl^-, which forms as the pH increases:

$$HOCl \leftrightarrow OCl^- + H^+ \tag{3.77}$$

Both HOCl and OCl^- are referred to as "free chorine" but HOCl is an order of magnitude more effective than OCl^- at killing pathogens. Therefore, disinfection requires that the pH stay below 7.5, the value at which HOCl is 50% of total free chlorine. Figure 6.1 in the chapter on chlorine measurement shows the relative concentration of hypochlorous acid and hypochlorite as a function of pH. Note that, for pH values less than 5, HOCl converts to chlorine gas according to the reaction:

$$2\ HOCl + 2\ H^+ \leftrightarrow Cl_2 + 2\ H_2O \tag{3.78}$$

Figure 6.1 does not show the conversion of HOCl to deadly chlorine gas at pH levels less than 5. The figure only shows the ideal pH range for maximum disinfection is 5.5 to 7.0. If maximizing HOCl concentration were the primary goal of water treatment, then water sent through the distribution would be within this range. However, minimization of corrosion necessitates a pH of greater than 7.5. The result of low pH values on lead service lines is leaching lead ions, as was made clear during the Flint water crisis of 2014.

8.10 Food and Dairy Production

Dairy production ranging from raw milk to cheese is highly pH dependent. Milk is typically maintained at pH 6.8. To control bacterial contamination, workers drop the pH to between 4.1 and 5.3. To make butter the pH of the cream must be between 6.70 and 6.85. Yogurt production requires a pH of the cultured milk of 5. When adding fruit, the pH of the yogurt must match the pH of the fruit.

Beer production is a fermentation process that is analogous to the denitrification of wastewater in an anoxic or anaerobic environment. The pH of crushed malt is 5.8. During fermentation, the pH changes and must be kept constant.

Grapes used for wine contain several carboxylic acids—compounds that contain the weak acidic group, COOH. The most prominent of these are tartaric and malic acids. Grapes grown in cooler climates have higher levels of these acids. One of the most important steps in winemaking is malolactic fermentation in which malic acid is converted to the less acidic lactic acid and CO_2. The final pH of wine ranges from 2.8 to 3.8. If the pH is higher than 4, then the wine becomes susceptible to bacterial growth. In general, sweeter wines are produced to a lower pH value so that the acidity balances out the sweetness.

The other by-product of grapes, jelly, requires a pH 3.3 to gel. At higher pH values, it will stay liquid, and at lower pH values, it becomes too stiff.

8.11 Metal Plating

Recall from Section 10.6 in Chapter 2 that ORP is monitored during the reduction of toxic Cr^{6+}, or Cr(VI) to the more harmless Cr^{3+}, or Cr (III). That reduction reaction is:

$$CrO_4^{2-} + 8\ H^+ + 3\ e^- \rightarrow Cr^{3+} + 4\ H_2O\ ;\ E^0 = 1.22\ V \qquad (3.79)$$

As the left side of Equation 3.79 shows, the reaction occurs under acidic conditions. The optimum pH is 2. In a subsequent treatment tank, the pH is raised to 8.5 where Cr (III) combines with OH^- to form $Cr(OH)_3$, a precipitate that settles to the bottom.

Cyanide (CN^-) is another highly toxic ion used in metal finishing that must be converted into harmless chemicals. In the first phase, CN^- is oxidized by sodium hypochlorite first to cyanate (CNO^-).

$$CN^- + OCl^- \rightarrow CNO^- + Cl^- \qquad (3.80)$$

Equation 3.80 can also be written with the anions replaced by their sodium salts:

$$NaCN + NaOCl \rightarrow NaCNO + NaCl \qquad (3.81)$$

Of course, the chemistry is the same. In the second stage, cyanate is further oxidized by hypochlorite into harmless sodium bicarbonate:

$$2\ CNO^- + 3\ OCl^- \rightarrow 3\ Cl^- + N_2 + 2\ HCO_3^- \qquad (3.82)$$

There are no H^+ ions involved in either of these oxidation reactions, so one would think that pH does not play a role. However, it is important that, during the oxidation expressed by Equation 3.81, the pH stays in the basic range (9.98 to 10.02) so that CN^- does not combine with H^+ to form deadly cyanide gas:

$$H^+ + CN^- \rightarrow HCN\ (g) \tag{3.83}$$

It is also important for Equation 3.82 to occur at the basic pH of 7.5 to 8.0 to prevent the hypochlorite converting to hypochlorous acid:

$$H^+ + OCl^- \rightarrow HOCl \tag{3.84}$$

9.0 REFERENCES

Blais, J. F., Djedidi, Z., Cheikh, R. B., Tyagi, R. D. & Mercier, G. (2008). Metal precipitation from effluents, Review. *Practical Periodical of Hazardous, Toxic and Radioactive Waste Management*, *12*(3), 135–149.

Cheng, K. L. (1989). pH glass electrode and its mechanism. Electrochemistry, past and present (Stock & Orna). American Chemistry Society Symposium Series.

Culp, R. L., & Culp, G. L. (1971). *Advanced wastewater treatment.* Van Nostrand Reinhold, 310.

Dey, D., Herzog, A., & Srinivasan, V. (2007). *Chemical precipitation: Water softening* [Thesis]. Michigan State University.

Grunditz, C., & Dalhammer, G. (2001). Development of nitrification inhibition assays using pure cultures of Nitrosomonas and Nitrobacter. *Water Research*, 35(2), 433–440.

Kang, J., Wang, M., & Xiao, Z. (2009). Modeling and control of pH in pulp and paper wastewater treatment process. *Journal of Water Resources and Protection, 2,* 122–127.

Lotfi, I., Ebrahimi, A., & Hajian, M. (2014). Comparison study of turbidity removal using synthetized poly-aluminum chloride-sulfate and poly-aluminum chloride in aqueous solutions. *International Journal of Environmental Health Engineering* *3*(3), 1–6.

Lowe, K., Tucholke, M. B., Tomaras, J. M. B., Conn, K., Hoppe, C., Drewes, J. E., McCray, J. E., & Munakata-Marr, J. (2010). Influent constituent characteristics of the modern waste stream from single sources. *Water Research Foundation.*

Monhemius, A. J. (2017). The iron elephant: A brief history of hydrometallurgists' struggles with element no. 26. *CIM Journal*, *8*(4), 197–206, Figure 1.

Morf, W. E. (1995). Lifetime of glass membrane electrodes: Theoretical model for the corrosion of silicate glasses. *Electroanalysis, 7*(9), 852–858.

Prawoto, Y., Ibrahim, K. M., & Wan Nik, W. B. (2009). Effect of pH and chloride concentration on the corrosion of duplex stainless steel. *Arabian Journal for Science and Engineering, 34* (2C), 115–127.

Ripp, K. M. (2000). Causes and cures of distribution system corrosion. *Opflow,* 26–28.

Radiometer Analytical. (2007). *pH theory and practice.*

Simoes, F., Vale, P., Stephenson, T., & Soares, A. (2018). The role of pH on the biological struvite production in digested sludge dewatering liquors. *Nature, Scientific Reports, 8*, 7225.

Stumm, W., & Morgan, J. J. (1981). *Aquatic chemistry: An introduction emphasizing chemical equilibria in natural waters*. Wiley.

Texier, P. (2008). Effect of acidification on sludge dewatering properties. *Semantic Scholar, 2007–2008*, 78–83.

Wett, B., Rostek, R., Rauch, W., & Ingerle, K. (1998). pH-controlled reject water treatment. *Water Science and Technology, 37*(12), 165–172.

Woods Hole Oceanographic Institute. (2021). https://www.whoi.edu/know-your-ocean/ocean-topics/ocean-chemistry/ocean-acidification/

4

Conductivity

1.0 WHAT IT IS AND WHY IT MATTERS

So far, this book has covered the workings of two probes that look virtually identical but measure two very different things. A pH sensor measures the concentration of a chemical species—H^+ ions—while an oxidation-reduction potential (ORP) sensor measures an aggregate property of water—its redox potential. Hopefully the reader, at this point, sees that the protons that a pH probe measures and electrons that an ORP probe measures are, in fact, highly correlated. Where there are electrons there are usually protons (H^+ ions).

Both pH and ORP sensors are potentiometric devices, which means that they generate a voltage but, ideally at least, no current. This chapter discusses another aggregate property—conductivity. This sensor measures neither a voltage nor a current. It measures the one parameter that connects them—the conductivity.

That water conducts electricity is no surprise to most. When James Bond, in the movie *Goldfinger*, tosses an electric fan into a bathtub to fry his enemy, he is relying on the high conductivity of bathtub water to kill his nemesis. What we may not know is what it is in the water that does the actual conducting of electricity.

Electricity is carried by charged particles. In the world in which we live those charged particles are usually electrons traveling over wires. In the world of water negatively charged anions and positively charged cations conduct current. In pure water, however, the only ions are the H^+ and OH^- ions that are present at very low concentrations of 10^{-7} moles/liter each. That's not a lot of charged particles to conduct electricity and makes pure water a good insulator. We discuss later the actual resistance of pure water later.

When one says that water conducts it's not the water itself that is conductive but the ions, or dissolved salts, in the water. Why is the conductivity of water such a big deal? The answer is that "dirty" water correlates with ion concentration—the purer the water the lower the ionic concentration and the lower the conductivity. In other words, conductivity is a surrogate measure for water purity.

Though this truism holds for most processes that occur in water it does not hold for all. One of my favorite demonstrations is to take two glasses of deionized water. In one I dissolve a packet of sugar and in the other I dissolve a pinch of salt. When I ask which glass of water is more conductive most people pick the glass of sugar. In fact, the sugar water is no more conductive than ordinary tap water whereas the pinch of salt drives the conductivity of its glass of water through the roof. The reason is that table sugar is not a salt. When it dissolves in water it does not break up into cations and anions as a salt does. It simply breaks up into individual sucrose molecules, which are electrically neutral. So, it is possible that a sample of water could show a low conductivity when, in fact, it is seriously contaminated with organic solvents or solutes. But that is the exception in the real world.

It bears repeating the explanation of why ions are prevalent in water but nonexistent in most other liquids. As first noted in Chapter 1, Section 3 water is a unique substance. The water molecule's two H atoms have a partial positive charge while its O atom has a balancing and negative charge. Overall, of course, the water molecule is neutral. But the charge separation that gives rise to it being a *polar* molecule allows anions and cations from a dissolved salt to remain as separate ions when, in the air or an organic solvent, they would recombine to form the neutral salt. It takes a lot of energy to create ions in the air. A collection of ions in the gas phase, or plasma, is characterized by a temperature in the thousands of degrees. Water does this at room temperature.

2.0 LET'S MAKE A MODEL

2.1 Ohm's Law for Aqueous Solutions

Compared to the mathematics behind pH and ORP, the formalism behind conductivity is remarkably simple. It starts with the most basic equation in electronics:

$$V = I\,R \text{ or } I = V/R \tag{4.1}$$

The simple expression is Ohm's law, and it makes more sense when written as the fraction in the second equality. A *voltage* (Volts or V) across a resistor drives a *current* (Amperes or Amps or A). The proportionality constant, R, that connects the voltage and current is the *resistance* (Ohms or Ω). Current is nothing more than either positive or negatively charged particles traversing along a pathway. This heralded equation usually means electrons pushed through a metal wire. For an aqueous solution Ohm's law looks more like Figure 4.1.

In the world of electronics resistance is a ubiquitous parameter. However, it's important to understand that the resistance (Ω) of a wire or a resistor or any other electronic component is not an intrinsic property of that component. For example, a 10-cm length of wire has a resistance that is twice that of a 5-cm length of wire. That approach works well when one is concerned about the resistance of a wire in an electrical circuit but not when one wants to measure the current carrying capacity of the metal that makes up the wire. If we want to compare, say, copper wire with aluminum wire, then we want to measure apples to apples—that is, we want to compare resistances between two wires of the same diameter and same length. We want to define a property that does not depend on the shape or dimensions of the wires. That intrinsic property is called the *resistivity*. Intuition tells us that the resistance is proportional to the length of the conductive path. A little more thought also tells us that it is inversely proportional to the cross-sectional area of the conductive path. After all one chooses a thicker

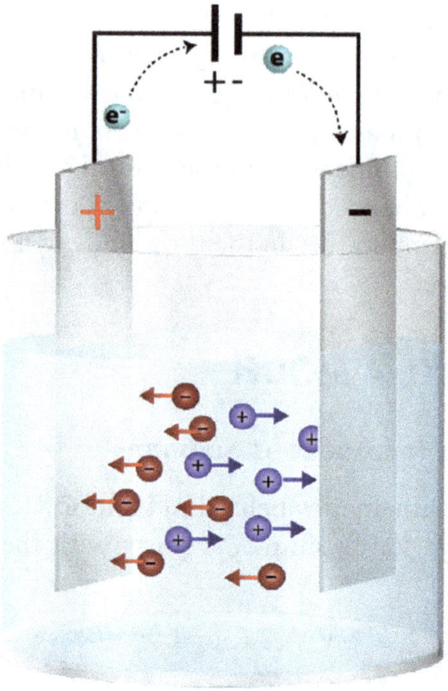

FIGURE 4.1 Ohm's law illustrated for an electrical circuit on the left and an aqueous solution of ions on the right. In both cases, the current equals the voltage across the conductive pathway divided by the resistance.

wire to carry more current. Putting these two logical constructs together generates a definition of resistivity as:

$$L = \frac{R}{l} \times A = R \times \frac{A}{l} \tag{4.2}$$

Simple multiplication and division of units shows that resistivity is in units Ω-cm. (The proper SI unit is Ω-m, but in the electrochemical world, it is more covenant to deal with centimeters than meters.) Now, determining whether aluminum or copper is more resistive is a sample matter of looking up their resistivity values—1.68×10^{-6} Ω-cm for copper and 2.82×10^{-6} Ω-cm for aluminum. This discrepancy explains why copper is still king in electrical wiring.

2.2 From Resistance to Conductance to Conductivity

Electrochemists have a different perspective of resistance than electrical engineers and electricians. They prefer to talk in terms of *conductance*, which is nothing more than the opposite—or, to be more precise, inverse—of resistance. Just as resistance is the inverse of conductance, the inverse of resistivity, an intrinsic property, is another intrinsic property, *conductivity* (σ)—that is:

$$S = \frac{1}{R} \text{ and } \sigma = \frac{1}{L} \tag{4.3}$$

As a clear demonstration of the reciprocal relationship between resistance and conductance the original unit defined for the latter was *mho*—that is, ohm spelled backward. Nowadays most use the term *Siemen* (S) though the two units are the same. Because actual conductance values in water can range into very low values, we are more likely to see resistivities in the thousandths of Siemens—milliSiemens or mS—or even millionths of Siemens—microSiemens or μS.

If we invert Equation 4.3 we get the following:

$$\sigma = \frac{S}{A} D = S \frac{D}{A} \tag{4.4}$$

It should come as no surprise that the units for conductivity are S/cm, or mS/cm or μS/cm. The conversion factor, D/A, in Equation 4.4 comes up so frequently that it gets its own name— the cell constant. Its importance becomes clearer in the upcoming discussion of the construction of a conductivity sensor.

2.3 Conductivity in the Real World of Electrolytes

The range of conductivity values that are measured in industrial settings spans six orders of magnitude.

On the low end is ultrapure water in which the only contributor to conductivity is the 10^{-7} M concentration of both H^+ and OH^- ions. Ultrapure water has a well-characterized value of 0.055 μS/cm. At this end of the conductivity scale users typically switch from units of conductivity to units of resistivity, which, for pure water, is 18.2 MΩ-cm. To be consistent with this convention a probe tailored toward this extreme end of the conductivity scale is called a *resistivity probe*. Ultrapure water is a staple in semiconductor manufacturing in which even trace concentrations of impurities can short out a conductive trace on an integrated circuit.

On the high end of the conductivity scale is brine, acids, and bases. The latter two are special cases of solutions that don't arise from dissolved salts. Brine is a concentrated salt solution that can be used in applications from pickling to the ocean to hydraulic fracturing. Conductivity values for brine solutions range from 100 to 500 mS/cm.

Somewhere in the middle is tap water and natural fresh water. Conductivity values vary from 200 to 400 μS/cm for the former, and 500 to 1000 μS/cm for the latter. Seawater is about 3% salt—mostly sodium and chloride but with a wide variety of metal cations and halogen anions. Its average conductivity value is 56 mS/cm.

Figure 4.2 shows the conductivity values for a range of common solutions—from ultrapure water to strong acids. Strong electrolytes—salts, acids, and bases—show a unique conductivity behavior. The concentration

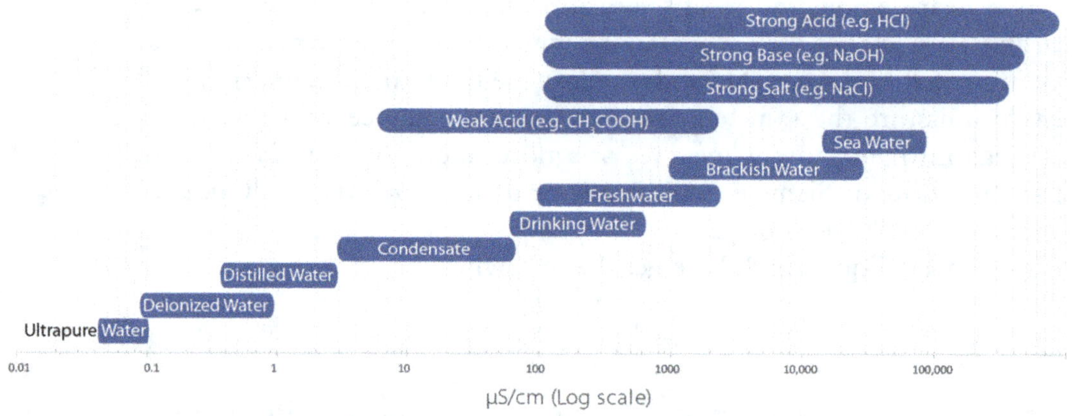

FIGURE 4.2 Typical conductivity *ranges* of common aqueous solutions. The horizontal axis is logarithmic.

range over which their conductivity increases with concentration is limited. At high enough concentrations, their conductivities actually decrease with increasing concentration.

2.4 Relating Ion Properties to Conductivity Values

Calculating conductivity values for any arbitrary collection of ions is, on paper, straightforward, though, in practice, anything but trivial. In contrast to electrical circuits consisting of electrons traveling through a metal conductor, the conductivity of aqueous solutions represents the sum of the individual conductivities of the many ions in the sample. To complicate matters further, the conductivity of one ionic species is not simply proportional to that ion's concentration. Interactions between ions—mainly repulsive ("Coulombic") forces result in the conductivity being less than that which one would predict if it were simply proportional to concentration. This is analogous to the concept of activity, discussed in Chapter 3, Section 6.1, as a parameter that is less than the concentration—especially so at high concentrations of H^+ ions.

To understand how this works we first need to continue the definition of molar conductivity that was first set forth in Chapter 3, Section 6.5. This parameter defines the conductivity parameter that is intrinsic to dissolved ions. For the sake of completeness, we redefine this parameter.

Recall that removing the dependence of the conductance on the dimensions of the aqueous solution allows one to go from this extrinsic property of conductance to the intrinsic property of conductivity. However, the conductivity of a sample is still dependent on the concentration of its ions' concentrations. To quantify the intrinsic conductivity of a salt that dissolves completely—aka "strong electrolyte"—one needs to remove the concentration dependence of the electrolyte from the conductivity of the solution.

To do so simply requires that one divide the conductivity of a solution, σ, contributed by the salt's concentration (C):

$$\Lambda_m = \frac{\sigma}{C} \tag{4.5}$$

The resulting parameter, Λ, is the *molar conductivity* (or *specific conductivity*) that was first defined in Chapter 3, Section 6.5. The subscript "*m*" denotes the ionic species. The units are in S m^2 mol^{-1} or μS cm^2 mol^{-1}.

As just mentioned, the molar conductivity is not a constant. The molar conductivity is proportional to the ionic species' concentration, but this "constant" of proportionality is not constant at all. It changes with concentration. The molar conductivity decreases as the ion's concentration increases as each ion increasingly feels the electrostatic repulsion from its neighboring ions. Its mobility is impeded to an increasing degree. The *limiting molar conductivity*, Λ_m^0, is the molar conductivity value at zero concentration.

The dissociation of a neutral compound into cations and anions behaves differently for a strong electrolyte, such as KCl, and a weak electrolyte such as acetic acid (CH_3COOH). The latter does not dissociate completely into its cation and anion constituents, as was described for buffers in Chapter 3, Section 6.4. Figure 4.3 shows typical molar conductivity values as a function of the electrolyte concentrations for the two types of electrolytes.

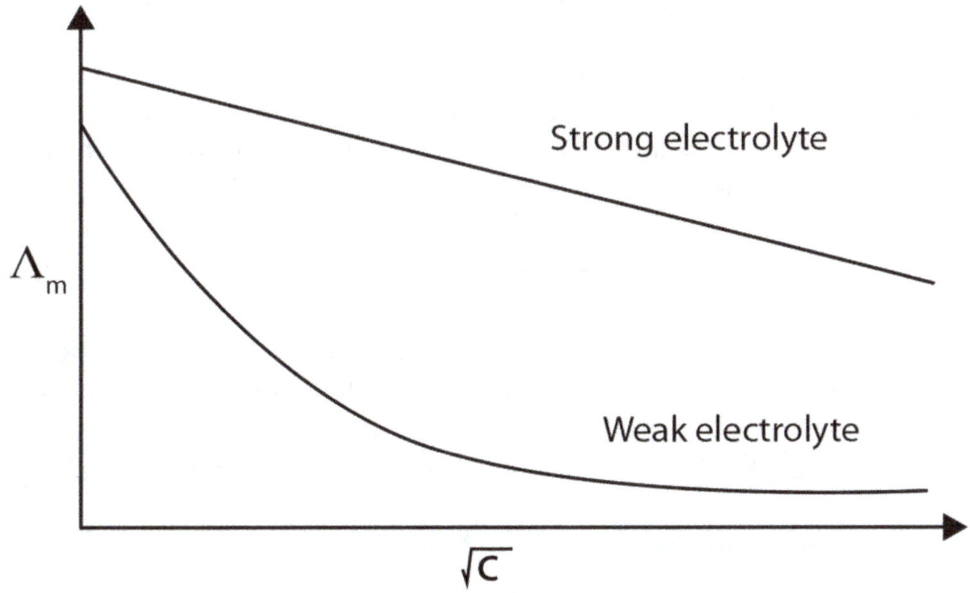

FIGURE 4.3 Molar conductivity (Λ) vs. the concentration.

Note that the conductivity in Figure 4.3 is plotted against the square root of the concentration. For a strong electrolyte, the relationship is linear when plotted this way:

$$\Lambda_m = \Lambda_m{}^0 - k\, C^{1/2} \tag{4.6}$$

A linear least-squares fit then gives a slope and intercept, the latter of which is $\Lambda_m{}^0$ and represents the molar conductivity at infinite dilution. Kohlrausch's law states that each charged species has a value for $\Lambda_m{}^0$ that is independent of other species. The total molar conductivity of a solution therefore is the sum of the $\Lambda_m{}^0$ values of each species.

We can go one level deeper: Every electrolyte consists of a cation and anion. A very good approximation is that the molar conductivity for an electrolyte is the sum of individual molar conductivities of the cation and anion—that is, the two are independent of each other:

$$\Lambda_m{}^0 = n_+\lambda_+ + n_-\lambda_- \tag{4.7}$$

The "+" subscript applies to the cation and the "−" subscript applies to the anion. The parameter λ is the molar conductivity of the ion, and n is the charge number. For instance, for the salt $CaCl_2$, n_+ is 1 and n_- is 2.

One can therefore assign molar conductivities to the constituent anions and cations. Equation 4.7 allows one to determine conductivity values for a wide range of salts. As long as one can look up the molar conductivities of the constituent ions, λ_+ and λ, one can determine the molar conductivity of the salt, $\Lambda_m{}^0$. Table 4.1 is a compilation of ionic conductivities.

For weak electrolytes, such as a weak acid or base, the relationship between the molar conductivity and concentration is more complicated. The following is Ostwald's dilution law and is the analogue of Kohlrausch's law for weak electrolytes.

$$\frac{1}{\Lambda_m} = \frac{1}{\Lambda_m{}^0} + C\frac{\Lambda_m}{(\Lambda_m{}^0)^2 K_a} \tag{4.8}$$

In Equation 4.8, K_a is the dissociation constant introduced in the discussion of buffers in Chapter 3, Section 4.8. According to Equation 4.8, a plot of $1/\Lambda_m$ against $C\Lambda_m$ yields a straight line. The intercept is $1/\Lambda_m{}^0$, and the slope is $1/(\Lambda_m{}^0)^2$Ka. Figure 4.4 shows a generalized example of this plot.

2.5 From Conductivity to Ion Mobility

Chapter 3, Section 6.5 introduced the parameter of ion mobility (u) to explain the diffusion potential of the liquid junction. This parameter is a

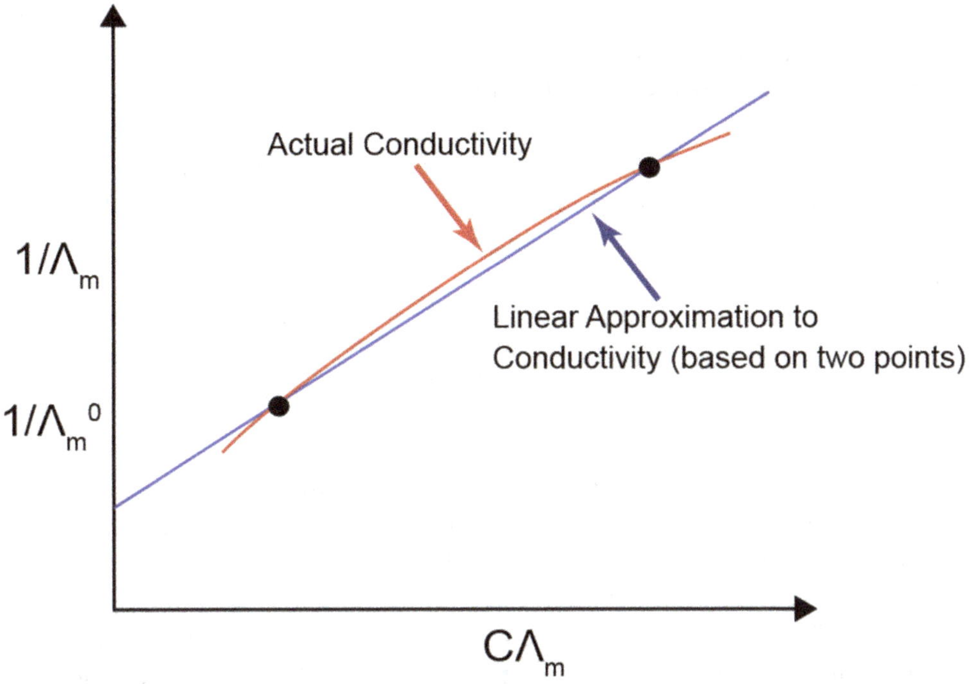

FIGURE 4.4 In accordance with Equation 4.8, $1/\Lambda_m$ is plotted against $C\Lambda_m$. The intercept is $1/\Lambda_m{}^0$, and the slope is $C/(\Lambda_m{}^0)^2K_a$ and gives a linear approximation to the actual conductivity.

measure of how fast an ion moves through a gas or, in this case, a liquid. It might seem that both the ion mobility and the limiting molar conductivity of an ion describe the same behavior. Indeed, the two parameters are proportional to each other. The proportionality between the molar conductivity and the ion mobility is given by:

$$\Lambda = F\, z\, u \tag{4.9}$$

Recall that z is the charge of the ion, and F is Faraday's constant. The units of ion mobility are the speed of an ion divided by the magnitude of the electric field, or $cm^2\ sec^{-1}\ volt^{-1}$. Ion mobility describes a single ion, whereas molar conductivity describes a mole of ions.

In terms of limiting conductivity values for the individual ions (u_+ and u_-):

$$\Lambda_m{}^0 = F(n_+z_+u_+ + n_-z_-u_-) \tag{4.10}$$

The additional parameter, n, is the number of ions in the undissociated salt. As an example, for $CaCl_2$, $n_+ = 1$, $z_+ = 2$, $n_- = 2$, and $z_- = 1$. Because all salts are neutral before they dissociate, $n_+z_+ = n_-z_-$.

The mobility of an ion is proportional to the ion's charge and inversely proportional to its radius. It is also inversely proportional to the solvent's (in this case, water's) viscosity. The meaning of the radius merits clarification. The polar water molecule sticks to anions with its positively charged H atoms and to cations with its negatively charged oxygen atom. This results in ions surrounded by an entourage of water molecules—that is, the ion is "solvated." The solvated ionic radius is much larger than the radius of the sole ion.

These three factors go into the determination of the mobility of an ion (i):

$$u_i = \frac{|z_i| e}{6\pi \eta R_i} \tag{4.11}$$

Here z_i is the absolute charge of the ion in units of electrons (i.e., for SO_4^{2-} it is 2), e is the elementary charge (i.e., the actual charge of an electron), e is the viscosity of the water, and R_i is the solvated radius.

To calculate the conductivity of a solution we just add up the product of the molar conductivities and the concentrations of all the constituent ions as given by Equation 4.9. To convert the molar conductivity to conductivity we use Equation 4.9; the sum of molar conductivities is:

$$\sigma = F \sum_i |z_i| u_i C_i = F \sum_i \frac{z_i^2 e}{6\pi \eta R_i} C_i \tag{4.12}$$

The real take-home lesson of Equation 4.12 is that the conductivity of a solution containing a mixture of ions is the sum of the conductivities of the constituent ions. Table 4.1 shows both Λ_0 and u values for common cations and anions. As the above discussion shows, either one is sufficient to calculate the conductivity of a solution of any mixture of ions.

2.6 Salinity—The Third Yardstick

There is still one more parameter that quantifies conductivity. *Total dissolved solid* (TDS) or, sometimes simply *salinity*, is often cited in field measurements. It is the easiest of all conductivity parameters to understand because it is simply the total concentration of dissolved ions, or salts. Often, it's the parameter we really want to know. It's more appropriate to think of conductivity as a measuring stick of salinity rather than the other way around. Salinity is also the easiest to measure if you have the time. All you need to do is take a one-liter sample solution, evaporate the water, and weigh the remaining salts. Its units are in mg/L or parts per million (ppm). Both units are identical as long as they refer to aqueous solutions.

TABLE 4.1 Ion mobility and limiting molar conductivity (at 25 °C). Note that the mobility values are proportional to the specific conductivity values.

Cation	Mobility 10^{-4} cm^2 s^{-1} V^{-1}	Specific Cond. S-cm^2 mol^{-1}	Anion	Mobility 10^{-4} cm^2 s^{-1} V^{-1}	Specific Cond. S-cm^2 mol^{-1}
H^{+C}	36.25	349.6	OH^-	20.84	199
Li^+	4.01	38.7	F^-	5.74	55.4
Na^+	5.19	50.1	Cl^-	7.91	76.4
K^+	7.62	73.5	Br^-	8.10	78.1
Rb^+	8.06	77.8	I^-	7.96	76.8
Cs^+	8.00	77.2	NO_2^-	7.44	71.8
Ag^+	6.41	61.9	NO_3^-	7.41	71.5
NH_4^+	7.62	73.5	ClO_3^-	6.70	64.6
Be^{2+}	9.33	90	ClO_4^-	7.08	67.3
Mg^{2+}	11.0	106	HCO_3^-	4.61	44.5
Ca^{2+}	12.3	119	$H_2PO_4^-$	5.91	57
Sr^{2+}	12.3	119	HSO_3^-	5.19	50
Ba^{2+}	13.2	127	HSO_4^-	5.19	50
Fe^{2+}	11.1	108	CH_3COO^-	4.24	40.9
Cu^{2+}	11.0	107	CO_3^-	14.3	137
Zn^{2+}	11.1	106	HPO_4^-	6.84	66
Pb^{2+}	14.7	142	SO_4^-	16.6	160
Al^{3+}	19.0	183	PO_4^-	21.5	207
Fe^{3+}	21.2	204	$Fe(CN)_6^{3-}$	31.4	303
La^{3+}	21.7	209	$Fe(CN)_8^{4}$	45.8	442

TDS measurements may be simple in theory, but evaporating water is a slow and energy-intensive process. Because we know from the previous section that conductivity is proportional to the sum of ionic concentrations, the opposite must be true: TDS values must be (at least roughly) proportional to conductivity values.

$$TDS = k\,\sigma \tag{4.13}$$

The only unknown is the value *k*—the so-called proportionality constant. Just as with ion mobility or molar conductivity, *k* is different for every ion. In the real world, the value of *k* typically varies from 0.5 to 1.0. *Standard Methods for the Examination of Water and Wastewater* (APHA et al., 2017) recommends a value between 0.55 and 0.7. For freshwater samples with salinity levels less than 5000 mg/L, the value is close to 0.5. For highly saline (>10 000 mg/L) and solutions of divalent ions, such as Ca^{2+} or SO_4^{2}, the *k* value is closer to 0.7.

It's very important to understand that analyzers that report measurements in TDS units assume an average value such as 0.50 or 0.67. Some allow the user to input k. It's also very important to remember the discussion in the previous section that conductivity is proportional to the ionic concentration, but the proportionality is not constant with concentration. For instance, the conductivity for NaCl at 38 ppm is 84 µS/cm. If *K* were constant and conductivity scaled linearly then, at 8532 ppm, one would expect the conductivity to be $\frac{8532\ ppm}{38\ ppm} \times 84\,\mu\frac{s}{cm} = 18\ 900\ \mu S/cm$. It is actually 15 000 µS/cm. In other words, *K* increases from 0.48 at 38 ppm to 0.56 at 8532 ppm. A higher value of *K* means a lower conductivity value per unit of TDS. Figure 4.5 shows experimentally determined *k* values for KCl and NaCl over a wide range of concentrations (Fondriest, 2021). As just noted, the ratio hovers around 0.5 for conductivity levels less than 10 mS/cm (about 5000 mg/L).

The take home lesson here is that, unless you can calibrate a TDS meter for the appropriate ion at the appropriate concentration, you can't place too much faith in a measurement that reports a TDS value. The higher the concentration, the larger the deviation from an average assumed *k* value.

When characterizing solutions with very high salt concentrations, some report the salt content as a percentage, such as 10% NaCl. When a solid is dissolved in water the percentage is always on a weight basis. For instance, a 1% NaCl solution means 1 gram of NaCl in 100 grams of water. Since 1 milliliter of water weighs exactly 1 gram, converting percentage by weight to mg/L or ppm is easy. In this example:

$$1\text{ g NaCl} / 100\text{ g } H_2O = 1000\text{ mg NaCl} / 100\text{ mL } H_2O$$
$$= 10\ 000\text{ mg/L (or ppm)}.$$

A concentration of 10 000 ppm is approximately equal to 15 000 µS/cm or 15 mS/cm. This is about 27% the salinity of seawater.

2.7 Conductivity Examples

Section 2.3 explained why conductivity is linear with concentration only in dilute solutions. Strong electrolytes, such as KCl, dissolve almost completely

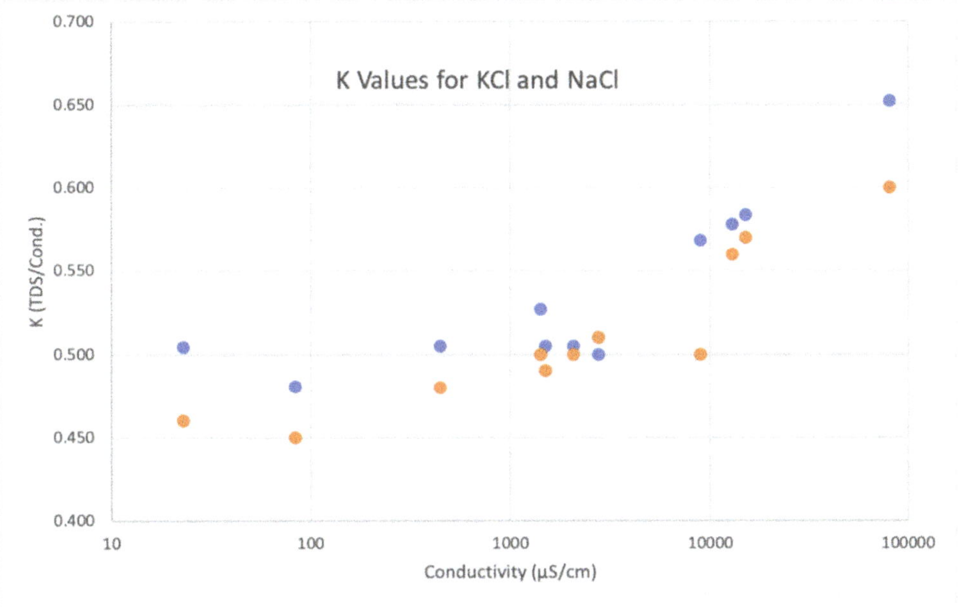

FIGURE 4.5 K values for KCl (blue) and NaCl (red) (Data from Fondriest, 2021).

and show approximately linear behavior throughout a wide concentration range, whereas weak electrolytes, such as acetic acid, show approximately linear behavior over a very narrow concentration range. Figure 4.6 shows the actual conductivity versus concentration behavior for solutions of two common salts (NaCl and $CaCl_2$), three strong acids (HCl, H_2SO_4, and HNO_3), one weak acid (CH_3COOH), and one strong base (NaOH). The conductivity for NaCl is nearly linear throughout its limited concentration range. The reason why the curve stops at about 30% weight concentration is simply because NaCl is 100% saturated at that point. The $CaCl_2$ conductivity rolls over and decreases with concentration. The strong acids and base reach a maximum in conductivity and decrease strongly toward zero as their concentration nears 100%. The conductdivity of the weak acid (acetic acid) is so small (maximum 1600 µS/cm) that it hardly registers on the graph.

The limiting molar conductivity for any of these species is the slope of curve in Figure 4.6 near the origin of the graph.

2.8 Temperature Dependence

No other parameter is harder to quantify nor affects conductivity values to a greater extent than temperature. Every ionic species has its own temperature dependence, and that temperature dependence also varies with concentration.

Figure 4.7 shows the conductivity of a 0.9% NaCl solution (9000 ppm) over a temperature range of −25 to 25 °C (Sauerheber & Heinz, 2015).

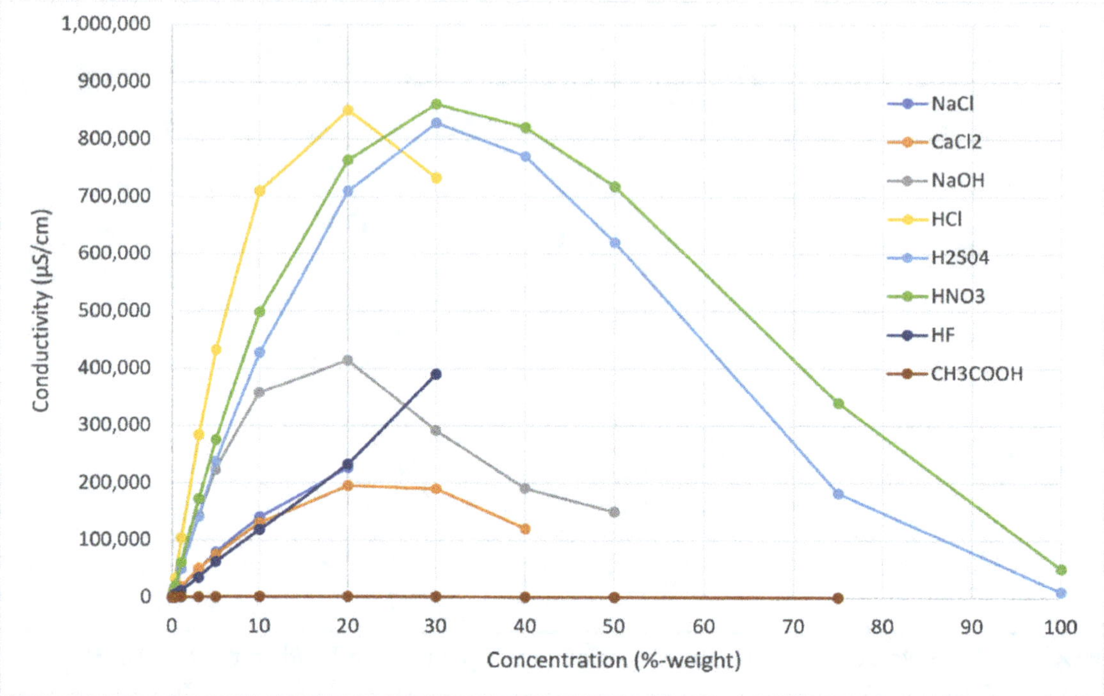

FIGURE 4.6 Conductivity as a function of concentration for two salts, three strong acids, one weak acid, and one base. At low concentrations the conductivity is proportional to concentration as evidenced by the straight line. At high concentrations the conductivity slope rolls over and—for the case of acids and bases—decreases.

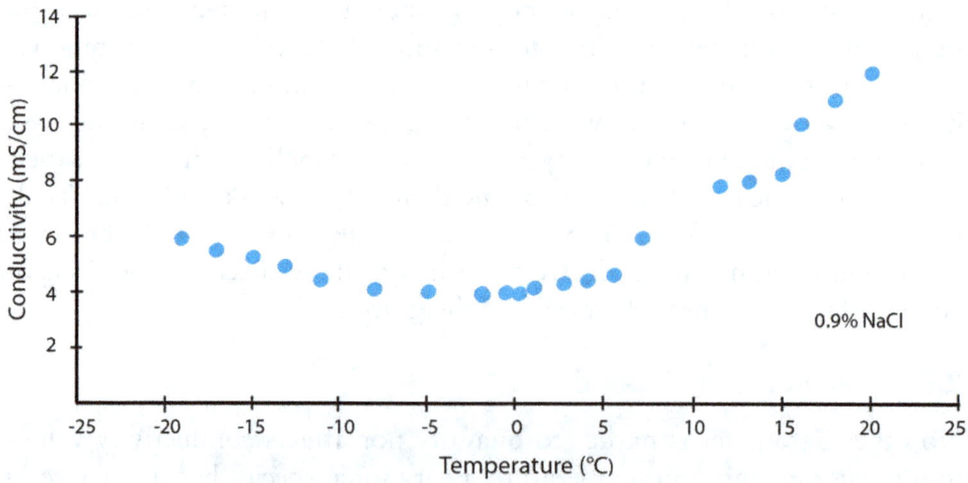

FIGURE 4.7 Conductivity as a function of temperature for 0.9% NaCl in water. (Adapted from Sauerheber & Heinz, 2015).

The conductivity actually decreases from −25 to 0 °C and then rises. From 5 to 25 °C the increase is approximately linear.

It would be most convenient to derive a correction for the conductivity of a solution from one temperature based on another. First, let's define a generalized correction factor α:

$$\sigma_T = \alpha(T, C)\sigma_{25} \qquad (4.14)$$

The correction factor, $\alpha(T,C)$, is written with T in parentheses to emphasize the fact that it is not necessarily a constant but changes with temperature as well as concentration. The challenge is to define the appropriate correction factor for the sample in question and over the temperature range and concentration range likely to be encountered. The approximate linearity in Figure 4.7 from 5 to 25 °C suggests that, over this temperature range in the "good-enough" world of industrial or munipcal measurements, a linear temperature correction is sufficient:

$$\sigma_T \approx \sigma_0[1 + \alpha(T - 25\ °\mathrm{C})] \qquad (4.15)$$

In this approximate formula, α_0 is the conductivity at 25 °C or the calibration temperature and the temperature correction factor, α, is in units of $°C^{-1}$ or K^{-1}.

Figure 4.8 shows conductivity values measured for tap water over a temperature rate of 10 to 40 °C (Water Analytics). From the uncorrected values (blue) we determined a temperature coefficient of 2.05% per degree Celsius. The red dots show conductivity values with this temperature correction. It's clear that the linear correction models the data to within the experimental error.

However, there are limitations to assigning a constant correction factor, α, for all temperatures, concentrations, and different dissolved salts. Figure 4.9 shows the change in conductivity of trace (ppb) concentration solutions of NaCl solutions over the temperature range of 0 to 100 °C (Light et al, 2005). The correction factor changes by a factor of 10.

All one can do is determine an approximate correction factor that is strictly valid only for one ion, over a defined and narrow range of both concentration and temperature. For this reason, conductivity measurements at temperatures other than 25 °C are limited in precision to no better than approximately 5%.

Because most processes of interest occur at temperatures between 10 °C and 50 °C, a linear temperature coefficient between 1% and 3% is "good enough." Table 4.2 lists approximate values of α for salt solutions, acids, and

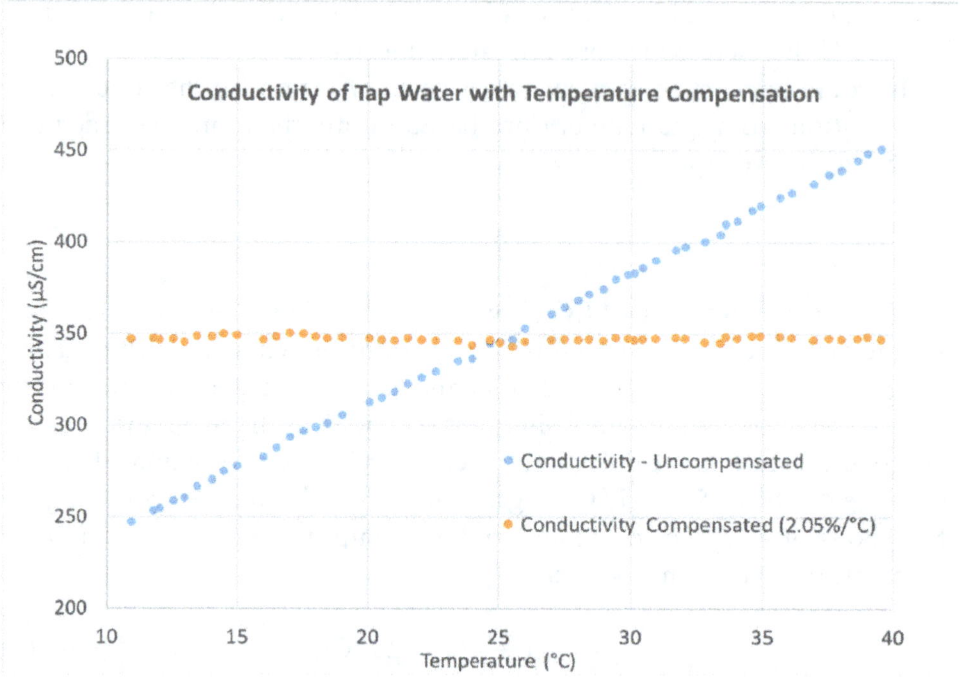

FIGURE 4.8 Temperature dependence of conductivity for tap water (Water Analytics).

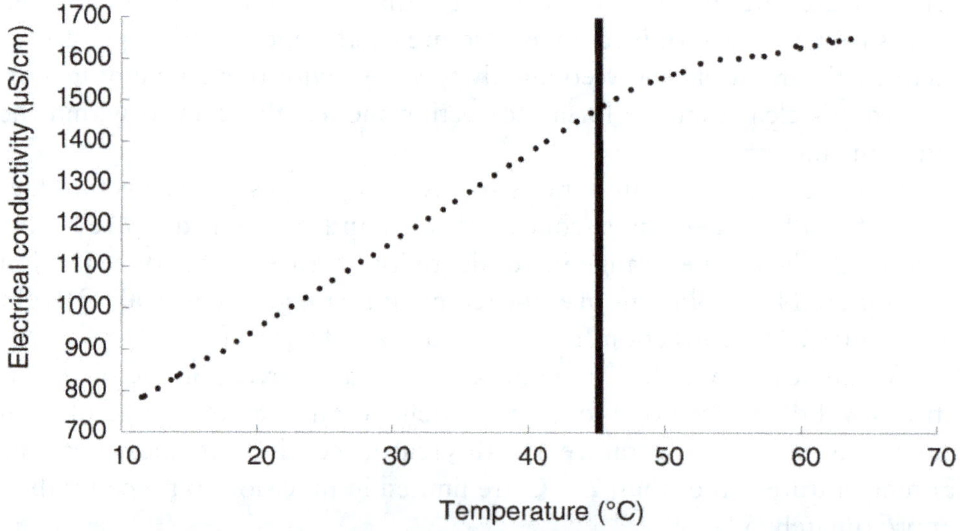

FIGURE 4.9 Water electrical conductivity increases with temperature but with a nonlinear behavior. Consequently, the fractional change per degree Celsius m_f is not constant with temperature. (Hermans et al., 2014).

TABLE 4.2 Approximate range of temperature coefficient (α) for salts, acids, and bases.

Solution	α (%/°C)
Acids	1.0–1.6
Bases	1.8–2.2
High Concentration Salt Solution	2.2–3.0
Low Concentration Salt Solution	1.8–2.5
Drinking Water	2.0
Ultrapure Water	5.2

bases. Because we most often measure dilute solutions, most conductivity analyzers use a value of 2%.

Simple arithmetic shows that, if the correct α is, say, 1.5% instead of the assumed 2%, then the error in the conductivity measurement increases at 1% for every two degrees removed from 25 °C. So, a measurement at 45 °C will be 10% higher than the actual value. Furthermore, a temperature measurement error of only 2 °C for an assumed temperature coefficient of 2%/°C generates a 4% error. In contrast a 2 °C error in a pH measurement at room temperature (298 K) generates an error of 298/296, or 0.6%. For this reason alone, the accuracy of conductivity measurements cannot be expected to be better than approximately 5%.

The measurement accuracy can be improved by discarding the linear relationship between conductivity and temperature and, instead, fitting the temperature dependence to a nonlinear curve containing multiple points. Though not common in controllers and transmitters, the multiple point curve fit option is possible in a supervisory control and data acquisition system.

For applications that use ultrapure water the temperature relationship is very nonlinear. As mentioned in Section 2.3 the 10^{-7} M of H^+ ions and 10^{-7} M of OH^- ions are responsible for the conductivity value, 0.055 µS/cm (at 25 °C), of ultrapure water. As the temperature increases the dissociation of H_2O molecules into H^+ and OH^- ions increases with temperature. Figure 4.10 (data from ASTM, 2005) displays conductivity values of ultrapure water as a function of temperature from 0 °C to 100 °C.

For the calculation of conductivity of surface and groundwater, the ISO/DIN 7888 standard proposes a fourth order polynomial to fit the temperature correction between 0 °C and 35.9 °C.

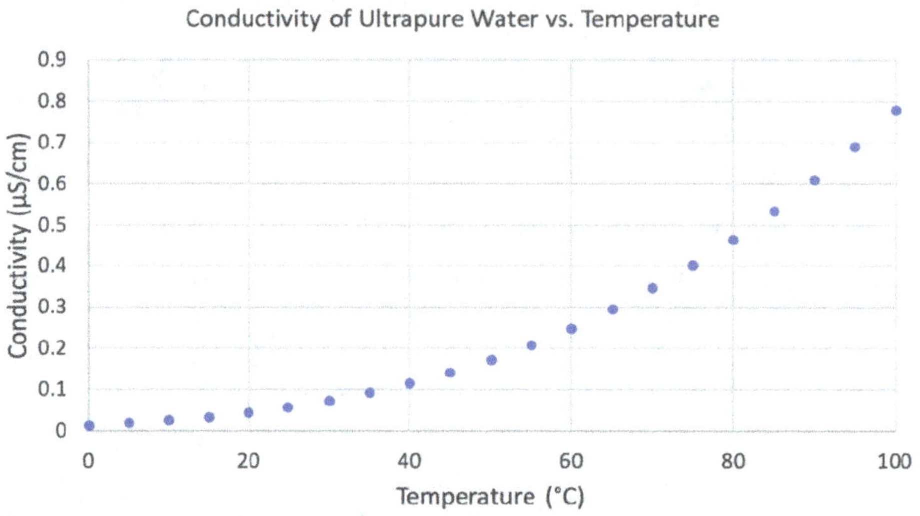

FIGURE 4.10 Conductivity of ultrapure water (Data from ASTM, 2005).

3.0 CONSTRUCTION OF A CONDUCTIVITY SENSOR

3.1 The Two Electrode Sensor

Figure 4.11 shows the design of the common two-electrode conductivity sensor. The design calls for a voltage source, two parallel plates, and a way to measure either the resistance across the volume defined by the distance between the electrodes (D) and the area of the plates (A). Note that the actual shape of the plates is irrelevant—only the surface area and the gap between the plates matter.

The moment the power supply applies the voltage, V, positive cations migrate toward the negatively charged plate and the negative anions migrate toward the positively charged plate. Current flows between the electrodes and the analyzer measures the resistance. But there's a problem. Unlike an electronic circuit that consists of only electrons flowing in one direction, the conductivity sensor consists of anions and cations migrating in opposite directions. After a short while, anions pile up at the positive plate and cations pile up at the negative plate, thereby effectively neutralizing the charge on both plates. This so-called *polarization effect* decreases the flow of current. Figure 4.11 shows schematically how polarization at the electrodes sets up two additional voltage sources in the sensor circuit.

The two polarization voltages add to the voltage drop across the sample, so the analyzer is unable to distinguish between the two. The most effective way to minimize the polarization of ions between the electrodes is to drive

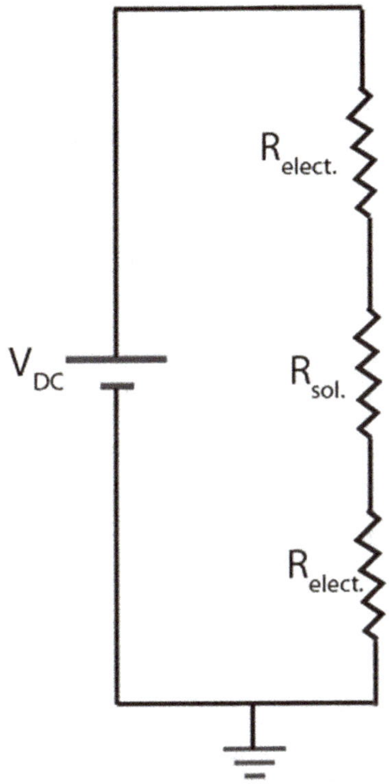

FIGURE 4.11 Schematic of a two-electrode conductivity sensor with a direct current voltage source. $R_{elect.}$ is the resistance from polarization at the surface of both electrodes. $R_{sol.}$ is the actual resistance of the solution.

the sensor with an alternating current voltage. Typical voltages range from 2 to 10 V at frequencies from 100 Hz to 50 KHz.

Figure 4.12 is an extension of Figure 4.11 for an alternating current driving voltage and shows schematically how the driving voltage minimizes the effect of the buildup of charge at the electrode surface on measurement. The buildup of charge also creates a boundary layer capacitance, thus creating an RC circuit in series with the solution resistance. Capacitors pass alternating current (AC). The buildup of charge on the electrodes gives a direct current voltage, which does not pass through the circuit. The alternating current frequency, ν. for which the current transmission is maximum is given by:

$$2\pi\upsilon = \sqrt{RC} \tag{4.16}$$

The important point illustrated in this equivalent circuit explanation is that the optimum frequency of the AC voltage is given by the conductivity of the solution. Many controllers use two or more frequencies with lower frequencies on the order of 100 Hz for conductivity levels below 100 μS/cm

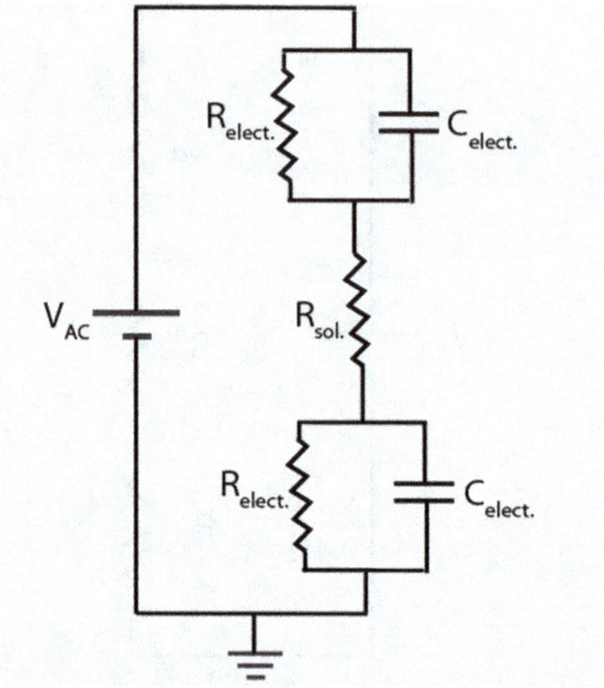

FIGURE 4.12 Schematic of a two-electrode conductivity sensor with an alternating current voltage source. The charge buildup on the surface of the electrodes creates RC circuits that can pass an AC voltage if its frequency is matched to the solution conductivity.

and higher frequencies on the order of 10 KHz for conductivity levels above 100 mS/cm.

A second method for reducing polarization is to use *platinized* electrodes—that is, electrodes that are coated with a layer of platinum black. Platinum black is powdered platinum. The orders of magnitude increase in the surface area of the platinum, enabling a corresponding decrease in the current density through the electrodes. The downside of platinized electrodes is that they foul easily, and as a result, the calibration, or cell constant, changes with time. The platinized surface easily abrades in water with suspended matter. For these two reasons sensors with platinized electrodes are relegated mostly to the laboratory.

To measure conductivity for a particular sample one needs to choose the cell constant tailored toward the conductivity range of the sample. Recall the definition of the cell constant, or K factor, stated in Equation 4.4. It helps to flip this equation to:

$$S = \sigma \frac{A}{D} = \frac{\sigma}{K} \tag{4.17}$$

Equation 4.17 states that, in order to keep the conductance (S) relatively constant, the cell constant (K) has to move in the same direction as the conductivity (σ). For instance, if we are measuring a low conductivity solution, then we use a probe with a small cell constant. To make a sensor with a small cell constant we make the distance between electrodes (D) relatively small and/or the surface area of the electrodes relatively large. If we are measuring a high salt solution that has a high conductivity, then we want a relatively insensitive sensor. We accomplish this my minimizing the surface area of the electrodes and/or maximizing their separation—hence a large D/A or cell constant.

Before leaving the parallel plate conductivity sensor, it is important to point out one very important characteristic of the cell constant: equating the cell constant to D/A is strictly valid only for the case of the conductive medium constrained to the same cross-section as the plates. This normally means a solid medium, like a copper wire, which naturally has a well-defined boundary. The electrons are constrained to move within the confines of the solid. However, when the conductive medium is an aqueous solution, there is no boundary to contain the flow of ions. The electric field bows out at the edges as shown in Figure 4.13. These edge effects increase the effective

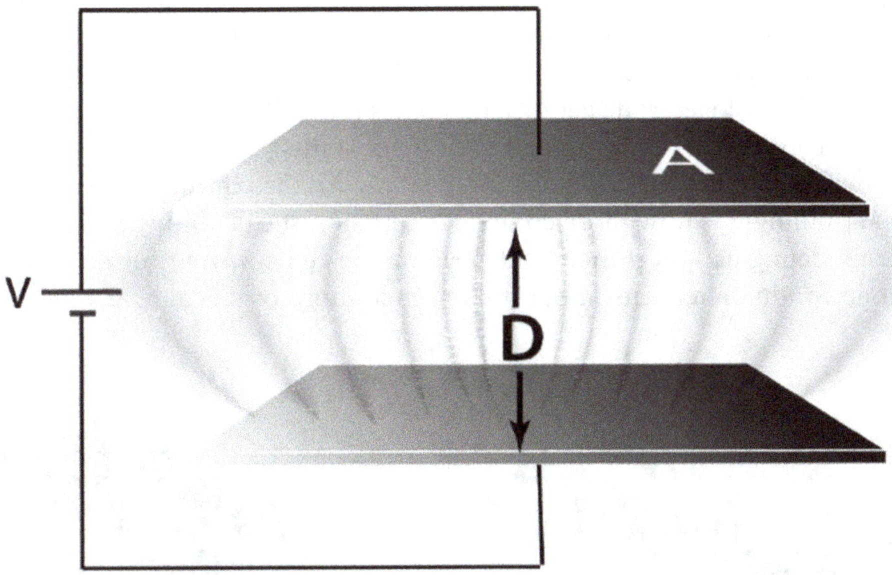

FIGURE 4.13 Schematic of conductivity sensor based on parallel plates. The cell constant is equal to the ratio of the separation between the plates to their surface area. The curved lines show the electric field. Note that the field lines extend beyond the edges of the plates means that the effective surface area is larger than the actual area. The units for K are cm^{-1} but it is typically written in the literature without the cm^{-1} appended.

surface area of the electrodes and, therefore, decrease the real cell constant from the simple D/A equation. Edge effects can affect not only the solution but also the container holding the sample. If the container is conductive (i.e., metallic), and the probe is close to the walls (by approximately one probe width), then the electric fields between conductors will extend through the walls. The result is a smaller effective cell constant and higher measured conductivity. This issue reappears later in this chapter in discussions of calibration and accuracy. The simple geometric relationship expressed in Equation 4.17 is accurate only for the case of infinitely large electrodes. The lesson to be learned is that, even for the simple parallel plate sensor, the only way to determine the cell constant is to measure it.

Conductivity sensors that use parallel plates are still in use in laboratory analyzers. However, this design does not make for rugged and compact probes for industrial use. For this reason, most designs use cylindrical electrodes that come in a variety of configurations. Figure 4.14 shows examples of probes using the more common concentric electrode configuration. The figure on left is an illustration whose outer shell—the outer electrode—is cut away to reveal the inner electrode. Both have a cell constant of 1 cm^{-1}. The probe on the right is an actual probe by Mettler Toledo. The electrode material is usually stainless steel, but it can be graphite that is impregnated with copper to make it more conductive. This latter can be used in acidic or highly oxidative solutions that would corrode stainless steel. Holes in the outer electrode facilitate the flow of water between the electrodes.

This variety of electrode geometries makes it possible to make industrial strength probes with a wide range of cell constants. The geometry of the electrodes determines the cell constant, just as with the parallel plate sensor. But determining it is now more complicated than just calculating D/A. Geometric modeling can, in principle, determine the cell constant for any two-electrode conductivity sensor, but it's so much more practical to determine

FIGURE 4.14 Conductivity sensors using concentric electrodes. Left: Cutaway illustration showing the inner electrode and hole for facilitating flow between the electrodes. The outer cylinder is the second electrode. Right: Probe from Mettler Toledo is based on this design (Courtesy Mettler Toledo).

it through actual measurement. The next section discusses several methods for accurately determining the cell constant for two-electrode sensors. The same methods are applicable to four-electrode and toroidal sensors as well.

3.2 Just What Does the Cell Constant Mean?

If there is one philosophical debate or a source of confusion in the world of conductivity sensors, it is likely to be the meaning of a cell constant. This is more than just an academic concern. Determining the actual cell constant for a probe is the same as calibrating it.

If one considers the origin of the cell constant to be the geometrical dimensions of the original parallel plate sensor (i.e., distance between plates and area of plates), then one naturally extends this simple geometric definition, as defined in Equation 4.18, into a more generalized geometrical calculation for more complicated electrode shapes. For contacting sensors with cylindrical electrodes, deriving the cell constant from the geometry of the electrodes is more complicated. The geometric modeling is much more sophisticated and requires computer modeling. But the principle remains the same. According to this interpretation, the determination of the cell constant is a purely mathematical exercise.

On the other hand, if one assumes that the cell constant is simply the proportionality between conductance and conductivity, then the determination of the cell constant is a purely empirical exercise: one measures the conductance of a calibration standard of known conductivity and divides the latter by the former to attain the cell constant. Calculating a generalized D/A for complex electrode configurations is no longer necessary or useful.

So, is the cell constant a calculated geometrical factor or an empirical proportionality constant? The debate becomes even more contentious for a toroidal sensor then it is for a contacting sensor (to be discussed in Section 3.7). The geometry of the electrodes for contacting sensors may be complex, but its geometric underpinnings are clear. The electrodes of a toroidal sensor consist of send and receive coils encased in a plastic shell, so a geometric interpretation of the constant is not appropirate. One determines the cell constant using calibration standards, which can be an aqueous standard or a precision resistor.

To measure the cell constant of a probe is straightforward. Equation 4.17 can be rearranged to yield the cell constant in terms of the known conductivity of a sample and its measured resistance. Since conductance (S) is the reciprocal of resistance (R), we get:

$$K = \frac{\sigma}{S} = R\,\sigma \tag{4.18}$$

So, to measure the cell constant we only need a conductivity calibration standard and a conductivity analyzer that will display the raw resistance. There are two high-precision ways of doing this.

3.2.1 The Wheatstone Bridge

The Wheatstone Bridge is a century-old device that is still the gold standard for measuring resistance. Because a conductivity measurement is a resistance measurement this is equivalent to balancing an unknown resistor to a known one. Figure 4.15 shows a schematic of a Wheatstone Bridge used to measure a probe's cell constant. It works by applying an alternating current voltage across two pairs of resistors in parallel. Two of the four resistors have fixed and precisely known resistance values. The third is a variable precision resistor (i.e., rheostat). The fourth, R_x, is the unknown resistance of the calibration solution measured by the probe. A voltmeter in the middle of the bridge measures a voltage imbalance between the two arms. When the voltage is zero the resistance of the unknown resistor is given by:

$$K = \frac{\sigma}{S} = R\,\sigma \tag{4.19}$$

Figure 4.15 shows the conductivity probe in the conductivity standard as the unknown resistance, R_x. It's important that the container dimensions be much larger than the probe dimensions to prevent the container's walls from influencing the measurement. It's also important that the choice of calibration standard be appropriate for the probe's nominal cell constant. The inverse of the measured resistance is the conductance, and the ratio of the calibration standard conductivity to the measured conductance is the cell constant. The accuracy of the Wheatstone Bridge measurement is limited by the accuracy of the calibration standard.

3.2.2 The Analyzer With a Resistance Readout

Many analyzers have a diagnostic menu that allows the user to view the raw resistance reading of the sensor. If so, then calculating the cell constant is as simple as sticking the probe in a known calibration standard and reading the resistance ®. Equation 4.18 yields the constant. To make the units come out right just use the following formulation of the simple equation for K:

$$K = 10^{-6}\,\sigma(\mu S\,/cm)\;R(\Omega) \tag{4.20}$$

As an example, if we use 1000 µS/cm calibration standard and the analyzer measures 950 Ω then $K = 10^{-6} \times 1000 \times 900\ \text{cm}^{-1} = 0.95$.

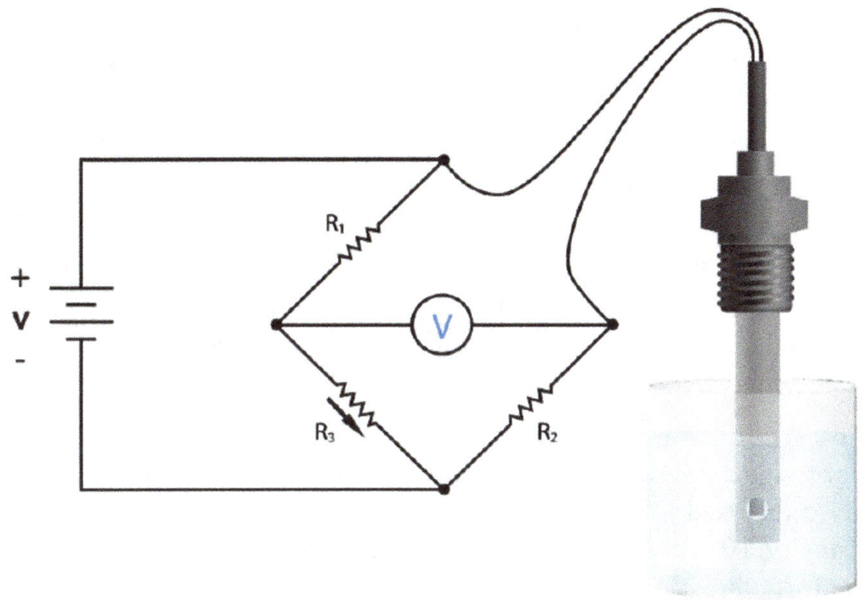

FIGURE 4.15 Wheatstone Bridge. Resistances R_1 and R_2 are fixed and precisely known. R_x is a resistor whose resistance is unknown. R_3 is a variable resistor that the user adjusts until the voltage across the bridge is zero.

3.2.3 *The Reference Resistor*

Wheatstone Bridges are expensive and not every analyzer displays the raw resistance reading. But here's a method that works and only requires a precision resistor and a beaker of calibration sandard. The idea behind it is that one first calibrates the analyzer to the resistor and then measures the conductivity of the calibration standard. To make the arithmetic easy we'll use a sensor with a nominal cell constant of 1 cm^{-1}, a 1 kΩ resistor, and 1000 µS/cm standard. The sensor will have 4 wires (5 if there is a shield wire) that connect to the analyzer.

There are four wires coming out of a conductivity sensor—five if there is a shield. Two are from the electrodes and two are from a temperature element—usually a thermistor or resistance temperature detector. There are two inputs into the analyzer for the conductivity leads of the sensor. Do not confuse these with the inputs for the temperature device.

1. Insert the two legs of the 1000 Ω resistor into the sensor inputs. The polarity does not matter. Because the temperature compensation for the analyzer may go haywire if it has no temperature input, either turn off temperature compensation (if possible) or connect the two temperature leads from the sensor to the two temperature inputs

2. Calibrate the analyzer by adjusting the reading to 1000 μS/cm. (The analyzer is dumb and has no idea it is connected to a resistor instead of a container of water.)
3. Remove the resistor and replace it with the two signal leads from the conductivity sensor.
4. Insert the sensor into the 1000 μS/cm calibration standard. (Avoid placing the sensor close to the walls.) Read the conductivity value.
   a. The cell constant is 1000 divided by the measured conductivity value. So, for example, if your standard reads 950 μS/cm, then your actual cell constant is 1000/950, or 1.05 cm^{-1}.

To carry out this exercise with other cell constants and with minimal mental fatigue, just use the same 1 kΩ precision resistor and a calibration standard that scales with the cell constant. For instance, a probe with a cell constant of 10 cm^{-1} calls for a 10 000 μS/cm calibration standard. The simple formula is:

$$K = \frac{\sigma_{calib.std}}{\sigma_{measured}} \tag{4.21}$$

The choice of cell constant is arguably the greatest source of error when choosing a conductivity probe. If the cell constant is chosen incorrectly, then the range of measured conductivity values will occupy only a fraction of the available range. For instance if a probe with a cell constant of 10 is used in a sample with a conductivity range of 50 to 100 μS/cm, then the actual measurements will occupy only about 5% of the total available 20 000 μS/cm range. That increases the uncertainty in the measurement more than an order of magnitude so that an uncertainty that would be ±2% in a probe of cell constant 1 cm^{-1} is greater than ±20 % in a probe with a cell constant of 10 cm^{-1}. In this case, we are not really introducing an error into the conductivity measurement. We are simply increasing its uncertainty. This is another example of how the accuracy remains unchanged, but the precision is worsened. In actuality, the non-linear response of the probe increases the error beyond this simple contraction of the measurement range.

3.3 The Two-Electrode Analyzer

The two-electrode sensor is by far the most common probe configuration. Figure 4.16 shows the schematic of a typical conductivity analyzer and probe. The analyzer contains a precision sense resistor in series, R_{sense}, with the resistance measured by the probe, R_{sol}. This circuit is a simple voltage

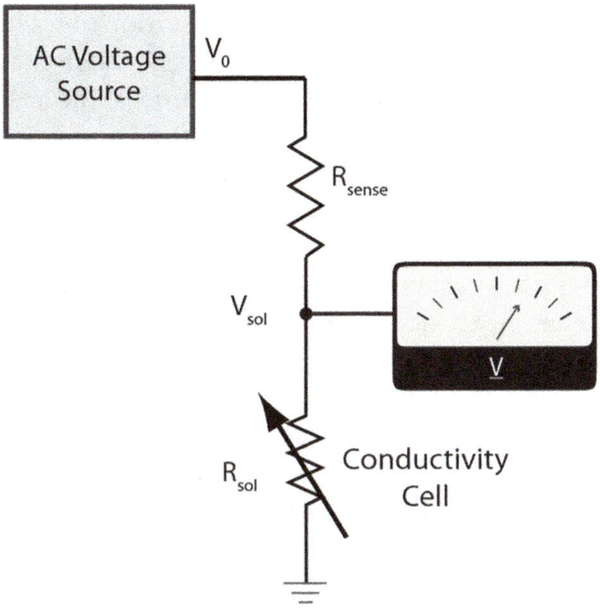

FIGURE 4.16 Simplified schematic of conductivity analyzer. This simple voltage divider measures the unknown resistance between the two electrodes by measuring the voltage across the precision sense resistor. The ideal measurement range occurs when the solution conductance is close to that of the resistor.

divider containing a driving voltage, V_{out}. The voltage across R_{sense} is V_{sense}. With Ohm's law and a little arithmetic, it's easy to show that:

$$R_{cell} = R_{sense}\frac{V_{out}}{V_{sense}} - 1 \ ; S_{cell} = \frac{1}{R_{cell}} \tag{4.22}$$

The precision resistor is on the order of 5 kΩ, which corresponds to a conductance of 1/(5 × 10^3 Ω), or 200 µS. From Equation 4.17 and a cell constant of 1, we get a conductivity range that is centered around 200 µS/cm. Typical electronics enable a measurement range of two orders of magnitude, or 100 to 1. So that means a conductivity sensor with cell constant of 1 should be able to cover a conductivity range of, say, 20 to 2000 µS/cm. At values below this range the lack of sensitivity reduces the measurement accuracy and, at values above the range, the probe response is less than linear, due to polarization.

Based on the discussion of cell constants in Section 2.2, the conductivity range is proportional to the cell constant. So, for example, a probe with a cell constant of 10 will cover a range 10 times larger—200 to 20 000 µS/cm. Note that we can design a conductivity analyzer that uses only one probe of cell constant 1 but switches between the precision resistor to cover different conductivity ranges. Some laboratory analyzers work this way.

A further variation of either circuit is the insertion of multiple sense resistors that allow the sense resistance to more closely match that of the sample.

Figure 4.16 represents a simplified explanation of the analyzer. In reality, this circuit would be subject to errors due to the finite input impedance of the voltmeter and output impedance of the voltage source. Actual analyzers are based on the usage of op-amps. The killer feature of the op-amp is that its input impedance is enormous, thus virtually eliminating impedance errors. Figure 4.17 is a schematic showing an op-amp configured for measuring an unknown impedance using an external voltage source, V_{source}, and a high-precision, reference resistor, R_{ref}. R_{cell} is the resistance of the sample measured by the probe.

Modeling an op-amp's output is easy once you understand the first commandment of this amazingly versatile chip: an op-amp does whatever is necessary to force the inverting input (−) voltage to equal the noninverting (+) input voltage. The result for this circuit is that:

$$V_{out} = -V_{source}\frac{R_{ref}}{R_{cell}}\;;\; S_{cell} = \frac{1}{R_{cell}} = -\frac{V_{out}}{V_{source}}\frac{1}{R_{ref}} \qquad (4.23)$$

Note that, if we switched R_{ref} with R_{cell}, the output would be proportional to the resistance instead of the conductance.

Before discussing the four-electrode sensor, it's important to note that a two-electrode sensor may have four electrodes. The four electrodes are actually two pairs of electrodes with those of each pair connected to each other. This is simply an alternate strategy for decreasing the cell constant

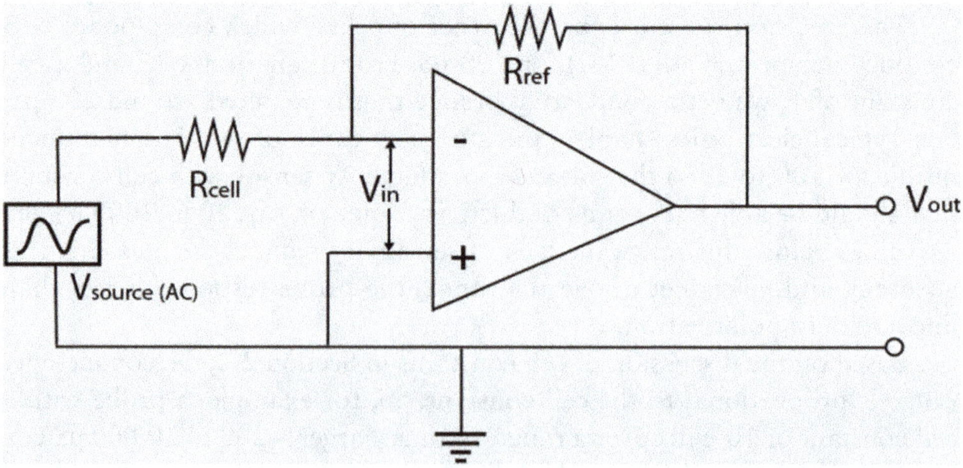

FIGURE 4.17 Op-amp configured for resistance or conductance measurement.

as opposed to one that uses longer electrodes or places the concentric electrodes in closer proximity.

The two-electrode sensor is the simplest, least expensive, and most common conductivity sensor. It does have several limitations:

1. The conductivity reading that the probe makes represents the sum total of all possible conductive pathways between the electrodes. As Figure 4.13 shows, the electric field set up by the electrodes extends beyond the actual electrodes. If the boundaries of the container in which the probe sits is close to the electrodes, then the reading will incorporate the conductivity of the container walls. If the electrodes actually touch the container the error can be greater than 100%.
2. At high conductivities, polarization at the electrodes occurs, the linear response rolls off and an artificially low conductivity reading results. To minimize polarization effects, most probe manufacturers increase the alternating current voltage frequency as the conductivity level increases. As mentioned in Chapter 3, Section 1, platinized electrodes minimize polarization, but they are too fragile to find widespread use outside the laboratory.
3. For any given cell constant the range of measurement spans two orders of magnitude. The range of conductivity values in the real world spans an enormous seven orders of magnitude, from 0.1 µS/cm to 1000 mS/cm. This means that at least 4 probe models having different cell constants are needed to cover this range. Typical cell constants of conductivity probes span a range from 0.01 to 100, with models every 10x, 5x, or 2x (e.g., 0.01, 0.1, 1.0, 10.0, 50).

Most applications cover a conductivity range that is less than three orders of magnitude, so the two-electrode probe suffices. For instance, tap water varies from 200 to 400 µS/cm. For this reason, in addition its low cost, the two-electrode sensor is the most common type of probe. The choice of a cell constant then becomes the only decision to be made. Table 4.3 gives the ranges of (approximately) linear responses for cell constant

The ranges are very approximate and depend on the frequency of the driving alternating current voltage and the resistance of the precision sense resistor. Polarization of the electrodes limits the upper limit of a probe's linear response range to less than 500 mS/cm.

3.4 Linearity and Calibration

Calibration of conductivity sensors is the usual mapping of the probe output to one or more solutions of known conductivity to the conductivity

TABLE 4.3 Conductivity ranges over which the two-electrode conductivity sensor is linear for common cell constants.

Cell Constant (cm^{-1})	Conductivity Range	Examples
0.01	0.2–20 μS/cm	Deionized water (0.1–50 μS/cm)
0.1	2–200 μS/cm	Rainwater (50–100 μS/cm)
1.0	20–2000 μS/cm	Drinking water (200–500 μS/cm)
10	200–20 000 μS/cm	Wastewater (1–10 mS/cm)
50	1–200 mS/cm	Seawater (50 mS/cm)

value(s). In other words, one places a conductivity sensor into a calibration standard, and the analyzer measures the response to that standard. We can "instruct" the analyzer to assume that a zero current corresponds to zero conductivity. (Some analyzers ask the user to hold the probe in air and measure the tiny current and some just assume that the current is negligibly small.) If we accept the approximation that conductivity is linear with concentration over a range less than two orders (which we know is not rigorously correct), then we can use this assumed zero offset point along with the measured probe current in the calibration standard to construct a straight-line calibration curve.

For routine calibration all that is needed is a calibration solution whose conductivity is in the range of the probe. For maximum precision it is important to choose a calibration solution close to the maximum conductivity where the response is linear, as listed in Table 4.3. A 10 μS/cm error in a 100 μS/cm solution is a 10% error but, in a 1000 μS/cm solution, is only a 1% error.

Calibration is nearly almost always done with KCl. Recall, from the discussion on equitransferent solutions in Chapter 3, Section 6.5, that the roughly equal mobilities of the K^+ and Cl^- ions means that they diffuse at roughly equal rates. This minimizes the diffusion potential that affects conductivity measurements in the same way that it affects pH and ORP measurements.

Making conductivity standards requires only a table of KCl molar conductivity values, distilled water, and high purity KCl. Table 4.4 lists conductivity values for three KCl dilutions. A common practice is to make up a concentrated solution of a calibration standard and create serial dilutions from it. In this case one might make up a 1 M solution of KCl and dilute it ten-fold to make a 0.1 M solution and then dilute that one to make a 0.01 M solution. It might be tempting to also assign conductivity values to the two dilutions by simply diving the 1 M value by 10 and 100, respectively. But we know that, strictly speaking, conductivity is linear with concentration

TABLE 4.4 Conductivity values for KCl solutions. (Pratt, Koch, Wu & Berezansky, 2001)

Temp (°C)	0.01 M	0.1 M	1 M
0	773	7117	63 488
5	891	8184	72 030
10	1014	9292	80 844
15	1141	10 437	89 900
20	1273	11 616	99 170
25	1408.23	12 825	108 620
30	1546.63	14 059	118 240
35	1687.79	15 316	127 970
40	1831.27	16 591	137 810
45	1976.62	17 881	147 720
50	2123.43	19 180	157 670

only for infinitely dilute solutions. Indeed, Table 4.4 shows that multiplying the conductivity value at 0.01 M by 10 to attain the conductivity at 1 M (or 0.1 M to 1M) will overshoot the actual conductivity by an amount on the order of 10%.

Standard conductivity values are reported at 25 °C. To minimize the error induced by temperature compensation one should calibrate as close to this temperature as possible. For measurement of pure water (<10 µS/cm), the calibration solution needs to be de-aerated. At ambient pressure, dissolved CO_2 will add 0.5 µS/cm to the conductivity level.

As discussed, the polarization of the two-electrode sensor places an upper limit on its linear measurement range. But, as also discussed and displayed in Table 4.4, even the linear range is not strictly linear. Figure 4.18 shows the measured conductivity versus the true conductivity of NaCl solutions over a wide concentration range and the conductivity as measured by a two-electrode sensor with a cell constant of 1. The plot on the left shows the response of the probe over the range of 0 to 100 000 µS/cm. Plotting the response this way removes the non-linear conductivity of the NaCl solution with concentration and allows us to just focus on the response of the probe. The cell constant 1 probe response over this very wide range is very non-linear but, if we narrow the range to 0 to 2000 µS/cm, the response appears linear.

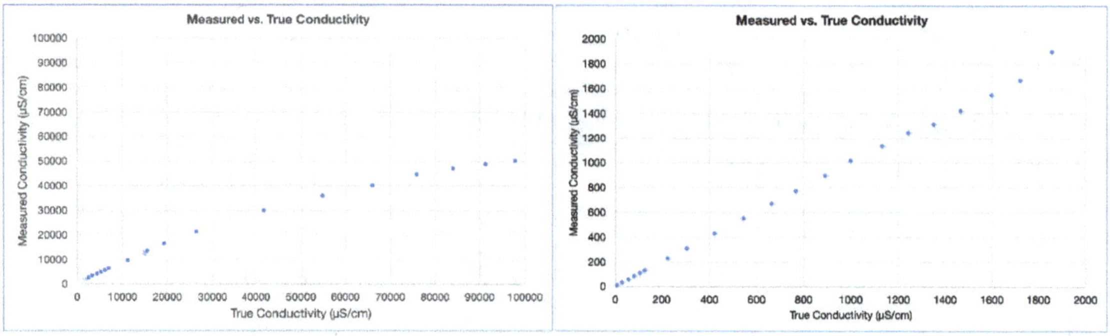

FIGURE 4.18 Measured conductivity of NaCl solutions versus actual conductivity using a probe with cell constant 1.0. The left-hand plot covers the range of 0–100 000 µS/cm. The right-hand plot confines the range to 0–2000 µS/cm, which is the linear range for the sensor (Water Analytics).

For measurements that span approximately two decades of conductivity and for which an uncertainty of ±5% is acceptable calibration can be done via a two-point calibration (with one point being an assumed zero) is all that is needed.

Moreover, if one extends the number of calibration points beyond the two needed for a linear calibration, then the measurement range can be extended significantly. Figure 4.19 shows the magnitude of the measurement error when measuring conductivity of KCl solutions using a cell constant of 1.0 cm^{-1} and one calibration point (red) versus seven points (blue). For a maximum calibration error of ±10%, the measurement range for the one-point calibration is 20 to 8000 µS/cm. For a seven-point non-linear calibration the range expands to 5 to 100 000 µS/cm.

Unlike the potentiometric probes, conductivity probes are prey to only one physical source of calibration error. Contaminants that coat the electrodes of a two-electrode sensor will result in a higher effective cell constant and result in a lower conductivity rating. Coating can be organic matter, suspended solids, oxidation of a metallic electrode (usually stainless steel), or scaling. Fouled electrodes can be cleaned with bleach. Stainless steel electrodes that have scaling can be cleaned in a weak acid. Plating, pulp-and-paper, beverage, and food processing present serious coating issues and call for a toroidal probe.

An alternative to calibrating a sensor with a standard solution is to use a second, previously calibrated sensor, also known as a *referee* sensor. The two sensors are in the same container but spaced at least one probe diameter apart, or they are both installed in a pipe—that is, inline. The latter configuration has the advantage for low conductivity measurements in which absorption of air can affect measurements of samples exposed to the atmosphere.

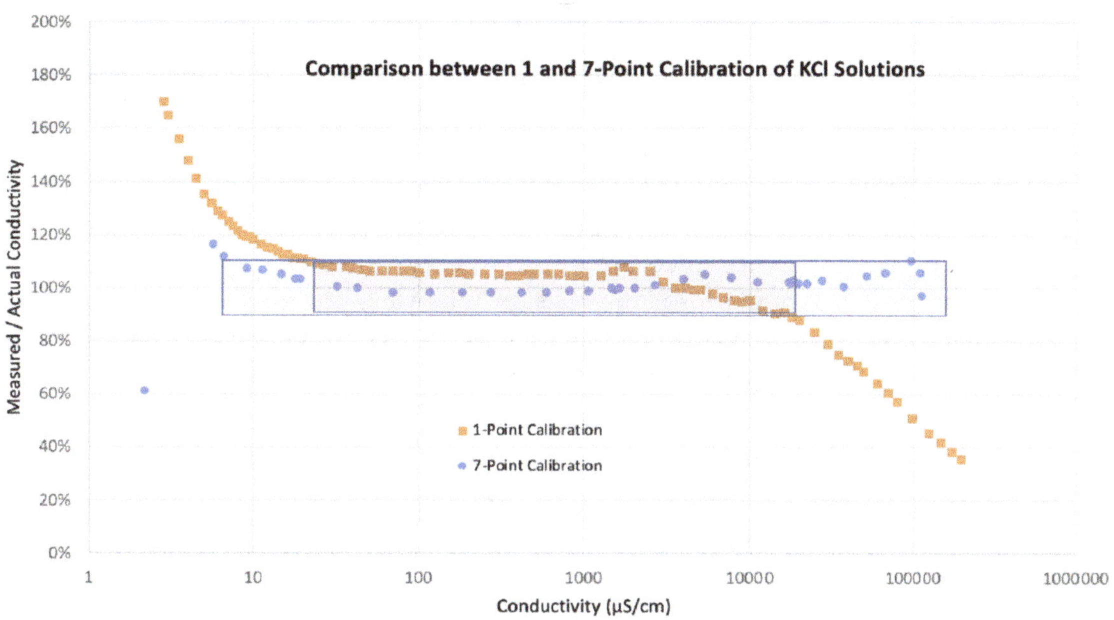

FIGURE 4.19 Accuracy for a 1-point calibration versus 7-point calibration. The cell constant is 1.0 cm^{-1}. The shaded rectangles display the range of conductivity for which the error is ±10%; red for 1-point calibration and blue area for 7-point calibration (Water Analytics).

3.5 Why the Cable Matters

The conductivity measurement is a resistance measurement. Unfortunately, all electrical conductors have resistance and, if the cable is long enough, its resistance will add to the measured resistance. Equation 4.24 gives the change in measured conductance in terms of the real solution conductance (S_s) and the cable resistance (R_c).

$$S_m = \frac{S_s}{1 + (R_c S_s)} \tag{4.24}$$

To gauge the magnitude of the cable resistance effect, consider the case of highly saline water at 100 mS/cm measured by a conductivity sensor with electrodes 1 cm apart. A typical cable length is 5 m and has a resistance of 0.15 Ω. The conductance between the electrodes is therefore 100 mS/cm × 1 cm, or 0.1 S. If we plug these numbers into Equation 4.23, then we get a measured conductance, or conductivity, that is low by 1.5%. For this reason, cable lengths are typically less than 5 m.

Cables also have capacitance. This effect is not as simple to estimate as resistance. It is only a factor for very pure water where the conductance is less than 4 µS.

3.6 The Four Electrode Sensor

The four-electrode conductivity sensor overcomes the limited measurement range of the two-electrode sensor. The left side of Figure 4.20 shows how it works. The right side shows an actual probe.

The four-electrode sensor consists of two pairs of electrodes. The drive circuit of the controller supplies an alternating current through the outer pair, the latter of which are appropriately called the *drive electrodes*. In the two-electrode sensor the resistance between the electrodes includes the resistance of the polarization layer that forms on their surfaces. In the four-electrode sensor a second set of electrodes, the *sense electrodes*, are set inside the outer electrodes. They measure the voltage drop between the drive

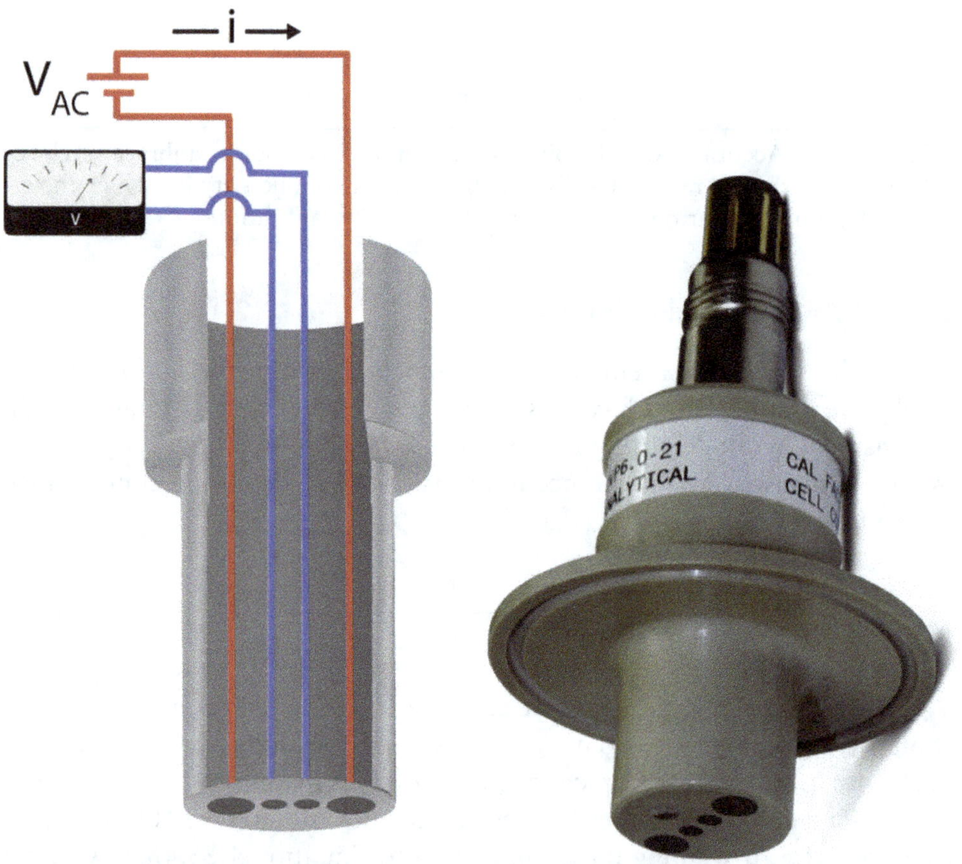

FIGURE 4.20 Left: Schematic of the four-electrode conductivity sensor. The outer two electrodes impose a fixed AC voltage, and the sensing electrodes measure the voltage drop. Right: Four-electrode conductivity sensor by Rosemount Analytical (Courtesy Emerson).

electrodes. Because they are not driven by a voltage source their surfaces are not polarized. This means that they can measure conductivity levels that are much higher than the two-electrode sensor. The conductance of the solution is equal to the current across the drive electrodes divided by the voltage measured between the sense electrodes. The measurement range is about five orders of magnitude—on the order of 1 μS/cm to 100 mS/cm—nearly as high as another type of sensor to be discussed next—the toroidal conductivity sensor. The only constraint on the four-electrode sensor is that capacitance effects at low conductivity values can alter the voltage measurement at the sense electrodes. Its use of two sets of electrodes also results in a larger diameter probe.

3.7 The Toroidal Conductivity Sensor

Like the four-electrode sensor the toroidal probe, also known as the inductive, electrodeless, or non-contacting probe, covers a conductivity range of roughly four orders of magnitude. Its operating principle, however, is completely different.

Chapter 3, Section 7.1.1 introduced the subject of magnetic induction when explaining ground loops. The principle behind induction dates to 1831 when English physicist Michael Faraday discovered that electricity and magnetism are intertwined. A changing magnetic field induces a voltage, also known as an electromotive force or EMF. If one changes the magnetic field inside a loop of wire, the resulting EMF generates a current through the wire. You can observe this simply by pulling a magnet through a wire loop connected to an ammeter. We observe induction in inductors, electric motors, transformers, and, unfortunately, ground loops.

As James Maxwell laid out in mathematical precision in his eponymous equations, the interplay between magnetic field and induced current is a two-way street. A changing magnetic field induces current in a coil and vice versa—a current in a coil induces a magnetic field. If, instead of one coil, one configures a wire into many coils like a spring, then the magnetic field created by the current is a loop that circulates through the coil, as shown on the left side of Figure 4.21.

To make a toroid we need only take this spring-like configuration and curve it around to form donut. Now the magnetic field takes on the circular shape of the doughnut, or *toroid*. To confine the field to the interior of the toroid we fill the space with a circular magnet—the *magnetic core*.

A manifestation of Maxwell's famous equations is that a charged species traveling perpendicular to a magnetic field experiences a force perpendicular to both the velocity direction and magnetic field lines. A circulating magnetic

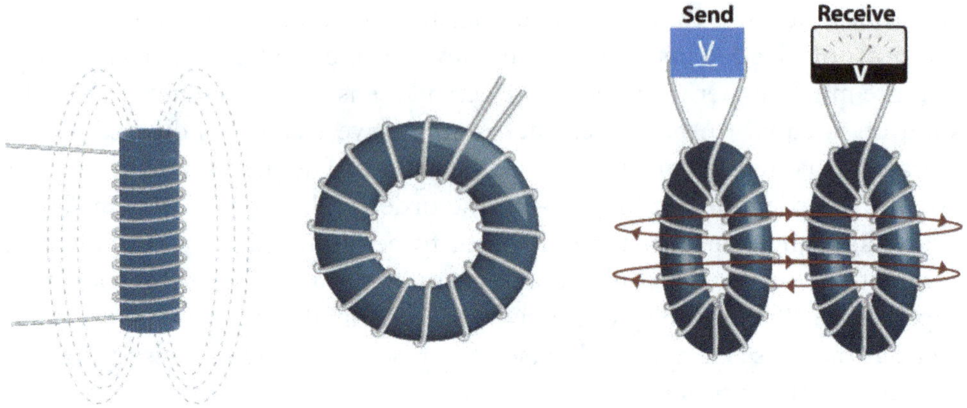

FIGURE 4.21 How to make a toroidal probe. Left: First make a coil so that the magnetic field inside the coil is aligned in a straight line. Middle: Next bend the coil into a toroid so that the magnetic field lines turn into a circle. Right: Then make another toroid and place the two parallel to each other (right side). Current flowing through the left coil creates a magnetic field that drives a current loop of ions that flow perpendicular to the plane of the first toroid. The ionic current creates a circular magnetic field in the secondary toroid, which creates a current in the toroid that can be measured.

field then creates a force that results in the flow of charged particles—that is, a current of ions, in an elliptical path, as shown on the right side of Figure 4.21. The current induced in the receive toroid is proportional to the conductivity. The analyzer measures the current flowing through the sense toroid—usually by dropping the current across a resistor and measuring the voltage across the resistor.

To model the operation of a toroidal sensor, each toroid can be thought of as a transformer. A transformer has two coils—primary and secondary—coupled by an induced magnetic field. Each coil has a number of windings, and the ratio of the windings determines the voltage change. One can think of toroidal sensor as consisting of two transformers—with one half (the primary) being an actual coil and the other half (the secondary) being the water itself. For the primary of the drive transformer the number of windings in the primary n_D and the number windings in the second—the solution—is 1. The sense transformer is the reverse: the number of windings in the primary coil—the solution—is 1 and the number of windings in the secondary coil is n_S.

Figure 4.22 shows the equivalent circuit for the coupled transformer model of a toroidal sensor.

The mathematical connection between the voltage in the sense toroid (V_S) is related to the drive voltage (V_D), number of windings in the drive

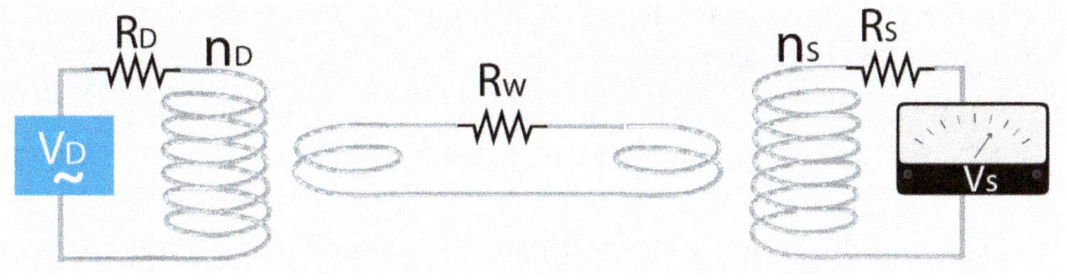

FIGURE 4.22 Equivalent circuit of a toroidal probe. The circuit consists of two transformers—a drive transformer and a sense transformer. The secondary of the drive transformer and the primary of the sense transformer is the solution itself, and both have the equivalent of one winding.

toroid (n_D), number of winding in the sense toroid (n_S), inductance of the sense toroid (L_s), drive frequency (ω), and the water resistance (R_W) by

$$V_S = \frac{n_S}{n_D} \frac{V_D}{1 + n_S^2 R_W \dfrac{1}{R_S} + \dfrac{1}{j\omega L_S}} \tag{4.25}$$

Let's look at the two terms inside the parentheses in the denominator. We'll start with the second term: The term including j (the imaginary number $\sqrt{-1}$) represents out of phase contributions to the output voltage resulting from capacitance and inductance (Panaggio, 2006). Remember that resistance is the real (in-phase) part of impedance. The spiral wound toroid is a very good inductor, so the imaginary component term is significant. The first term is the inverse of the resistance of the sense resistor that converts current in the sense toroidal to a voltage.

It's clear that there is no linear relationship between the probe output voltage and the sample conductivity or the windings. We can look at the limit where the product, ωL_S, is much larger than R_S. In this case, the out-of-phase inductive component is negligible compared to the in-phase resistive component and Equation 4.25 simplifies to:

$$V_S = \frac{n_S}{n_D} \frac{1}{1 + n_S^2 \dfrac{R_W}{R_S}} V_D \tag{4.26}$$

To make ωL_S as large as possible, the driving frequency for the toroidal sensor is chosen to be as high as practical. Components for driving and sensing AC voltages increase in cost as the frequency increases so there is a trade-off in sensor design between measurement range and cost. Typical driving frequencies are between 100 and 1000 KHz.

If the conductivity of the water is high, R_W is much less than 1 and Equation 4.26 collapses to:

$$V_S \approx \frac{n_S}{n_D} V_D \tag{4.27}$$

From this discussion we expect that the output voltage of the toroidal sensor is linear with conductivity for high enough conductivity values and is proportional to the ratio of windings, n_S/n_D. Indeed Figure 4.23 shows this behavior (Parra et al., 2015). The probe's output is linear with conductivity values above about 20 mS/cm.

A fifth order polynomial covers the entire conductivity range to within a few percent.

Figure 4.24 shows a typical toroidal conductivity probe. Like the four-electrode sensor, it measures a conductivity range of approximately 4 orders of magnitude. Its one principal advantage over the four-electrode probe is that the coils that do the actual measuring are encased in a polymer such as PEEK, PTFE (Teflon), or polypropylene. The controller drives an alternating current through the drive toroid and the sense toroid measures the induced alternating current or voltage. The frequency is in the 10 KHz range and is optimized for the greatest signal-to-noise.

Field effects resulting from the geometry and composition of the container or pipe are more significant for toroidal sensors than they are for contacting probes. The electric field that the circulating magnetic field sets up and that drives the ionic current extends well beyond the confines of the enclosed toroids. If the sensor is in the vicinity of a metallic container or pipe wall, the conductive surface increases the electric field, drives a

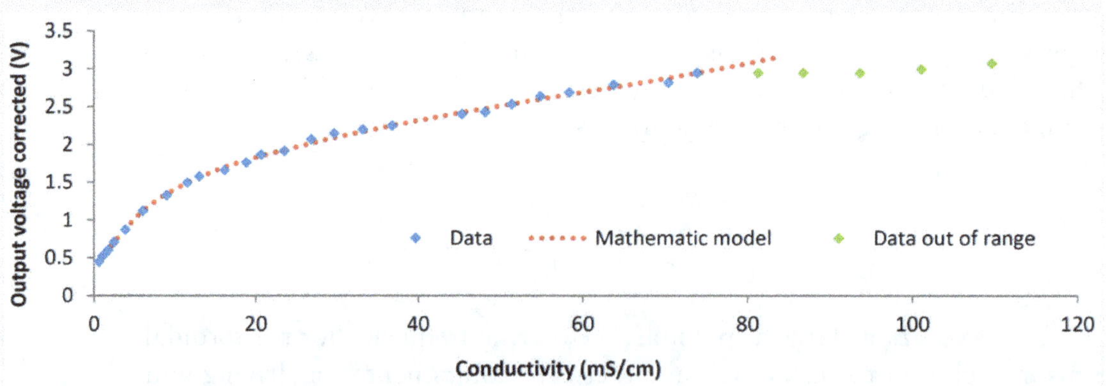

FIGURE 4.23 Output voltage of a toroidal sensor with conductivity. Note that the output is linear for conductivity values greater than about 20 mS/cm (Parra et al., 2015).

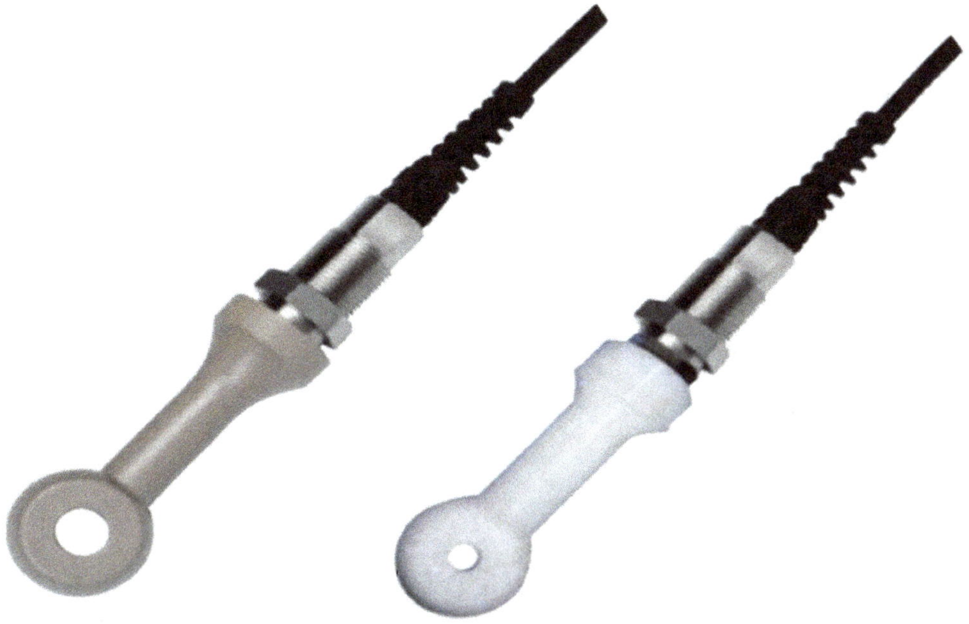

FIGURE 4.24 Toroidal conductivity probes. Note that the assembly of drive and sense coils are encapsulated by a Teflon polymer, which renders the sensor to be "non-contacting." The probe on right has a smaller effective cell constant (Courtesy Yokogawa).

stronger ionic current, and results in an artificially high reading. The presence of a nonconductive surface, such as a PVC pipe or glass container, has the opposite effect. In either case the effect is equivalent to changing the cell constant. Figure 4.25 (Rosemount Analytical, 2010) shows the increase in conductivity when the probe face is parallel to the walls of a metallic container (red line) and when it is perpendicular (blue line).

It's easy to understand this difference when looking at the magnetic field lines in Figure 4.21. The field lines loop through the cores, so interjecting an external surface perpendicular to them interrupts the field. If one were to calculate the cell constant through the traditional calibration exercise by placing the probe less than an inch from the sides of the container, one would arrive with a lower value of the cell constant t han that which would have been obtained if the probe were more than two inches from the side. As the figure shows the shift in conductivity or measured cell constant can be greater than 10% and underscores the need to calibrate the sensor in the container in which measurements are being made.

Section 3.2 elaborated on how to calculate a cell constant by using a precision resistor or measuring the raw resistance in a calibration standard. A similar exercise can be done for a toroidal sensor. One loops a precision resistor connected to a single wire through the hole in the sensor. The

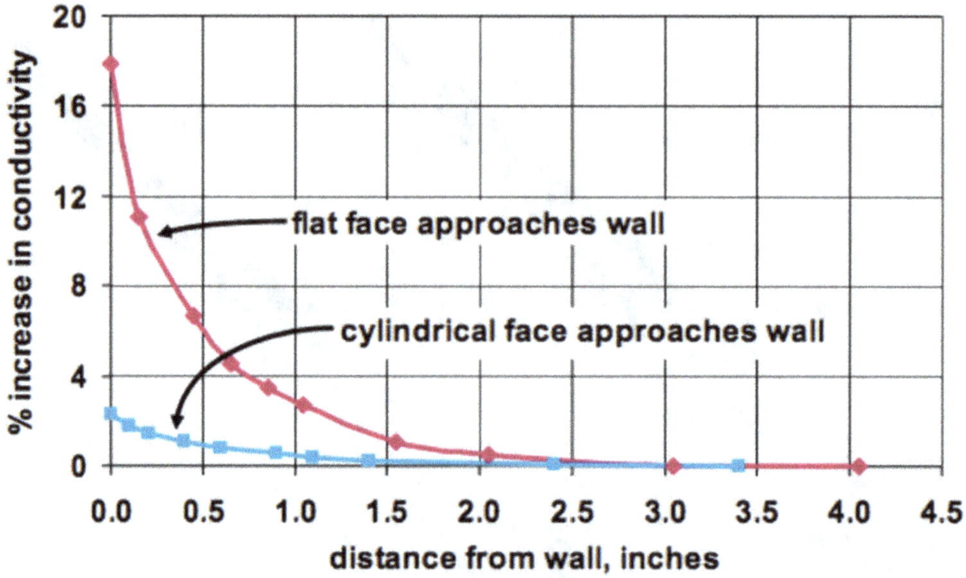

FIGURE 4.25 Change in conductivity due to orientation of toroidal probe to a metallic container wall. The maximum change occurs when the toroidal probe is parallel to the wall (red line) (Courtesy Rosemount).

measured resistance divided by the resistance of the resistor is the cell constant. For example, suppose the resistor is 1000 Ω, and the analyzer reads 2500 Ω. The actual cell constant is 2500/1000 or 2.5.

For the ideal sensor—contacting or toroidal—zero conductivity yields zero sensor response. However, this is not true for the toroidal sensor, and a finite sense voltage is nearly always present in the absence of any conductive medium. The cause is the presence of ambient voltages in the vicinity of the sensor. The size of this voltage offset is typically between 1 and 50 mV. The output of a typical toroidal sensor is in the range of several volts. The error resulting from not accounting for the offset varies from less than 0.1% in highly conductive solutions to as much as 10% in pure water. If one uses a toroidal probe to measure conductivity values less than 100 µS/cm—a dicey proposition for most toroidal sensors—one can take a zero-point measurement by holding the probe in air and measuring the offset.

In addition to field effects, other factors affect the sensitivity and accuracy of the toroidal sensors. The factor that most affects the lower limit is the hysteresis of the magnetic core. The magnetic field circulating through the toroid induces a magnetization in the magnetic core. The resulting magnetization appears to have a memory; it persists after the magnetic field turns off. The magnitude of this "memory," or *hysteresis*, determines the detection limit of the sensor. Other factors include the coil resistance, shielding, amplifier noise, and thermal noise.

4.0 REAL WORLD APPLICATIONS

4.1 Water Purity

As we said in the beginning of this chapter, conductivity is a simple and accurate measurement of water purity. The major market for pure water is manufacturing.

The semiconductor industry requires ultrapure water—that is, water with <1 μS/cm conductivity. Any ions in the water during processing can result in electrical shorts on the nanometer sized conductive paths of integrated circuit chips. Ultrapure water that is exposed to air absorbs CO_2, forms carbonate ions, and increases the conductivity. To seal an ultrapure water sample from air, one uses a flow-through cell—a glass jacket that surrounds the sensor.

Boilers have particularly challenging needs because the high temperature of the water accelerates chemical reactions. The most important reaction is the formation of calcium carbonate, or *scaling*, in the boiler's heat exchangers.

Filtration of water is often the final step in water treatment in both industrial environments and water treatment. We classify filtration by four levels of increasing filtering efficiency:

1. **Microfiltration** removes colloidal particles and suspended particles. Microfiltration does not remove dissolved organic matter, salts, and viruses. Typical pore size ranges from 0.1 to 10 microns.
2. **Ultrafiltration** uses a high-pressure flow against a filtration medium to block large organic solutes and multivalent ions. It does allow low molecular weight organic solutes and salts to pass through. It is used in mostly dairy, food, and biomedical industries. Typical pore size is 0.01 micron.
3. **Nanofiltration** removes organic solutes and allows only monovalent (+1 or −1) salts to pass through. Typical pore size is 0.01 micron (or 10 nm).
4. **Reverse osmosis** uses high pressure to drive water through a membrane that allows only water to pass through. The pores in the membrane reject even the ions of dissolved salts. Typical pore size is 0.1 nm.

Regardless of the filtration method used, making conductivity measurements before and after filtration is standard operating procedure to ensure that the filter is operating normally. A break in the filtration medium shows up as a spike in conductivity in the filter effluent.

4.2 Pure Water

The measurement of pure water is special case. Ultrapure water has a conductivity of 0.055 μS/cm (18 MΩ-cm). Upon exposure to the atmosphere, CO_2 dissolves to form carbonic acid and increases the conductivity to 0.75 μS/cm. So, to measure conductivity for demanding applications such as semiconductor manufacturing, the process water must be sealed from the environment. To do so one uses a flow cell that seals the sensor from the outside atmosphere and minimizes the entrainment of air. Figure 4.26 shows two examples of flow cells by WTW: one of stainless steel and one of glass.

A non-platinized probe with a cell constant of 0.01 and a low driving frequency is used for very low conductivity measurements.

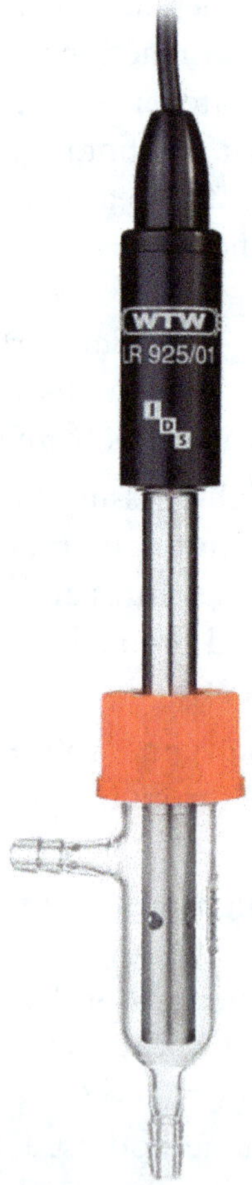

FIGURE 4.26 Flow cell for conductivity measurements of pure water. The sensor is sealed from the atmosphere so that air cannot leak into the cell and contaminate the sample with carbon dioxide (Courtesy Xylem Analytics Germany, WTW).

Many workers report conductivity measurements in units or resistivity, or MΩ-cm. Conductivity measurements can detect concentrations of impurity ions at concentrations as low as 100 parts per thousand (ppt, or 0.1 ppb). Figure 4.27 shows the resistivity for a sample salt (NaCl), acid (HCl), and base (NaOH) at concentrations between 1 ppt and 10 ppb (Bevilacqua, 1988).

4.3 Cooling Towers

To cool large-scale air conditioning units, refrigerators, chillers, and any other equipment in an industrial environment, water circulates through these units. To cool the circulating water requires a second, large reservoir of water, known as a *cooling tower*. When water evaporates, the water that remains as liquid cools. (Water molecules that are warmer than their neighbors evaporate and leave behind their cooler neighbors.)

Evaporation of water in a cooling tower leaves behind salts that increase in concentration. Some of these salts—carbonates, sulfates, and phosphates of the divalent cations—calcium and magnesium—form scaling in the cooling tower. The higher the *evaporative load* the faster the drawdown of the cooling tower water and the faster the buildup of salts. To keep the concentration of salts from precipitating out at scaling, the tower undergoes *blowdown* in which a portion of the water is discharged and replaced with fresh makeup water. The makeup water must be high enough in volume to also make up for the water that has evaporated.

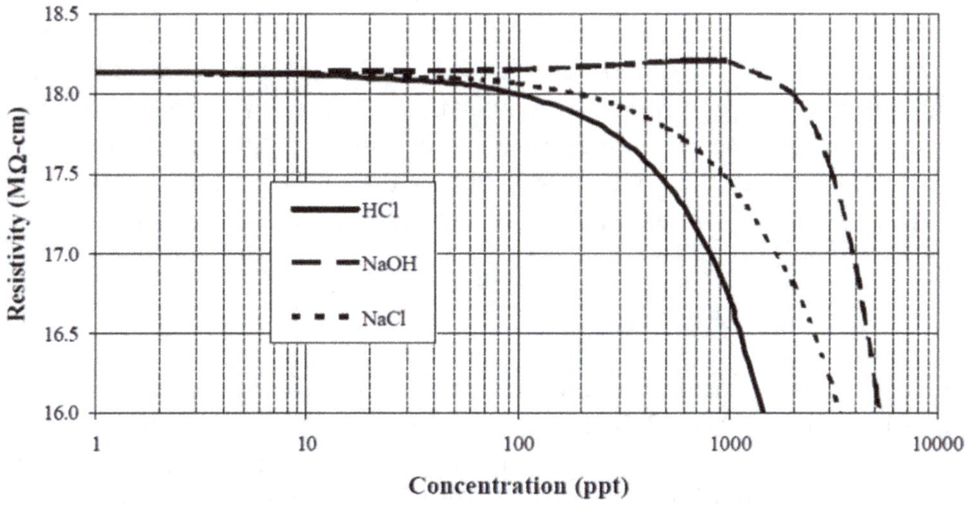

FIGURE 4.27 Resistivity for three sample ions at very low concentrations (Bevilacqua, 1988) (Reprinted with permission from the author).

To minimize the amount of water lost through blowdown the latter is down when the conductivity of the cooling tower water reaches a critical point. This set point is typically at a conductivity value between 1000 and 2000 µS/cm. Makeup water is typically between at 100 and 400 µS/cm. Water that is too low in conductivity accelerates corrosion. Makeup water is often mixed with acids to prevent scaling, corrosion inhibitors and algaecides to minimize biofouling.

4.4 Boiler Feedwater

The water that feeds boilers has far more stringent requirements than that for cooling towers. As a boiler generates steam, any dissolved solids in that water are left behind and, just as with cooling towers, build up in concentration. Scaling on the insides of a boiler reduces heat transfer efficiency and can be downright catastrophic if the scaling restricts flow in the pipes that carry steam.

The steam-powered electric industry has a particular need to measure the total concentration of ions. Corrosion at the high temperatures inside a boiler is a serious problem and is caused mainly by cations, such as Na^+, Ca_2^+, Mg^{2+}, and NH_4^+, and anions such as Cl^-, SO_4^{2-}, and HCO_3^-. (Ammonia or amines are added to raise the pH of feedwater and minimize corrosion.) Plant engineers insert a cation exchange system into the measurement loop. The cation exchange resin is charged with HCl and exchanges the cations listed above for H^+ ions. What is leftover are H^+ ions plus the anions, all of which pass through unaffected. The reason for swapping all cations for H^+ ions is that the conductivity of H^+ ions is many times that of the alkaline cations. It thus amplifies the conductivity signal of these trace ions. The resulting measurement is termed *cation conductivity*, which is very misleading because the conductivity measurement is that of H^+ ions plus anions. A better term is cation-exchange conductivity. The conductivity analyzer uses a special temperature correction for HCl dissolved in pure water to calculate the conductivity of this relatively pure water. Workers strive to keep the conductivity of boiler feedwater for steam generation no greater than 0.25 µS/cm.

4.5 Chemical Concentration

A pH probe can measure pH values between 0 and 14 but, at values less than 1 or greater than 12, interferences that were discussed in the last chapter can affect the accuracy of the measurement. Furthermore, strong acids can have pH values lower than 0 and strong bases can have values greater than 14. Add to these two constraints the third factor of shorter probe lifetimes at either end of the pH scale and you can see that a pH probe is not a long-term solution for measuring strong acids and bases. A conductivity sensor is a much better solution because it can be constructed from materials that resist a strong acid or base. In particular, graphite electrodes are chemically inert against either.

The conductivity probe can come in very handy in these environments. As Figure 4.6 shows, the conductivity of acids and bases vary nonlinearly with concentration. Over a very limited range one can approximate the relationship between concentration and conductivity with a constant but, for accuracy better than a few percent, this approximation breaks down.

Measuring an acid or base's concentration by its conductivity is just one example of converting a conductivity measurement to a TDS one. The units of TDS, mg/L, are proportional to another frequently used unit of acid/base concentration—percent by weight—that is, grams of the acid or base per gram of water. Since 1 liter of water has a mass of 1000 g it's easy to show that one simply divides the TDS value (mg/L) by 10 000 to get the percent by weight. For example, a solution whose TDS concentration is 1000 mg/L is 0.1% by weight. Specifying the concentration of an acid or base by its molarity is also common. To go from the molar value to the percent-by-weight value to molar values multiply the former by the molecular mass of the chemical and divide by 10. For example, HCl has a molecular mass of 36 g/mole so a 1 M HCl solution is a 3.6% (by-weight) solution.

The various units one may choose to express concentrations differ from each other by simple constants. They don't, however, change the linear relationship of the concentration to conductivity relationship. Section 2.6 stated that the constant of proportionality between conductivity units (µS/cm) and TDS units (mg/L) was typically between 0.5 and 1.0—that is, 1 µS/cm is equivalent to 0.5 to 1 mg/L. To convert conductivity values of an acid or base, one needs to construct a nonlinear relationship. The simplest way to do so is to add a quadratic term to the linear one:

$$C = a_0\sigma + a_1\sigma^2 \qquad (4.28)$$

Equation 4.28 is non-linear but it is still monotonically increasing with concentraton. As Figure 4.6 shows, the conductivity actually rolls over and decreases with increasing concentration. Figure 4.28 shows such a relations for phosphoric acid (H_2PO_4 on quantitative terms).

The rollover response of a conductivity with concentration means that, if one were to measure, say, 150 mS/cm, and had no prior knowledge of the H_3PO_4 solution, Figure 4.28 shows that the concentration can be 22% or 60%. It would be impossible to distinguish between the two based on the conductivity measurement alone. The quadratic approximation embodied by Equation 4.28 only works in sections of the concentration versus

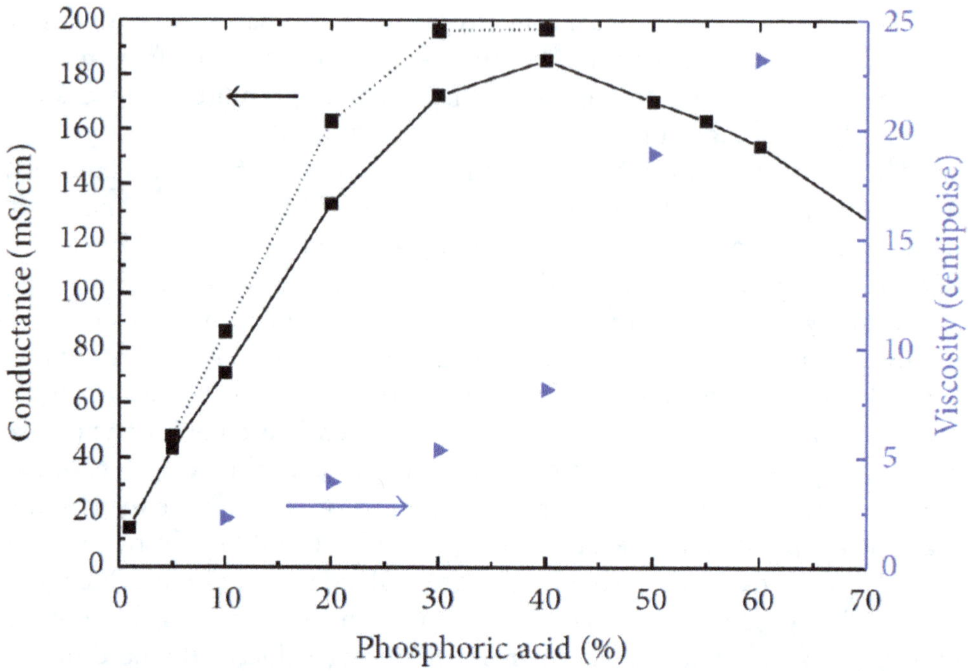

FIGURE 4.28 Concentration versus conductance for phosphoric acid. This is a more detailed depiction of the roll-over behavior of conductivity in acids and bases first shown in Figure 4.6 (Machorro et al., 2013).

conductivity curve, where the curve is increasing. Most analyzers that perform this function constrain themselves to fitting just the rising part of the curve in Figure 4.28 to Equation 4.28. One then needs to ensure beforehand that the concentration of the acid or base in question is within the range defined by the rising part of the conductivity curve.

For guidance selecting a probe with the correct cell constant for an acid or base, Table 4.5 lists conductivity values for concentrations of common acids and bases that are within the rising portion of the conductivity versus concentration range.

One application well suited to conductivity measurements is the use of clean-in-place (CIP) equipment in the food and beverage industry. Caustic, or a strong oxidizing agent, is typically stored at high concentrations, diluted as needed and dispensed throughout the plumbing to be cleaned. A conductivity controller ensures that the concentration of NaOH is held constant throughout the cleaning process. When the line is flushed after cleaning with clean water the conductivity analyzer also ensures that all caustic has been removed.

TABLE 4.5 Conductivity values for common acids and bases (mS/cm). Temperature is 25 °C, unless otherwise noted.

Acid	1%	5%	10%	15%	20%	25%	30%
HCl	112	430	700	820			
H_2SO_4 (25 °C)	50	221	440	595	721		
HNO_3	63	281	500	672			
HF	12	59	121	190	265	325	305
H_3PO_4	10	32	64	92	123		
CH_3COOH	0.64	1.43	1.70	1.82			
Base	**1%**	**5%**	**10%**	**15%**	**20%**	**25%**	**30%**
NaOH	60	220	365	410			
NaOH (50 °C)	76	320	530	650			
NaOH (50 °C)	100	495	850	1110	1310		
KOH	38.5	178	356	495	525		

4.6 Conductometric Titration

As with pH, a conductivity measurement is an aggregate one that cannot distinguish one ion from another. Nonetheless Section 4.5 shows how one can use a conductivity analyzer to determine the concentration of a dissolved salt or acid whose chemical composition is known but whose concentration is not. There are three types of chemical species for which a conductivity-based (or *conductometric*) titration is applicable: strong acids, weak acids, and salts.

For a strong acid, say, HCl, one adds a strong base, say NaOH, of known concentration one drop at a time and measures the conductivity with each drop. Recall that a strong acid or base dissociates completely into its constituent ions. As the NaOH is added, the Na^+ and OH^- ions tie up the H^+ and Cl^- ions from the HCl to make NaCl and water.

$$H^+ + Cl^- + Na^+ + OH^- \rightarrow NaCl + H_2O \quad (4.29)$$

The conductivity drops until the amount of NaOH added, in moles, is exactly equal to the amount of HCl. As one adds more NaOH, there is no more HCl with which to combine, and the conductivity rises. The bottom of the titration curve—the *end point*—gives the concentration of HCl. An idealized plot of the process juxtaposed with an actual titration is shown in Figure 4.29.

For a weak acid, say, acetic acid (CH_3COOH), that does not dissociate completely, one adds a weak base, say, ammonium hydroxide (NH_4OH):

$$\begin{aligned} CH_3COOH + NH_4OH &\leftrightarrow CH_3COO^- + H_2O + NH_4^+ \\ &\leftrightarrow NH_4CH_3COO + H_2O \end{aligned} \tag{4.30}$$

Equation 4.30 is broken into two reaction steps, containing a double arrow to make clear the incomplete dissociation of both acid and the base. The conductivity of the acetic acid solution prior to adding the NH_4OH is much lower than that of HCl because of the incomplete dissociation. As one adds NH_4OH, the total concentration of NH_4^+ and acetate (CH_3COO^-) increases thus increasing the conductivity. A fraction of the ammonium and acetate will combine to make ammonium acetate. When the added NH_4OH concentration equals that of the CH_3COOH there is no more acid to neutralize and the surplus NH_4OH concentration increases. As it does the conductivity increases more slowly. The end point is at the point at which the slope of the conductivity curve levels off. Figure 4.30 illustrates this type of titration. The midpoint of the inflection as the rising curve plateaus is the equivalence point and the correspoinding acid concentration As an

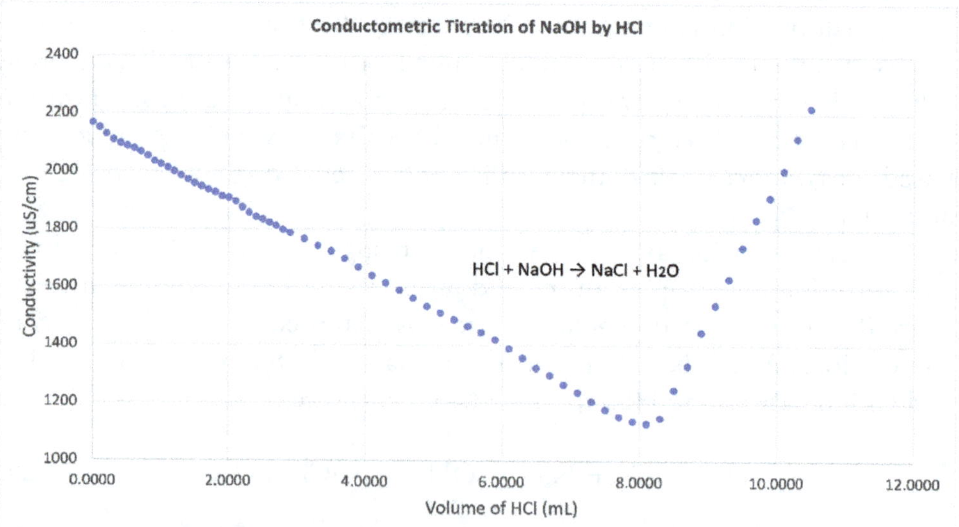

FIGURE 4.29 Conductometric titration of an unknown concentration of NaOH by a solution of known concentration of HCl, a strong base. At the minimum conductivity, the molar concentration of HCl equals the molar concentration of NaOH (Water Analytics).

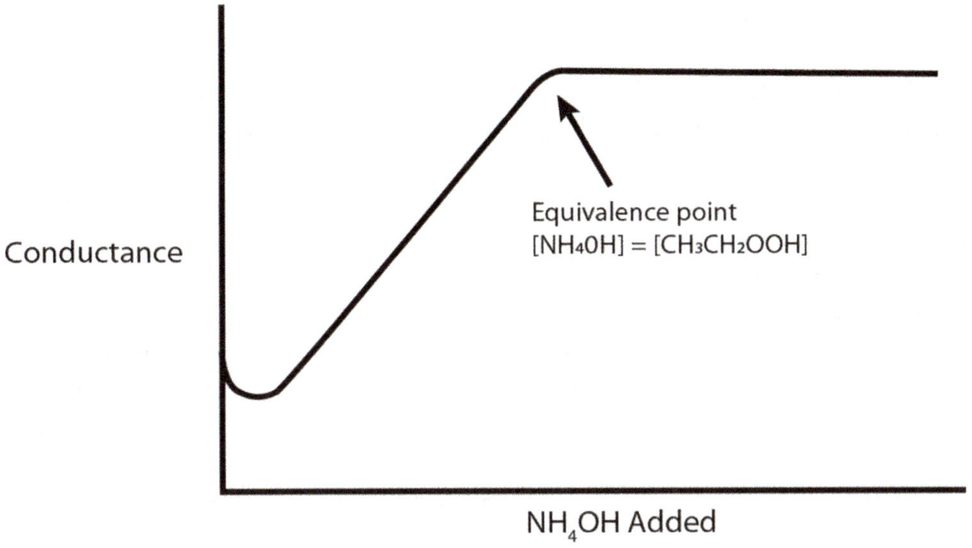

FIGURE 4.30 Conductometric titration of a weak acid, acetic acid (CH_3CH_2COOH), by a weak base, ammonium hydroxide (NH_4OH).

alternative one can add a strong base. The difference is that the leveling off will become a sharp increase.

The same titration procedure can be used to determine the concentration of a salt. Take the example of silver nitrate, $AgNO_3$. One can add a second salt, say, NaCl, that will combine to form a salt with low solubility—in this case AgCl:

$$Ag^+ + NO_3^- + Na^+ + Cl^- \rightarrow AgCl\ (s) + Na^+ + NO_3^- \quad (4.31)$$

Because most of the AgCl precipitates out, the conductivity drops just as it did with a strong acid. After the end point the added salt increases the conductivity. The plot of conductivity versus added salt looks just like the one for a strong acid.

4.7 Leak Detection

One unique use of a conductivity cell is as an indicator for the presence of water either in an area that might flood or intrusion of water inside a container that contains a nonaqueous liquid. In this case a simple conductivity sensor without temperature compensation serves as a switch. A typical example is a tank of solution (e.g., caustic) that is surrounded by a concentric low barrier to contain any spill. A conductivity sensor placed at floor level will conduct if any solution escapes the tank and fills the enclosed pool. It then closes a relay and/or sends an alarm so that remediation can take place.

For this application the sensor is nothing more than two parallel electrodes connected to a low voltage device. This type of conductivity sensor works simply as an on/off switch and is called a *liquid level controller*. Two parallel rods are connected to a direct current power supply hanging down from the top of the tank. One typically reaches the bottom of the tank, and the other can be adjusted to a level that must be maintained. The power source also powers one or two relays. As long as both electrodes are immersed, the circuit is complete, and the relay remains in its normally open or closed state. When the liquid level drops below the adjustable electrode, the circuit opens, and the relay changes its state to activate a pump that fills the tank to the set level. To maintain a dead band, the liquid level controller has a third electrode connected that hangs down to the upper end of the dead band.

5.0 REFERENCES

American Public Health Association, American Water Works Association, & Water Environment Federation. (2017). *Standard methods for the examination of water and wastewater* (23rd ed.). American Public Health Association.

ASTM. (2005). *Standard test methods for electrical conductivity and resistivity of water*, D 1125–1195.

Bevilacqua, A. C. (1998). Ultrapure water—The standard for resistivity measurements of ultrapure water. *1998 Semiconductor Pure Water and Chemicals Conference.*

Fondriest. (2021). https://www.fondriest.com/environmental-measurements/parameters/water-quality/conductivity-salinity-tds/

Hermans, T., Nguyen, F., Robert, T., & Revil, A. (2014). Geophysical methods for monitoring temperature changes in shallow low enthalpy geothermal systems. Energies, 2014(7), 5083–5118.

Light, T. S., Licht, S., Bevilacqua, A. C., & Morah, K. R. (2005). The fundamental conductivity and resisitivity of water,. *Electrochemical and Solid-State Letters*, *8(*1), E16–E19.

Machorro, J. J., Olvera, J. C., Larios, A., Hernandez-Hernandez, H. M., Alcantara-Garduno, M. E. & Orozco, G. (2013). Electrodialysis of phosphates in industrial-grade phosphoric acid. *ISRE Electrochemistry, 2013*(865727), 1–12.

Panaggio, A. (2006). *The development of a conductivity sensing network based on wireless transmission* [Master thesis]. GIM—Instituto de Telecomunicacion.

Parra, L., Sendra, S., Lloret, J. & Bosch I. (2015). Development of a conductivity sensor for monitoring groundwater resources to optimize water management in smart city environments. *Sensors*, *15*, 20990–21015.

Rosemount Analytical. (2010). *Calibrating toroidal conductivity sensors*. Application Data Sheet ADS 43-025.

Sauerheber, R., & Heinz, B. (2015). Temperature effects on conductivity of seawater and physiologic saline, mechanism and significance. *Chemical Sciences Journal,* 6(4), 1–5.

5

Dissolved Oxygen

Everything discussed in the book so far has dealt with bulk properties of water. The oxidation-reduction potential (ORP) of a solution is the sum of the individual redox values of all species in the water. Likewise, the conductivity is the sum of the individual conductivities of all ionic species. The pH of a solution represents the concentration of H^+ ions, but most still think of pH as being a bulk property because this parameter has such a wide-ranging effect on the course of chemical reactions.

This chapter and the next address measured parameters that refer to a specific chemical species. Both are gaseous molecules at room temperature that consist of two atoms of the same element. One—dissolved oxygen—is a key parameter in wastewater treatment. The other—chlorine—is the key component of disinfection, though its chemical makeup, when dissolved in water, is very different from that in its gaseous form.

Both oxygen and chlorine share a common property that is far more important than their formulation as diatomic molecules—they are both strong oxidizing agents. The principal chemistry of dissolved oxygen in wastewater treatment is as an oxidizing agent of ammonia and nitrite—albeit inside the cell walls of bacteria. The chemistry of free chlorine is like an oxidizing agent that penetrates the cell walls of bacteria and oxidizes their protective cell membrane. It also oxidizes metals to form solids that precipitate out of solution. Because both chemicals are oxidizing agents, both can be measured indirectly by ORP sensors.

These next two chapters describe sensors that measure the two oxidants directly. This chapter focuses on oxygen. Unlike Cl_2, O_2 remains intact as it dissolves in water. Like chlorine, it is the reduction of oxygen on a metal anode that results in a current whose magnitude is proportional to its concentration. This chapter also delves into a measurement technology that is not electrochemical in nature—this one involving light—to describe a second technique for measuring dissolved oxygen that has become the prevalent mode of detection.

1.0 THREE PARAMETERS THAT AFFECT DISSOLVED OXYGEN CONCENTRATIONS

Before starting a discussion on instrumentation for dissolved oxygen it's important to establish an understanding of parameters in the environment that can affect dissolved oxygen concentrations. There are three: atmospheric pressure, temperature, and salinity. A dissolved oxygen analyzer responds to the partial pressure of oxygen in solution. To get from this often-misunderstood parameter to the actual dissolved oxygen concentration in parts per million (ppm) requires the analyzer to know how these three parameters work. Some users prefer to see dissolved oxygen values expressed as percent-saturation, but it's important to understand that 100% saturation means different absolute concentrations under different conditions. Dissolved oxygen units are a source of confusion, so our first business is to set the record straight.

1.1 From Oxygen in the Air to Oxygen in the Water

Let's start with the basics—*Dalton's law*. We live at the bottom of an ocean of air and that ocean has a mass and, therefore, a pressure at Earth's surface due to gravity. The total pressure of the atmosphere can be expressed in several units. The one unit that is arguably the easiest to remember is the atmosphere, or *atm*. Another unit that is almost as easy to understand is the *bar*. One bar, or 1000 *mbar*, equals one atm. Other units are Pascals, torr, psi (pounds per square inch), inches, and centimeters (the latter two refer to the level of mercury in a column).

The total pressure measured on the ground is the sum of many *partial pressures*, in which each partial pressure is the pressure for one constituent gas. The concept of partial pressure is a consequence of Dalton's law that states that, for a gas "i," in a mixture with other gases:

$$p_i = f_i \, p_{total} \tag{5.1}$$

In Equation 5.1, p_i is the partial pressure of gaseous species i, f_i is the molar fraction (i.e., the fraction of molecules) of the gas, and p_{total} is the total pressure of the atmosphere. The law dates to 1801 to English scientist John Dalton and is just a mathematical formulation of the statement that the total is equal to the sum of the parts. In the case of gases, the total pressure is equal to the sum of pressures of each constituent gas. Earth's atmosphere is approximately 78% nitrogen and 21% oxygen (and 1% everything else). Therefore, the partial pressure of nitrogen is approximately 0.78 atm, and oxygen's partial pressure is approximately 0.21 atm.

Dalton's law relates to gases. To determine how much gas dissolves in a liquid, one must invoke another law based on another English scientist, William Henry, who formulated the following law in 1803:

$$p_i = K_{H,i,}C_i \;;\; C_i = \frac{p_i}{K_{H,i}} = \frac{f_i \, p_{total}}{K_{H,i}} \tag{5.2}$$

Henry's law simply states that the partial pressure of a gas and its concentration as a species dissolved in liquid are proportional to each other. In the second equality of the above equation, Equation 5.1 is used to replace p_i with the total atmospheric pressure. *Henry's law* simply states that the partial pressure of the gas i, p_i, in the atmosphere above the liquid is proportional to the concentration of the dissolved gas in a liquid, C_i (moles/liter). The two terms are related through a proportionality constant called *Henry's constant*, K_H. Because we want to know the inverse—the concentration of the dissolved gas as a function of its partial pressure, we simply flip the equation around as shown in the right side of Equation 5.2. One just needs to remember that the constant of proportionality becomes $1/K_{H,i}$. In fact, some formulations of Henry's law are written in this order so that Henry's constant is the inverse of the one defined in Equation 5.2. To distinguish between the two cases, pay close attention to the units of the constant. For the formulation of Equation 5.2, the units are pressure/concentration—most often atm/M (1 M = 1 mol/L), or L-atm/mol. Other units can be atm-m^3/kg or atm-L/mg or any other combination of volume and pressure units. For the inverted definition, the most used units are mol/L-atm or M/atm.

Every gas has a unique value for Henry's constant. For oxygen, the constant is 769.23 atm/M or 1.28×10^{-3} mol/L-atm, at 25 °C. The dependence of K_i on temperature has very important consequences. As water heats up, the concentration of dissolved oxygen decreases. With this all-important constant, one can calculate how much oxygen dissolves in pure water at 25 °C under 1 atm of air pressure.

$$X_{D.O.} = 0.209 \times 1 \text{ atm} \times 1.25 \times 10^{-3} \text{ mol L}^{-1}\text{atm} \times 32.0 \text{ g mol}^{-1} = 8.6 \text{ mg/L} \tag{5.3}$$

The 32 g/mol term in Equation 5.3 is the molecular weight of O_2. Remember that units of mg/L are identical to units of ppm. It is common to see values as low as 8.2 mg/L in the literature. The discrepancy is due to assumption of 0% water vapor in Equation 5.3. For most applications in a typical moist temperate environment, 8.3 mg/L is the most used value. More on humidity contributions to dissolved oxygen is in Section 1.5.

One way to look at a gas like oxygen dissolving in a liquid like water is to view the gas as exerting partial pressure in the atmosphere as well as in the liquid medium. Both are in equilibrium with each other. In other words, at equilibrium, the partial pressure of the oxygen dissolved in water equals the partial pressure of oxygen in atmosphere. If the partial pressure of O_2 in solution is greater than that in the air, then some O_2 will leave the solution until balance is restored. If the partial pressure of O_2 in the air is greater than that in solution more O_2 will dissolve in the water until the partial pressures are equal.

If you were able to see individual molecules, you would see O_2 molecules flying around the atmosphere, and some would strike the surface of the water and dissolve. If you could see below the surface, you would also see O_2 molecules zipping around water molecules, and some would strike the surface of the water from below. Those molecules would exit the water and become part of the air above it. Pressure is force per area, and the partial pressure of a gas like oxygen is the result of molecules striking the surface and exerting force against any solid surface they encounter. Because the partial pressure of the oxygen in solution must equal its partial pressure in air, the former is constant as long as atmospheric pressure above it is.

The ramification of this seemingly simple statement is quite startling. Suppose there are two aqueous solutions—one consisting of pure water and the other, a salt solution—both saturated with oxygen. The salt solution will have a lower concentration of dissolved oxygen than the pure water, but both will have the same partial pressure of dissolved oxygen. It is very important to understand this distinction because dissolved oxygen sensors respond to partial pressure—not concentration. Both give the same sensor reading.

The approximate value of 8.3 mg/L of dissolved oxygen in water-saturated air is a number that every person who measures dissolved oxygen should know. A measurement of dissolved oxygen in pure water at 25 °C and sea level—that is, 1 atm—ought to yield 8.3 mg/L. If it does, one says that the water sample is 100% saturated with oxygen. If the actual dissolved oxygen concentration is, say, 7 mg/L, then the sample is at 84% saturation. It is common for dissolved oxygen concentrations that are less than 100% saturation. As this section has hinted, and the following sections show, 100% saturation is a relative measure of concentration. It shows how much oxygen has dissolved compared to how much can be dissolved.

Under different conditions of atmospheric pressure, temperature, and salinity, 100% saturation means different absolute values of dissolved oxygen as expressed in mg/L or ppm units. For this reason, many operators prefer the absolute units of mg/L. However, as will be discussed later, a dissolved oxygen cannot measure absolute concentration. It can only calculate it based on the probe reading, atmospheric pressure, temperature, humidity, and salinity.

1.2 Variation of Dissolved Oxygen Concentration With Air Pressure and Sea Level

Equation 5.3, which combines Henry's and Dalton's laws, states that the dissolved oxygen concentration in water is proportional to atmospheric pressure. There are two ways in which air pressure can change: by going higher in altitude or when a weather front moves through an area and the *barometric pressure* changes. Changes in barometric pressure are slight deviations from the average 1 atm.

At sea level the average atmospheric pressure is 1 atm. I say "average" because variations in air pressure exist everywhere and give rise to high-pressure areas, associated with fair weather, and low-pressure areas, associated with precipitation. Barometric pressure changes are within 3% (hurricanes notwithstanding). Air pressure also changes with altitude, as shown in Figure 5.1.

The pressure at Denver, the "Mile High City," averages 0.841 atm, which equates to 7.2 mg/L dissolved oxygen at 100% saturation. It's clear that changes in altitude have a much greater influence on atmospheric pressure than changes in the weather. It's important to understand that "100%" means different concentrations at different ambient pressures. For a dissolved oxygen analyzer to report the percent-saturation of a sample, it must know the ambient pressure.

1.3 Variation of Dissolved Oxygen Concentration With Temperature

The last section already mentioned that the amount of gas dissolved in solution changes with temperature. We all experience firsthand how cold water holds more dissolved gas than warm water. That's why a warm can of soda "explodes" when you open it, and why bubbles appear in water as it heats up—before it boils. That oxygen shows decreasing solubility with increasing temperature stands in opposition to the behavior or solids and

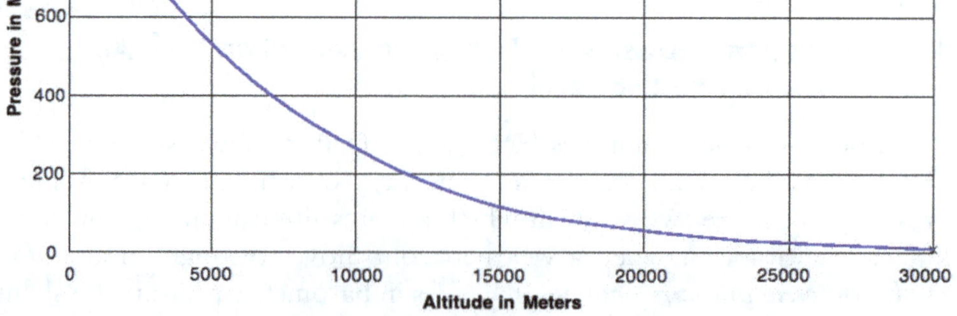

FIGURE 5.1 Air pressure as a function of altitude (height above sea level). Dissolved oxygen concentrations vary in the same proportion.

most liquids that dissolve to a greater degree with increasing temperature. This paradoxical behavior is a manifestation of the change in free energy in a solution in which a substance dissolves. When a solid, such as a salt, dissolves, energy is removed from the liquid in the form of heat to break up its chemical bonds as it mixes. When water, or any liquid, heats up, it gains free energy to a degree proportional to the rise in temperature. Hence, a rising temperature allows water to dissolve more solids. When a gas like oxygen dissolves, it releases energy—that is, the dissolution step is an exothermic reaction. In accordance with LeChatelier's principle, adding more heat drives the dissolution reaction to the left back toward gaseous oxygen.

$$O_2\ (gas) \rightarrow O_2\ (aq.) + heat \qquad (5.4)$$

The dependence of the concentration of a dissolved gas on temperature is embodied in Henry's law constant. This behavior makes a mockery of the term *constant*, so Henry's constant is more accurately called *Henry's coefficient*.

The Van't Hoff equation is a thermodynamic expression that relates the change in Henry's law constant and temperature. For the case of an ideal gas, it gives the simplified relationship:

$$K_H(T) = K_H(298\ K)\ \exp\ [C_H\ \frac{1}{T} - \frac{1}{T_0}\ ] \qquad (5.5)$$

In Equation 5.5 the temperature is expressed as in Kelvin, which is the value in degrees Celsius (°C) plus 273.15. K_H(298 K) is Henry's coefficient at 298 K, or 25 °C. This second constant, C_H, is, to a good approximation, a constant and has a value of 1700 K for dissolved oxygen. The blue line in Figure 5.2 shows the concentration of dissolved oxygen as a function of temperature and the exponential behavior one expects from Equation 5.5.

An ideal gas is one in which its bulk properties are completely independent of its concentration. The more dilute the gas, the more ideally it behaves. The deviation of behavior as gas pressure increases is analogous to the Nernst equation where activity of an ion equals its concentration only at low concentrations.

Gases that consist of molecules with more than one atom behave ideally only at very low concentrations. For the real world of dissolved oxygen, a more accurate formulation is required. Figure 5.2 compares the dissolved oxygen concentration using the approximate Equation 5.5 with the actual concentration determined by measurement. The difference is as much as 1 mg/L for temperatures between 5 °C and 20 °C.

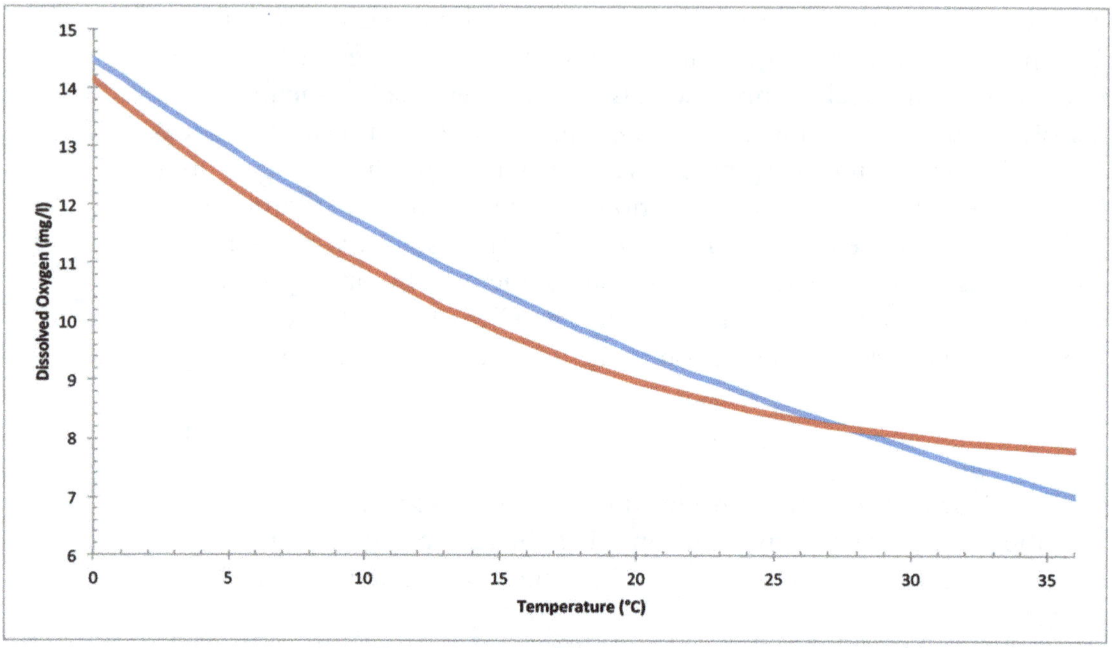

FIGURE 5.2 Dissolved oxygen concentration as a function of temperature for two models. The red line shows the relationship given by the Van't Hoff equation. This equation assumes that oxygen is an ideal gas. The blue line is the actual concentration.

All dissolved oxygen analyzers perform temperature compensation. They can use look-up tables that store concentration values for discrete temperatures, or they can use a mathematical relationship such as the one in the next section.

A membrane-based dissolved oxygen sensor must account for a second temperature effect on the dissolved oxygen measurement. As the temperature increases, the permeability of the membrane increases, and more oxygen diffuses across the membrane. The result is an increased reading that the analyzer must correct. Section 3.4 addresses this correction.

1.4 Variation of Dissolved Oxygen With Dissolved Salts

One final physical parameter that affects the concentration of a dissolved gas is the concentration of dissolved salt. As a salt dissolves it dissociates into cations and anions. Both ions form strong bonds with water molecules and create ions encased in cages of H_2O molecules. The formation of these so-called solvated ions results in less space between water molecules in which dissolved gases can fit. Seawater holds about 20% less oxygen than fresh water. Table 5.1 is a compilation of dissolved oxygen concentrations over a range of salinity and temperature values.

There have been several empirical and theoretical relationships that correlate the concentration of dissolved oxygen with the concentration of dissolved salt (X_{salt}). The simplest is the simple *Sechenov equation*:

$$X_i = X_{i,o}\, e^{-K_{sol}\, C_{salt}} \tag{5.6}$$

In this equation $X_{i,0}$ is the concentration of the gas (O_2 in this case) in pure water, C_{salt} is the molar fraction of the dissolved gas, and K_{sol} is the Sechenov constant.

Equation 5.5 is a simple theoretical expression for the temperature dependence of dissolved oxygen concentration, Equation 5.6 a simple expression for correcting dissolved oxygen concentration for salinity. There have been empirical equations that correct for both temperature and salinity. One of the most commonly used is (Benson & Krause, 1980).

$$\ln C = -139.34 + \left(1.5757 \times \frac{10^5}{T}\right) - \left(6.6432 \times \frac{10^7}{T^2}\right) + \left(1.2438 \times \frac{10^{10}}{T^3}\right) - \left(8.6219 \times \frac{10^{11}}{T^4}\right) - S\left(0.017674 - \frac{10.754}{T} + 2.1407 \times \frac{103}{T^2}\right) \tag{5.7}$$

C is in units of mg/L, T is in degrees Kelvin, and S is in units of parts per thousand (ppt). (1 ppt = 1000 ppm.)

Figure 5.3 shows values, calculated using Equation 5.7, of dissolved oxygen concentrations for a range of temperatures and two total dissolved solids (TDS) values: 0 and 35 ppt (35 000 ppm). The latter is the average TDS value of ocean water.

Some analyzers include a conductivity sensor to measure salinity (or TDS). Recall from Chapter 4, Section 2.6 that one can only derive an accurate value of salinity from a conductivity measurement if one knows the identities and concentrations of the salts that make up the solution. However, approximations of the salinity are sufficient for dissolved oxygen concentrations that are accurate to within 1% in most water or wastewater environments. Analyzers that don't include a conductivity sensor for automatic salinity concentration usually offer the option of manually entering the salinity value.

1.5 Variation of Dissolved Oxygen With Relative Humidity

The effect of water vapor pressure in the air is frequently overlooked. If one starts with an atmosphere of 1 atm, a temperature of 25 °C, and zero water vapor and then adds water vapor to achieve a relative humidity of 100%, then the water vapor adds 0.031 atm to the 1 atm. Dalton's law

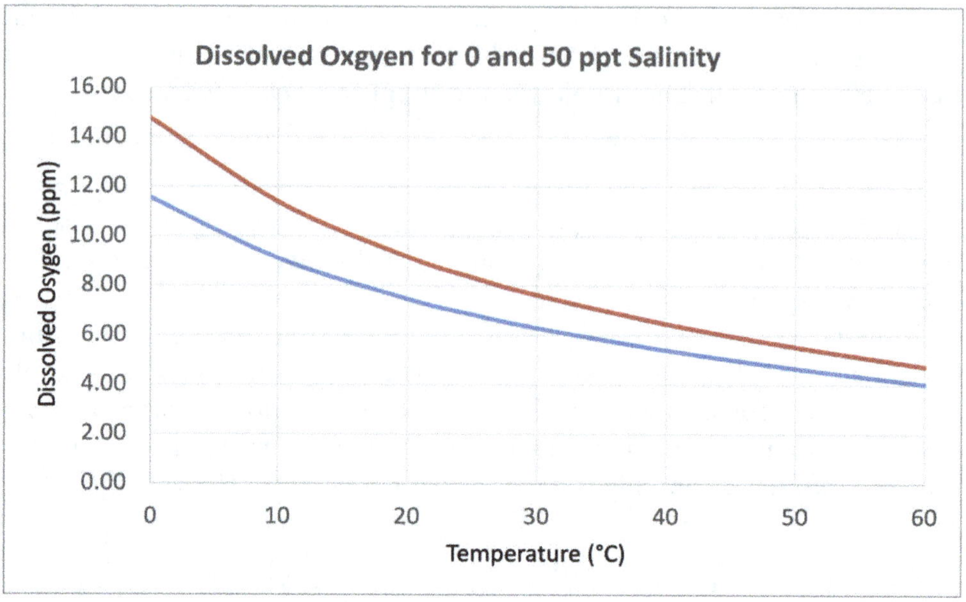

FIGURE 5.3. Dissolved oxygen concentration as a function of temperature for a TDS of zero (red squares) and 35 ppt (35 000 ppm) (blue circles). The latter is the salinity of ocean water.

implies that gases that make up a gas mixture (i.e., the atmosphere) behave independently of each other. This means that the partial pressure of dissolved oxygen in solution does not change with the addition of humidity. However, because the total pressure is now 1.031 atm, the fraction of air that is now oxygen is lower than the fraction in dry air. The increase in p offsets the decrease of f in Dalton's law, Equation 5.1 so that the partial pressure of O_2 remains constant, as it must.

Conversely suppose one takes a measurement of dissolved oxygen in an atmosphere saturated with water vapor and a total pressure of 1 atm. The corollary of the above argument states that the fraction of O_2 in the atmosphere, f, is less than the common value of 20.95% that is commonly associated with the atmosphere. In fact, the 20.95% value is now 20.95/1.031 = 20.32%. The concentration of dissolved oxygen as predicted by Henry's law also decreases by 1/1.031.

The following empirical relationship can be used to calculate the dissolved oxygen concentration as a function of total atmospheric pressure and temperature (APHA et al., 2017).

$$\text{For } 0\ ^\circ\text{C} < T < 30\ ^\circ\text{C};\ C_{DO}\ (\text{mg/L}) = 0.678\frac{(p_{total} - p_{water\ vapor})}{35 + T} \quad (5.8)$$

$$\text{For } 30\ ^\circ\text{C} < T < 50\ ^\circ\text{C};\ C_{DO}\ (\text{mg/L}) = 0.827\frac{(p_{total} - p_{water\ vapor})}{49 + T}$$

Figure 5.4 shows the dissolved oxygen concentration in water for the two limiting cases of 0% and 100% relative humidity.

It's important, when looking at this graph, to understand that the dissolved oxygen concentration doesn't really decrease as humidity in the air above increases. It means that if there are two air samples—one with 0% and the other with 100% relative humidity—and both have a total pressure of 1 atm, that the former will have more oxygen in the air and therefore more dissolved oxygen in solution. Whereas temperature and salinity physically affect the concentration of dissolved oxygen in water the effect of atmospheric water vapor is really an arithmetic artifact.

This discrepancy of approximately 3% at 25 °C is the reason why some sources state that the dissolved oxygen concentration in water at 25 °C is 8.25 mg/L and some state 8.56 mg/L. It is also the reason why manuals that describe calibration of a dissolved oxygen sensor in air should be done in 100% saturated air. It could just as easily be done in dry air but maintaining a 0% relative humidity atmosphere is much more difficult than maintaining a 100% one. Implicit in this prescription for calibration is that the analyzer's effective Henry's constant is that for water saturated air.

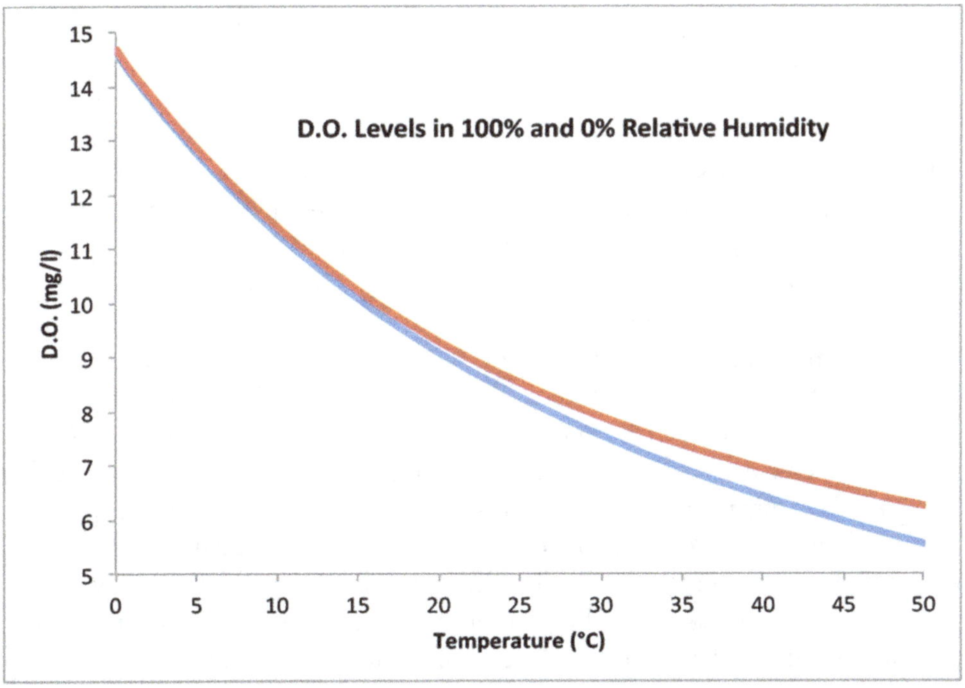

FIGURE 5.4 Dissolved oxygen concentrations at 100% and 0% relative humidity (R.H.). (Red line = 0% R.H., Blue line = 100% R.H.) The atmospheric pressure is 1 atm for both cases. The real reason for the divergence is not physical but the lower concentration of oxygen in water saturated air at 1 atm versus dry air at 1 atm.

1.6 A Dissolved Oxygen Sensor Measures Partial Pressure, Not Concentration

Arguably the most confusing fact underlying how a dissolved oxygen sensor—electrochemical or optical—works is that the sensor measures partial pressure, not concentration. This principle applies to all dissolved gases and the sensors that measure them.

Recall from Section 1.1 that the partial pressure of oxygen dissolved in water must equal the partial pressure of oxygen in the atmosphere. It seems counterintuitive that a gas dissolved in a liquid indeed has a pressure just as it does in the atmosphere. However, one can think of the aqueous environment as being fundamentally the same as the column of air above it—just denser and more viscous. Pressure, in any medium, is the result of molecules striking a surface and delivering their momentum to the surface. In a dissolved oxygen probe, the oxygen molecules strike the probe surface, and the probe output is proportional to the number of molecules striking it. That, in turn, is proportional to the partial pressure of the dissolved gas—not its concentration. A highly concentrated solution of dissolved oxygen molecules will give a zero signal if the molecules were somehow stationary and did not move.

A dissolved oxygen sensor responds to partial pressure. A solution of water at 25 °C, at sea level and at 100% saturated with oxygen would give a dissolved oxygen reading equal to the partial pressure of oxygen in the atmosphere—or 0.21 atm. However, a dissolved oxygen analyzer does not display the 0.21 atm or any other unit of pressure. We humans don't know what to make of dissolved gas readings in pressures. So, rather than reporting the output of a dissolved oxygen sensor in pressure units, the analyzer divides that pressure by the pressure it expects—0.21 atm in this simple example of sea level, zero salinity, and 25 °C. The fraction is the *percent saturation*. This parameter—not concentration—is the most direct measurement of a dissolved oxygen analyzer. The only other parameters needed are the ambient air pressure and, to a lesser degree, the relative humidity.

Yet every analyzer also reports values in absolute concentrations (mg/L or ppm) because we users demand it. As explained earlier, the connection between concentration and partial pressure is simply Henry's constant which, as previously explained, is dependent on temperature and salinity. Dissolved oxygen analyzers experience only minor variations in pressure—whether due to altitude or weather—but temperatures can be wildly different from sample to sample. Therefore, we are approximately correct in stating that a dissolved oxygen sensor's native measurement is percent saturation. Admittedly it's not the whole truth but it's not far from it. Figure 5.5 illustrates this point as well as the input of temperatue and salinity values to convert %-saturation values to concentration ones.

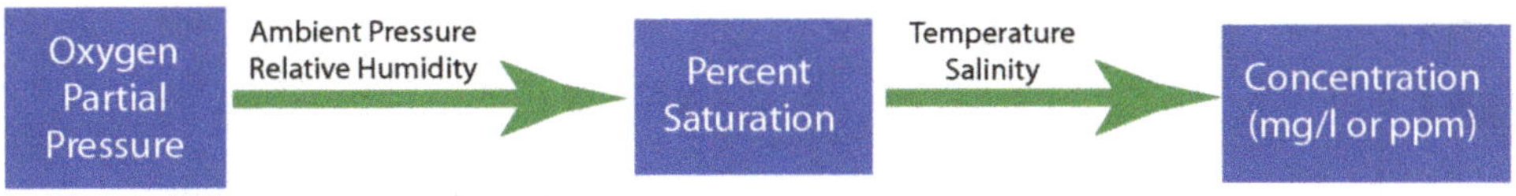

FIGURE 5.5 Converting the native measurement of partial pressure to the derived values of percent saturation and absolute concentration. While it is incorrect to say that the percent saturation is a native measurement, it is derived from the probe signal using only atmospheric pressure and humidity. The conversion to absolute concentration is a more complicated function of temperature and salinity (or TDS).

2.0 THE GRAND UNIFIED THEORY OF ELECTROCHEMICAL SENSORS

The first two sensors that this book explored—those that measure pH and ORP—are potentiometric. They produce a voltage but—ideally—no current. This chapter, and the next, focuses on amperometric sensors. An external voltage drives reactions at the anode and cathode and produces a current through a circuit that connects the two electrodes. These two types of sensors operate on different principles, but they are just two modes of operation under one framework.

The human mind seems to be wired to find simple, underlying principles that explain a wide collection of facts. Albert Einstein went to his grave feverishly working on a model that would explain gravity and electromagnetism with one master theory. He called it the Grand Unified Theory, and more than 65 years after his death, physicists are still working on a solution.

Fortunately, there is a "unified" model that explains all electrochemical sensors—potentiometric, amperometric, and voltametric. Credit for this explanation of this model goes to Professor Jiri Janata, who's textbook on chemical sensors (Janata, 2009) has been a staple in graduate courses for decades. The material in this section is typically found only in electrochemistry textbooks, but understanding it yields a stronger grasp of the workings of amperometric and potentiometric sensors. Nonetheless you may skip to Section 3.0 with a minor loss of understanding.

Potentiometric sensors generate a voltage at zero current. Amperometric sensors work at a fixed potential and produce a current proportional to the species whose concentration we want to measure. These two modes of operation seem diametrically opposed but can be represented by one graphical interpretation. Figure 5.6 shows this representation. Note that positive voltage is on the left part of the x-axis, and negative voltage is on the right.

Figure 5.6 depicts a hypothetical electrochemical cell, or a "thought experiment." In this cell, which exists only in the mind, one of the two

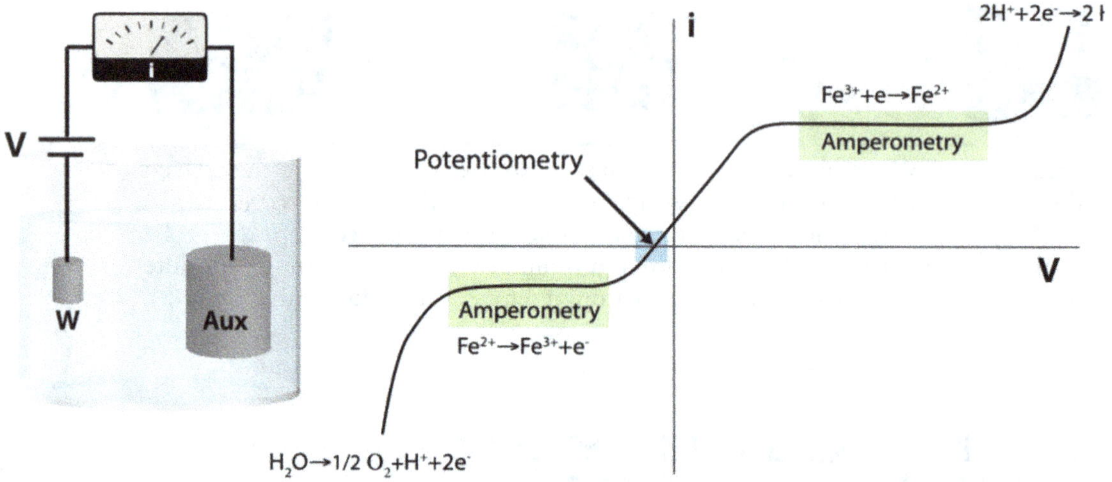

FIGURE 5.6 Current—Voltage (*i-V*) curve for the simple redox couple of Fe^{+2} and Fe^{+3}. The point at which the current is zero defines the domain of potentiometry. The region where the current is relatively flat defines the domain of amperometry (Redrawn from Janata, 2009).

electrodes is a large *auxiliary* electrode (Aux) that supplies any current desired for the *working* electrode (*W*). It is at the working electrode where a half-cell reaction takes place. The figure, called an *i-V* curve, shows the current from a reduction or oxidation reaction at the working electrode as a function of the voltage supplied by the auxiliary electrode. The meter, *M*, measures the current. The polarity of the auxiliary electrode can switch to make the working electrode a cathode for reduction reactions or an anode for oxidation reactions. Reduction takes place on the right side of the graph where the potential at the working electrode is negative. The flow of electrons into the cathode where reduction takes place is called the *cathodic current*. Oxidation takes place on the left side of the plot where the potential at the working electrode is positive. The flow of electrons out of the anode where oxidation takes place is called the *anodic current*.

For purposes of illustration, we consider the very simple redox couple: ferrous ion, Fe^{2+}, and ferric ion, Fe^{3+}. The two half-reactions that convert one to the other are:

$$Fe^{+2} - Fe^{+3} + e^- \tag{5.9}$$

$$Fe^{+3} + e^- - Fe^{+2} \tag{5.10}$$

This pair of oxidation and reduction reactions involve the exchange of one electron between the two. The current changes as the voltage changes, which demonstrates the ease with which the Fe^{3+}/Fe^{2+} redox couple exchange a single electron. The facile exchange yields a large change in current at the working electrode—into the electrode for the reduction of Fe^{3+} and out of the electrode for the oxidation of Fe^{2+}.

If we were to plot the *i-V* curve for a two-electron exchange, say, the oxidation of water into O_2 and H^+ and the reduction of H^+ to H_2, then the curve would look very different. The slope of the curve would be nearly flat in the middle of the plot ($i = 0$) and change rapidly at the ends to merge with the steep slopes shown in the graph. The exchange of two electrons and rearrangement of atoms in the water reactions is slow and manifests itself in a small current exchange at the electrode. Only when a critical voltage is reached does the reaction "take off" and the curve's slope drastically increase.

Let's take a closer look at the reactions between ferrous and ferric ions. The *i-V* curve shows two distinct types of behavior as exemplified by the two very different slopes. When the applied potential is close to zero, the change in current with voltage is linear—that is, a straight line. For instance, as the voltage becomes increasingly negative the rate of reduction of Fe^{+3} increases and, hence, the cathodic current does too. But at a certain potential the current levels off, indicating that the reaction rate does not change upon further increase in the magnitude of the negative potential. The reason for this is simple: Let's take the case of the reduction of Fe^{+3} though the same argument holds for the oxidation of Fe^{2+}. At a certain voltage Fe^{+2} ions are reduced faster than they can be the replenished. In other words, the reaction of ferrous ions at the electrodes results in the concentration of these ions being zero on the surface of the electrode. Only *mass transport*—diffusion, convection, or flow—from the bulk of the liquid to the electrode can replenish the depleted ions. When reaction at the electrode is so fast that mass transport cannot keep up with it, the electrode is said to be operating in a *mass-transport limited* regime. This is the world in which amperometric probes reside. Increasing the magnitude of the voltage won't increase the current and the *i-V* curve plateaus. This enables us to make a probe that outputs a steady current over a range of voltages. Increasing the voltage past a certain point will eventually cause the current to sharply increase but that is due to new reactions coming "online." In this case the new reactions are the reduction of H^+ at the cathode or the electrolysis of water at the anode. The range of potentials over which the current is stable defines the operating range of the amperometric sensor.

Now look at the point on the *i-V* curve where the current is zero. This is the point at which Fe^{+2} and Fe^{+3} ions are at equilibrium with each other. The oxidation reaction rate of Fe^{+2} to Fe^{+3} occurs at the same rate as the reduction rate of Fe^{+3} to Fe^{+2}. Consequently, the anodic current and the cathodic current are equal in magnitude but opposite in direction so that they cancel each other out. This zero crossing of the *x*-axis is the point at which a potentiometric sensor operates. For the redox reaction in which both species are at the same concentration that voltage is at 0.77 V—the standard redox potential is given in Table 2.1. This voltage is also referred to as the *equilibrium potential* (E_{eq}) for the electrode and redox species concentrations. When the species are at the stoichiometric ratio (equal in this case) the equilibrium potential equals the standard potential. If the concentrations of species were not equal one can calculate the equilibrium potential using the Nernst equation. It is important to restate the point made in Chapter 2, Section 4: the zero, net current in a potentiometric sensor does not mean that no redox reactions are taking place. It means that the anodic and cathodic currents cancel each other and result in an *exchange current* of zero.

This exercise of looking at the operation of a potentiometric sensor from a kinetic viewpoint is very different from the thermodynamic viewpoint that led to the Nernst equation in the first place. Thermodynamics treats a chemical species as one homogenous substance that has bulk properties while kinetics treats a chemical species as a collection of individual atoms, molecules or ions. Ultimately, they must give the same result. The following exercise shows that they do.

From the kinetic theory of electrodes comes the equation for the *net current*—that is, the difference between the cathodic and anodic currents:

$$i = nFk_0\,[C_0\,(0)e^{-[\alpha nF/RT](E\,-\,E0)} - C_R\,(0)e^{-[(1\,-\,\alpha)nF/RT](E\,-\,E0)} \quad (5.11)$$

This equation consists of two parts: the cathodic current is the first term in brackets and the opposing anodic reaction is the second term. The magnitude of the cathodic current is proportional to the oxidized partner (C_0) and the anodic current is proportional to the reduced partner (C_R). The (0) after the concentration terms denotes the concentration of each species at the surface of the electrode. There is a potential barrier that can make one half-reaction proceed faster than the other. It's reflected in the *symmetry coefficient*, α. When α is ½ both reactions are equal. When α > ½ the reduction reaction proceeds faster, and vice versa. The standard redox potential is E^0 and the applied potential is E. The reaction rate constant, k_0, is equal for both reactions; the asymmetry factor makes them different. The parameters *n*, *F*, *R*, and *T* have the same meaning as they do in the Nernst equation: *n* is the number of electrons exchanged, *F* is Faraday's constant, *R* is the gas constant, and *T* is temperature in Kelvin.

For the case in which the oxidized and reduced species are approximately equal (or at their stoichiometric ratio in the more general case) and Equation 5.11 reduces to:

$$i \approx nFk_0C_0[e^{-[\alpha nF/RT](E\,-\,E0)} - e^{-[(1\,-\,\alpha)nF/RT](E\,-\,E0)}] \quad (5.12)$$

Because the two concentrations are approximately equal C_0 is the concentration of either species. It's important to realize that this approximation works only when the potential of the electrode is close to the standard state potential of the redox reaction. As with Equation 5.12, the first term in the brackets is the cathodic current, and the second term is the anodic current. If the symmetry coefficient is ½—that is, the energy barriers to the oxidation and reduction reactions are equal—then two currents are equal, and the net current is zero.

The sought-after consequence of this kinetic model of the electrode occurs when the applied potential (E) equals the equilibrium potential (E_{eq}). The latter is the potential at which the net current equals zero, which is the point at which the i-V curve crosses the x-axis. When this condition occurs, one can set the cathodic (first) term of Equation 5.12 equal to the anodic (second) term. A little simplification leads this equality to:

$$E = E^0 + \frac{RT}{nF} \ln \frac{C_0}{C_R} \tag{5.13}$$

Equation 5.13 is, of course, is the Nernst equation. It is very satisfying that two different approaches to the potentiometric sensor—thermodynamics and kinetics—both arrive at the exact same conclusion.

However, the Nernst equation cannot explain the magnitude of the current that flows into or out of an electrode during an oxidation or reduction reaction. Only the kinetic theory introduced in this section can do that. The current magnitude becomes a critical factor in an amperometric sensor like the dissolved oxygen probe, as the next section sensor makes clear.

Figure 5.6 illustrates the distinction between a potentiometric sensor and an amperometric sensor. A potentiometric sensor works at the applied potential at which there is no net change of oxidizing or reducing reactants—at the $i = 0$, or x-axis, of the i-V curve. An amperometric sensor works at an applied potential at which consumption of the analyte of interested occurs faster than the mass transport (usually diffusion) can replace the reacting species. Both modes of operation allow one to measure the concentration species of the analyte. The potentiometric sensor does so through the Nernst equation. The amperometric sensor does so as a function of the current.

Discussion has been centered on the imaginary electrochemical cell depicted in Figure 5.6. It's important to bridge the discussion to a real probe. The fictitious auxiliary electrode that supplies the desired voltage is close to the real thing in an amperometric probe. For the Clark dissolved oxygen sensor, whose discussion follows, the auxiliary electrode—the silver anode—relies on an external potential to drive the oxidation of Ag to Ag^+ to supply the supply of electrons needed by the cathode. At the cathode, O_2 molecules are reduced to OH^- ions.

In its potentiometric mode, the auxiliary electrode in the imaginary cell adjusts the potential so that current at the working electrode is zero. In the actual potentiometric probes of this book—pH and ORP—the auxiliary electrode is the Ag-AgCl reference electrode. Inserting a high-impedance voltmeter ensures that a net current of zero flows between the electrodes.

Figure 5.6 contains one very useful piece of information that merits a brief discussion. The graph is a plot of current between the electrodes and the voltage applied to the working electrode. The slope of the *i*-*V* plot is *i*/*V*, which, according to Ohm's law, has the units of an inverse resistance (1/R = *i*/*V*). This means that one can associate with the working electrode an effective resistance. Because the current in via ions in solution and electrons at the electrodes, this effective resistance is termed *charge transfer resistance*, R_{CT}.

If we were to plot the *i*-*V* characteristics of a complex reaction, like the dissociation of water as mentioned above, the resulting curve would be nearly flat up to both ends of the plot. The working electrode can be said to have very high resistance. Water molecules and ions adsorb onto the surface and create a charge layer. This phenomenon is equivalent to the layer of ions that form on the electrodes of a conductivity sensor as discussed in Chapter 4, Section 3.1. The charge layers form a double layer capacitor—hence the very small slope and the correspondingly high impedance. Just as in any capacitor, as soon as a threshold voltage is reached, the capacitor breaks down and current increases drastically with additional voltage. Another way of describing this mechanism is to invoke the concept of an energy barrier introduced in Chapter 1, Section 6.0. The free energy change to make a redox reaction go may be negative (downhill), but the reaction still may not go unless the voltage is high enough to overcome the energy barrier, as depicted in Figure 1.4.

But that's not what happens with the Fe^{+3}/Fe^{+2} redox couple. Around the origin of the graph the *i*-*V* curve has a steep slope. Current changes very strongly with voltage. Ferrous and ferric ions encounter very little resistance toward releasing or picking up electrons. The redox reactions, Equations 5.9 and 5, occur at low resistance. The electrical engineer recognizes that this working electrode has a capacitor in parallel with a low-resistance resistor, and the resistor shorts out the capacitor.

Modeling sensors as electrical circuits with a resistance and capacitance gives us greater insight into the workings of potentiometric and amperometric sensors.

As discussed above, potentiometric sensors operate at zero current—along the *x*-axis of the figure. To be exact: at the voltage generated by the probe, the anodic current corresponding to oxidation is equal in magnitude but opposite in direction to the cathodic current corresponding to reduction. The magnitude of the current varies according to the reaction type and can be modeled by an electrical circuit containing a resistor and capacitor in parallel. A simple electron exchange reaction, like the iron redox reactions of Equations 5.9 and 5.10 occur at a relatively high current because the equivalent resistor in this reaction has low resistance and shorts out the capacitor. A complex reaction involving rearrangement of atoms and more than one electron, like the reactions of reduction of oxygen drives the dissolved oxygen sensor. Its *i-V* curve is characterized by a resistor of much higher resistance, so the capacitor (formed by the double layer of ions) plays a dominant role. Its curve is much flatter around the origin. This further explains the observation, first made in Chapter 2, Section 8.3, that an ORP probe responds in seconds to a calibration solution of iron complexes but responds in tens of minutes to tap water containing molecular redox species like oxygen and HOCl.

Amperometric sensors operate at a voltage higher than the equilibrium potential—the point at which the anodic and cathodic currents are equal and opposite. The potential applied to the cathode to drive reduction just past the equilibrium potential predicted by the Nernst equation is called the *overpotential*. The overpotential takes into account the charge transfer resistance and the resistance of the electrolyte. The transfer resistance depends strongly on the electrode material and the condition of the electrode surface (Amine & Mohammadi, 2019). The additional potential needed to overcome the resistive elements in the full cell goes into creating heat. This is why batteries warm up as they discharge and do work.

In their range of operation, the rate of reaction, as manifested in the cathodic and anodic currents, is much faster than the mass transport (mostly by diffusion). This is true even for very slow reactions like the reduction of oxygen. Amperometric probes with membranes are mass transport limited.

3.0 THE DISSOLVED OXYGEN AMPEROMETRIC PROBE

The previous section presented a framework that explains the theory behind potentiometry and amperometry. The key differentiator between the two is the voltage driving the electrodes. Now we are in a better position to understand why one or the other is used to carry out an electrochemical measurement.

The ORP sensor represents a good jumping off point for transitioning from a potentiometric sensor to an amperometric one. This sensor is an indiscriminate one—it measures the redox potential of everything in the water. The potential generated at the platinum electrode equals the sum of all the individual redox reaction potentials present in solution. It's positive if the bulk of the sum of the redox potentials of all redox active species (weighted by their concentration) add up to create a net oxidation potential and negative if the bulk of the reactions add to create a net reduction potential.

The reference electrode of the ORP sensor supplies the half-reaction that is complementary to the half-reaction at the process electrode. Recall the reaction at the ORP reference electrode for saturated KCl:

$$AgCl(\text{sat'd KCl}) + e^- \leftrightarrow Ag^+ + Cl^-; E = 0.197\ \text{V} \qquad (5.14)$$

If the overall redox reaction at the measurement electrode is an oxidation reaction, then Equation 5.14 occurs just as written—that is, from left to right. It consumes the electrons generated by the reduction. If the overall reaction is a reduction, then Equation 5.15 goes in reverse—that is, from right to left.

The placement of the reference electrode in its own separate compartment filled with a buffer solution means that the only way the probe can have a complete electrical circuit is to include the sample solution in its pathway. Otherwise, the two electrodes would constitute an open circuit. The reference solution or salt bridge supplies the ions from its KCl contents needed to complete the circuit.

The design of the ORP, or pH, sensor is ideally suited for a potentiometric sensor that requires the current be as close to zero as possible. It does that simply by inserting a very high impedance voltmeter in the circuit. But it won't work for an amperometric sensor that is driven by a voltage and produces a measurable current. The only practical design that generates a stable current is to place both electrodes in one cell. Recall that the imaginary electrochemical cell described in the previous section did consist of its two electrodes immersed in one container. The operator configured it for potentiometry by adjusting the potential of the auxiliary electrode to zero out the current on the working electrode. For amperometric operation the operator simply dialed up the potential on the auxiliary electrode to produce a current that was stable over a range of potentials. If the potential on the working electrode is positive—that is, reducing—the latter becomes the cathode, and the auxiliary electrode becomes the anode, and vice versa. This is an ideal arrangement of electrodes for an amperometric probe.

In an amperometric sensor one electrode is always the anode, at which anions or neutral molecules give up electrons to the circuit, and the other is always the cathode, at which cations or molecules take up electrons. The two electrodes are immersed in an electrolyte solution that acts as a reservoir for redox reactants and products. The applied voltage drives the reactions that produce ions at both electrodes—cations at the anode and anions at the cathode. The oppositely charged ions balance each other to maintain the electrical neutrality of the electrolyte. Electrons flow from the anode to the cathode via a conductor and an ammeter inserted into the connecting wire to measure the current. (Remember that current flow is opposite to electron flow. We can thank Benjamin Franklin for that.)

The current that flows from the cathode to the anode is proportional to the sum of the oxidation reaction at the anode and reduction reaction at the cathode. If one can keep the reaction rate at the anode stable, then any changes in current are due to reactions at the cathode. This is an essential condition to all amperometric sensors. One way of satisfying this condition is to make the anode surface larger than the cathode surface so that the reduction of the desired species is the limiting reaction.

The amperometric probe looks like the galvanic cell pictured in Figure 2.4 except that there is no salt bridge or membrane separating the electrodes. The potential generated by the oxidation reaction at the anode adds to the potential generated by the reduction reaction at the cathode to create one overall voltage of the cell. If the sum of the half-cell potentials is positive, then the half-cell reactions will proceed spontaneously. This is the basis of the battery or fuel cell. It's also the basis for the Galvanic dissolved oxygen probe we discuss in Section 5.3.3. If the sum of the half-cell potentials is negative, then an external voltage must be supplied to make the half-cell reactions go. This is the basis of the electrolytic cell and the Clark dissolved oxygen probe.

3.1 The Clark Cell

Consider a cell in which oxygen molecules reduce at a gold or platinum cathode to form OH^- anions and the silver atoms in a pure silver anode oxidize to form Ag^+ cations. The half-reactions and their associated potentials come from Table 2.1. The reduction reaction is:

$$O_2 + 2\ H_2O + 4\ e^- \rightarrow 4\ OH^-\ ;\ E^0 = +0.401\ \text{V} \qquad (5.15)$$

In an acidic solution, the reduction reaction occurs with a larger free energy change:

$$O_2 + 4\ H^+ + 4\ e^- \rightarrow 2\ H_2O\ ;\ E^0 = +1.229\ \text{V} \qquad (5.16)$$

Note that the gold or platinum that comprises the cathode does not play a role in the reduction reaction. The cathode is (mostly) inert and simply provides a surface on which reduction takes place.

The reaction at the anode is the familiar reaction at the Ag-AgCl reference electrode, which is Equation 2.24 in reverse:

$$Ag + Cl^- \rightarrow Ag\ Cl + e^-\ ;\ E\ (\text{sat'd KCl}) = -0.199\ \text{V} \qquad (5.17)$$

In contrast to the gold or platinum cathode, the silver anode itself plays the leading role in the oxidation reaction. As it oxidizes it loses material.

Adding Equations 5.16 and 5.17 gives the total redox reaction:

$$4\ Ag + 4\ Cl^- + O_2 + 2\ H_2O \rightarrow 4\ AgCl + 4\ OH^- \ ; E = +0.202\ V \qquad (5.18)$$

An alternate formulation includes the dissolved, spectator K^+ cation:

$$4\ Ag + 4\ KCl + 4\ O_2 + 2\ H_2O \rightarrow 4\ AgCl + 4\ KOH \qquad (5.19)$$

The AgCl product is an insoluble solid that coats the Ag anode. The reference electrode in the potentiometric sensor is a silver wire coated with AgCl. But in the case of the potentiometric sensor the Ag of the wire electrode is in equilibrium with the Ag^+ in solution. In the absence of a net reaction there is no loss of material. In the case of the amperometric sensor the coating limits the anodic reaction and must eventually be removed, or the anode replaced.

Equation 5.18 generates a positive voltage so it should proceed spontaneously in the forward direction. But an amperometric sensor made from it will not produce a sizable current. The reason is the nature of the reduction of oxygen at the cathode.

Equation 5.15 (or 5.16) is a complex reaction involving the exchange of four electron, which contrasts with the one-electron exchange of the Fe^{2+}/Fe^{3+} redox pair. As described in the previous section, we can model this reduction electrically by assigning it a high capacitance. The capacitance is due to the formation of a double layer of charge at the surface of the electrode. Alternatively, we can model the reduction chemically by assigning it a high free energy barrier.

Furthermore, there are resistive elements in the redox circuit that decrease the current. This includes the accumulation of AgCl on the surface of the anode and the plating of silver ions to become silver atoms on the cathode. The very high charge transfer resistance plus the resistance of the electrolyte plus the resistance of the AgCl surface leads to a considerable voltage demand. The potential applied across the electrodes will have to be greater than the sum of the equilibrium potentials predicted by the steady state potentials to create a positive net current between the electrodes. This is the overpotential defined at the end of Section 2. The overpotential is the voltage the analyzer supplies to push the itAl curve onto the plateau as displayed in Figure 5.6, at which the cathodic current is limited by mass transport. This is the territory of the amperometric sensor. Figure 5.7 is a schematic of an amperometric dissolved oxygen sensor.

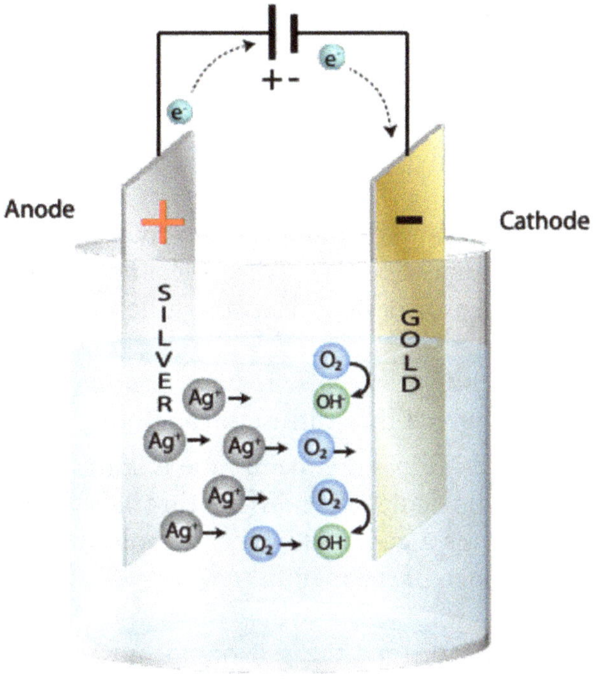

FIGURE 5.7 The principle behind the amperometric ("Clark") cell. An external potential is inserted between the anode and cathode to drive Equation 5.18 and produce a measurable current.

The electrochemical cell in which O_2 reduces to OH^- is the basis of operation for the first electrochemical dissolved oxygen sensor invented. It can be traced to 1923 (Heyrovskya, 1924), but the cathode upon which oxygen reduced was open to the sample and contaminated easily. This first generation of dissolved oxygen sensors used a mercury drop cathode. Leland Clark pioneered the oxygenation of blood in one of the first heart-lung bypass machines. He developed his sensor to address criticism that he had not measured the oxygen content of blood introduced by his machine, thereby casting doubt on the efficacy of the innovation. Clark isolated the cathode from the blood environment with a permeable polyethylene membrane.

Clark termed his probe "polarographic." Today this term strictly refers to a voltametric sensor using a mercury drop for the working electrode. This book follows this etymology and avoids use of the term *polarographic* in connection with the Clark cell. However, the term, *polarize*, is commonly accepted to denote the application of a voltage across the anode and cathode to drive the redox reactions at their surfaces.

The membrane is permeable to dissolved oxygen and protects the electrodes from contamination. Though there is no membrane that is totally selective only to O_2 several come close. The most common is a thin sheet of polyethylene or polytetrafluorethylene, aka PTFE or Teflon. To complete the list of synonyms for the amperometric dissolved oxygen sensor, the name *membrane-based* or *membrane-covered* sensor is often used to denote the critical role that the selective membrane plays. Figure 5.8 shows the schematic of a Clark cell.

The membrane does more than impart selectivity to the sensor. Without it the rate of diffusion would change over time, and the signal would vary. Let's assume that the sensor design meets the following two conditions:

1. The rate at which O_2 molecules reduce at the cathode is much faster than the rate at which they diffuse through the membrane.
2. The flow of water across the membrane is sufficient to replenish dissolved oxygen molecules that diffuse through the membrane—that is, the flow ensures a constant concentration of dissolved oxygen at the exterior surface of the membrane.

Under these two conditions the sensor output resulting from reduction of dissolved oxygen is *diffusion-limited* between the inside surface of

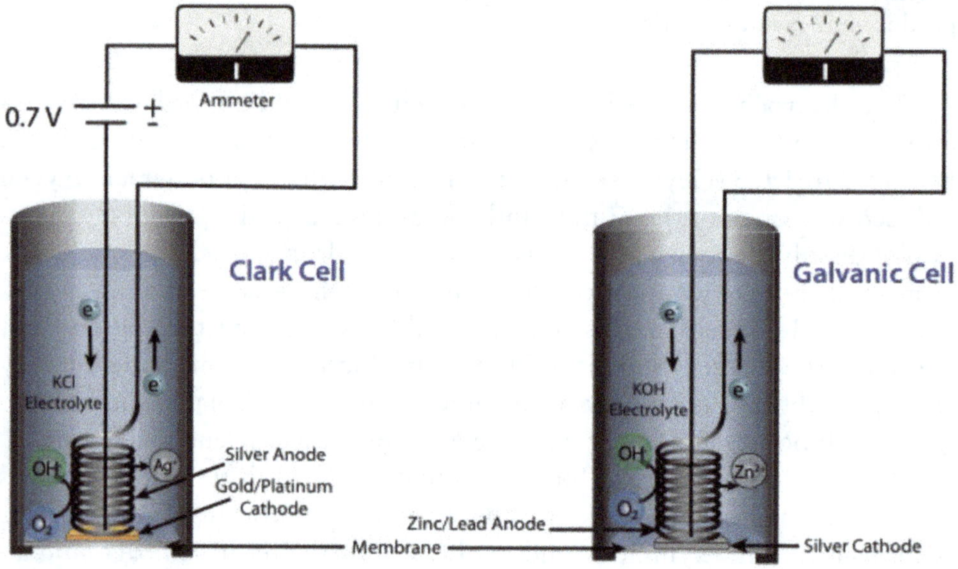

FIGURE 5.8 The left side is a schematic of an amperometric ("Clark") cell for measuring dissolved oxygen. The anode is typically an inert metal like gold, and the cathode is an inert metal—usually silver, gold, or platinum. A voltage of 0.7 to 0.8 V is applied to the anode to drive the reduction of O_2. The right side depicts a Galvanic cell. The anode is usually zinc or lead. The same redox reactions occur but without the need for an external voltage.

the membrane and the surface of the cathode. The second condition states that flow of water supplies oxygen molecules at a rate faster than diffusion inside the membrane. Our discussion of the voltametric analyzer in Chapter 8, Section 3.1 dives a little more deeply into the three mechanisms of mass transport for an amperometric setup that does not include a membrane.

With the two conditions stated above, the constant rate of diffusion produces a stable and reproducible signal that determines the current output of the sensor. Section 3.4 discusses the choice of materials and their thickness for constructing a membrane having the optimum combination of response time and signal stability. The primary decision that a Clark cell engineer must make is to dial in the voltage across the electrodes that ensures that the above two criteria are met.

The selection of the optimum voltage is an exercise that was already discussed in Section 2 and depicted in Figure 5.6. The minimum voltage is one at which the rate of reduction is faster than the rate of mass transport. The maximum is the potential at which other reduction reactions start to occur—mainly the hydrolysis of water. The optimum voltage is therefore somewhere in the middle of this range.

The left side of Figure 5.9 shows actual *i*-*V* graphs for the reduction of oxygen at the cathode and oxidation of silver at the anode (Fatt, 1982). These plots figure prominently in the discussion of voltammetry in Chapter 8, Section 3. Each graph in Figure 5.9 is plotted for one dissolved oxygen concentration. At dissolved oxygen levels above 5% saturation, there is a steep increase in current as the polarization voltage increases from 0 to 0.3 V. (To be clear, these voltages are negative because they are for reduction.) From 0.3 V to 0.7 V the current is nearly constant. This is the diffusion-limited range of operation. At potentials greater than 0.7 V the current increases very rapidly due to the hydrolysis of water:

$$2\,H_2O + 2\,e^- \rightarrow H_2 + OH^- \qquad (5.20)$$

Examination of this family of graphs shows that the optimum operating voltage of a Clark cell is any potential between 0.4 and 0.7 V. In practice most amperometric probes work between 0.6 and 0.7 V. The right side of Figure 5.9 is a plot of the current at 0.7 V for each of the oxygen concentrations. The resulting current versus percent saturation dissolved oxygen serves as a calibration curve for the sensor. The magnitude of the current is on the order of 100 nA per ppm of dissolved oxygen. The linear plot shows that zero current at zero applied voltage, but that is not strictly correct. There is always a *residual current* due to the reduction of impurities that make their way into the electrolyte. Studies of my company's Clark cell showed no measurable (<1 nA) current at zero dissolved oxygen when

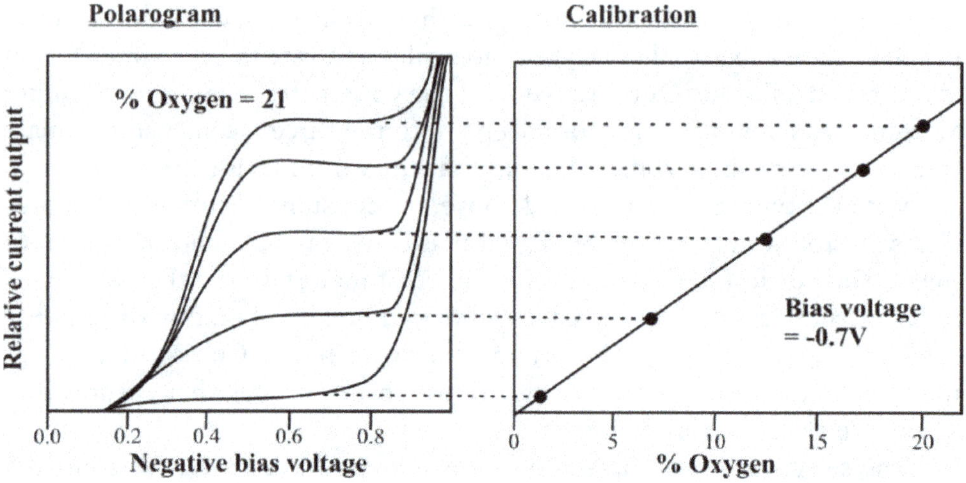

FIGURE 5.9 Clark cell current as a function of polarization voltage for different dissolved oxygen concentrations. The right-hand graph takes the values of the current at 0.7 V and plots these currents as a function of the dissolved oxygen concentration (Fatt, 1982) (Reprinted with permission from Krieger Publishing).

the electrolyte and membrane are fresh. Others have reported residual currents between 0.01% and 5% of the 100%-saturation point, though it's probable that the higher levels are due to contamination of the electrolyte. Figure 5.10 shows two typical Clark cell probes. In the exploded view of the Mettler Toledo sensor one can clearly see how much larger the anode is compared to the cathode. As discussed Section 2 this ensures that the anode supplies sufficient current to the cathode so that the reduction of oxygen at the cathodic current is proportional to the dissolved oxygen partial pressure.

3.2 Variations on the Clark Theme

There are two variations of the two-electrode amperometric probe that address two shortcomings of the conventional two-electrode amperometric sensor.

The depletion of oxygen at the cathode as it reacts creates a concentration gradient between the cathode and the membrane. For maximum signal, the gap between the two is very small, resulting in a concentration gradient that extends beyond the membrane and into the sample. To maintain the local concentration of dissolved oxygen there must be continuous flow of water across the membrane. This is the main drawback of the electrochemical dissolved oxygen sensor that the optical dissolved oxygen sensor circumvents.

The job of the anode is two-fold. The description of the electrochemical cell in Section 2 characterizes the anode as the "auxiliary" electrode to

FIGURE 5.10 Clark cell dissolved oxygen probes. Top: YSI probe: The gold cathode is in the center of the front-end. The silver anode is inside the tip behind the cathode. The membrane covers the front of the cartridge. Bottom: Eploded view of a Mettler-Toledo probe: The cathode is at the tip of the white cylinder, the anode is the concentric cylinder to the left of the white cylinder, and the membrane cap covers the front of the black cylinder (Courtesy YSI and Mettler-Toledo).

denote its function as supplying a sufficient flow of electrons to ensure that the electrode does not limit the reaction rate at the "working" electrode. Hence the large anode shown in Figure 5.10. The other job of the anode is to maintain a stable voltage between it and the cathode to ensure that the operating point of the sensor lies on the plateau of the *i-V* curve. The three-electrode design addresses this constraint by assigning two electrodes for each function. It is found more commonly in free chlorine sensors but can occasionally be found in dissolved oxygen sensors. The next chapter on free chlorine describes its design more fully. The driving force for the 3-electrode dissolved oxygen sensor is to ensure accurate dissolved oxygen measurements over a wide range of concentrations, enabled by the stable working potential of the cathode regardless of dissolved oxygen concentration.

3.3 Galvanic Sensor

The Clark electrode requires an external ("polarizing") voltage to drive the reduction of oxygen at the anode. Though the two half-reactions proceed with a positive free energy change, the energy barrier at the cathode surface and resistive losses on both electrodes, require an overpotential to drive the combined redox reaction. To construct an amperometric sensor that works without an external voltage simply requires an oxidation that is greater than approximately 0.3 V to overcome the overpotential.

Two common metals, zinc (Zn) and lead (Pb), do oxidize with a positive potential. In the presence of an electrolyte containing OH^-, the following oxidation reactions are very energetically favorable:

$$Zn \rightarrow Zn^{2+} + 2\,e^- \;;\; E^0 = +0.762 \text{ V} \tag{5.21}$$

$$Pb \rightarrow Pb^{2+} + 2\,e^- \;;\; E^0 = +0.126 \text{ V} \tag{5.22}$$

Zn^{2+}, or Pb^{2+}, react with OH^- to form hydroxide and the unstable hydroxide converts to the oxide.

$$Zn^{2+} + 2\,OH^- \rightarrow Zn(OH)_2 \;;\; Zn(OH)_2 \rightarrow PbO + H_2O \tag{5.23}$$

$$Pb^{2+} + 2\,OH^- \rightarrow Pb(OH)_2 \;;\; Pb(OH)_2 \rightarrow Zn + H_2O \tag{5.24}$$

The reduction reaction at the gold or platinum cathode is the same as that of the Clark cell:

$$O_2 + 2\,H_2O + 4\,e^- \rightarrow 4\,OH^- \;;\; E^0 = +0.40 \text{ V} \tag{5.25}$$

We combine the potential for either the Zn or Pb anode half-reaction combined with the formation of their respective oxide and the reduction of oxygen at the cathode. The result is a very energetic overall reaction:

$$2\,Zn + O_2 + 4\,OH^- \rightarrow 2\,ZnO + 4\,OH^- \;;\; V = +1.65 \text{ V} \tag{5.26}$$

$$2\,Pb + O_2 + 4\,H_2O \rightarrow 2\,PbO + 4\,OH^- \;;\; V = +0.98 \text{ V} \tag{5.27}$$

The voltage of this electrochemical cell is much larger than the 0.175 V of the Clark cell standard potential and is enough to supply the overpotential needed to make the reaction go spontaneously. This configuration is essentially a galvanic cell but without a salt bridge or membrane separating the half-cells.

In the Clark cell the Ag^+ cations react with the Cl^- anion from the KCl electrolyte to form an insoluble AgCl film that adheres to the anode. In a galvanic sensor the operative anion is OH^-. The most common electrolyte used is KOH. The K^+ and OH^- products are highly soluble and persist as dissolved ions. In the galvanic sensor, the reaction products PbO or ZnO are insoluble. They flake off from the anode and form a milky white suspension in the KOH electrolyte. Maintaining the galvanic sensor therefore requires more frequent replacement of the electrolyte. On the other hand, because the zinc or lead oxide flakes off the surface of the anode, one does not need to clean the surface as one must do with the silver anode of the Clark cell. Figure 5.8 illustrates the difference in construction between a Clark and galvanic probe.

Another major difference between the two is that the probe current of the galvanic probe at zero dissolved oxygen levels. As long as an operating voltage powers the Clark cell, there is always a residual current. Figure 5.11 shows an example of a galvanic probe made by ATI.

The galvanic sensor, like the Clark electrode, has been largely displaced in this country by the optically based probe. However, as a gas sensor for the measurement of oxygen in hazardous atmospheres or the fuel mixture in a vehicle, it is still the predominant technology (Changet al., 1993). Because of its zero residual current it also excels at measuring ppb levels of dissolved oxygen.

3.4 The Key Role of Membranes

In a membrane-based probe, Clark or galvanic, the cathode membrane is stretched tightly over the cathode, resulting in a microscopically thin layer of electrolyte between the two.

FIGURE 5.11 Galvanic dissolved oxygen sensor. This model by ATI uses a replaceable membrane and electrolyte solution (Courtesy ATI, Division of Badger Meter Company).

We can use the long-established laws of diffusion to model the diffusion of dissolved oxygen across the membrane. As shown in Figure 5.12 the three properties of a membrane that affect the rate at which oxygen diffuses through it are the surface area of the cathode, membrane permeability, and membrane thickness.

The probe current as a function of these parameters can be derived from Fick's law of diffusion:

$$I = nFJ = \frac{nFAD}{W} C(O_2) \tag{5.28}$$

In the first equality, the current density equals the product of n, the number of electrons per number of molecules of oxygen (4) reduced at the cathode, F is Faraday's constant (the charge in Coulombs of a mole of electrons), and J is the flux of oxygen molecules as they impinge on the cathode. The second equality equates the flux, J, to A, the surface area of the membrane, W, the membrane thickness, D, the membrane diffusion coefficient for dissolved oxygen (also known as the permeability coefficient) and $C(O_2)$, the concentration of dissolved oxygen. We know that the latter is proportional to the partial pressure for a given temperature and salinity. We return to this formulation in Chapter 8, Section 3.2 on voltammetry. Table 5.1 (Lighton, 2019) lists common membrane materials.

All these membranes share a common property. They are hydrophobic—that is, they repel water, other polar molecules, and ions. Oxygen is nonpolar,

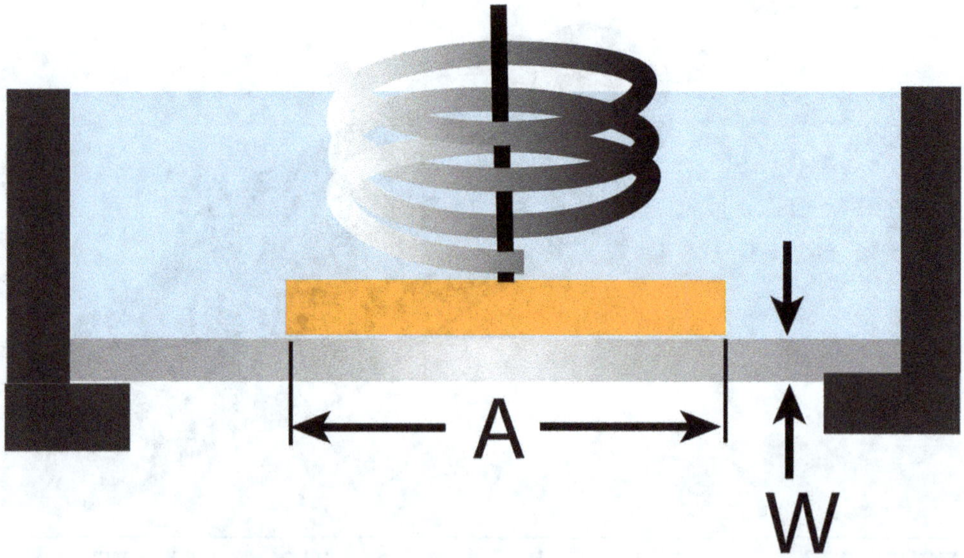

FIGURE 5.12 Schematic of a generalized electrochemical dissolved oxygen sensor showing the cathode (gold rectangle), anode (coil), and membrane.

TABLE 5.1 Common membrane materials (Lighton, 2019). The more permeable (larger diffusion coefficient) the faster the response but the more dependent on flow and the lower the maximum dissolved oxygen concentration.

Membrane Material	Diffusion Coeff ($cm^2\ s^{-1}$) $\times 10^{15}$
PTFE (Teflon)	2.54
PFA	2.69
FEP (Teflon FEP)	1.73
Tefzel	1.34
PTME	3.43

so it can easily slip through the membrane. Other constituents, including all salts, are polar or charged, so they are not able to diffuse through. The hydrophobicity of the sensor membrane, along with the voltage between the anode and cathode give the dissolved oxygen sensor its selectivity.

The Cottrell equation, also derived from the diffusion theory, quantifies how the probe current decays in time (t) if there is no flow.

$$I(t) = nFA\sqrt{\frac{D}{\pi t}}\ C(O_2) \tag{5.29}$$

From these two equations we can draw some conclusions:

1. The probe signal (or current) is proportional to the area of the cathode.
2. The signal is proportional to the square root of the diffusion coefficient (D), which is proportional to the membrane permeability and inversely proportional to the membrane thickness.
3. The signal decays as the inverse of the square root of time (t).

The second point embodies a highly advantageous characteristic of a membrane-based sensor. The use of a membrane renders the probe current limited by diffusion through the membrane. This results in a signal that is linear with the partial pressure of oxygen. If there were no membrane, then the signal would be limited by mass transport, including diffusion through the aqueous medium. This is a much faster process and would result in the probe current saturating at a low dissolved oxygen partial pressure. The signal would no longer be linear in $p(O_2)$. A thinner membrane means a lower detection level but also a lower saturation level.

The last point warrants a brief discussion. One can change the permeability by choosing a different membrane material, using Table 5.1 as a guide. But it's easier to change the thickness, as permeability is proportional to thickness. For example, if the signal decreases 90% of its initial value for a 2-mil thick membrane in 20 seconds it will decrease 90% for a 0.5-mil thick membrane in 10 seconds. The consequence is that there is a trade-off between signal stability and response. A thinner membrane gives a larger signal but is more susceptible to changes in flow. In other words, a thicker membrane requires less flow to yield a stable signal. In general, a flow of several cm/s is sufficient to replenish the diffusion of dissolved oxygen through the membrane.

One final note is that the above diffusion-based model of the probe signal is very simple and does not account for variables such as the three-dimensional geometry of the probe or the electrolyte layer thickness. In any case, changing the membrane and geometry of the probe changes the probe response but calibration takes these factors into account so that no other compensation is necessary.

3.5 Temperature and Membranes

Several physical parameters of the Clark or galvanic sensor affect the current between electrodes. Understanding how these parameters affect the output current is necessary for two reasons: to derive an accurate dissolved oxygen concentration value from the current and to guide the user into selecting the right components of the sensor.

There are two effects of temperature on the dissolved oxygen reading of a membrane-based dissolved oxygen analyzer. One is real while the other one is a measurement artifact.

As discussed in Section 1.3, the solubility of oxygen decreases as temperature increases. However, the partial pressure of dissolved oxygen remains constant. Instead, Henry's coefficient decreases in value so that the dissolved oxygen concentration also decreases. This means that two identical probes in two solutions of differing temperature will give the same output, but the probe in the colder solution will have more dissolved oxygen on an absolute basis—that is, when expressed as mg/L. The dissolved oxygen analyzer must measure the correct temperature and transform the current signal into a concentration value using a correction factor, either in the form of a look-up table or formula such as Equation 5.6.

The second effect of temperature is the change in membrane permeability with a change in temperature.

The change in membrane permeability with temperature is approximately 4% for temperatures near 25 °C. An empirical relationship between the membrane diffusion (D) and temperature is (Seutheramanet al., 2009):

$$D_{memb}\,(T) = A\ e^{-B/T} \tag{5.30}$$

Both A and B are constants that are determined by fitting experimental data to Equation 5.30. They are different for every membrane material and dissolved gas.

Experimental data gathered in our labs gave the nearly linear relationship over a temperature range of approximately 20 °C to 40 °C.

$$K_{memb}\,(T) = K^{0}_{memb}\,(T_0)(T - T_0)^{1.03} \tag{5.31}$$

There is a solution for correcting membrane permeability for temperature that is as ingenious as it is simple. That is to insert a thermistor into the electrical circuit in the analyzer between the cathode and anode. Recall that the signal being measured is a current. A thermistor is a resistor whose resistance decreases with temperature. This means that, as the temperature increases the membrane permeability increases, the cathode current increases and the resistance of the thermistor decreases. The thermistor is in parallel with the circuit between the cathode and analyzer's op-amp input. The result is a decrease in current to the op-amp that increases with temperature and a lower reading. This analog temperature compensation of the membrane temperature effect works well for temperatures close to room temperature but deviates to as much 20% or greater at higher temperatures.

With the advent of more powerful processors, modern dissolved oxygen analyzers resort to a temperature reading and an algorithm, such as Equations 5.30 or 5.31, to compensate for the changing membrane permeability. Since the probe already incorporates a temperature element (resistance temperature detector, RTD, or thermistor) to enable the analyzer to convert partial pressure to concentration, the correction for the membrane requires no additional hardware.

3.6 Clark or Galvanic?

The argument over which electrochemical sensor is superior continues to this day. As one would expect, manufacturers of one type of electrochemical sensor present persuasive arguments as to why that type is superior to the other. In an effort to be as objective as possible, this section compares performance and maintenance of each.

A Clark cell only works when the external voltage is switched on—that is, "polarized." As with any electrochemical device it takes a while for the redox reactions that take place on the anode and cathode to reach equilibrium. That time is between 10 and 20 minutes for a Clark cell. Because a galvanic sensor is "on" all the time by virtue of its "self-polarization," that warm-up time is effectively zero. This is a significant advantage for galvanic probes, which makes the preferable for portable instruments.

The current output of any amperometric sensor increases with the surface area of the cathode and the membrane permeability. Both parameters being equal, the Clark electrode typically produces a larger current—on the order of several μA's—by virtue of the applied external voltage. The galvanic sensor produces a smaller current—on the order of 1 μA—by virtue of it being self-polarizing (Wei et al., 2019).

For both sensors the electrolyte is the component that is consumed most quickly and must be replenished. In the Clark cell the KCl electrolyte provides Cl^- anions to react with produced Ag^+ cations to form AgCl. The formation of the coating on the anode adds a resistive component that grows over time and diminishes the signal. Removing the coating requires chemical polishing. A common method is to soak the silver in ammonia. Ammonia reacts with the surface AgCl as follows:

$$AgCl + 2\ NH_3 \rightarrow Ag(NH_3)_2Cl \tag{5.32}$$

The resulting complex salt does not adhere to the silver as the AgCl does and precipitates into the cleaning solution. A 2- to 3-minute dip in a 14% NH_4OH solution is a faster solution. The temptation to etch the silver with a weak acid is ill-advised, as it will only convert more silver to a salt (e.g., more AgCl with HCl). Cleaning the electrode necessitates replacing the KCl electrolyte, which is quick and cheap.

The accompanying depletion of chloride ions in the electrolyte necessitates changing it at least every 2 months.

As mentioned, the surface of the anode of a galvanic sensor, whether it is zinc or lead, does not coat with the metal oxide. The oxidation products, ZnO or PbO, flake off the anode and go into suspension in the electrolyte. The white precipitate gives a milky white appearance to the KOH electrolyte, so the built-in visual aid offers a visual aid for determining when to change the electrolyte. The highly visible suspension of lead or zinc hydroxide precipitate adds a resistive element to the galvanic circuit, thereby increasing its impedance.

The voltage across the electrodes is fixed, so an increase in this contributor to the voltage subtracts from the overpotential that drives the reduction reaction. The amperometric sensor operates in a diffusion-limited regime so a diminishing overpotential does not have a significant effect on the current until the reaction rate reaches the critical point where it is less than the diffusion rate. At that point the current output drops precipitously as shown in Figure 5.13 (Warburton et al., 2001). In other words, the effect of increasing concentration of oxidation products is the same as turning down the voltage driving the electrochemical cell.

This resistive element occurs in the electrolyte as opposed to the surface of the anode as with the Clark cell. Because polishing of the anode is not needed in the galvanic sensor, maintenance is simpler than that for the Clark cell. Add another point to the galvanic side.

The cathode for both the Clark and galvanic sensor is a noble metal that does not undergo reduction. Therefore, it is the consumption of the anode and electrolyte through oxidation that determines the lifetime of the sensor. However, the cathode will plate with silver from the Ag^+ ions in the electrolyte and need to be cleaned.

The always-on nature of the galvanic cell means that it undergoes electrochemistry continuously and so will require maintenance more often than the Clark cell. The latter only undergoes chemistry when it is powered. However, most process control applications require a continuously powered

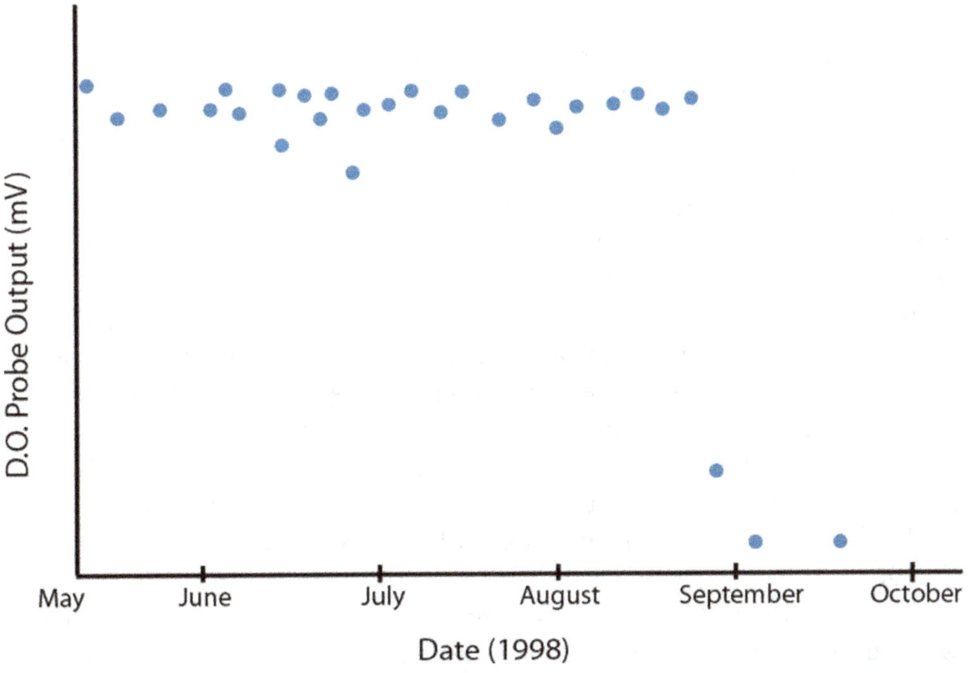

FIGURE 5.13 Drop in current in a galvanic probe as the cathode and electrolyte accumulate increasing amounts of lead oxide. The point at which the output current drops represents the point at which the rate of oxygen reduction drops below the rate of diffusion (Data from Warburton et al., 2001).

dissolved oxygen sensor so the Clark cell's lifetime advantage as well as its warm-up time disadvantage are non-issues.

The same is true for the cathode. Both types of sensors utilize a noble metal cathode—gold or platinum. The cathode does not participate in the reduction reaction. However, electroplating of the anode metal—silver, lead, or zinc, onto the cathode does occur over time—silver plating for the Clark cell and lead or zinc plating for the galvanic sensor. One can observe the plating on a gold cathode as a loss of the gold color. Again, because of the always-on operation of the galvanic sensor, it may need to be replaced sooner than the amperometric sensor. However, this advantage disappears when the sensor is in continuous operation.

The sensitivity of both types of membrane-based probes is comparable though galvanic probes have an edge. Both can achieve detection levels down to sub-ppm levels. When we talk about current at zero levels of dissolved oxygen we must differentiate between the *residual current*—the current that results when there is zero dissolved oxygen but impurities that reduce at the cathode—and the *zero current*—the background current that the external power supply generates in the absence of any reduction reactions. Both probes will clearly experience the same residual current. The galvanic probe has no external power so there is zero current background when there is zero dissolved oxygen. In contrast the Clark cell is externally powered so there is always some current. Therefore, the galvanic probe can measure lower levels of dissolved oxygen, which is particularly useful for measuring ppb levels in boilers. Experiments with our company's Clark cell in our lab show that the zero current in oxygen-free water, is on the order of nA's.

Last, there is the question of stability. The Clark cell operates at a very precisely controlled external voltage applied between the cathode and anode, whereas the voltage between the electrodes in a galvanic sensor is dependent on the redox reactions that occur spontaneously. As the Nernst equation predicts, voltage is dependent on the concentration of Zn^{2+} or Pb^{2+} ions in solution. The result is that the galvanic sensor is subject to voltage changes, which results in current changes. That point for more stable signal ought to go to the Clark cell but, as with detection limits, I have yet to find actual data to show that this advantage manifests itself in any observable manner. It's likely that real fluctuations in dissolved oxygen overrides any minute change in probe voltage.

The membrane is also common to both Clark and galvanic probes. It will foul over time and need to be replaced. Fouling is particularly problematic in wastewater in which dissolved oxygen measurements are ubiquitous. The consequence of membrane fouling is a decrease in signal or—equivalently—a lesser calibration slope. Most manufacturers provide an optional cleaning attachment that injects pressurized water or air at the membrane.

In addition to fouling, *membrane tension* diminishes over time. As the membrane stretches, its permeability changes. If a new membrane is mounted by stretching it, over time it loses its elasticity and the tension diminishes. To minimize signal changes as the membrane ages one should stretch it before mounting it. Membranes that are packaged as part of a cap that screws onto the probe end have already been stretched and show minimal changes in tension.

3.7 Interferences

As amperometric devices the Clark and galvanic sensor will respond to any chemical that can be reduced at the cathode. One major interfering compound is free chlorine, HOCl or OCl^-. As the next chapter shows, an amperometric free chlorine analyzer is very similar to a dissolved oxygen analyzer, so it only makes sense that both oxidizing agents interfere with each other. A second major interfering dissolved compound is hydrogen sulfide (H_2S), which is commonly found in an anaerobic part of a wastewater stream. As with pH and ORP sensors, sulfides react with silver to form silver sulfide (Ag_2S) and poison the anode.

In this chapter we have focused on Clark and galvanic sensors for measuring dissolved oxygen. Though they have been largely replaced (except for physiological applications) by the optically based dissolved oxygen probe we are about to explore, the preceding discussion is still a very useful explanation of the principles of amperometry. We will see the same mode of operation for the chlorine probe, which is the subject of the next chapter.

4.0 OPTICAL DISSOLVED OXYGEN SENSORS

4.1 The Weird World of Fluorescence

Up to this point this book has dealt exclusively with electrochemical sensors. The amperometric dissolved oxygen sensor has been a tried-and-true tool since 1954 when Leland Clark developed the eponymous probe. In medical and biological applications, the membrane based dissolved oxygen sensor is still the predominant oxygen probe, by virtue of the ease with which the design can me miniaturized. However, a sensing technology based on optics has become the predominant technology for measuring dissolved oxygen in the drinking water and wastewater measurement markets.

To understand how an optical dissolved oxygen sensor works one needs to understand a phenomenon that is rooted in the world of quantum mechanics—the world of objects that are roughly smaller than a micron. The predominant feature of quantum mechanics is that the energy levels of quantum systems—typically atoms, molecules, and ions—are quantized; that is, they exist in discrete levels rather than a continuum as in classical mechanics. Discrete energy levels comprise their rotational, vibrational, and electronic motions. A change from discrete energy level to another is accompanied by the absorption or emission of discrete bundles of energy—often in the form of light.

Electrons around the nucleus change their orbits by absorbing or emitting light that is typically in the visible or ultraviolet range. The discrete electronic energy levels of an atom or molecule can be portrayed in an *energy level diagram*. Figure 5.14 shows a generalized example. The absorption of light (aka a *photon*) promotes an electron from a lower level to a higher—usually from a ground state to an excited state. Every electronic energy level of a molecule or molecular ion can be further broken down into a series of vibrational energy levels that result from the molecules quantized vibrations of its atomic framework. The molecule then loses some of that energy through dropping down to a lower vibrational state of the excited energy level and then emitting a photon that, due to energy conservation

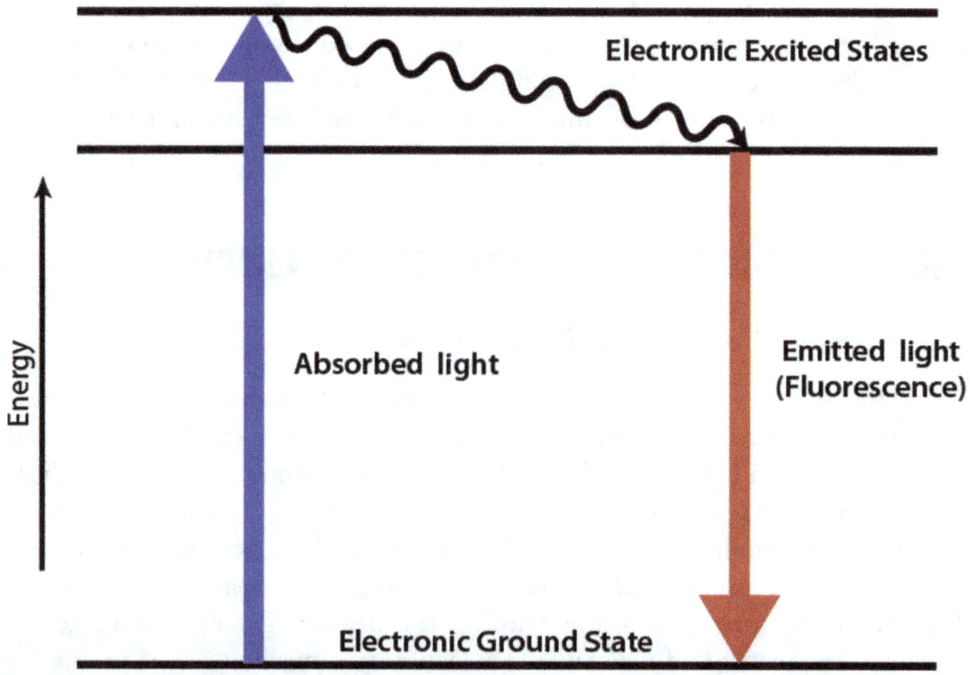

FIGURE 5.14 Generalized energy level diagram. The absorbed light is colored blue to denote that it has higher energy. The red light is at a lower frequency (longer wavelength) because it comes from a lower energy state. The multiple lines represent sublevels of the electronic state that have differing amounts of vibrational energy.

has less energy than the photon that was absorbed. Hence the blue color of the absorbed photon and red color of the emitted photon.

One of the early triumphs of quantum mechanics, due in large part to the work of Einstein and fellow German Max Planck, is that photons—discrete bundles of energy—have no mass but do have energy. The energy of a photon is given by:

$$E = \frac{hc}{\lambda} \tag{5.33}$$

Here, λ is the wavelength of light, c is the speed of light (2.998×10^8 cm/s), and h is Planck's constant (6.626×10^{-27} cm^2 g/s). Planck's constant is one of the most important numbers in all of physics as it ties together energy and the wavelength of light. The salient feature of Equation 5.33 is that the energy of a photon—and, hence, of light—varies inversely with the wavelength. Our understanding of light changed monumentally with this relationship.

The emission of photons from a higher energy level to a lower one is called *radiative transfer*. The loss of energy from a vibrational level to lower ones is called *non-radiative transfer*. Sometimes the non-radiative transfer dumps 100% of the energy absorbed with the result that the molecules heat up. Sometimes radiative transfer dumps 100% of the energy. When this happens the light that comes out of the molecules is exactly the same wavelength and energy as the light that went in. This is the phenomenon that makes the sky blue. Oxygen molecules in the atmosphere absorb sunlight toward the blue end of the spectrum and reemit the blue light in all directions.

This phenomenon of emitting a lower energy photon after absorbing a higher energy photon, as shown in Figure 5.14, is called *fluorescence*. It is one of two possible modes for a molecule or atom to shed a photon to release energy. The other mode is *phosphorescence*. In this mode, there is another excited electronic state that is at a lower energy than the initial one that is populated by the absorbed light. That second state, for reasons given by quantum mechanics, does not connect strongly to the electronic ground state. It therefore takes a lot longer—from milliseconds to hours—for that second state to emit light and jump down to the ground state. This is the process behind glowing watches, ceiling stars, and glow sticks. The glowing light persists long after one turns off the lights because of the so-called *forbidden* transition from the second excited state to the ground state. The term *luminescence* refers to both fluorescence and phosphorescence. Unfortunately, the terms *fluorescence* and *luminescence* are often used loosely when describing optical dissolved oxygen sensors.

One can think of the fluorescence signal as a marker for the excited molecule as it dumps some of its energy into vibrational motion of the initially excited electronic state and then sheds the bulk of its excess energy in the form of a photon. Some molecules emit a photon instantly before they transfer any energy while others cascade down the energy staircase giving off packets of vibrational energy along the way. The fluorescence signal represents the dynamics of the entire ensemble of molecules and, therefore, gives an average picture of those dynamics. The rise and decay of the fluorescence signal mirrors the progression of molecules as they emit photons and end up back in the ground state.

Figure 5.15 illustrates the time resolved uptake of light energy and the subsequent release of it. The blue dotted line represents the light that molecules absorb to promote them from the electronic ground state to the excited electron state. The light that promotes a molecule from the ground state into an upper electronic state is termed the *excitation* light. The red line represents the fluorescence light as it decays. Fluorescence in most atoms and molecules occurs within less than a microsecond.

Though the timescale differs, the shape of the fluorescence decay is the same in all molecules and atoms. It is a shape found in many natural phenomena and is a *decaying exponential* that follows the following relationship.

$$I(t) = I_0 e^{-t/\tau} \tag{5.34}$$

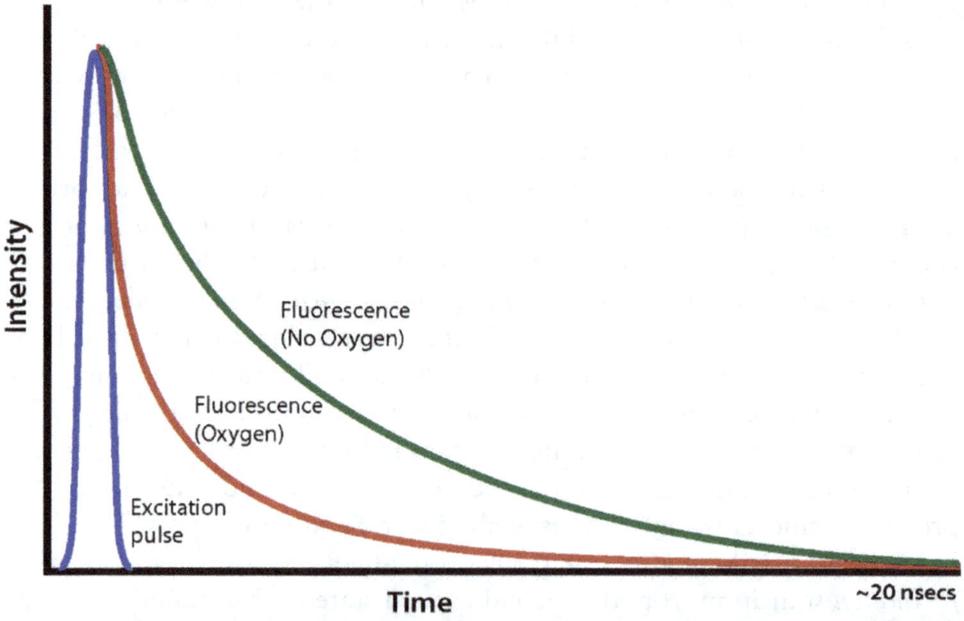

FIGURE 5.15 Schematic of fluorescence. The narrow blue pulse that excites absorbing molecules from the electronic ground state to an excited state. The green line is the fluorescence that excited molecules emit as they jump back down to the ground state. The shape of the fluorescence decay is given by Equation 5.34. The red line shows the same fluorescence process affected by another, quenching species.

$I(t)$ is the intensity of the fluorescence at any time (t), I_0 is the intensity at the start of fluorescence, and τ is the *time constant*. After an interval of one time constant the signal decays to 37% of its initial intensity. Its units are seconds.

The time constant is a quantitative measurement of how fast the fluorescence decays—the smaller the constant the faster the decay. It is key to the operation of an optical dissolved oxygen sensor. The time constant of a molecule is a fixed constant under the condition that it does not collide with other molecules that can knock it out of its excited state. This collision-free lifetime is notated as τ_0. It is most often referred to as the *fluorescence lifetime*. The green line in Figure 5.15 shows the fluorescence. When other molecules, atoms, or ions do collide with the fluorescing molecule, the latter drops from its excited state into a lower one—usually the ground state. This pathway to the ground state is non-radiative. As more and more excited molecules are knocked out of the fluorescing state two things happen:

1. Fewer fluorescing molecules means less fluorescence. This means that the fluorescent light intensity decreases.
2. The added pathway to remove energy shortens the average lifetime of the ensemble of excited molecules. That means the fluorescence decay that one measures occurs faster. We say that the fluorescence lifetime is shorter—that is, τ decreases. The phenomenon of collisions with other molecules decreasing the quantity of fluorescence is called *quenching*. The interfering chemical species quenches the fluorescence of the radiating species. The red line in Figure 5.15 shows the shorter fluorescence lifetime resulting from quenching.

The shortening of the radiative lifetime or the decrease of the fluorescence intensity follows a simple relationship:

$$\frac{1}{\tau} = \frac{1}{\tau_0} + k_Q[M] \; or \; \frac{1}{I} = \frac{1}{I_0} + k_Q[M] \qquad (5.35)$$

In Equation 5.35, k_Q is the quenching rate constant and [M] is the concentration of the quenching species, M. For the equation on the left, τ is the fluorescence lifetime in the presence of a quenching species and τ is the fluorescence lifetime in the absence of quenching. For the equation on the right, I is the fluorescence intensity in the presence of the quencher and I_0 is intensity in its absence. Either equation is a prescription for determining oxygen concentration. If one knows the values for τ and k_Q then a measurement of $1/I$ or $1/\tau$ determines [M]. A calibration exercise, outlined in the next section, allows one to determine either τ or k_Q. Both expressions in Equation 5.35 are formulations of the very widely used *Stern-Volmer equation*.

4.2 Ruthenium–The Secret Ingredient

One chemical species figures prominently in the optically based dissolved oxygen sensor. It's a very large cation that consists of a ruthenium metal atom in the center surrounded by 15 aromatic (benzene based) rings. It's chemical name is [Ru(II)-Tris(4,7-diphenyl-1,10-phenanthroline)]$^{2+}$. The double positive charge of the entire molecular complex denoted by the "2+" surrounding the brackets means that it exists as a salt. Two Cl^- anions shown in the center of the complex balance the positive charge. Figure 5.16 shows this beautifully symmetric molecular complex. There are three shorthand names: Ru-$(Ph_2phen)_3{}^{2+}$, the more compact abbreviation $Ru(ddp)_3{}^{2+}$ and the even more compact Ru(ddp).

Ru(dpp) is the active ingredient of the optical dissolved oxygen sensor because it absorbs light at 455 nm, which is blue, and fluoresces at 613 nm,

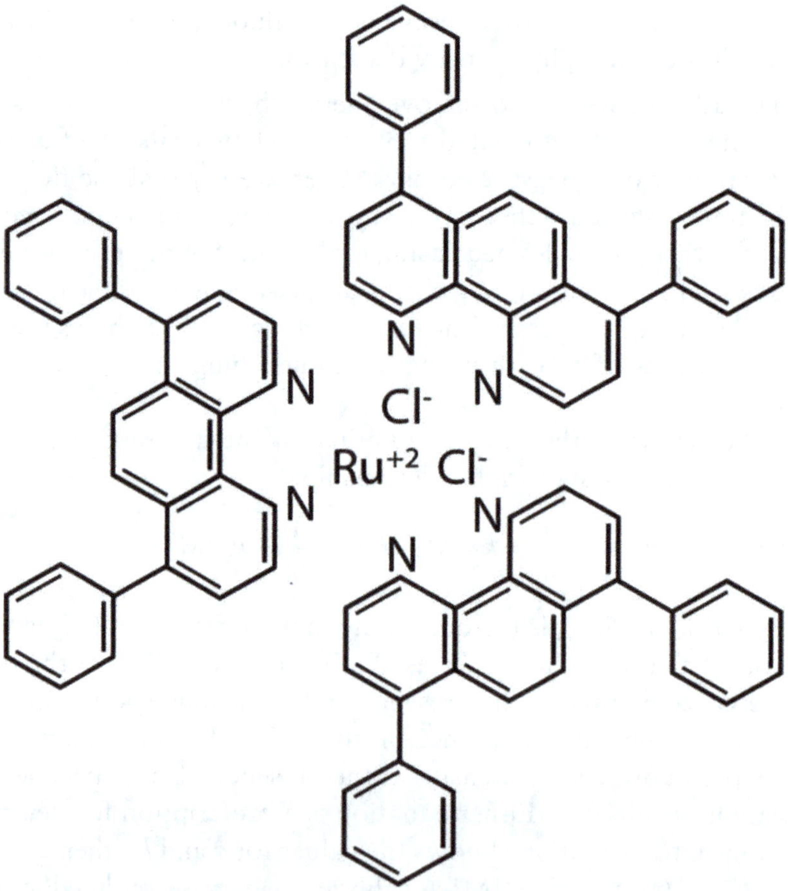

FIGURE 5.16 The molecular structure of the fluorescing compound $Ru(dpp)_3{}^{2+}$. The two Cl^- anions do not bond to the complex but, as with any salt, balance the +2 charge of the complex.

which is red. Its lifetime in the excited electronic state is 650 ns—in the absence of oxygen (Montalti et al., 2006). Both the relatively long lifetime of the fluorescence and the very different wavelengths of the absorbing and emitting light are uncommon for a fluorescing molecule. The reason for this unusual, but very convenient, behavior is that the excited electronic state is fundamentally different from the ground state. The fluorescing state is a *triplet* state and the ground state is a *singlet* state. This makes the fluorescence technically phosphorescence as described above. Because of this difference, the $Ru(dpp)_3^{2+}$, molecule takes a long time to emit its red photons—long enough for another chemical species to collide with it and knock it into an electronic state that loses energy through a nonradiative pathway. Not just any other molecule can interrupt the fluorescence process. One is particularly effective. And that molecule is oxygen.

When O_2 molecules collide with $Ru(dpp)_3^{2+}$ molecules, the fluorescence at 613 nm decays faster, and the intensity of emission decreases—just as discussed in the last section. The more O_2 collisions, the shorter the fluorescence lifetime and the lower the fluorescence intensity. By measuring the quenching in accordance with Equation 5.35 one can determine the oxygen concentration.

4.3 Making a Difficult Measurement Easy

To make an optical dissolved oxygen sensor one only needs an excitation light source at 455 nm to irradiate the Ru(ddp) lumiphore and collect the fluorescence at 613 nm. To measure the concentration of dissolved oxygen requires making two measurements:

1. Measure the fluorescence intensity or lifetime in the absence of oxygen.
2. Measure the fluorescence intensity or lifetime in the presence of oxygen.

Measuring fluorescence intensity sounds easy enough but is, in fact, difficult. Any fluctuations in the light source, detector, or collection optics will throw off the measurement. Furthermore, the change in intensity is very small and must be measured against a very large background. Measuring the fluorescence lifetime makes a lot more sense because its value is totally independent of the 455 nm light intensity. Thus, the measurements one wants to make are those illustrated in Figure 5.15. All one needs to do is to fit the decay curves to the exponential in Equation 5.34. From the curve fit one gets the lifetime, τ, for both conditions. In practice, one measures the decay curve over a range of known dissolved oxygen concentrations and constructs a calibration curve from the experimentally determined values of τ.

This approach of measuring the fluorescence decay is a sound approach, but it has drawbacks. For one, we would need to come up with a 455 nm light source that emits a pulse of duration less than a few nanoseconds. A pulsed laser can easily meet this requirement, but it would cost tens of thousands of dollars and its size and complexity would relegate it to a laboratory. We also need a detector that can measure the fluorescence on a timescale of nanoseconds. Again, this is achievable but at a high cost.

There is another method of measuring fluorescence lifetime without making a time-based measurement at all, and it's ripped from the pages of a basic electronics textbook. The method uses a continuous light source that is modulated. *Modulation* is a common electronics technique that makes radio possible whether it is modulation of the carrier frequency signal for FM radio or modulation of the amplitude of the carrier frequency for AM radio. The optical dissolved oxygen sensor does the latter. Modulation is any method that takes a steady signal and imposes upon it a cyclic, or periodic, pattern. That pattern can be a sine wave (sinusoidal) pattern, a saw tooth pattern, a square wave pattern, or any other repeating pattern. Most of the time, including this case, it is a sinusoidal wave as shown in Figure 5.17.

Instead of pulsing the 455 nm excitation light source optical, dissolved oxygen sensors use a continuous light source. A 20 kHz modulation source converts the continuous light into a modulated light source. A 20 kHz signal has a pulse width of 50 000 nsec, which is about 100 times longer than the

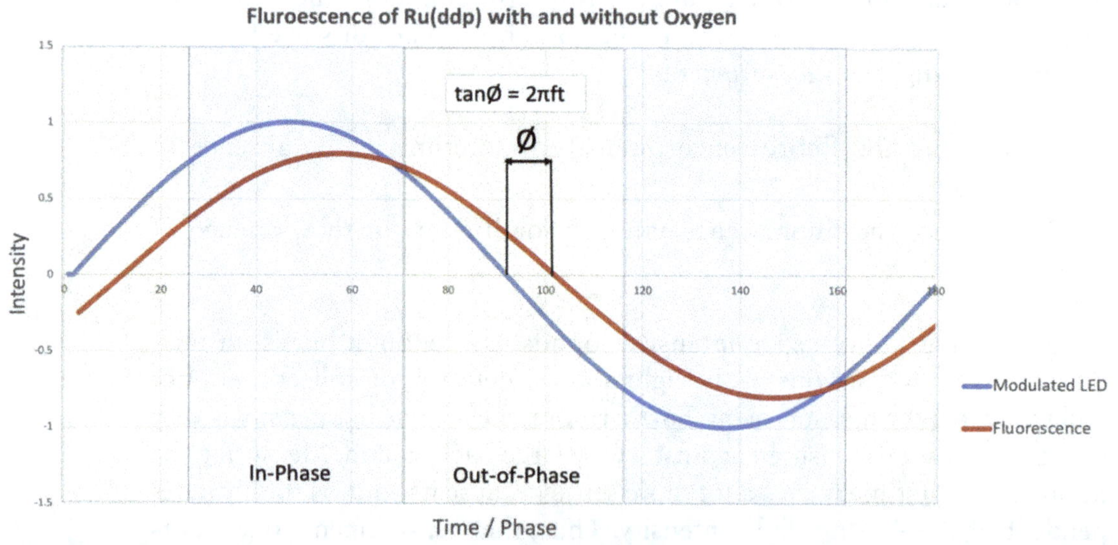

FIGURE 5.17 Synchronous detection scheme to measure the phase shift of the fluorescence. In addition to the phase shift there is a decrease in intensity of the fluorescence with respect to the incident 455 nm light source in accordance with Equation 5.35.

fluorescence decay of the Ru complex. The difference in time scales means that the fluorescence appears to be modulated at the same frequency of the 455 nm light source. However, because the fluorescence is delayed with respect to the excitation the modulated fluorescence is shifted in time from the modulated excitation. This so-called *phase shift* is a sideways shift of the entire sinusoidal curve. The frequency doesn't change—only the position along the time axis of the oscillating signal. Figure 5.17 shows the modulated excitation light and the phase shifted fluorescence light. Notice that the intensity of the modulated fluorescence is less than that of the excitation source. This is in accordance with the right side of Equation 5.35.

Relating the phase shift to the decay constant is a mathematical exercise, which I summarize. Figure 5.17 shows the sinusoidal modulation of the incident light source at frequency f and the subsequent modulation of the fluorescence at the same frequency. A boxcar averager, a type of synchronous detector, first integrates the combined oscillations during the in-phase half period of the incident wave, as shown in Figure 5.17, and then switches to the out-of-phase component.

The ratio of the out-of-phase integrated signal to the in-phase integrated signal is related to the phase delay and fluorescence lifetime (τ) by:

$$\tan\varphi = 2\pi f\ \tau = \frac{I_{out\text{-}of\text{-}phase}}{I_{in\text{-}phase}} \tag{5.36}$$

Those with a background in electronics will recognize that Equation 5.36 and Figure 5.17 describe the response of a sinusoidal electrical signal as it passes through an RC circuit (i.e., one containing a resistor and capacitor) and undergoes a phase shift.

In accordance with the right side of Equation 5.35 one can also measure the fluorescence lifetime by measuring the ratio of the modulated fluorescence intensity amplitude to the modulate excitation intensity amplitude.

Measuring the phase shift of the fluorescing light enables one to measure the lifetime of the fluorescence. All that remains is to relate the lifetime to the O_2 concentration. The Stern-Volmer equation, Equation 5.35, gives that relationship. The fluorescence lifetime or intensity is that of Ru(dpp), and oxygen is the quenching species, M. With the amperometric probe it's all about the current. With the optical probe it's all about the phase.

To determine the quenching rate constant, k_Q, one plots $1/\tau$ (or $1/I$) as a function of known concentrations of dissolved oxygen. The left side of Figure 5.18 shows the results of this calibration exercise. The slope of the best-fit line through the points is k_q, as given by Equation 5.35.

Equation 5.35 is an approximation. The right-hand side of Figure 5.18 shows the nonlinear behavior of the $1/\tau$ over a much wider concentration.

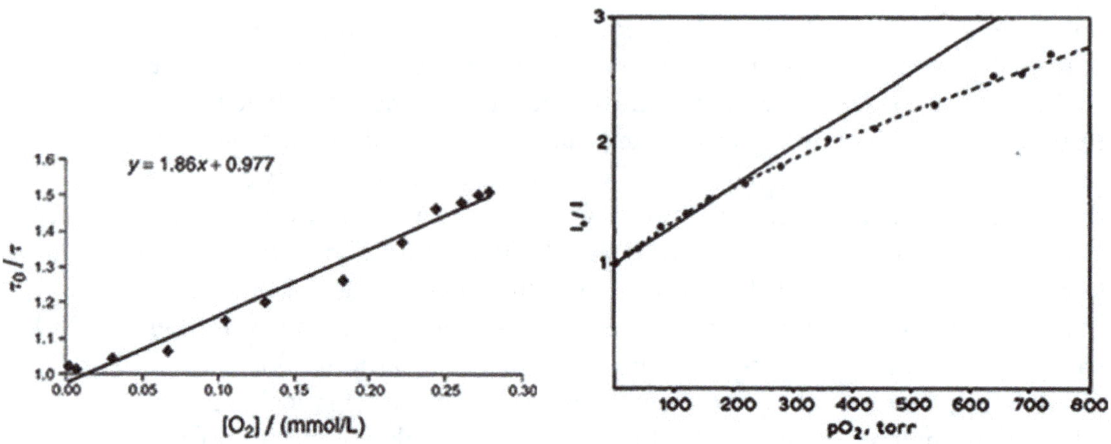

FIGURE 5.18 Extracting the dissolved oxygen concentration from fluorescence lifetime measurements. Left: Stern-Volmer plot of the fluorescence lifetime as a function of the dissolved oxygen concentration. The concentration range is approximately 0% to 120% saturation. The slope of the plot gives the quenching coefficient, k_q, which can then be used to calculate the dissolved oxygen concentration from the linear Equation 5.35 (Rusak et al., 2006). Right: Actual calibration showing nonlinear relationship. The wider concentration range of 0% to 500% saturation shows a nonlinear relationship that is fit by the empirical relationship Equation 5.37 (Mills, 1997).

range of approximately 0% to 500% saturation. The nonlinear behavior has been attributed to the kinetics of quenching and solubility of oxygen in the membrane. Equation 5.38 can be adapted to the nonlinear behavior by adding additional terms. The following is an empirical equation that fits the right-hand plot (note that it plots the inverse of the fluorescence intensity, which is proportional to the inverse of the lifetime) (Mills, 1997).

$$\frac{1}{\tau} = \frac{1}{\tau_0} + k_{1,Q}[M] + \frac{k_{2,Q}[M]}{1 + k_{3,Q}[M]} \tag{5.37}$$

There remains just one technical challenge to making a working sensor. The phase shift of the fluorescence signal with respect to the excitation LED signal is relatively easily measured. However, there is also the phase shift due to the temperature effect on the measurement electronics. To compensate for this important source of error, the optical dissolved oxygen probe almost always includes a second LED source that emits a wavelength of light—in the red (590 nm)—that is not absorbed by the Ru complex. It is modulated at the same frequency and in phase as the excitation blue light. The sensor typically alternates pulses, of a duration on the order of a second, of the blue excitation LED and then, the red LED.

Figure 5.19 is a schematic of an optical dissolved oxygen sensor.

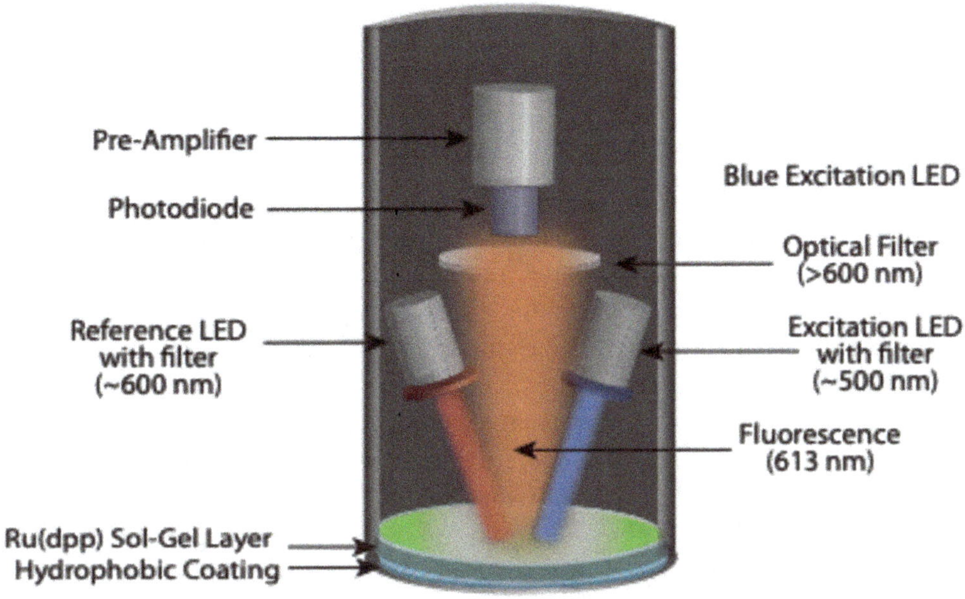

FIGURE 5.19 Schematic of an optical dissolved oxygen sensor. The $Ru(dpp)_3^{2+}$ is encapsulated in the sol-gel coating. The Reference LED pulses a red light that is not absorbed by the Ru complex. The blue LED pulses a blue light that is absorbed by the Ru(dpp) and fluoresces.

The Ru(ddp) lumiphore is encapsulated in a special polymer, called a sol-gel, at the tip of the probe. The permeable membrane allows O_2 to diffuse through it. There are two LED light sources, as explained above. The reference LED pulses a red light that is modulated on the order of 100 kilohertz. The modulation frequency is a compromise: As the frequency increases the phase sensitivity increases but the signal-to-noise decreases (McDonagh et al., 2001). The two LEDs are matched as closely as possible with respect to their temperature dependence. A photodiode picks up the light that scatters off the ruthenium but whose phase of modulation is unaltered. Some sensors use an infrared reference source. The sample LED pulses a modulated blue light at the same frequency that is absorbed by the the Ru(ddp). The ruthenium complex fluoresces at the longer 613 nm wavelength, which is detected by the same photodiode. The mixing of the reference and modulated signals enables the electronics to determine the phase difference due to both fluorescence and electronics. The permeable membrane coating and fluorescing ruthenium sol-gel complex are both deposited on a cap, which is easily replaced. A black coating on the front of the sol-gel seal prevents ambient light from entering the probe and adding noise to the detector signals. This picture makes it clear why optical dissolved oxygen sensors are attractive. There is no electrolyte, cathode, nor anode to worry about. About once per year the user simply unscrews the cap containing the lumiphore and screws on a new one.

4.4 Why Calibration of the Optical Dissolved Oxygen Sensor Is a One-Point Exercise

A look back at Figure 5.18 shows how calibration of an optical dissolved oxygen sensor works. The slope of the fluorescence lifetime versus O_2 concentration is equal to the quenching constant, k_q. Calibration means measuring the slope of the straight line that connects the points on the graph.

The calibration to determine k_q is done once. Routine calibration of the optical dissolved oxygen sensor during use is a one-point exercise. That's because the sensor is self-referencing. Every time the red-light source irradiates the lumiphore it establishes a zero condition—to be exact, the zero-phase modulated signal as shown in Figure 5.17. The phase difference between the fluorescence and reference signals is proportional to the oxygen partial pressure. (The phase difference caused by the temperature effect on the electronics is compensated by the reference LED.)

As with the electrochemical sensors, either water saturated air or air saturated water may be used to peg the 100% saturation phase difference. Also, like the electrochemical sensor, the analyzer uses the temperature and salinity to convert percent-saturation point to an absolute concentration in units of mg/L or ppm.

The effect of temperature on the measurement electronics and, consequently the measured phase, is compensated for by the reference LED. The remaining temperature effect on the phase angle is due to the ruthenium fluorescence and the diffusion of oxygen through the membrane. Figure 5.20 (McDonagh et al., 2001) show that the change in phase between 5 °C and 30 °C is between 3 and 5 degrees, which, as Figure 5.20 shows, corresponds to a change in measured dissolved oxygen on the order of 10%.

The simple empirical corrections, Equations 5.30 and 5.31, used to correct the amperometric dissolved oxygen probe were based only on the diffusion of oxygen through the membrane as a function of temperature. The additional dimension of fluorescence renders a temperature correction for the membrane of an optical probe more complicated. Empirical correction equations have been developed (Li et al., 2015) but invoke a more complicated power series in two variables (dissolved oxygen partial pressure and temperature) that are beyond the scope of the book. One common approach is to supply a power series (cubic) to fit the phase versus dissolved oxygen concentration over a discrete range of temperatures. Another is to simply provide a look-up table for the membrane temperature correction over discrete dissolved oxygen and temperature values.

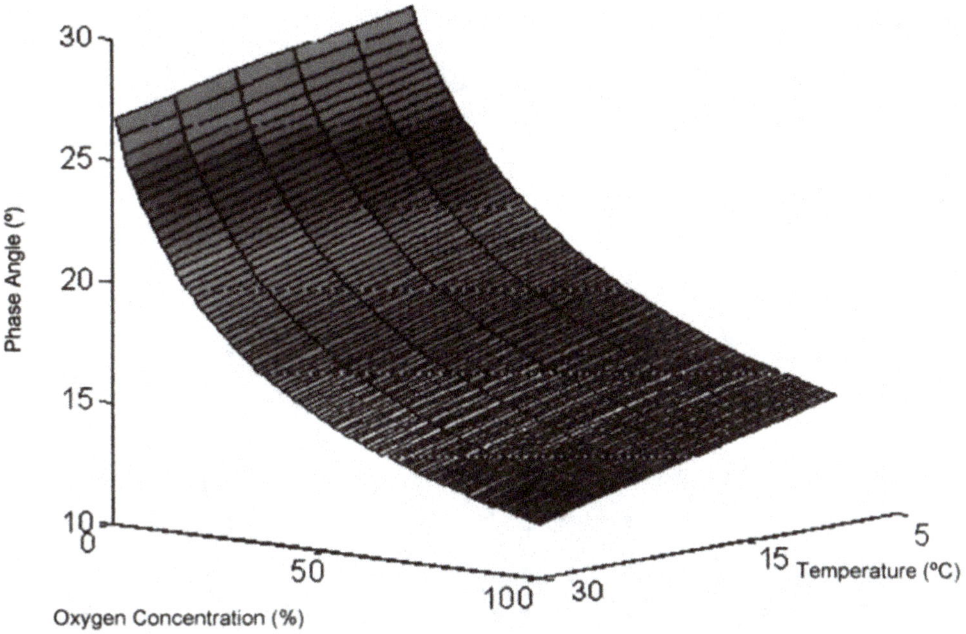

FIGURE 5.20 Variation of phase angle on the dissolved oxygen concentration and temperature (McDonagh et al., 2001).

4.5 Comparison Between Electrochemical and Optical Dissolved Oxygen Sensors

Like any amperometric probe, the Clark cell and galvanic sensors respond within seconds to changes in dissolved oxygen. The measurement duration depends on the time it takes for the redox reactions at the cathode to reach steady state. The rate-limiting step is diffusion of oxygen through the membrane. Typical response times for either type of electrochemical sensor is between 10 to 20 seconds. A thicker or less permeable membrane will result in a longer response time but a more stable signal.

The optical dissolved oxygen sensor has a response time that varies from 20 to 30 seconds. The difference is, therefore, not significant. Though flow does not affect the measured dissolved oxygen value, it does reduce the response time by a few seconds. The limiting step for making measurements is the lag between the pulsing of the red reference LED and that of the blue excitation LED through several cycles to make the measurement. Figure 5.21 (YSI, 2009) shows dissolved oxygen readings from three sensors as they are moved from a 100% saturation water sample to a 0% one (YSI, 2009). The sensors are: Clark cell with a 1.25-mil polyethylene membrane (orange dots), Clark cell with a 1.0-mil Teflon membrane (blue dots), and optical sensor (green triangles).

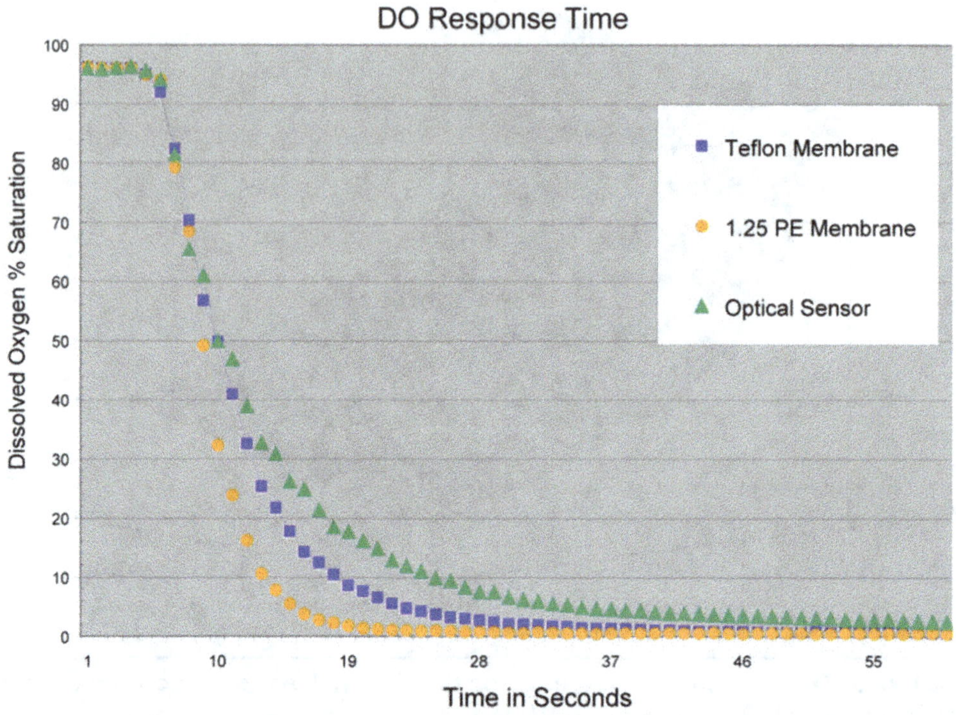

FIGURE 5.21 Response times for two amperometric sensors and an optical sensor. The two Clark cells differ in the membrane choice of Teflon and polyethylene. The measurements are for the sensors in air-saturated water that are placed in zero oxygen water (YSI, 2009).

For this important metric of measurement range, optical and electrochemical sensors are very comparable. A more precise way of phrasing the question is to seek the range over which the measurement range is linear. Both optical and electrochemical probes are capable of measuring ppb levels on the low end and 500% saturation on the high end. One adjusts the measurement range of an electrochemical probe to lower levels by using a thinner membrane and/or increasing the size of the electrode. To change the range of the optical probe the critical "knob" is the modulation frequency, which can vary more than an order of magnitude from approximately 20 KHz to 200 KHz, the latter for ppb level detection.

Figure 5.22, summarizes a study (Johnston & Williams, 2006), which placed two Clark cells and three optical dissolved oxygen sensors in the Tualatin River in Oregon and made continuous measurements over a 3-week period in 2005. The authors cleaned and recalibrated all five probes twice during the measurement time frame and made corrections for the drift due to both fouling and falling out of calibration. The data shown in Figure 5.22 show the corrected data. As the figure shows, the two sensor types tracked

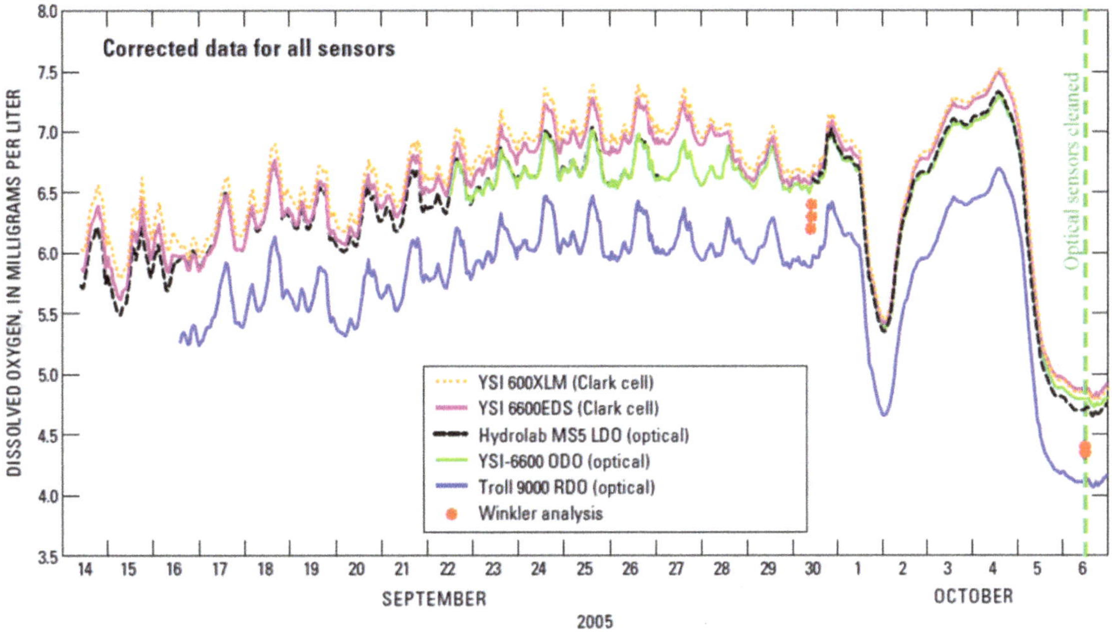

FIGURE 5.22 Comparison of 2 Clark cells and 3 optical dissolved oxygen sensors over 3 weeks in field measurements (Johnson & Williams, 2006).

each other well and deviated by about the same amount from the laboratory-based Winkler method. (The next section reviews the Winkler method.)

Drifting caused by fouling was 0.17 to 0.37 mg/L for the Clark cells and 0 to 0.02 mg/L for the optical dissolved oxygen sensors. Calibration drift ranged from -0.22 to 0.03 mg/L for the Clark cells and −0.09 to 0.02 mg/L for the optical sensors.

In the category of calibration, the optical sensor has a clear advantage over the amperometric probe. Figure 5.23, also taken from Johnson & Williams, shows the same results for two of the six sensors as Figure 5.22 with one difference: the data is shown both in its raw form and as corrected for calibration. The top graph is for the YSI 600XLM Clark cell while the bottom is for the Troll 9000 R optical probe. Clearly correction for the optical sensor is within 1% over the 3-week period whereas the drift for the Clark cell is about 4% over the same period.

The optical probe needs calibration every several months—as long as 1 year in clean water. The electrochemical sensors require calibration at a frequency of several days to a month, depending on the accuracy requirement and environment. Yes, calibration is fast and easy, requiring no calibration standards and only one point. But the drastically reduced calibration requirement for optical sensors makes them highly desirable.

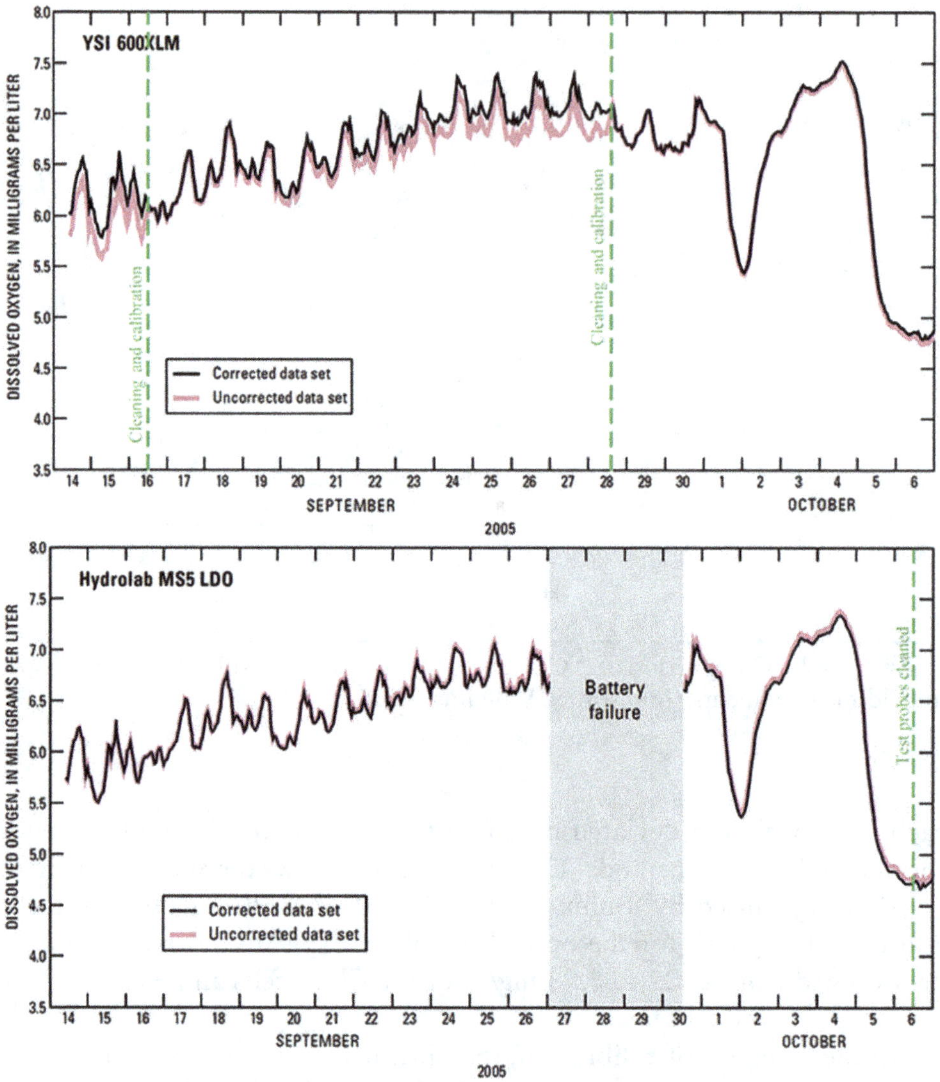

FIGURE 5.23 Comparison of stability of a Clark cell sensor (a) to an optical sensor (b). The red line is the raw dissolved oxygen signal uncorrected by calibration. The black line represents the calibration corrected readings. Clearly there is more drift in the electrochemical cell (Johnson & Williams, 2006).

Interfering compounds for electrochemical sensors are oxidizing agents that reduce on the cathode. This includes mainly free chlorine and other chlorine alternates (chloramine, chlorine dioxide, permanganate, and peracetic acid. For optical sensors interfering are compounds are those that quench the fluorescence of the Ruthenium lumiphore. Explaining the mechanism requires greater knowledge of spectroscopy, which is beyond the scope of the book. The major interfering dissolved gases are sulfur dioxide (SO_2)

and chlorine. Other interfering compounds include nitric oxide (NO), sulfur dioxide (SO_2), and bromine (Br_2).

For an excellent comparison of Clark, galvanic, and optical dissolved oxygen sensors, I highly recommend the review article by Wei et al. (2019).

5.0 CALIBRATION

Calibration requires a minimum of two points. If one assumes that zero oxygen produces zero output, then only one point is needed, and that point is water 100% saturated with oxygen. Thanks to Henry's law one can make the 100% calibration measurement either in water that is oxygen saturated or in air is 100% saturated with oxygen.

Calibration in water only requires an air injection device that can be as simple as an air pump with an air stone. These items are available at any pet or fish store. To calibrate an electrochemical sensor an inexpensive aquarium power head will ensure that there is sufficient flow. Because a pump will inject air at greater than 100% atmospheric pressure, the calibration point will be slightly higher ($< 2\%$) less than 100%.

To calibrate in water-saturated air, one encloses a saturated sponge in a sealed container such as a BOD bottle. When taking the sensor out of water it is important to allow enough time for the temperature sensor(s) to equilibrate to the ambient temperature. Otherwise, both the absolute temperature reading and the correction for the membrane permeability will be incorrect. Keep in mind that the permeability of a dry membrane is approximately 1% less than a wet one.

It bears repeating that a dissolved oxygen sensor measures partial pressure. Knowing the ambient pressure and, to a minor degree, the relative humidity, allows the analyzer to convert its current output to a percent saturation measurement. Measurement of the temperature and salinity allows the analyzer to convert its output current into an absolute concentration.

For precise measurement, the small, but finite, current at zero dissolved oxygen concentration requires that this second point be measured. This can be achieved by bubbling nitrogen gas through a container of pure water or by adding a reducing agent, such as sodium sulfite (Na_2SO_3), to an enclosed container of water at a concentration that is in excess of the dissolved oxygen. The oxygen is consumed oxidizing the Na_2SO_3 to sodium sulfate:

$$2\ Na_2SO_3 + O_2 \rightarrow 2\ Na_2SO_4 \tag{5.38}$$

It takes several minutes for the sodium sulfite to consume all dissolved oxygen. Stirring is necessary so a magnetic stirrer is advisable. Some users add cobalt chloride ($CoCl_2$) to speed up the process even more. Figure 5.24 shows such a two-point calibration for a Clark or Galvanic probe. The current at zero dissolved oxygen was less than 1 nA so that it does not register in the graph.

Recall that the membrane's permeability changes with temperature. Even though most analyzers compensate for changes in temperature, the accuracy of calibration is maximized by calibrating at the same temperature as the expected sample temperature.

No discussion of calibration would be complete without mention of the Winkler first developed by Lajos Winkler in 1888 and delineated in a U.S. Environmental Protection Agency (U.S. EPA) procedure (U.S. EPA, 2007). The only water parameter measurement kit more complicated is the biochemical oxygen demand (BOD) method discussed in Section 6.2. Four reagents are required.

Manganese sulfate, $MnSO_4$, reacts with KOH to form manganese hydroxide, $Mn(OH)_2$:

$$Mn^{2+} + 2\ OH^- \rightarrow Mn(OH)_2 \tag{5.39}$$

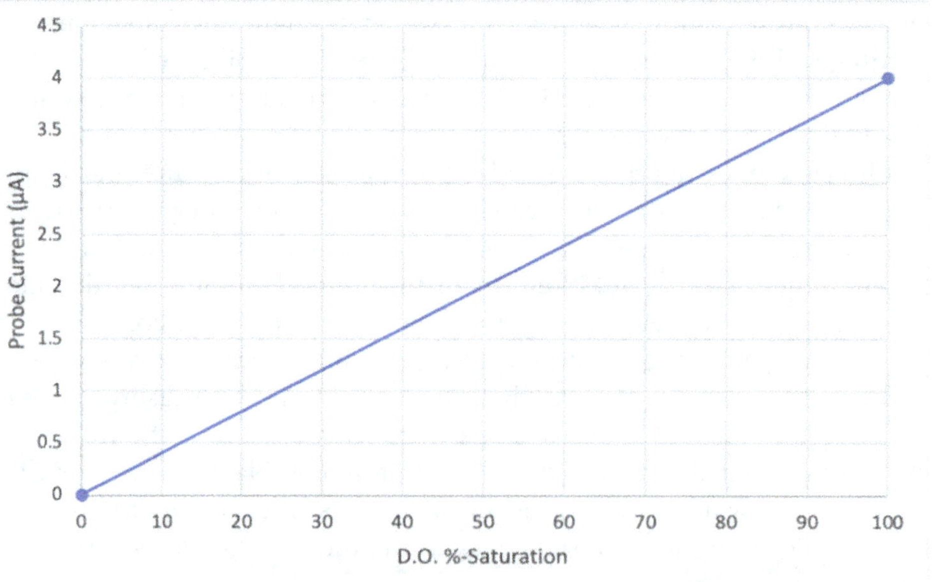

FIGURE 5.24 Calibration of Clark dissolved oxygen sensor. The (0,0) point was measured to be < 1nA. For most applications, not measuring trace levels of dissolved oxygen nor requiring very high precision, the user can assume the first point to be (0,0) (Water Analytics).

O_2 oxidizes $Mn(OH)_2$ to form a brown precipitate, manganic hydroxide, $MnO(OH)_2$:

$$2\ Mn(OH)_2 + O_2 \rightarrow 2\ MnO(OH)_2 \qquad (5.40)$$

Sulfuric acid (H_2SO_4) converts $Mn(OH)_3$ to manganic sulfate ($MnOSO_4$):

$$2\ MnO(OH)_2 + 2\ H_2SO_4 \rightarrow 2\ Mn(SO_4)_2 + 6\ H_2O \qquad (5.41)$$

Manganic sulfate oxidizes potassium iodide to elemental iodine (I_2):

$$2\ Mn(SO_4)_2 + 4\ KI \rightarrow 2\ MnSO_4 + 2\ K_2SO_4 + 2\ I_2 \qquad (5.42)$$

A colorimetric titration with sodium thiosulfate, $Na_2S_2O_3$, and starch as a color indicator:

$$2\ I_2 + 4\ Na_2S_2O_3{}^{2-} \rightarrow 2\ NaI + 2\ S_4S_4O_6 \qquad (5.43)$$

From the above sequence of equations, it takes 4 moles of sodium thiosulfate to react with 1 mole of oxygen. The starch indicator turns from blue to indicate when Equation 5.43 is complete. Thus, by knowing the concentration of the thiosulfate solution one can easily calculate the dissolved oxygen concentration.

The uncertainty of the method is approximately 0.1 mg/L. In practice it can be much higher. It's small wonder that the Winkler titration is rarely used. On the other hand, 100% water saturated air or 100% air saturated water requires no reagents and takes less than 5 minutes.

6.0 DISSOLVED OXYGEN PROBES IN THE REAL WORLD

6.1 Biological Nutrient Removal

By far, the most prevalent use of dissolved oxygen analyzers is to monitor and control the biological nutrient removal in wastewater treatment: nitrification, denitrification, and phosphorous removal. Chapter 2, Section 10.2 covered the chemistry of nitrification and denitrification in wastewater treatment (U.S. EPA, 1993). Figure 5.25 shows the profile of ammonium ($NH_4{}^+$) and dissolved oxygen along a plug flow reactor (Nagel & Haworth, 1969). (A plug-flow reactor is one in which the contents of the tank move in unison—that is, there is no horizontal mixing.) Therefore, a plot of ammonium and oxygen over the length of the reactor looks just like a plot over time of a thoroughly mixed system.

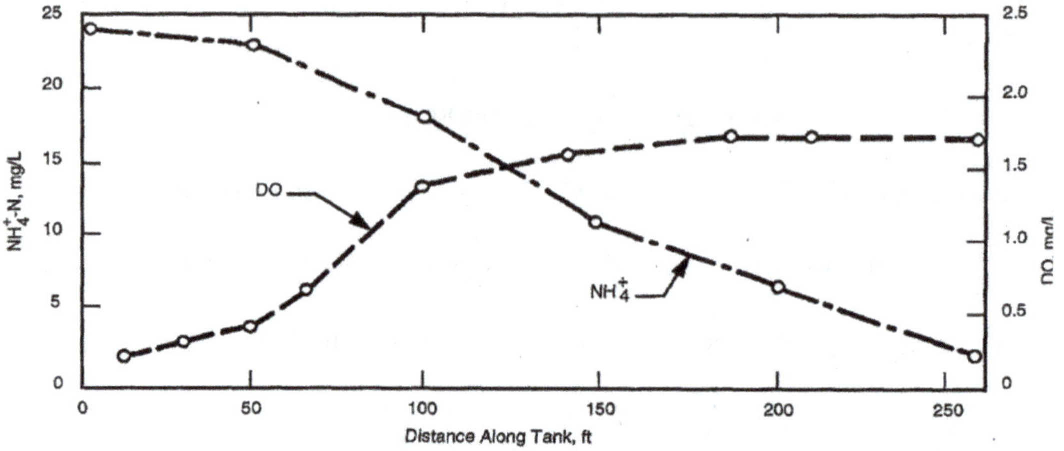

FIGURE 5.25 Ammonium and dissolved oxygen correlation in a nitrification tank (Nagel & Haworth, 1969).

The nitrification stage has traditionally called for a minimum dissolved oxygen concentration of 2 mg/L. However recent work has shown that nitrification can occur efficiently at dissolved oxygen concentrations as low as 0.5 mg/L (Bellucci et al., 2011). This has been attributed to the growth of highly efficient ammonia oxidizing bacteria as well as a large gradient of dissolved oxygen concentration in a biofilm from the outer surface to the inner surface (Gomez et al., 2002). The aeration machinery (blowers or diffusers) that controls the dissolved oxygen concentration is the single largest energy consumer in the wastewater facility. Achieving the correct dissolved oxygen concentration is not only important for maximizing nitrogen removal but for minimizing operating costs. Chapter 2, Section 10.2 discussed the approach of using ORP to monitor nitrification and denitrification. Nonetheless, measuring dissolved oxygen concentrations is almost always used as the main monitoring measurement.

The anoxic denitrification requires a dissolved oxygen concentration of less than 0.5 mg/L (van Haandel & van der Lubbe, 2017). However, for reasons stated in the previous paragraph, this condition can occur on a microscopic scale even in an aerated chamber in which the bulk dissolved oxygen concentration is much higher.

6.2 BOD Measurement

Most of the engineering that goes into a wastewater treatment facility is aimed at breaking down organic matter and lowering the concentration of the two major nutrients, nitrogen and phosphorous. As the saying goes, "You can't treat what you can't measure." Organic matter is particularly difficult to measure because it consists of a wide range of chemical species, all of which have carbon as a molecular backbone. Chapter 9 dives into the measurement of organic matter. The "gold standard" for measuring organic matter is to oxidize every organic compound all the way to carbon dioxide and measure the CO_2 that is driven off as gas. Organic matter measured this way is termed total organic carbon, or TOC. Long before the very convenient though very expensive TOC analyzer, the BOD_5 test was a mainstay of the treatment facility. In fact, the test dates all the way back to Victorian England when the Royal Commission on Sewage was established in 1898. The commission enacted the test 10 years later. The 5-day incubation period of the BOD_5 test that is still widely used today represents the average time it takes for pollution at a source in London to travel down the Thames River to the ocean.

The BOD_5 test is time consuming, cumbersome, and prone to large uncertainties. Nonetheless, it is codified as an accepted U.S. EPA method, so its continued use is assured. The basic premise behind the test is that the amount of oxygen that bacteria consume to break down organic compounds is proportional to the amount of these compounds. The full procedure for the U.S. EPA Standard Method 5210B BOD test can be found in many references, including *Standard Methods* (APHA et al., 2017). The following is a summary:

1. Prepare 1 phosphate buffer solution, 3 salts ($MgSO_4$, $FeCl_3$, and $CaCl_2$), one acid solution (H_2SO_4), and one base solution (NaOH).
2. Make up nutrients for bacteria with a mixture of glutamic acid and glucose.
3. Grab a wastewater sample.
4. Prepare 5 dilutions of the sample with distilled water to ensure that at least one of them starts and ends with BOD readings that are within range of the measurement.
5. Gather a source of seed bacteria from wastewater effluent or mixed liquor.
6. If the test is for carbonaceous BOD, or CBOD, add a nitrification inhibitor to kill off nitrifying bacteria. The difference between BOD and CBOD is actually negligible for incubation time less than 10 days.
7. Seed the sample or dilution water with nutrients.
8. Check the pH of the samples.
9. Prepare one sample of pure dilution water seeded with the nitrifying bacteria at the same amount. This is the control sample.
10. Measure the dissolved oxygen concentrations for all 5 samples and the control sample.

11. Incubate all 5 samples and control sample at 20±1 °C for 5 days.
12. Come back in 5 days and measure the dissolved oxygen concentrations again.
13. Calculate the BOD_5 from the difference in dissolved oxygen readings:

$$BOD_5 = \frac{D.O._0 - D.O._5}{D} \tag{5.44}$$

$D.O._0$ is the dissolved oxygen concentration (mg/L) measured immediately after preparation of the samples, *dissolved oxygen*$_5$ is the concentration after 5 days and *D* is the fraction of sample in the dilution—that is, the dilution factor.

14. If the BOD_5 calculation for the control sample is more than 0.2 mg/L, throw everything out and start over.

To say that the BOD_5 test is labor-intensive, time-intensive, and material-intensive would be an understatement. Furthermore, the numerous reagents and conditions involved result in measurement uncertainties as high as several hundred percent at low BOD levels (Young et al., 2003).

Some labs measure chemical oxygen demand (COD) in which chemical oxidizing agents break down carbonaceous matter. Using chemicals rather than bacteria makes the test much simpler, but the results from BOD tests are directly transferable to determining the amount of oxygen needed in wastewater treatment. As mentioned above, TOC measurements are gaining in popularity. Although the initial cost for the analyzer is much higher, the manpower and chemical needs of BOD tests reduce the long-term cost of the TOC analyzer (Gariepy & Turgeon, 2002). In practice, a lab makes measurements of TOC and BOD_5 and calculates the ratio between the two. Going forward it can then use this calibration to transform TOC measurements into BOD_5 ones. Chapter 9, on the measurement of organic matter, will dive more deeply into TOC.

Nonetheless, the measurement of BOD is still a mainstay in wastewater facility laboratories. Some manufacturers package their dissolved oxygen sensors into a sampling system that automates dissolved oxygen readings and calculations based on Equation 5.44.

6.3 Boiler Water Deaeration

Chapter 2, Section 10.7 discussed briefly the role of oxygen in boiler water. Because high temperatures accelerate most reactions, dissolved oxygen at

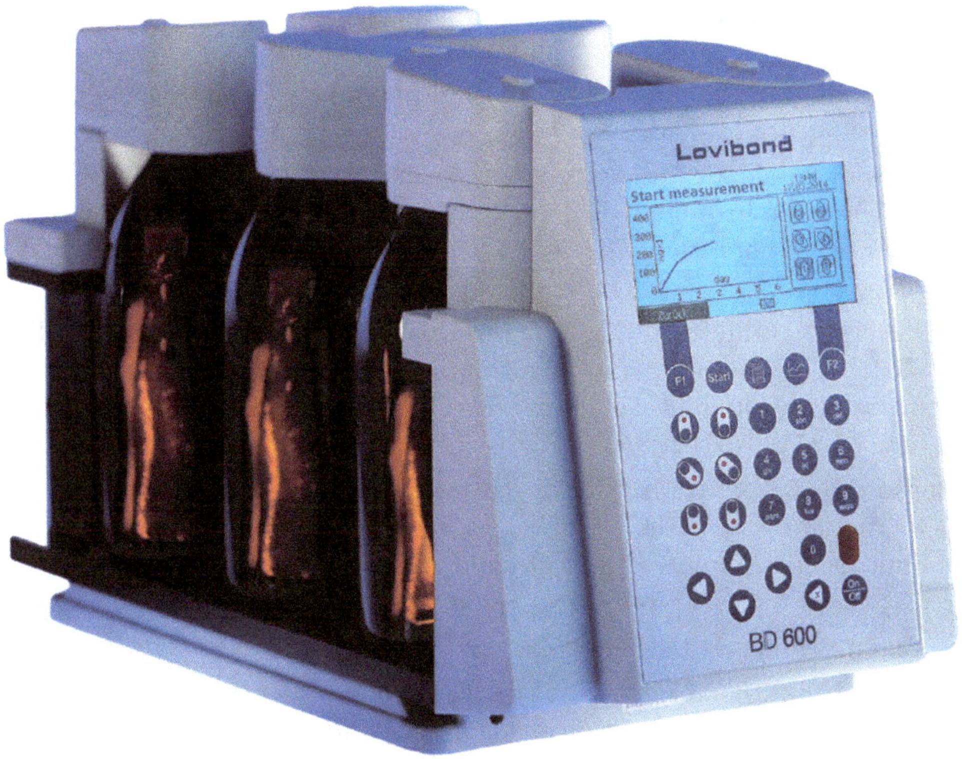

FIGURE 5.26 The Lovibond BD 600 automates the measurement of dissolved oxygen in multiple samples and calculates the BOD_5 values (Courtesy Lovibond).

boiler temperatures rapidly corrodes stainless steel and, if left unchecked, can cause a catastrophic failure in the boiler or the associated piping. To minimize corrosion, operators maintain a dissolved oxygen concentration to below 10 ppb. Dissolved oxygen is removed through mechanical and/or chemical treatment. Mechanical treatment typically involves a deaerator that heats the feedwater and converts it to steam. It then condenses the steam. To remove trace levels of oxygen, a reducing agent—hydrazine (N_2H_4) or sodium sulfite (Na_2SO_3)—is added to the feedwater. As mentioned in Chapter 2, Section 10.7, a trace level of 5 ppb of dissolved oxygen is sometimes considered desirable as it forms an oxide layer that protects the stainless-steel surface.

The ORP chapter made the case for measuring dissolved oxygen through ORP measurements though the standard practice is to directly measure the oxygen concentration by either electrochemical or optical sensors. Both electrochemical (mostly galvanic) and optical probes have improved to the point where they can measure ppb levels.

7.0 REFERENCES

Amine, A., & Mohammadi, H. (2019). *Amperometry, encyclopedia of analytical science* (3rd ed.), 85–98.

American Public Health Association, American Water Works Association, & Water Environment Federation (2017). *Standard methods for the examination of water and wastewater* (23rd ed.). American Public Health Association.

Bellucci, M., Ofiteru, I. D., Graham, D. W., Head, I. M., & Curtis, T. P. (2011). Low-dissolved-oxygen nitrifying systems exploit ammonia-oxidizing bacteria with unusually high yields. *Applied Environmental Microbiology, 77*(21), 7787–7796.

Benson, B. B., & Krause, D., Jr. (1980). The concentration and isotope fractionation of gases dissolved in freshwater in equilibrium with the atmosphere. *Limnology and Oceanography*, *25*(4), 662–671.

Chang, S. C., Stetter, J. R., & Cha, C. S. (1993). Amperometric gas sensors. *Talanta, 40*(4), 461–477.

Fatt, I. (1982). *Polarographic oxygen sensor: Its theory of operation and its application in biology, medicine and technology*. Krieger Publishing.

Gariepy, D., & Turgeon, M. (2002). TOC and BOD_5 correlations for biodegradable organic matter in the pulp and paper industry. Conference of the American Paper & Pulp Association.

Gomez, M. A., Hontoria, E., & Gonzalez-Loopz, J. (2002). Effects of dissolved oxygen concentration on nitrate removal from groundwater using a denitrifying submerged filter. *Journal of Hazardous Materials, B90*, 267–278.

Heyrovskya, J. (1924). The processes at the mercury dropping cathode. Part II. The hydrogen overpotential. *Journal Transactions Faraday Society, 19*, 785–788.

Janata, J. (2009). *Principles of chemical sensors* (2nd ed.). Springer. [Text is modified from original Janata material with permission of the author].

Johnston, M. W., & Williams, J. S. (2006). Field comparison of optical and Clark cell dissolved oxygen sensors in the Tualatin River, Oregon (USGS Open File Report 2006-1047).

Lakowitz, J. R. (2006). *Principles of fluorescence spectroscopy*. Springer.

Li, F., Wei., Y., Chen, Y., Li, D., & Zhang, X. (2015). An intelligent optical dissolved oxygen measurement method based on a fluorescent quenching mechanism. *Sensors*, *15*, 30913–30926.

Lighton, J. R. B. (2019). *Measuring metabolic rates. A manual for scientists* (2nd ed.). Oxford Press.

Manole, A., Neacsu, I., Apostu, M. O., & Melnig, V. (2006). Membrane covered polarographic oxygen sensor manufacturing theoretical considerations. *Analele Stiintifice Ale Universitatii Al. I. Cuza IASI*, Tomul II, 25–34.

McDonagh, C., Kolle, C., McEvoy, A. K., Dowling, D. L., Cafolla, A. A., Cullen, S. J., & MacCraith, B. D. (2001). Phase fluorometric dissolved oxygen sensor. *Sensors and Actuators B, 74*, 124–130.

Mills, A. (1997). Optical oxygen sensors, utilizing the luminescence of platinum metals complexes. *Platinum Metals Review, 41*(3), 115–127.

Montalti, M., Cedi, A., Prodi, L., & Gandolfi, M. T. (2006). *Handbook of photochemistry* (3rd ed.). CRC Press.

Nagel, C. A., & Haworth, J. G. (1969). Operational factors affecting nitrification in the activated sludge process. *42nd Annual Conference of the Water Control Federation*, Dallas.

Rusak, D. A., James, W. H., Ferzola, M. J., & Stefanski, M. J. (2006), Investigation of fluorescence lifetime quenching of ru(bpy)32+ by oxygen using a pulsed light-emitting diode. *Journal of Chemical Education, 83*(12), 1857–1859.

Sethuraman, V. A., Khan, S., Jur, J. S., Haug, A. T., & Weidner, J. W. (2009). Measuring oxygen, carbon monoxide and hydrogen sulfide diffusion coefficient and solubility in nafion membranes. *Electrochimica Acta, 54*(27), 6850–6860.

U.S. Environmental Protection Agency. (1993). *Process design manual: Nitrogen control.* Washington, D.C.

U.S. Environmental Protection Agency. (2007). *Standard operating procedure for dissolved oxygen micro method, Winkler titration* (Publication LG501).

van Haandel, A., & van der Lubbe, J. (2017). *Handbook of biological wastewater treatment* (2nd ed.) IWA Publishing.

Warburton, P. R., Sawtelle, R.S., Watson, A., & Wang, A. Q. (2001). Failure prediction for a galvanic oxygen sensor. *Sensors and Actuators B, 72*, 197–203.

Wei, Y., Jiao, Y., An, D., Li, D., Li, W., & Wei, Q. (2019). Review of dissolved oxygen detection technology: from laboratory analysis to online intelligent detection. *Sensors, 19*:3995, 1–38.

Young, L. G., Liptak, B. G., & Tatera, J. F. (2003). Chapter 8.8: Biological oxygen demand, chemical oxygen demand and total oxygen demand. In B. Liptak (Ed.), *Instrument engineers handbook—Process measurement and analysis* (pp. 1224–1234).

YSI. (2009). *The dissolved oxygen handbook.*

6

Free Chlorine

1.0 HYPOCHLOROUS ACID AND HYPOCHLORITE

Chapter 2 covered oxydation-reduction potential (ORP) sensors as a universal measurement tool for the most ubiquitous class of reactions that occur in water—oxidation-reduction, or redox, reactions. Chapter 5 covered sensors that measure the concentration of the most important oxidizing agent on earth—oxygen. Oxidation by oxygen drives so much of the chemistry on earth from the formation of minerals to forest fires to providing energy to cells comprising plants and animals. An important example of the latter is oxidation that bacteria use to oxidize ammonia to nitrite and nitrite to nitrate in wastewater.

Recall that chemical elements on the right side of the Periodic Table—up to the right-most column containing the inert noble gases—make effective oxidants. These are the elements that are only one or two electrons short of filling up their outer orbital. Oxygen is short by two electrons, so it is moderately reactive. Chlorine and its fellow *halogen* gases in the second-to-the-right-most column are missing only one electron, so they are highly reactive. Within this column (the second to last on the right of the Periodic Table), the smaller the atom the more reactive it is. Accordingly, fluorine, at the top of the column, is the most reactive. Next in line is chlorine. In chemical parlance, redox reactions involving chlorine oxidation occur with large free energy decreases.

It is chlorine's effectiveness as a strong oxidizing agent that treatment facilities use to kill microorganisms in treated water. As already discussed in Chapter 2, Section 10.1, Cl_2 is not really the disinfecting agent. What is called *free chlorine* is *hypochlorous acid* (HOCl) and its anion, *hypochlorite* (OCl^-). Hypochlorous acid forms when chlorine dissolves in water:

$$Cl_2 + H_2O \rightarrow HCl + HOCl \tag{6.1}$$

HOCl is a weak and unstable acid. It is weak because it dissociates into H^+ and OCl^- ions.

$$HOCl \leftrightarrow H^+ + OCl^- \tag{6.2}$$

The double arrow emphasizes the point that the reaction does not go strongly from left to right but results in an equilibrium mixture consisting of both species.

Both HOCl and OCl^- are effective oxidants. However, as Table 2.1 shows, the standard reduction potential of HOCl is 1.49 eV, whereas that of OCl^- is 0.90 eV. This makes HOCl a more potent oxidizing agent. HOCl and OCl^- do their work as disinfectants by oxidizing proteins and lipids inside bacterial cells. Another factor that plays into HOCl's relative strength is its neutral charge, which enables it to slip through neutral, hydrophobic cell membranes more easily than a charged species like the hypochlorite anion. The human body itself makes HOCl as a bactericide. The body's trained assassins, white blood cells, engulf pathogens and inject hypochlorous acid into the invaders.

In the realm of water treatment, chlorine's strong oxidizing potential is primarily directed toward disinfecting. However, it offers additional benefits. It oxidizes two major metal contaminants in water—iron and manganese. Both give water an unpleasant metallic taste. As dissolved ions, they exist in oxidation states of +2 for iron (Fe^{+2}) and +3 for manganese (Mn^{+3}). Free chlorine oxidizes them to Fe^{+3} and Mn^{+4}, which then form ferric hydroxide

[$Fe(OH)_3$] and Manganese dioxide (MnO_2), respectively, that precipitate out. Other contaminants that free chloride removes through oxidation include humic compounds, algae, and unwanted tastes and odors.

What is known as bleach is a dilute (3% to 6%) solution of sodium hypochlorite, NaOCl. Passing chlorine gas through caustic solution produces hypochlorite solution:

$$Cl_2 + 2\ NaOH \rightarrow NaOCl + NaCl + H_2O \tag{6.3}$$

The instability of hypochlorite solutions results from one or both of the following decomposition reactions:

$$NaOCl \rightarrow 2\ NaCl + O_2 \tag{6.4}$$

$$3\ NaOCl \rightarrow 2\ NaCl + NaClO_3 \tag{6.5}$$

Both reactions are characterized by rates that increase with hypochlorite concentration, temperature, pH values less than 10, light, and metal impurities such as copper, iron, nickel, and cobalt. It is for this reason that some water treatment facilities choose to make hypochlorite solutions fresh on location by mixing chlorine and caustic at the facility.

Because the storage of chlorine gas comes with safety risks, some facilities generate chlorine gas and HOCl on-site by electrolysis of a NaCl solution. At the anode chloride oxidizes to chlorine:

$$2\ Cl^- \rightarrow Cl_2 + 2\ e^- \tag{6.6}$$

At the cathode water is reduced to hydrogen gas:

$$2\ H_2O + 2\ e^- \rightarrow H_2 + 2\ OH \tag{6.7}$$

The combined reaction is:

$$2\ H_2O + 2\ Cl^- \rightarrow 2\ HOCl + 2\ OH \tag{6.8}$$

As the double arrow in Equation 6.2 denotes, HOCl is a weak acid. The addition of acid changes the ratio of OCl^- to HOCl. As the pH decreases, the fraction of HOCl in the mixture increases.

To calculate that fraction, all that is needed is the Henderson-Hasselbach equation, Equation 3.38.

$$pH = pK_a + \log \frac{[A^-]}{[HA]} \tag{6.9}$$

A^- is the conjugate base of the weak acid, or OCl^-. *HA* is HOCl. The dissociation constant, K_a, is a function of temperature and salinity. A little more rearrangement of Equation 3.38 gives us the fraction of free chlorine ($[HOCl]+[OCl^-]$) that is just [HOCl]:

$$\frac{[HOCl]}{[HOCl]+OCl^-} = \frac{1}{1+10^{(pH-pK_a)}} \tag{6.10}$$

Figure 6.1 shows the fraction of [HOCl] as a function of pH, at 25 °C, calculated from Equation 6.10. Note that pK is 7.58 (at 25 °C).

Not included in the above analysis or in the figure is the evolution of chlorine gas from HOCl for pH values less than 5—the reverse of Equation 6.1 (which is why mixing bleach with vinegar can be fatal).

HOCl is the more effective disinfectant, so the effectiveness of free chlorine decreases as the pH increases. This ratio will become very important to the design of an amperometric chlorine analyzer that this chapter reveals. Note that decreasing the pH below 7 creates an acidic condition that increases the rate of corrosion, thus setting up a Faustian choice for treatment facility operators between disinfection and corrosion control.

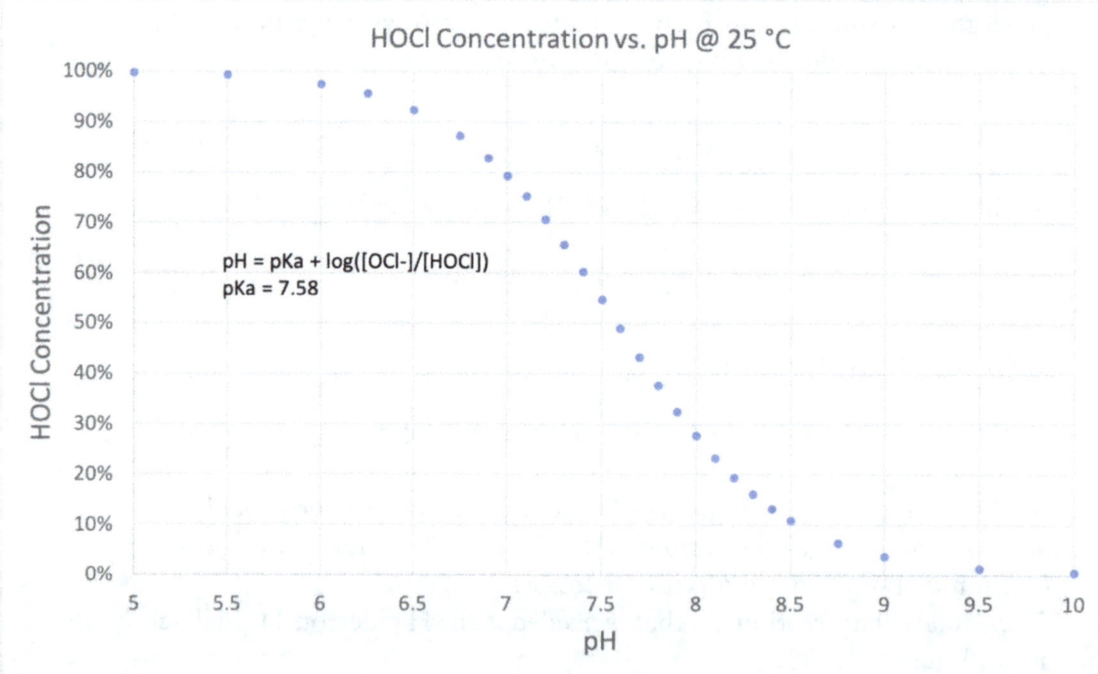

FIGURE 6.1. The fraction of free chlorine ($HOCl + OCl^-$) as HOCl at 25 °C calculated by the Henderson-Hasselbach equation and pK = 7.58. Not included is the evolution of Cl_2 at pH < 5.

A close relative to sodium hypochlorite is calcium hypochlorite, also known as *bleaching powder* or *chlorinated lime*. It's made by passing chlorine through lime [$Ca(OH)_2$]:

$$2\ Cl_2 + 2\ Ca(OH)_2 \rightarrow Ca(OCl)_2 + CaCl_2 + 2\ H_2O \qquad (6.11)$$

As the chlorinated lime's molecular formula shows, it contains two hypochlorite anions, thus making it a more effective oxidizing agent in dilute solutions. It is also more stable.

2.0 BREAKPOINT CHLORINATION

The large reduction potential of chlorine is a double-edged sword. It enables it to react with a lot more than just bacteria and viruses. Organic compounds, ammonia, and metals all are readily oxidized to form chlorides. Though the term *disinfection by-product*, or DBP, applies to any compound resulting from the oxidation of impurities by a disinfectant, the major culprit includes the most problematic ones—chlorinated methane and chlorinated ammonia—chloramines (NH_2Cl, $NHCl_2$, and NCl_3). Monochloramine, NH_2Cl, is in fact, a desirable disinfectant itself. The reaction of free chlorine with other chemicals to create DBPs, especially chloramine, presents a barrier to a treatment facility trying to maintain a minimum level of free chlorine. The strategy of overcoming the barrier is a process called *breakpoint chlorination.*

Breakpoint chlorination ensures that there is enough free chlorine to clear out the reducing compounds that consume it so that only free chlorine is left. The process can be found in water treatment, wastewater treatment, pools, and spas. Figure 6.2 (Reynolds & Richards, 1996) shows conceptually how this works. This breakpoint chlorination curve consists of four stages as the free chlorine is added:

1. Between points 1 and 2, all free chlorine is destroyed (i.e., reduced) by easily oxidized species, including iron (Fe^{2+}), manganese, hydrogen sulfide (H_2S), and some organic compounds. The chlorine residual is zero in this first region.
2. Once this "low hanging fruit" of reducing agents has been consumed, the free chlorine can turn to oxidizing ammonia. Monochloramine is the first DBP to form. The section of the graph between points 2 and 3 follows the increase in NH_2Cl. If chlorine is being consumed, why does the curve in stage 2 increase? Because monochloramine is an effective disinfectant and is included as part of the *total chlorine* budget.

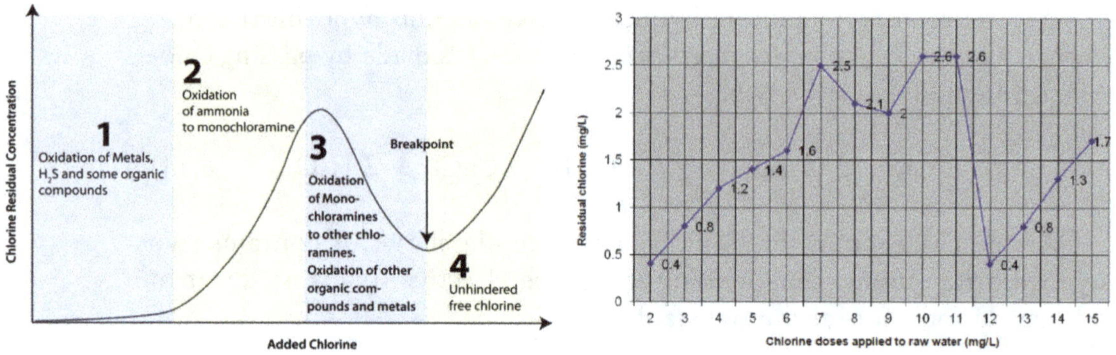

FIGURE 6.2 Breakpoint chlorination. Left: Conceptualization of the breakpoint process. Right: Actual data from a treatment facility in India in which the source of water is the River Yamuna (Kumar, 2013). The author hypothesized that the additional dip in chlorine residual at 9 mg was due to numerous DBPs and/or microorganisms.

3. After HOCl and, to a lesser degree, OCl^-, has converted all ammonia to monochloramine, it then oxidizes NH_2Cl to $NHCl_2$ and then $NHCl_2$ to NCl_3. The free chlorine then turns to remaining organic compounds and harder-to-oxidize compounds and metals. The consumption of HOCl, OCl^-, and NH_2Cl leads to a downward trend in free chlorine. That is the region between 3 and 4. Eventually all constituents that can be oxidized are oxidized.
4. Further addition of HOCl or OCl^- can now go "straight to the bottom line." The total curve swings upward. The point at which the free chlorine survives as chlorine residual—that is, point 4—is the *breakpoint*. The chlorine that persists in the distribution system is *residual chlorine*.

Treatment facilities, pools, and spas that use HOCl or OCl^- for residual chlorine adjust the addition of free chlorine to ensure that the measured chlorine is somewhere past the breakpoint to maintain the required residual chlorine level needed before leaving the plant.

3.0 THE DPD ANALYZER

3.1 The Wondrous Wurster Red Dye

N,N-diethyl-p-phenylenediamine, or DPD, is an organic compound whose molecular structure is shown in the left side of Figure 6.3. Two nitrogen atoms attach on opposite sides of the benzene molecule. A nitrogen atom usually bonds to three atoms via a pair of electrons per bond (*single bond*). However, there is a pair of electrons (*lone pair*) on the N atom that is left over. Even though the molecule is neutral, this local negative charge density

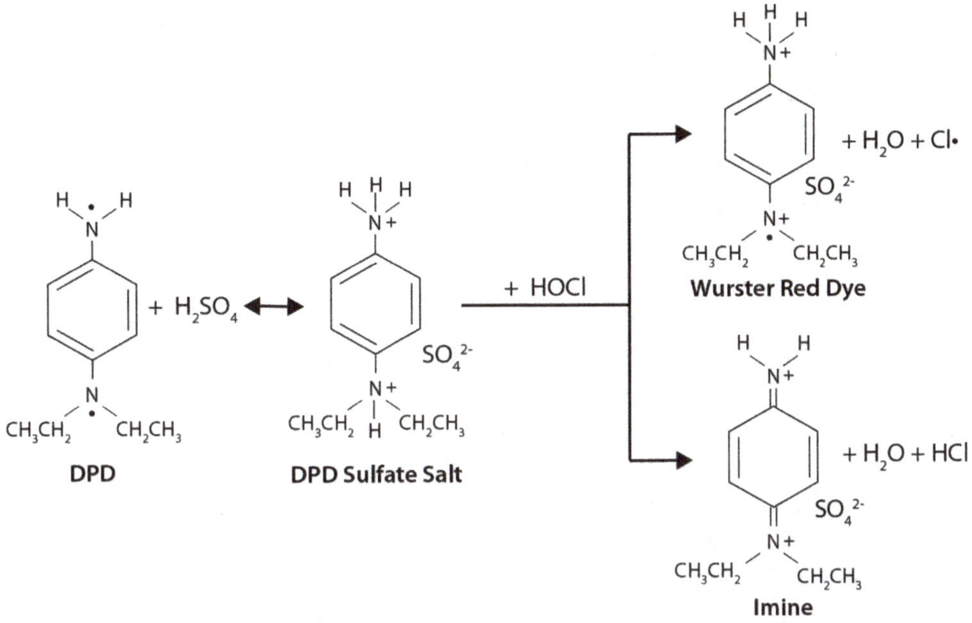

FIGURE 6.3 DPD forms a salt with H_2SO_4 to make DPD^{2+}-SO_4^{2-}. The salt reacts with free chlorine (or other oxidants) to form Wurster Red Dye, a radical cation. Further oxidation of Wurster Red Dye by HOCl creates an imine. *Note: I am unaware of a single text that shows the fate of the HOCl molecule in the oxidation of DPD. The species shown in this figure are the result of my own conjecture and an online discussion.*

attracts a H^+ ion, though doing so adds a positive charge. This is why NH_3, in the presence of an acid environment, will grab an H^+ ion to make NH_4^+. Ditto for both N atoms attached to the two C atoms on the top and bottom of the benzene ring in the figure.

One sulfuric acid molecule (H_2SO_4) can donate two H^+ ions (*protonate*) to the two N groups (*amines*). Doing so gives the DPD molecule two positive charges, the SO_4^{2-} anion balances the charge and creates a *double-protonated complex,* or a salt, that is, on the whole, balanced. The sulfate anion is typically ignored so that the DPD cation part of the sulfate salt has a +2 charge.

Now let's add a strong oxidant—in this case HOCl (but it can be Cl_2, Br_2, I_2, Br_2, $KMnO_4$, or O_3). It pulls off one electron, along with an H^+ ion (a proton) from the two-electron bond between the bottom N atom and H atom. In sum, it loses a neutral H atom. This leaves behind one lonely, unpaired electron on the N atom. The proton reacts with the HOCl to make a neutral water molecule and a neutral Cl atom. The resulting DPD product has a strong magenta color and is called *Wurster's Red.* A similar compound to DPD has methane groups attached to both N atoms and oxidizes to form *Wurster's Blue.*

A lone electron on the *Wurster's Red* dye molecule renders it a free radical—a neutral molecule with an unpaired electron. Free radicals are common in gases but rare in an aqueous environment in which polar water molecules like to associate with ions. The discussion of advanced oxidation processes at the end of the chapter introduces the powerful free radical hydroxyl oxidant, OH•.

The departure of both a proton and electron from the bottom N atom leaves that part with its positive charge. Together with the top N atom's positive charge, by virtue of the attached H^+ ion, the dye ion retains its +2 charge. The highly unusual Wurster's dye free radical owes its existence to the presence of the two positive charges on the nitrogen atoms, which render it an ion and a radical all in one—a *radical cation*. The Wurster's Red radical cation is stable because the lone electron actually spends much of its time in the cloud of electrons belonging to the benzene ring.

The Wurster's Red molecule balances its +2 charge with the SO_4^{2-} anion. Below the Wurster's Red molecule in Figure 6.3 is a second oxidation product. The Wurster's Red molecule is the result of the DPD molecule shedding one proton and electron. A second oxidation repeats the process of removing the single electron on the bottom N atom and a proton from the top N atom to make another H atom. Electrons in the benzene ring and the C-N bonds rearrange to create the molecule, N,N-diethyl-p-quinonediimine. This compound is a type of imine—a group or compound with a carbon atom doubled bonding to a nitrogen atom. As a diimine it has two such C = N bonds. The diimine is colorless and plays no role in the measurement of chlorine. The conversion of Wurster Red to the clear diimine limits the measurement range of DPD to approximately 5 mg/L. After the free chlorine limit exceeds this limit, the pink color fades as more Wurster dye oxidizes to the imine. To extend this measurement range. one can dilute the sample.

Keep in mind that all oxidants react with DPD. It is a generic oxidant indicator. Because free chlorine includes both HOCl and OCl^-, this universal colorimetric titrating reagent allows one to measure total free chlorine. Therefore, a solution containing mostly HOCl will give the same colorimetric reading as a solution containing mostly OCl^-, but the former is more effective as a disinfectant. This brings up the perpetual question of just how useful the measurement of total chlorine is. More on this question later.

To measure free chlorine, the user mixes a specified quantity of DPD reagent, waits for the magenta color to emerge, and compares the shade of pink to a chart of shades. The shade of pink is a function of the concentration of Wurster's Red and, by extension, that of the free chlorine. Figure 6.4 shows the comparator in a typical test kit. Note that there are color charts for free bromine (HOBr and OBr^-) and pH. Color changes are read through transmitted light and the comparator ensures that the same intensity of light is transmitted through the sample as through the color swatches.

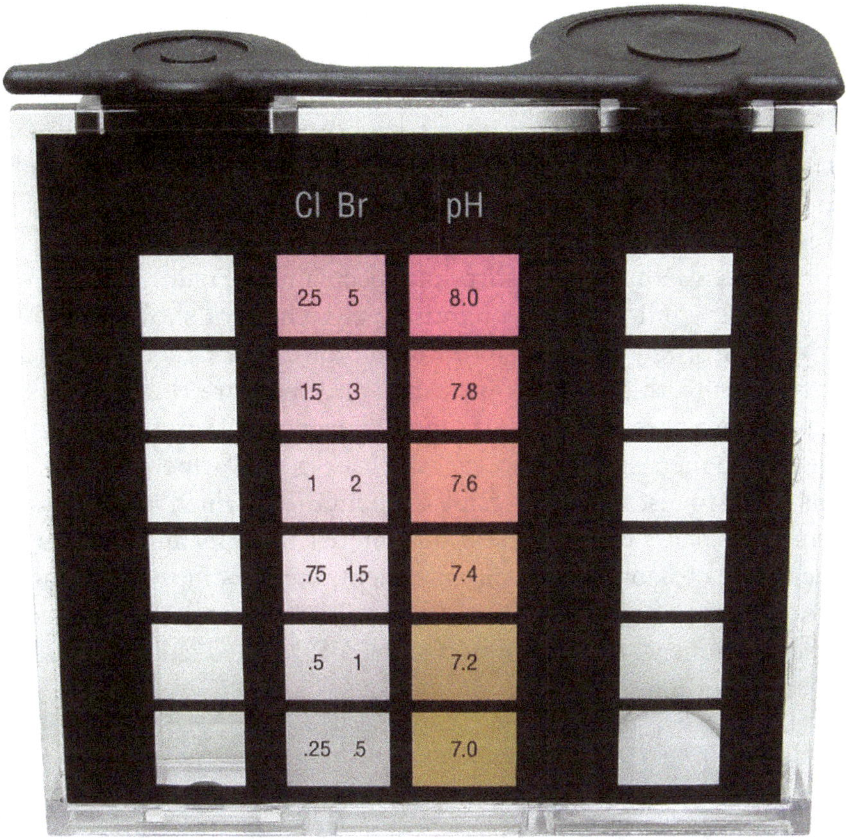

FIGURE 6.4 The comparator from a DPD test kit with color swatches that are read with the same transmitted light as the sample itself. This kit measures low values of free chlorine and bromine. Others measure higher levels (Courtesy Taylor Technologies).

To increase its accuracy, one can convert this simple colorimetric measurement into a titration measurement. After the free chlorine in the sample reacts with the DPD, one adds ferrous ammonium sulfate—$Fe(NH_4)_2(SO_4)_2{\bullet}6H_2O$ or FAS. FAS reduces the Wurster Red dye back to DPD. The number of drops required to turn the pink color of the dye to the colorless DPD is proportional to the free chlorine concentration. The sudden change of color makes titration measurements more precise than colorimetric ones. This is yet another example of titration. The first one was a conductometric titration of an acid by a base (or vice versa) in Chapter 4. We will encounter more.

To use a colorimetric test kit like DPD requires the ability to discern slight differences in the pink shade. This perceptual requirement limits both the range and accuracy of the test. For greater accuracy, one can substitute a spectrophotometer for the human eye to measure the intensity of the color. A spectrophotometer is a spectrometer that measures the absorption of light over a defined wavelength range that matches either the absorption of the analyte of interest or, more commonly, the absorption of a reagent that reacts with the analyte.

To understand exactly how a spectrophotometer works to measure the Wurster's Red concentration, let's turn to the absorption spectrum in Figure 6.5. Color theory states that a dye shows its characteristic color because it removes that color's complement. The reason DPD is magenta is because it absorbs green light. The absorption spectrum in Figure 6.5 (Kim et al., 2011) shows this information quantitatively. The plot shows that Wurster Red absorbs strongly between 510 nm and 550 nm. One can tune a spectrophotometer to this wavelength range and measure the light absorbed. There is another absorption band centered around 325 nm and can be used to increase the measurement accuracy (Moberg & Karlberg, 2000).

This book focuses on process instrumentation, which means those that make measurements in real time. Colorimetric test kits and sensors do not lend themselves to continuous measurement, but by simply automating the test process in a robotic fashion, a colorimetric analyzer can become a real-time instrument. Most common is the DPD-based chlorine analyzer as shown in Figure 6.6.

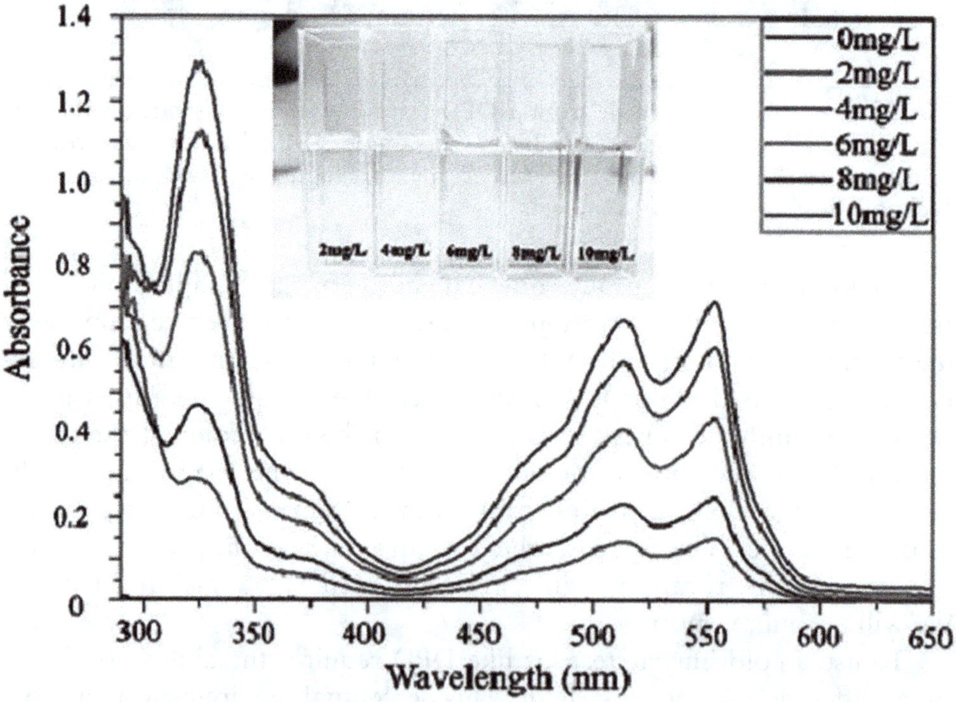

FIGURE 6.5 Absorption spectrum of DPD. One can tune a spectrophotometer to one of two wavelengths—both in the blue end of the spectrum. Another absorption band is at 325 nm—to the left of the graph (Kim et al., 2011).

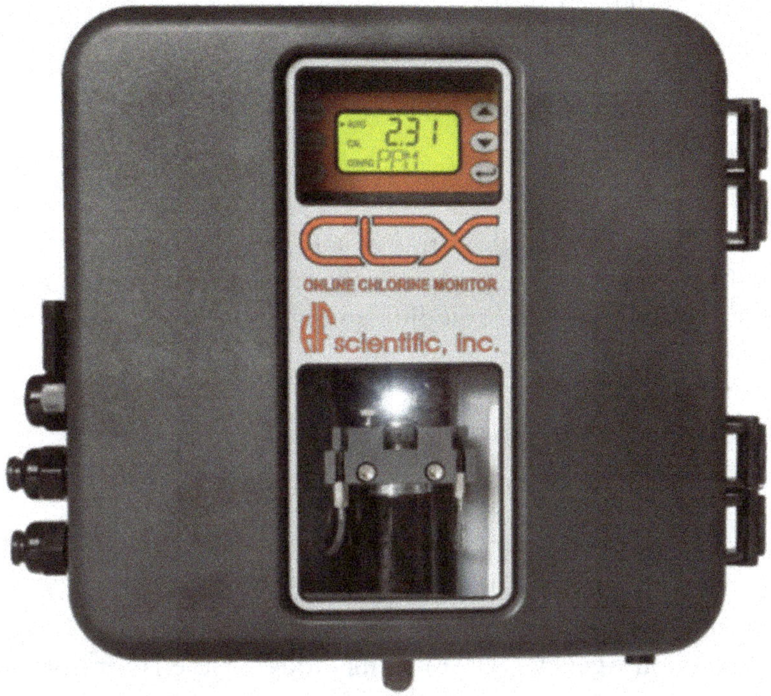

FIGURE 6.6 This colorimetric free chlorine analyzer by HF Scientific automates the determination of chlorine concentration by the DPD colorimetric test (Courtesy HF Scientific).

Reacting free chlorine with DPD and measuring the intensity of the oxidized DPD is the most commonly used reagent-based method. It's simple, fast, and inexpensive. But it's not the only method. There are methods that use either colorimetry or amperometry to measure the reaction limit of the reagent and free chlorine. For a more complete description of these techniques and their relative merits, refer to Daniel Harp's review article (Harp, 2002).

Sophisticated analyzers that measure a wide range of species such as phosphorus, ammonia, nitrate, arsenic, manganese, and many other metals use an array of reagent bottles and liquid dispensers. The measurement principle is the same. When an optical sensor does the colorimetric measurement, the technique is referred to as *spectrophotometry*.

3.2 To Measure Total Chlorine Just Add Iodide (and Starch)

Total chlorine consists of free chlorine—HOCl and OCl^-—and combined chlorine. The latter is mainly monochloramine, NH_2Cl, but also the less desirable $NHCl_2$ and NCl_3. To measure the amount of total chlorine, one simply adds iodide, I^- (usually as the salt, KI). This ubiquitous colorimetric agent reacts with HOCl and monochloramine, under acidic conditions (pH < 4), to make triiodide (I_3^-).

$$HOCl + 3\,I^- + H^+ \rightarrow H_2O + Cl^- + I_3^- \tag{6.12}$$

$$NH_2Cl + 3\,I^- + H_2O + H^+ \rightarrow NH_4OH + Cl^- + I_3^- \tag{6.13}$$

The triiodide concentration thus equals the sum of free and combined chlorine. (Similar reactions occur for the other chloramines, $NHCl_2$ and NCl_3.) One can think of the triiodide anion as the sum of an iodine anion and iodide molecule.

Some texts ignore the accompanying I^- and replace I_3^- in chemical reactions with I_2; for example:

$$HOCl + 2\,I^- + H^+ \rightarrow H_2O + Cl^- + I_2^- \tag{6.14}$$

For this reason, the term *triiodide* may be switched with *iodine.*

As the above reactions show, adding iodine converts the DPD test for free chlorine into a DPD test for total chlorine. There are five U.S. Environmental Protection Agency (U.S. EPA) methods for the laboratory measurement of total chlorine—330.1 through 330.5. All the methods, except 330.5, involve titration—a measurement in which a reducing reagent is slowly added, consumes the analyte, and changes the color of the analyte. The color of triiodide is the red color of iodine, and it fades to clear as it is reduced.

One of the most popular chemistry demonstrations is adding a solution of iodine (I_2) and iodide salt (KI) to a potato and watching the starchy vegetable turn dark blue. The triiodide anion nestles into a coiled molecular form of starch called amylose. Iodine is deep red, but the triiodide-starch complex is deep blue. The disappearance of the blue color to a pale yellow is considered easier to visualize in a titration than the change of the iodine red color to clear. Whether through iodine or starch-iodine, the titration of triiodide is known as an *iodometric titration.* For simplicity we will assume that the iodine-starch complex is in play in the following discussion of iodometric titration.

Sodium thiosulfate ($Na_2S_2O_3$) dissolves to create the thiosulfate anion, which oxidizes as follows:

$$2\,S_2O_3^{2-} \rightarrow S_4O_6^{2-} + 2\,e^- \tag{6.15}$$

The titration consists of adding thiosulfate to the triiodide, so that the latter reduces back to iodide and the thiosulfate oxidizes to $S_4O_6^{-2}$:

$$I_3^- + 2\,S_2O_3^{2-} \rightarrow 3\,I^- + S_4O_6^{2-} \tag{6.16}$$

As the titration proceeds, the iodine-starch blue color fades. The amount of thiosulfate needed to turn the solution colorless yields the I_3^- and, hence, the HOCl concentration. The stoichiometry in Equation 6.16 tells us that 2 moles of $Na_2S_2O_3$ react with 1 mole of triiodide. Equation 6.12 shows that 1 mole of HOCl or OCl^- is titrated by 2 moles of $Na_2S_2O_3$.

This titration is called a *forward titration.* Because it converts free and combined chlorine to I_3^- it measures total chlorine. Its main limitation is that other reducing compounds in the water can react with the triiodide and, hence, give a free chlorine reading that is low. This is more of an issue with wastewater, which contains more impurities than clean water that can be oxidized.

To overcome this limitation, one uses *back titration.* In this scheme, the user reverses the order of triiodide and reducing agent. One first adds a known amount of thiosulfate in excess of the free or combined chlorine to the sample:

$$HOCl + 2\ S_2O_3^{2-} + H^+ \rightarrow H_2O + Cl^- + S_4O_6^{2-} \quad (6.17)$$

$$NH_2Cl + 2\ S_2O_3^{2-} + 2\ H^+ \rightarrow NH_4OH + Cl^- + S_4O_6^{2-} \quad (6.18)$$

In the second step, the user adds triiodide to the sample to oxidize the leftover thiosulfate. The reaction is identical to Equation 6.16:

$$I_3^- + 2\ S_2O_3^{2-}(excess) \rightarrow 3\ I^- + S_4O_6^{2-} \quad (6.19)$$

This second reaction, the back titration, determines the amount of excess thiosulfate. By subtracting the excess thiosulfate from the initial thiosulfate added in Equation 6.17 or 6.18, the user determines the amount of thiosulfate consumed by the free chlorine. The validity of this approach rests on the assumption that the thiosulfate reacts with the free chlorine and not other oxidizing agents.

Because of U.S. EPA requirements to measure and report total chlorine, the replacement of free and combined chlorine with triiodide is standard practice. In fact, the DPD colorimetric test that was discussed first can be modified as a back titration. In this variation, described by U.S. EPA Method 330.4, one first prepares a buffered solution of DPD and adds KI. Next, one adds the sample. The KI converts free and combined chlorine to triiodide. The user titrates the triiodide with the reducing agent ferrous ammonium sulfate, or FAS, whose chemical formula is $Fe(NH_4)_2(SO_4)_2{\bullet}6H_2O$. The titration endpoint occurs when the magenta color disappears. The amount of FAS added is proportional to the total chlorine.

Method 330.5 is the oddball. It is the spectrophotometric measurement using DPD. All others are titration methods. Only DPD method 330.5 allows the option of adding KI to measure total chlorine rather than free chlorine. All other methods mandate it.

Turbidity makes the visual or colorimetric detection of the titration endpoint difficult. For spectrophotometric DPD measurements, the solution is simply to measure the transmission of a sample without DPD and subtracting the resulting baseline measurement from subsequent measurements. For other methods, filtration can remove suspended solids that interfere with color measurement.

3.3 Titration Goes Better With Current Than Color

The forward and back titrations just discussed use the disappearance of color in the sample to determine the end point of the titration. As impressive as the eye is, it is, nonetheless, a subjective detector. In view of the discussion of the DPD colorimetric test replacing the eye with a photometer is the "upgrade" needed to replace the subjective eye with an objective measuring device. But, in the case of titration, there is a tool with greater precision than an optical measuring device.

That tool is a variation of the familiar amperometric sensor for measuring dissolved oxygen or chlorine, the latter of which is covered next. Recall that this technique works by imposing a voltage across two electrodes and measuring the current that results from the analyte of interested being reduced. The amperometric probe in the titration analyzer can be of several configurations, but the most common one consists simply of a pair of platinum electrodes and lacks the membrane that most amperometric dissolved oxygen and chlorine probes contain. The cathode at which chlorine reduces is the usual inert metal—typically platinum. The reduction half-reaction under acidic conditions is:

$$\mathrm{HOCl} + \mathrm{H}^+ + 2\,e^- \rightarrow \mathrm{Cl}^- + \mathrm{H_2O};\ E^0 = 1.49\ \mathrm{V} \tag{6.20}$$

At even lower pH values the reaction is:

$$\mathrm{HOCl} + 2\,\mathrm{H}^+ + 2\,e^- \rightarrow \mathrm{Cl_2} + \mathrm{H_2O};\ E^0 = 1.63\ \mathrm{V} \tag{6.21}$$

The analogous reduction of OCl^- to Cl^- is:

$$\mathrm{OCl} + 2\,\mathrm{H}^+ + 2\,e^- \rightarrow \mathrm{Cl}^- + \mathrm{H_2O};\ E^0 = 0.81\ \mathrm{V} \tag{6.22}$$

Also, as noted in the previous section, one can measure free and combined chlorine by converting these species to triiodide, the latter of which reduces at the cathode

$$\mathrm{I_3}^- + 2\,\mathrm{H}^+ + 2\,e^- \rightarrow 3\,\mathrm{I}^- + \mathrm{H_2O} \tag{6.23}$$

We refer to all three possible chlorine-related species—HOCl, OCl,$^-$ or I_3^-—as the analyte. At the platinum anode, the exact opposite occurs. The reduced analyte, Cl^- or I^-, is oxidized back to the starting analyte. Why a platinum anode instead of the usual Ag anode where Ag oxidizes to Ag^+? Because we want to restrict the oxidation reaction to that of Cl^- and not Ag. This "recycling" of chlorine or triiodide results in its stable concentration over time and, therefore, a stable current. (Some titrators use an Ag-AgCl electrode to maintain a very stable potential between the cathode and anode. However, the titration endpoint is not as sharp.)

The sensor sits in the sampling container to which a titrant is added. Adding a titrating agent, or *titrant*, reduces the analyte, removes it from solution, and causes a drop in current. The titrant can be the thiosulfate that was used in the back titration. However, the current practice is to use phenyl arsine oxide (PAO) or C_6H_5AsO because it is stable over a wide pH range and gives a sharper titration endpoint than thiosulfate. In fact, the U.S. EPA Method 330.2 mandates the use of PAO for this technique. The arsenic atom in this compound is in the +3-oxidation state and gives up two electrons to convert to the +5 state, which then binds to two OH^- anions to become then neutral phenyl arsenic acid. The oxidation half-reaction of PAO by the analyte is:

$$C_6H_5AsO + 2\ H_2O \rightarrow C_6H_5AsO(OH)_2 + 2\ H^+ + 2\ e^- \qquad (6.24)$$

Note that this is yet another example, as with DPD, of a redox reaction that converts a neutral molecule into another neutral molecule, as opposed to a cation. Oxidation of PAO releases two electrons plus two H^+ ions and thus maintains electrical neutrality.

The combined redox reaction is, for HOCl:

$$C_6H_5AsO + HOCl + 2\ H_2O \rightarrow C_6H_5AsO(OH)_2 + 2\ H^+ + Cl^- \qquad (6.25)$$

The corresponding reactions for total chlorine converted to triiodide:

$$C_6H_5AsO + I_3^- + 2\ H_2O \rightarrow C_6H_5AsO(OH)_2 + 3\ I^- + 2\ H^+ \qquad (6.26)$$

A typical burette can dispense the PAO in volumes as low as 0.001 mL, which, means that the resolution of the titration is on the order of 0.001 mg/L (1 ppb). After the addition of each aliquot of PAO, a molar equivalent of analyte is consumed and the current drops. At the point at which all analyte has been consumed, further addition of PAO does not result in further change in current. The number of moles of PAO added up until this point is equal to the number of moles of free or total chlorine in solution. Figure 6.7 shows an automatic titration analyzer by Hach and an actual titration using

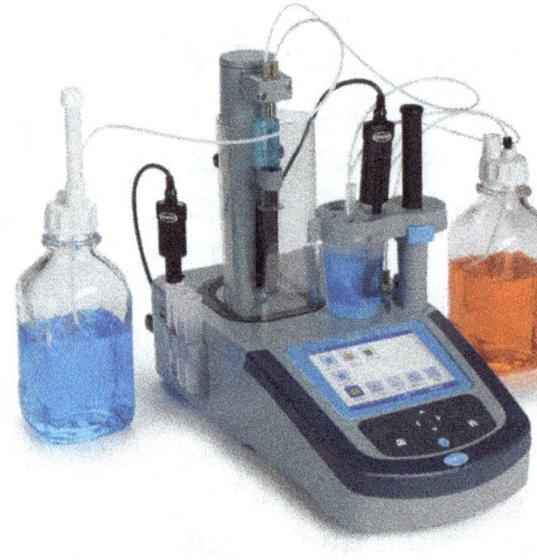

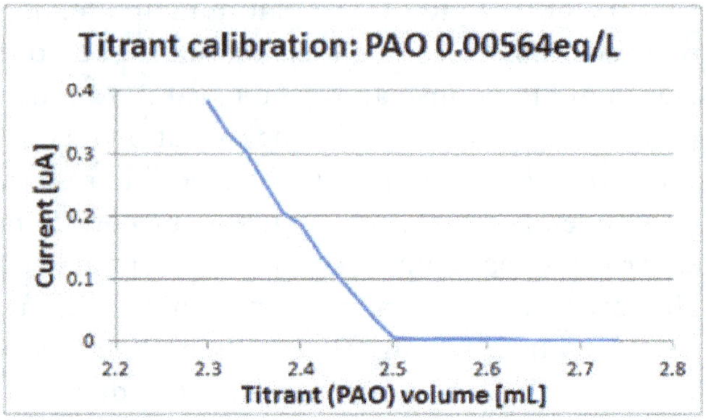

FIGURE 6.7 Amperometric titration to accurately determine concentrations of chlorine (and other oxidizing agents). Left: Hach AT1000 automatic titration analyzer. Right: Actual amperometric titration curve showing the complete reduction of residual chlorine to determine the concentration of residual chlorine with a resolution of 5 ppb (Courtesy Hach).

the analyzer. The extremely high sensitivity of the amperometric titrator and the precision of this measurement makes it capable of measuring chlorine concentrations down to 5 ppb. When it comes down to accuracy, it's the gold standard of titrations. However, for real-time monitoring, it is impractical.

One economical variation of the amperometric titrator is to use an ORP sensor, effectively converting the analyzer from an amperometric device to a potentiometric one. A more complete discussion of these methods, including the potentiometric version of titration, can be found in the review article by Hernlem and Tsai (2000).

As noted in Section 3.2, back titrations are useful for minimizing reactions between chlorine, or triiodide, and other reducing agents found in wastewater. This involves adding PAO as the reducing agent first, and in excess, so that all free and total chlorine is exhausted, and there is no amperometric current.

$$HOCl + C_6H_5AsO\ (excess) + 2\ H_2O \rightarrow Cl^- + C_6H_5AsO(OH)_2 + H^+ \quad (6.27)$$

The user then adds triiodide to oxidize the excess PAO:

$$I_3^- + C_6H_5AsO\ (excess) \rightarrow 3\ I^- + C_6H_5AsO(OH)_2 \quad (6.28)$$

As the user dispenses HOCl or triiodide, the current stays level, or zero, until the point at which all excess PAO has been consumed. Further addition of I_3^- enables its reduction at the anode and the current rises. The initial rise in the amperometric current marks the endpoint of the titration. In addition to minimizing interferences from other reducing agents, back titration is used when there is a delay between sample collection and measurement, resulting in evaporation of chlorine. Adding the PAO locks up the chlorine species and gives the user time to come back to complete the two-step procedure.

Although amperometric titration is the gold standard for laboratory chlorine measurement, it is subject to sources of error, mainly in the form of interferences from other redox species. Other oxidizing agents include oxygen containing anions, *oxyanions*, such as chlorine dioxide (ClO_2), ozone (O_3), hydrogen peroxide (H_2O_2), and potassium permanganate (K_2MnO_4). Bromide (Br^-) in seawater reacts with chlorine species to create brominated analogues such as HOBr and NH_2Br. Reducing agents such as sulfide and nitrite and low oxidation state metals such as Mn^{2+} and Fe^{2+} consume chlorine, but they can be ameliorated via back titration (Methods 330.2 and 330.4).

Table 6.1 summarizes the five U.S. EPA 330 methods. For a more thorough discussion of variations of colorimetric and amperometric techniques, see the excellent review article by Engelhardt and Markov (2013).

4.0 THE AMPEROMETRIC CHLORINE ANALYZER

4.1 The Two-Electrode Sensor

Reagent-based analyzers have been around for decades, and they will continue to be staples of water and wastewater facilities. For measuring free and total chlorine over a range of pH values they are the most accurate analyzers.

TABLE 6.1 Summary of EPA Methods 330.X. Back titration methods (2 and 4) are preferred when measuring total chlorine in wastewater. All but 330.5 are titrations.

EPA Method	Technique	Notes
330.1	Amperometric	PAO or $Na_2S_2O_3$ for amperometric detection.
330.2	Back Iodometric or Amperometric	Excess PAO or $Na_2S_2O_3$. Amperometric or starch indicator. Back titrate with I3-. Wastewater.
330.3	Iodometric with Starch Indicator	PAO or S_2O_3 for colorimetric (iodometric) titration.
330.4	DPD–FAS	DPD followed by FAS to titrate until clear.
330.5	DPD Spectrophotometric	DPD test measured photometrically.

They do have one constraint—reagents. The purchase of chemicals and the disposal of the waste stream both add to the total cost of ownership. The mechanics of adding reagent and measuring a colorimetric or amperometric change means that real-time measurement is limited by the rate at which the discrete measurements can be made. The use of the electrochemical chlorine sensor has been growing since 1970 when the first gas phase chlorine sensor was patented in 1973 (Oswin & Blurton, 1973).

With a couple modifications, the amperometric (Clark) dissolved oxygen sensor can be converted into an amperometric free chlorine sensor.

Like the Clark, cell the cathode of the chlorine sensor is a noble metal—gold (Au) or platinum (Pt) upon which the reduction of HOCl occurs:

$$HOCl + 2e^- \rightarrow Cl^- + OH^-;\ E^0 = 1.49\ V \tag{6.29}$$

Equation 6.21 is the analogous reaction for the reduction of OCl^-. The hypochlorite anion will convert to HOCl under acidic conditions—thus paving the way for measuring free chlorine without correction for pH. The oxidation reaction at the anode involves the oxidation of a metal that is usually silver and, occasionally, copper. The following discussions assume an Ag anode though the reactions for a Cu electrode is analogous. The oxidation is the familiar:

$$Ag + Cl^- \rightarrow AgCl + e^-;\ E^0\,(\text{sat'd KCl}) = -0.199\ V \tag{6.30}$$

Some texts leave out the role of the chloride ion in the anode reaction:

$$Ag \rightarrow Ag^+ + e^-;\ E = -0.800\ V \tag{6.31}$$

While this gives an accurate working picture of the sensor, it is not exactly correct. The standard state potential of Equation 6.31 is −0.800 V versus −0.236 V for Equation 6.30, so the presence of Cl^- ions at the anode favors enables the latter. As with the Clark cell, the electrolyte is usually KCl. The combined redox reaction is:

$$HOCl + Ag + Cl^- \rightarrow Ag^+ + 2\ Cl^- + OH^-;\ E = 1.29\ V \tag{6.32}$$

Including K^+ from the KCl electrolyte the redox reaction can also be written as:

$$HOCl + Ag + K^+ + 2\ Cl^- \rightarrow AgCl + KCl + KOH \tag{6.33}$$

As with the Clark electrode, the positive potential of 1.29 V ought to enable the chlorine sensor to proceed spontaneously. In fact, early chlorine sensors developed during the 1970s worked without an external voltage. However, the current was weak. The AgCl and KOH on the right-hand side of 6.33 are both relatively insoluble and precipitate out. Solid AgCl adheres to the anode while KOH forms a milky white suspension in the electrolyte (as it does in a galvanic dissolved oxygen probe). Both increase the resistance in the electrochemical pathway. As with the Clark electrode, the electrolyte resistance plus the anode surface resistance plus the charge transfer resistance add up to a voltage drop that necessitates an overpotential to drive Equation 6.32.

The optimum voltage for reducing the HOCl, or any other species of interest (including chlorine alternative), is typically chosen from performing voltammetric scans. Chapter 8, Section 3 will describe this technique. As noted in Chapter 5, Section 2, the potential varies with different cathode materials and cathode surface conditions. Most chlorine sensors use a gold cathode, whereas most Clark cells use a platinum cathode

The 1.29 V standard potention for chlorine reduction is much higher than the 0.20 V for the dissolved oxygen reduction, so one would expect that a lower polarization voltage is needed for the chlorine sensor. Indeed, the polariziation volteage for the amperometric chlorine probe is approximately 0.3 V whereas the voltage applied to a Clark cell is approximately 0.7 V. At this lower voltage, the chlorine sensor is minimally sensitive to dissolved oxygen. Similar to the Clark cell, the magnitude of current between electrodes is on the order of 100 nA per ppm chlorine.

The reduction of HOCl is the driving force behind the amperometric chlorine probe but any redox reaction will work. For instance, HOCl can be oxidized, given a large enough anodic potential:

$$6\ \mathrm{HOCl} + 3\ \mathrm{H_2O} \rightarrow 2\ \mathrm{ClO_3}^- + 4\ \mathrm{Cl}^- + 12\ \mathrm{H}^+ + \frac{3}{2}\mathrm{O_2} + 6\ e^- \qquad (6.34)$$

This is actually the principle of operation behind a chlorine probe with a reverse polarity (Kodera et al., 2004).

A schematic of the free chlorine sensor is shown in Figure 6.8. As the figure shows, the principle of operation is the same as that for the amperometric dissolved oxygen sensor. HOCl takes the place of oxygen as the analyte being reduced.

The membrane for a typical chlorine sensor is hydrophobic in order to reject water and OCl^- but pass the neutral HOCl. Most membrane materials consist of fluorinated hydrocarbon chains and include PTFE (Teflon) membrane, PP (polypropylene), PVDF (Kynar), and FEP (Teflon FEP). The common use of fluorinated polymers in amperometric sensors such as chlorine and dissolved oxygen means that these sensors are subject to interferences from all oxidizing agents, including chlorine, oxygen, and chlorine

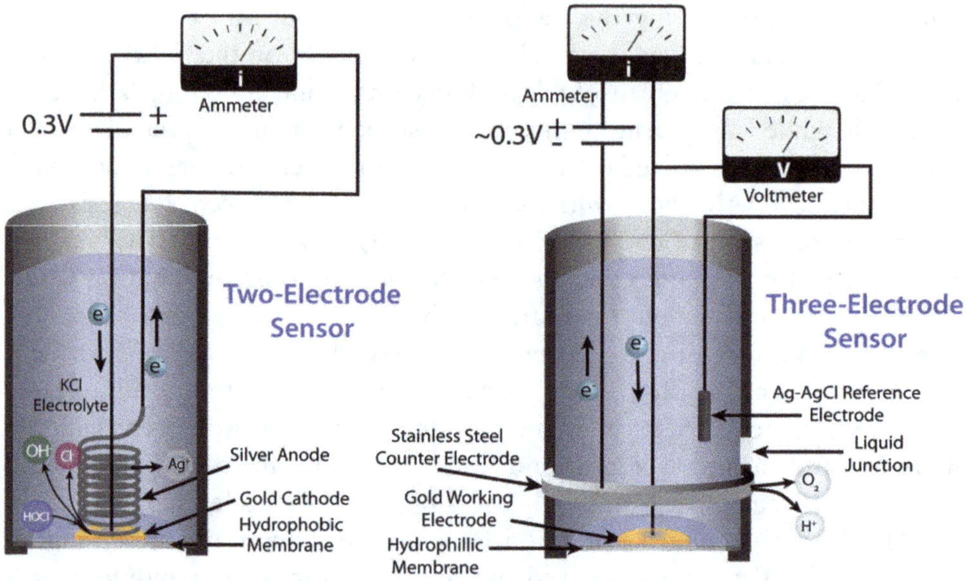

FIGURE 6.8 Schematic of amperometric free chlorine sensor. Left: is a two-electrode sensor.

dioxide. As with all membranes, fouling reduces the probe response over a period of weeks.

The pressure on the membrane affects its tension, which, in turn, affects the diffusion of HOCl through the membrane. To minimize signal changes due to pressure changes, most free chlorine analyzers add one additional component—a constant pressure *flow cell*, often in tandem with a pressure regulator. A flow regulator that is usually a variable area flow meter maintains a constant flow through the cell. The chlorine sensor sits in a hole drilled in the cell. Water flows across the sensor membrane and out of the cell into a drain. The pressure on the membrane is therefore close to atmospheric pressure (1 bar). Figure 6.9 shows the flow dependence of a typical probe for two membrane types. Figure 6.10 shows an amperometric probe in a flow cell with a pH probe for compensation and analyzer.

Free chlorine is defined as the sum of HOCl and OCl^-. Yet an amperometric chlorine probe only measures the redox-active HOCl species. The hydrophobic membrane ensures that only HOCl reaches the cathode. The fraction of HOCl that comprises the free chlorine concentration is a strong function of pH as shown in Figure 6.1. For pH values less than 5, that fraction is nearly 1. However, as the pH increases, the fraction of HOCl decreases. At a pH of 7, the fraction is about a half and, at a pH of 9, the fraction is close to zero. (Recall that pK_a is 7.58 at 25 °C.)

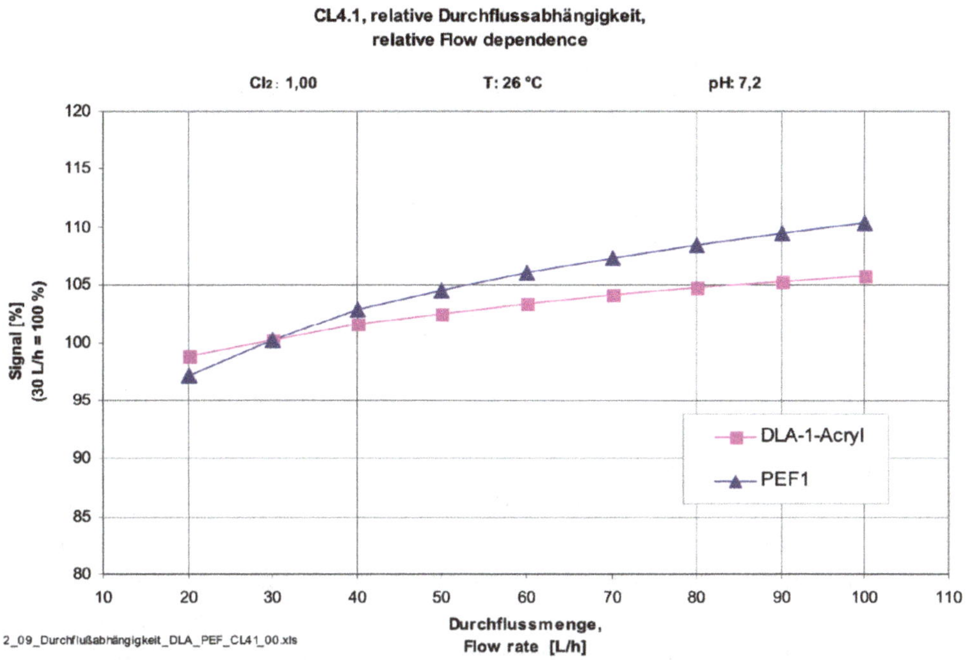

FIGURE 6.9 Flow dependence of chlorine sensor output. The sensor is a two-electrode probe with two different membranes, FEP (Courtesy Dr. Reiss).

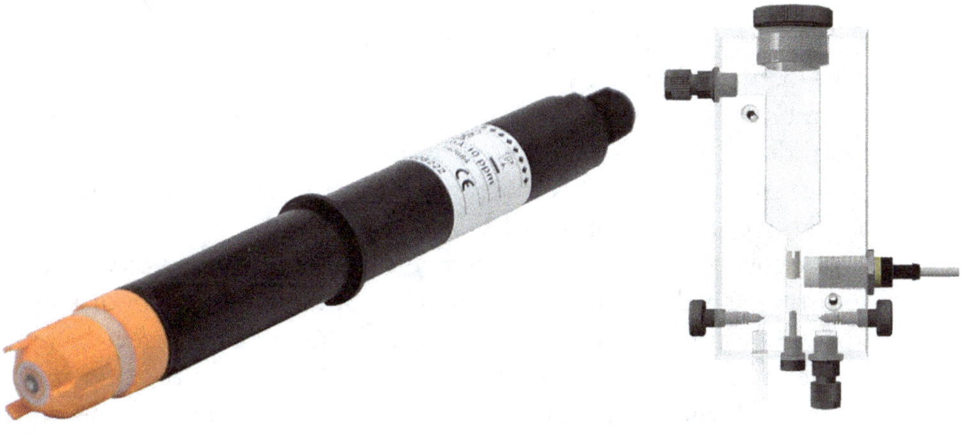

FIGURE 6.10 Left: Amperometric free chlorine sensor by Prominent. Right: Flow cell for maintaining constant pressure on membrane (Courtesy Prominent, ATI).

For pH values greater than 6, the analyzer must correct for the pH if we wish to determine the free chlorine concentration. A rearrangement of Equation 6.10 provides the correction we need:

$$[\mathit{Free\ Chlorine}] = [\mathrm{HOCl}] + [\mathrm{OCl}^-] = [1 + 10^{(pH-pKa)}][\mathrm{HOCl}] \quad (6.35)$$

Most chlorine analyzers allow the user to either manually enter the pH value or accept the input of an actual pH probe. In drinking water facilities, the pH for any stage of treatment is very stable, and the former is often sufficient.

As Equation 6.10 or Figure 6.1 imply, the compensation for pH becomes wildly inaccurate when the pH is greater than 9. For instance, at pH 9.5, the correction factor that converts the probe signal to free chlorine is a factor of 100. An alternative strategy for accounting for the hypochlorite concentration is to simply acidify the sample with a buffer that maintains a pH of 5 or less. This ensures that all OCl^- is converted to HOCl and that the reading of the analyzer is of the total free chlorine. The drawback to this approach is the need for a constant supply of acid.

Instead of conditioning the water sample with acid, some manufacturers condition the electrolyte in the probe. By switching the membrane from hydrophobic to hydrophilic, both HOCl and OCl^- pass through the membrane, and all the latter is converted to the former. The Rosemount 498CL probe uses a polyethersulfone membrane and saturated succinic acid ($C_4H_6O_4$) as the electrolyte. This weak acid has two pK_a values—4.2 and 5.6. Recall, from Section 4, that the pK_a value is the pH value at which the weak acid is most stable. The two pK_a values ensure that the pH remains below 6. According to the manufacturer, the probe's signal decreases less than 4% per pH unit from pH 6.5 to 10. The drawback of these probes is that the electrolyte passes through the membrane and needs to be replaced frequently—between 2 and 3 months. The membrane is also not as robust as hydrophobic ones need to be replaced frequently. JUMO claims that a probe signal that varies less than 1% variation in HOCl concentration over the pH range of 6.5 to 10. This sharply reduces the pH dependence of the signal on pH over a range of 4 to 12 (Schleicher, 2008).

In our discussion of the effect of temperature on dissolved oxygen sensors, we distinguished between the real effect of temperature on dissolved oxygen probe output due to Henry's law and the artificial effect of temperature due to membrane diffusion. Because it operates on the same principle, the free chlorine probe should show the same real and temperature dependence. But it doesn't. Oxygen in the atmosphere dissolves in water in accordance with Henry's law. The law is valid as long as the oxygen in the air is in equilibrium with the oxygen dissolved in water. There is very little HOCl in the air (hopefully) so Henry's law does not apply. The HOCl dissolved in water is a "sometime thing." Leave a glass of chlorinated water out in the open and, eventually, all the HOCl will "outgas" in a futile attempt to reach equilibrium with HOCl in the air

The major effect of temperature on the free chlorine probe is the dissociation of HOCl to OCl^- . As the temperature increases the dissociation increases, [HOCl] decreases and the probe output drops. The dissociation constant (Ka) is 1.5×10^{-8} at 0 °C, 2.9×10^{-8} at 25 °C and 3.44×10^{-8} at 35 °C. With a simple calculation using Equation 3.36, we find that the concentration of HOCl decreases by about 50% from 0 to 35 °C (Morris, 1966). To my knowledge, the typical analyzer does not account for this temperature effect as it does for the pH induced change in [HOCl].

As we saw in Chapter 5, Section 3.4, an increasing temperature results in an increasing membrane permeability that must be corrected. As the sensor temperature increases, the permeability increases. As with the dissolved oxygen sensor, the probe design balances time response against signal signal stability.

A third variation of the amperometric probe is the open sensor—that is, a sensor without a membrane. Without a membrane, the electrodes respond faster to chlorine concentration changes. Response time to reach 90% of its final value (T_{90}) is about 30 seconds. The response time for membrane sensors varies from 30 seconds to 2 minutes. The open sensor can work at pressures up to 8 bars. Without a membrane, the sensor responds to all oxidants. Open cell probes usually use copper anodes to generate sufficient anodic current in the sample. Copper readily forms oxides, which can be cleaned with a 10% nitric acid solution for about 5 minutes. Some manufacturers use a polishing mechanism in which small balls made of ceramic or plastic circulate around the anode. The gentle abrasion of the balls against the electrodes removes the oxide coat. At least one manufacturer (Dr. Kuntze) uses a self-cleaning mechanism in which the polarity on the electrodes is reversed and the resulting electrolysis produces hydrogen and oxygen gas.

The simplicity and fast response make the open sensor an attractive option when there are few other oxidizing agents other than free chlorine. However, one significant drawback is that, without the membrane, the process is in constant contact with the electrodes, which can then foul them over a short period (hours to days). Without a membrane to contain the KCl solution, the Ag-AgCl anode no longer works.

The common underlying principle of operation for the Clark dissolved oxygen and the free chlorine sensor prompts the question: Why not a galvanic chlorine sensor? This is especially pertinent because polarization voltage for the free chlorine sensor is less than that for the dissolved oxygen sensor. Perhaps the polarization voltage is close enough to zero that the distinction is minor. To date, I have encountered one manufacturer who makes a galvanic probe but, at this point, I cannot give an adequate explanation. The other open question is why the chlorine sensor output is so sensitive to flow and pressure on the membrane that a flow cell is necessary, but it is not so with the Clark or galvanic dissolved oxygen sensor.

4.2 Total Chlorine

Section 3.2 discussed adding KI and acid to a sample to convert HOCl, OCl^-, and combined chlorine (chloramines) to the single oxidizing agent, triiodide (I_3^-). This allows the determination of total chlorine in a single titration measurement. The same strategy works for converting a free chlorine sensor to a total chlorine sensor. There are two ways to make a total chlorine analyzer:

1. Use a standard free chlorine sensor but inject an acidic KI solution into a so-called conditioning system that precedes the sensor. I^- reacts with chlorine species to create triiodide (I_3^-). Vinegar is a low-cost effective acid that is frequently used. The cathode of the free chlorine probe will then oxidize the triiodide and generate a current that is proportional to the total chlorine concentration.

$$I_3^- + 3\,e^- \rightarrow 3\,I^- \tag{6.36}$$

   The Rosemount TCL analyzer shown in Figure 6.11 is an example of this system.

2. Modify a free chlorine sensor by adding the acidic KI reagent directly into the electrolyte of the sensor. The cathode reduction reaction is the same as that of the conditioning system. This eliminates the need for a conditioning system but does require a more expensive electrolyte refill.

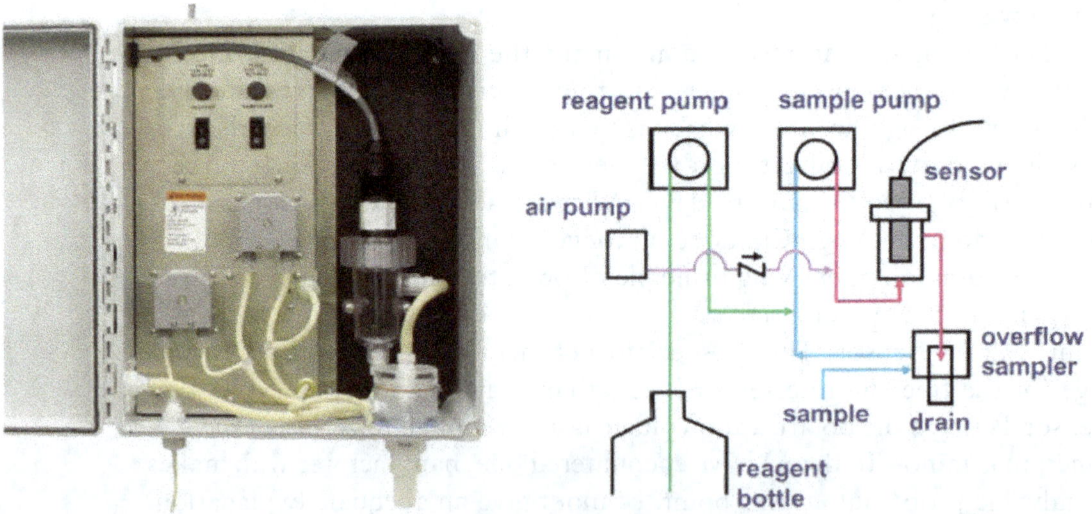

FIGURE 6.11 The Rosemount TCL total chlorine analyzer. The reagent bottle houses a KI-vinegar solution that is combined with the sample prior to being pumped into the overflow sampler (Courtesy Rosemount Analytical).

Calibration of a sensor can be a simple exercise with a DPD colorimetric test kit or a more accurate affair using the U.S. EPA titration methods summarized in Table 6.1. Both have variations for free, combined, or total chlorine. The DPD-1 test measures free chlorine. DPD-3 adds an acid and KI to the DPD reagent to measure total chlorine. Subtracting the DPD-1 free chlorine measurement from the DPD-3 total chlorine measurement yields combined chlorine (DPD-2). As discussed in Section 3, back titration with sodium thiosulfate increases the accuracy and minimizes interference with other oxidizing agents.

4.3 The Potentiostatic (Three-Electrode) Sensor

The amperometric sensor based on an anode and cathode has been the standard model for detecting oxidizing compounds. The anode has two jobs:

1. The anode supplies the electrons needed by the cathode for reducing the analyte.
2. The potential of the anode must remain constant so that the potential on the cathode is also stable. The anode maintains the potential of an Ag-AgCl reference electrode as long as the current is not large enough to perturb the potential from its steady state potential (0.199 V in saturated KCl) as predicted by the Nernst equation.

These two requirements appear to be at cross purposes. One solution to maintaining a constant potential at the anode is to simply make the anode as large as possible and, thereby, reduce its current density while keeping the total current constant. That approach is typically followed in two-electrode sensors, as shown in the Mettler Toledo Clark dissolved oxygen probe in Figure 5.10.

Another solution to the conflicting roles of the anode is to split it into two electrodes—one to provide current for the cathode and the other to provide a stable potential against which the cathode can operate. This is the *raison d'être* of the three-electrode sensor. The *counter electrode* provides the current, and the *reference* electrode provides the reference voltage (and draws no current). The cathode is usually called the *working* electrode. The working electrode is not always the cathode, and the counter electrode does not always source electrons. Though they are not covered in this book, there are electrochemical sensors that measure reducing compounds, such as hydrogen sulfide. In this case, the anode becomes the working electrode, and the cathode becomes the counter electrode. The resulting three-electrode sensor is often called a *potentiostatic* sensor. Be aware, however, that some writers use this term incorrectly for two-electrode sensors.

The right side of Figure 6.8 shows, conceptually, how the potentiostatic sensor works. The reference electrode (RE) draws no current. The potentiostatic circuit maintains a constant voltage between the reference and working electrode. It does that by adjusting the voltage that drives current into (or out of) the counter electrode. What might change the potential of the working electrode? The reduction of HOCl (or OCl^- or I_3^-) on its surface, which is proportional to its concentration. The voltage required to keep the potentials of the working electrode and reference electrode equal is therefore proportional to the free chlorine concentration.

There will always be a finite potential difference between the working electrode and reference electrode, due to the resistance of the electrolyte between them. To minimize the difference, the two electrodes need to be as close together as possible. The counter electrode must be large enough to supply the current needs of the working electrode. It is typically a ring electrode on the outside of the probe, as shown in Figure 6.12. Doing so necessitates using a hydrophilic membrane, which allows both HOCl and OCl to pass through the membrane and be reduced. This yields the added benefit of minimizing the pH dependence of the probe output—in contrast

FIGURE 6.12 Close-up of a three-electrode (potentiostatic) sensor. The counter electrode is on the outside of the probe and has a much larger surface area than the working electrode (the tip of the probe). This arrangement necessitates a hydrophilic membrane over the working electrode to complete the circuit between the working and counter electrodes (Courtesy Dr. Reiss).

to the hydrophobic membrane of the two-electrode sensor. The disadvantage is that the electrolyte fouls more quickly and must be changed.

Because the voltage between the reference electrode and, by extension, the working electrode, is more tightly controlled than that of the two-electrode sensor, the three-electrode sensor can maintain its accuracy over a wider range of both free chlorine concentration and pH range. At present, I have been unable to find measurement data that demonstrates this advantage but I have seen that one manufacturer states that its two-electrode sensor can maintain ±0.1 ppm accuracy over a range of 0–10 ppm and pH range between 5 and 9, whereas its three-electrode sensor can maintain that accuracy over a range of 0–20 ppm and pH range between 4 and 12. Another advantage is that the counter electrode is constructed of an inert metal, such as platinum or stainless steel. The electrons that supply the working electrode come from the power supply to the circuit in contrast to the silver anode of the two-electrode sensor, which continuously sheds silver in as Ag^+ ions to create a flow of electrons.

For applications, such as municipal water treatment, in which the pH and free chlorine ranges are tightly controlled and free chlorine rarely exceeds 8 ppm, the cost of the three-electrode sensor may not justify the added expense. For applications in which the chlorine concentration varies widely or for which accuracy is paramount, the three-electrode sensor is a good choice.

5.0 INTERFERENCES AND OTHER SOURCES OF ERROR

As with all membrane-based sensors, cross sensitivities to interfering species results in redox reactions adding to that of HOCl. At the cathode, three reduction reactions involve common oxidizing agents:

$$ClO_2 + e^- \rightarrow ClO_2^- \qquad (6.37)$$

$$O_2 + 2\ H_2O + 4\ e^- \rightarrow 4\ OH^- \qquad (6.38)$$

$$O_3 + 2\ H_2O + 2\ e^- \rightarrow O_2 + 2\ OH^- \qquad (6.39)$$

Not specifically listed above are free chlorine alternatives, such chlorine dioxice and monochloramine.

At the anode, or working electrode, one common interfering oxidation reaction is that of sulfate:

$$SO_3^{2-} + H_2O \rightarrow SO_4^{2-} + 2\ H^+ + 2\ e^- \qquad (6.40)$$

As with all amperometric sensors, the chlorine probe has four components that need maintenance: the cathode, anode, electrolyte, and membrane.

Replacement of the electrolyte is typically done every several months. It's usually convenient to replace the membrane at the same time. The gold or platinum cathode is inert and requires little or no maintenance. Household vinegar will remove most deposits. The silver anode oxidizes to form AgCl and leaves a fresh surface as it does so. Occasional cleaning with household ammonia usually suffices.

Table 6.2 summarizes the most common issues for the two-electrode probe with pH compensation.

6.0 HOW DO ORP MEASUREMENTS STACK UP AGAINST FREE CHLORINE MEASUREMENTS?

Chapter 2, Section 10.1 discussed the feasibility of using ORP measurements to measure the disinfection level in a water or wastewater treatment facility. Because the U.S. EPA mandates a minimum residual of 0.2 mg/L, there is

TABLE 6.2 Common problems affecting amperometric free chlorine sensors. This table does not include problems found specifically in open cell or three-electrode sensors. It also assumes pH compensation is operable.

Problem	Cause	Solution
Signal drifts on a short time scale	Membrane pressure/flow is not constant	Measure flowrate using a graduated container; adjust flow
Sensor reads high compared to DPD test	Other oxidants (e.g., O_3, H_2O_2, ClO_2) inflate the sensor reading	Restrict interfering oxidants
Sensor reads low compared to the DPD test	Calibration done at lower pH than actual measurements	Repeat calibration with known pH
	Cathode is coated	Clean cathode with fine abrasive
Very slow response	Fouled membrane (biofouling, grease, or scale)	Change membrane. For scaling, soften water
	Flowrate too low	Increase flow; ensure that flow outlet is open to atm
Slow response and/or difficult to calibrate	Contaminated electrolyte: Contains oxidants or depleted in KCl	Run probe in change electrolyte
	Surfactants (e.g., detergents) convert hydrophobic membrane to hydrophilic one	Remove source of surfactant

no chance that ORP sensors will supplant all chlorine analyzers. Nonetheless it is worth revisiting the strategy of using ORP measurement, wherever possible, as a surrogate for chlorine measurement.

As Figure 6.10 shows, there is a correlation between the ORP potential and the free chlorine concentration. If chlorine is the major oxidizing agent in the water, ORP will correlate to chlorine concentration. If the value is right—and that usually means greater than 650 mV—disinfection will be effective. Free chlorine measurement cannot distinguish between the OCl^- anion and the more potent HOCl.

The correlation between ORP and free chlorine depends on the pH and temperature. Even if the temperature is fixed at the standard 25 °C, there are two variables—pH, free chlorine (and, to a lesser degree, temperature)—upon which the ORP value depends. This means that plotting ORP versus free chlorine concentration has to be done as a family of curves, with each curve representing a different pH value. Figure 6.13 (based on data by Lambrou, 2012) shows a family of plots of ORP versus free chlorine concentration (ppm) for pH values between 7.0 and 8.2.

Constructing the curves shown in Figure 6.13 gives a set of calibration data that can then be input into the firmware of an ORP analyzer, thus turning the ORP analyzer into a de facto chlorine analyzer. This is the basis of

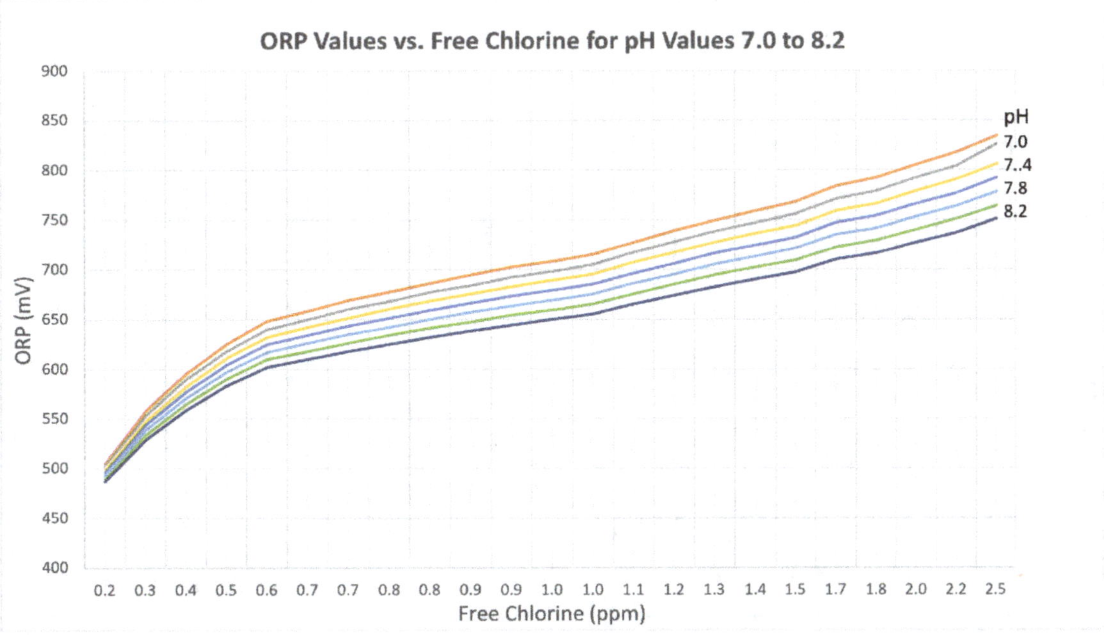

FIGURE 6.13 Correlation between ORP, free chlorine concentration, and pH (Data from Lambrou, 2012).

low-cost chlorine analyzers that are based on ORP measurement. They are typically portable units that also measure pH.

It's important to stress that an equivalent-free chlorine measurement should never be construed as a real free chlorine measurement. As discussed in Chapter 2, Section 7.2, ORP measurements are typically accurate only to within approximately 10 mV. A second look at Figure 6.13 shows that a chlorine analyzer can display accurate readings over a wide range of chlorine concentration from 0.2 to more than 10 ppm. The corresponding range of an ORP sensor is a much more compressed one of 500 to 800 mV. A precision of approximately 10 mV translates to, at best, 3% precision in the disinfectant level. The precision of an amperometric sensor is approximately 1%. The root cause of the diminishing accuracy of ORP sensors with increasing chlorine concentration is simply that ORP varies logarithmically with concentration.

As we noted in Chapter 2, Section 7.3, ORP measurements are slow and can take more than 20 minutes to complete. An amperometric chlorine sensor has a typical response time of 20 to 30 seconds. This may not be critical in most water or wastewater treatment facilities but is still a disadvantage. For measuring the residual chlorine in the distribution system before and during flushing, the response time of the ORP sensor is a serious problem.

To validate whether ORP or chlorine measurements are more effective for disinfection, one can take measurements of bacterial concentrations at varying concentrations of chlorine and take simultaneous ORP measurements. At a wastewater treatment facility in Kankakee, Illinois, Kim and Hensley (1997) made the measurements that are displayed in Figure 6.14.

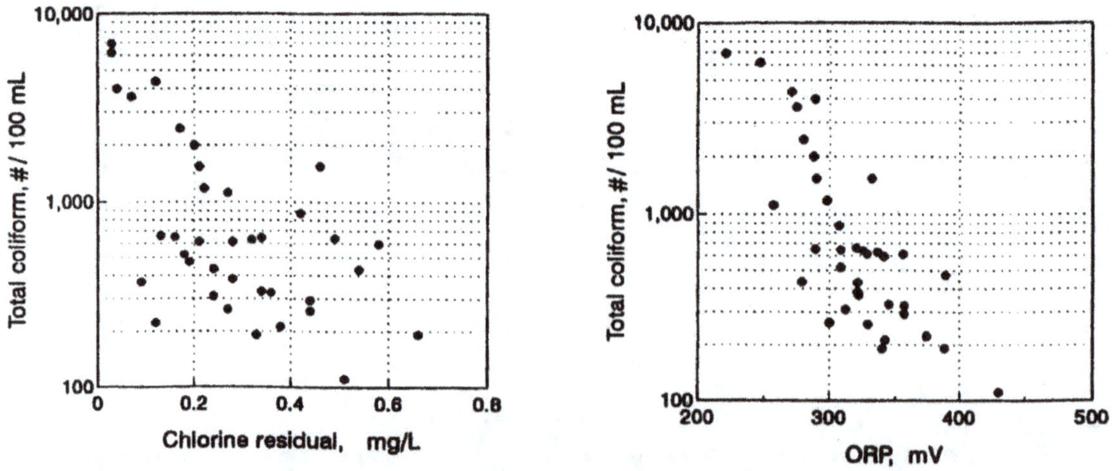

FIGURE 6.14 Disinfection efficiency (as measured by total coliform counts) versus free chlorine concentration and ORP. Comparison of the two shows a stronger correlation with ORP than chlorine (Kim & Hensley, 1997).

The pair made amperometric titration measurements as described in Section 3.3 alongside standard ORP measurements. The total coliform count is a high-level U.S. EPA–mandated method for assessing bacterial contamination. It consists of counting the number of colonies of coliform bacteria in a culture medium after incubation. When total coliform counts are plotted against the free chlorine concentration, as shown on the left side in Figure 6.14, the correlation is very weak. However, when total coliform counts are plotted against ORP values, they fall into a more recognizable downward sloping line. The authors reported reduced costs of chlorination by 47% and dichlorination by 62% compared to dosing of these chemicals based on chlorine measurement.

Not all such experiments have yielded such a sharp contrast between ORP and chlorine measurements. Diane Phelan measured fecal coliform counts at an unidentified 4.5 mgd wastewater treatment facility (Phelan, 1996). Fecal coliform counts are a subset of total coliform counts and contain only those bacteria species that are found in human and animal guts. This study differed from that of Kim and Hensley in that Phelan did not vary the chlorine concentratin and measure the resuling ORP and coliform counts. Instead, she measured all three parameters during the normal course of operation. Therefore, these experiments can't draw any definitive correlations between chlorine measurement or ORP and disinfection. What it did show is that the chlorine measurements were more stable than the ORP measurements. The explanation offered was that there were variations in pH and other redox species not accounted for.

Last, Victorin et al. (1972) measured ORP values and concentrations for seven different disinfectants. Out of the seven, only hypochlorite and monochloramine are in common use today. For both disinfectants, the correlation between the kill rate of *E. Coli* bacteria correlated more strongly with the ORP value than the disinfectant concentration.

The bottom-line answer to the question posed at the beginning of this section is this: an ORP measurement is an effective substitute measurement of free chlorine if:

1. The time response of less than 20 minutes is not critical.
2. The pH is stable.
3. The chlorine concentration is lower than approximately 10 ppm.
4. The number of species that can be easily oxidized is limited and constant.

Oxydation-reduction potential measurements are especially effective indicators of disinfecting efficacy if the pH of the water is likely to rise above 7.5 at which point most of the free chlorine is in the form of hypochlorite.

To use ORP measurements as a disinfectant measurement tool, one would need to calibrate the ORP sensor(s) against a standard such as a DPD measurement or a calibrated amperometric one and do so at a specific pH and temperature. With all this said, regulatory requirements, at least for U.S. public treatment facilities, are likely to require direct free chlorine measurements, thereby relegating ORP measurements as supplementary.

7.0 CHLORINE ALTERNATIVES

7.1 Chlorine's Achilles Heel

This chapter has dealt with free chlorine as the primary disinfectant in water treatment. Though it is used for disinfection by 83% of surface water treatment facilities and 86% of groundwater treatment facilities (U.S. EPA, 2002), free chlorine either as a primary or secondary disinfectant has one major drawback. It reacts to form undesirable disinfection by-products, or DBPs. Free chlorine reacts with methane to form a series of three chlorinated compounds in a series of reactions that are analogous to the formation of chloramines:

$$CH_4 + HOCl \rightarrow CH_3Cl + H_2O \qquad (6.41)$$

$$CH_3Cl + HOCl \rightarrow CH_2Cl_2 + H_2O \qquad (6.42)$$

$$CH_2Cl_2 + HOCl \rightarrow CHCl_3 + H_2O \qquad (6.43)$$

$$CHCl_3 + HOCl \rightarrow CCl_4 + H_2O \qquad (6.44)$$

Bromine reacts with methane in the same way and one can revise the chemical reactions above simply by substituting one or more Br atoms for Cl atoms. Although brominated methane compounds are part of the family of DBPs, it is the chlorinated compounds that are a problem for water treatment facilities because of the use of chlorine as a disinfectant. The family of methane compounds in which Cl and/or Br atoms—collectively called halogen atoms—substitute for H atoms is called *trihalomethanes* or THMs. The name is a misnomer because the "tri" implies that all the H atoms are substituted by halogen atoms, which is not always the case.

Free chlorine also reacts with acetic acid (CH_3COOH) to form haloacetic acid along reactions that are analogous to those for the formation of chlorinated methane listed above:

$$CH_3COOH + HOCl \rightarrow ClCH_2COOH + H_2O \qquad (6.45)$$

$$ClCH_2COOH + HOCl \rightarrow Cl_2CHCOOH + H_2O \qquad (6.46)$$

$$Cl_2CHCOOH + HOCl \rightarrow Cl_3CCOOH + H_2O \qquad (6.47)$$

There are analogous reactions in which Br atoms substitute for Cl atoms. These DBPs are called *haloacetic acids*, or HAAs.

All these DBPs carry varying degrees of adverse health effects, and some are considered to be carcinogenic. The Disinfection By-product Rule—Stages 1 and 2—represent the U.S. EPA's attempts to limit their concentrations in drinking water systems and strengthen reporting requirements. To minimize the formation of DBPs during disinfection, water treatment facilities minimize the amount of organic matter that reaches this late stage of treatment. Water that leaves the treatment facility with free chlorine residual can still form DBPs from organic matter in the distribution system.

The handling of chlorine gas to produce free chlorine is also fraught with the serious risk of a gas release. For these reasons alternatives to chlorine have been expanding in use. Because all of them (except for UV) are strong oxidants, they can all be measured amperometrically. The major adjustment an amperometric sensor makes to tune its sensitivity to the oxidant in question is the cathode potential. It correlates to the reduction potential but is always higher due to the overpotential needed to overcome constraints brought up in Chapter 5, Section 3.0.

7.2 Monochloramine

An increasing number of facilities are switching to monochloramine (MCA) as a primary of residual disinfectant. Not only does it not form the common DBPs of THMs and HAAs; it is also stable for a long time, which is necessary for maintaining a residual concentration of disinfectant throughout a distribution system. It has other benefits compared to free chlorine: it penetrates biofilms more effectively, has lower operating costs, and doesn't generate taste and odor problems. Among WRRFs that draw upon surface water, the vast majority add monochloramine as the residual disinfectant.

Monochloramine's major drawback, relative to free chlorine, is that it is not as reactive as free chlorine. Its standard reduction potential ranges from 1.45 V to 0.74 V, depending on pH, as shown in Equations 6.51 and 6.52. The pH dependent reduction potentials holds for all the disinfectants that follow. MCA's disinfection efficacy is only about 5% of that of HOCl. It does produce other DBPs, including N-nitrosamines. The decomposition of MCA to ammonia or the incomplete reaction between ammonia and hypochlorite can result in high levels of ammonia and require higher levels of nitrification.

Passing ammonia through hypochlorite solution creates monochloramine:

$$NH_3 + OCl^- \rightarrow NH_2Cl + OH^- \text{ OR } NH_3 + HOCl \rightarrow NH_2Cl + H_2O \qquad (6.48)$$

If these two ingredients are in equal proportion, Equation 6.48 results in only monochloramine. As previously described in breakpoint chlorimaniation (Section 2) a surplus of hypochlorite will substitute one or both H atoms of NH_2Cl with more Cl atoms:

$$NH_2Cl + OCl^- \rightarrow NH_2Cl + OH^- \qquad (6.49)$$

$$NHCl_2 + OCl^- \rightarrow NCl_3 + OH^- \qquad (6.50)$$

Dichloramine, $NHCl_2$, is the product in Equation 6.49. Further chlorination as shown in Equation 6.50, creates trichloramine, NCl_3, also called nitrogen trichloride. These two unwanted amines create the familiar "chlorine" odor in swimming pools and is an indicator that there is more ammonia than free chlorine. Both reactions, in conjunction with LeChatelier's Principle, show that lowering the pH drives the reactions to the right and exacerbates the production of these unwanted species. Conversely, raising the pH maintains the maximum fraction of chloramine species existing as monochloramine.

Water treatment facilities that use chloramine as a residual disinfectant maintain as close to a one-to-one ratio of ammonia to free chlorine to ensure that monochloramine is the predominant chloramine species. The total concentration of chloramines is called *combined chlorine*. *Total chlorine* is the sum of free and combined chlorine. The total concentration of free chlorine and monochloramine in the distribution system is *chlorine residual*. If this is confusing, just remember that combined chlorine is "good" as long as it is most or all monochloramine.

The most common method for measuring monochloramine is to measure total chlorine through the conversion of free chlorine and combined chlorine to triiodide, as discussed in Section 3.2. The difference, as you will recall, is combined chlorine, which is mostly MCA.

There are amperometric probes which measure MCA directly. The reduction of MCA, in an acidic medium, at the cathode is (Rajasekharan et al., 2007):

In a neutral or basic medium, the reaction is:

$$NH_2Cl + H_2O + 2e^- \rightarrow NH_3 + OH^- + Cl^-;\ E^0 = 0.74\ \text{V} \qquad (6.51)$$

$$NH_2Cl + 2\ H^+ + 2e^- \rightarrow NH_4 + Cl^-;\ E^0 = 1.45\ \text{V} \qquad (6.52)$$

The operating voltage is appprxomately 0.3 V. This is the same polarization voltage used to detect HOCl, so the probe can only work in the absence of free chlorine. At other potentials, the reduction of oxygen strongly interferes with MCA detection (Lee et al., 2015).

The polarization voltage for MCA is in the same range (0.2 to 0.5 V) for all the disinfectants mentioned in this section so there are cross sensitivities between them. The actual polarization voltage for each disinfecting species is a trade secret to probe manufacturers and it has proven to be very difficult to ascertain the precise potential for each.

There are a small number of manufacturers that manufacture amperometric probes that measure MCA directly. Some sensors, for example, the Rosemount MCL sensor, are two-electrode amperometric sensor that subsistute the electrolyte from KCl to KBr, buffered by acetic acid. The anode oxidizes the bromide (Br^-) anions provided by the KBr to bromine (Br_2) in the presence of H^+ ions supplied by the acetic acid:

$$NH_2Cl + 2\ Br^- + H^+ \rightarrow NH_3 + Br_2 + Cl^- \quad (6.53)$$

The Br_2 reduces back to bromide at the cathode. The KBr is an example of a *redox mediator*. This strategy extends the concentration range, sensitive or electrolyte stability of other oxidant measuring sensors.

7.3 Chlorine Dioxide

A third disinfectant is chlorine dioxide, ClO_2. It has traditionally been used for bleaching paper pulp, but it has found increased use in water treatment and food disinfection. There are three possible chemical reaction pathways that generate ClO_2, all of which start with sodium chlorite, $NaClO_2$:

$$2\ NaClO_2 + Cl_2 \rightarrow 2\ ClO_2 + 2\ NaCl \quad (6.54)$$

$$2\ NaClO_2 + HOCl \rightarrow 2\ ClO_2 + NaCl + NaOH \quad (6.55)$$

$$5\ NaClO_2 + 4\ HCl \rightarrow 4\ ClO_2 + 5\ NaCl + 2\ H_2O \quad (6.56)$$

This gas has several advantages over free chlorine:

1. It has a reduction potential of about 1.6 V, which is 0.2 V higher than that of HOCl. Its reactivity and resulting disinfection efficacy are equal to that of free chlorine.
2. ClO_2 attacks biofilms more aggressively than free chlorine. This makes it very effective against Legionella and Giardia.

3. ClO_2 is effective at oxidizing Fe^{2+} and Mn^{2+}, so that their oxides can be removed as precipitates.
4. ClO_2 is stable over a wide range of pH values.
5. In the presence organic matter, it forms lower concentrations of distribution by-products—trihalomethanes and haloacetic acids.

Chlorine dioxide's major drawback is that the gas is explosive, so it must be generated on-site. The cost of generated on-site generation is another drawback. Because ClO_2 is such an effective oxidizing agent, it can only be used in finished water that has low concentration of organic carbon. This limits many treatment facilities from using it. Chlorine dioxide decomposes to chlorate, a carcinogen, when exposed to sunlight.

Like all oxidizing agents, chlorine dioxide can be measured by DPD colorimetric tests. Probes to measure ClO_2 are based on the typical amperometric design. The reduction reaction of ClO_2 at the cathode at a neutral pH is:

$$ClO_2 + e^- \rightarrow ClO_2^- \tag{6.57}$$

Under acidic conditions the reduction reaction is

$$ClO_2 + 4\,H^+ + 5\,e^- \rightarrow Cl^- + 2\,H_2O;\ E^0 = 1.6\ V \tag{6.58}$$

A variation of chlorine dioxide targeted to be used as a spray or dipping bath for food is *acidified sodium chlorite*. Sodium chlorite is dissolved along with a food grade acid, such as citric, phosphoric, hydrochloric, or malic acid, to generate chlorous acid, $HClO_2$.

$$NaClO_2 + H^+ \leftrightarrow HClO_2 + Na^+ \text{ or } ClO_2^- + H^+ \leftrightarrow HClO_2 \tag{6.59}$$

The second formulation, along with the double arrow, makes it obvious that the pH-dependent ClO_2^- and $HClO_2$ equilibrium is analogous to that for OCl^- and HOCl. At a pH 2.3, 31% of the mixture is chlorous acid (Rao, 2002).

7.4 Peracetic Acid

Another disinfectant gaining popularity in food processing and, more recently, in wastewater treatment is peracetic acid, or peroxyacetic acid (PAA). Both PAA and hydrogen peroxide (H_2O_2) are peroxides, which means they contain two oxygen atoms joined by a single pair of electrons (*single bond*). Peroxides are energetically disposed to reacting with other compounds that can break the O-O bond. In contrast, the oxygen molecule's two O atoms are joined by two pairs of electrons (*double bond*), which makes it much more stable. PAA is prepared on-site by the two peroxides:

$$CH_3CO_2H + 2\ H_2O_2 \leftrightarrow CH_3CO_3H + H_2O \qquad (6.60)$$

An excess of H_2O_2 ensures that the final mixture contains both peroxides, both of which are oxidizing agents. PAA can be made commercially and sold in concentrations of approximately 15% PAA and 10% to 30% H_2O_2. Alternately acetic acid and hydrogen peroxide can be mixed onsite for convenience and savings.

Peroxyacetic acid's actual disinfecting agents are the free radicals: hydroxyl (OH•) and hydrogen peroxy (HO_2•). Both work by lysing bacterial cell membranes. We'll return to the hydroxyl radical in Section 7.6.

Peroxyacetic acid acid is effective against a very wide range of bacteria, viruses, and fungi and penetrates biofilms more effectively than chlorine. It penetrates cell membranes more quickly than H_2O_2 and breaks down sulfur bonds in proteins (Bettenhausen, 2020). A double-barreled approach combines PAA with UV radiation—the latter which damages DNA and RNA in pathogens and, at high doses, creates hydroxyl radicals from the residual H_2O_2 to introduce advanced oxidation processes (Wittbold & Kim, 2021).

Because PAA does not form DBPs and decomposes into acetic acid, oxygen, and water, it is finding increasing use in wastewater treatment. In addition, it works when UV radiation is ineffective due to absorption by colored wastewater. Other markets, besides food processing and wastewater treatment, include industrial cleaners, agriculture, and algae control. On the horizon is the simpler performic acid, HCO_3H.

Peracetic acid has a reduction potential of 1.8 V, which is 0.4 V higher than that of HOCl. Its reduction reaction is, at pH 7 (Zhang et al., 2018):

$$CH_3CO_3H + e^- \leftrightarrow CH_3CO_3H + H_2O;\ E^0 = 1.39\ V \qquad (6.61)$$

In acidic medium, the reaction is:

$$CH_3CO_3H + 2\ H^+ + 2\ e^- \leftrightarrow CH_3CO_2H + H_2O;\ E^0 = 1.75\ V \qquad (6.62)$$

7.5 Ozone

Ozone (O_3) is the jack-of-all-trades chemical for disinfection and other oxidation processes. It is used prior to oxidation, prior to filtration or as a final disinfectant. As the oxidant with the highest reduction potential, it is the most effective at breaking down organic matter. Ozone does not oxidize all organic compounds all the way to CO_2, but it does break them down to smaller compounds that are more easily filtered by biologically active filters (sand and activated carbon). Doing so lessens the load on filters and enables cleanings to be done on a less frequent basis. The smaller organic compounds ozone produces also coagulate more effectively and are more easily removed during flocculation and settling.

Cryptosporidium and *Legionella* are protozoa that form shells (oocysts) that are resistant to chlorine. Only two disinfectants—ozone and UV radiation—are effective against these two pernicious threats to drinking water, swimming pools, and HVAC systems. Ozone oxidizes organic compounds of concern such as DBPs, pesticides, pharmaceuticals, and compounds causing taste-and-odor issues. Ozone oxidizes metals such as iron, manganese, copper, and zinc so that they precipitate out as oxides or hydroxides. Ozone does not create DBPs.

Ozone has a lifetime of approximately 1 hour. Therefore, it is always produced on-site by feeding oxygen into an electrical discharge:

$$3\ O_2 \rightarrow 2\ O_3 \tag{6.63}$$

Ozone generators use either compressed oxygen cylinders or air compressors that remove particulates and water vapor. Ozone has a low solubility in water, so it is typically introduced via Venturi injection. Due to its short lifetime, it does not persist in finished water and cannot be used as a residual disinfectant in drinking water. Ozone does react with bromide to create bromates. Ozone's decomposition back to oxygen makes it suitable as a final disinfectant in wastewater treatment.

The reduction reaction at that cathode at pH 7 (Koppenol, 1982):

$$O_3 + e^- \rightarrow O_3;\ E^0 = 1.6\ \text{V} \tag{6.64}$$

The reaction under acidic conditions is:

$$O_3 + 2H^+ + 2\ e^- \rightarrow O_3 + H_2O;\ E^0 = 1.8\ \text{V} \tag{6.65}$$

The use of bromide, as KBr, in place of a chloride (as KCl) yields the following reaction at the anode:

$$Br^- + 2\ OH^- \rightarrow BrO^- + H_2O + e^- \tag{6.66}$$

KBr in the electrolyte is another case of redox mediator, introduced in Section 7.2. There are stoichiometric variations of Equations 6.65 and 6.66 that can be fine-tuned by changing the polarization voltage. Changing the voltage is one strategy to minimize interferences from other oxidants.

The voltage is in the vicinity of 0.5 V (Stanley & Johnson, 1979). Combining Equations 6.67 and 6.66 shows that the use of KBr minimizes the pH change in the electrolyte.

Ozone's strong oxidation potential confers an additional advantage in that it oxidizes metal ions, especially Fe^{2+} and Mn^{2+}, to higher oxidation state metals that, in turn, form hydroxides and precipitate out. It can also be used in predisinfection to oxidize sulfides and organic compounds and thus lower the demand for free chlorine in the final disinfection stage.

Ozone absorbs strongly between 230 and 280 nm so measurement is possible through direct ultraviolet absorption—typically via the absorption of 254 nm light emitted by a mercury lamp.

7.6 UV and Advanced Oxidation Processes

No list of alternate disinfectants would be complete without at least mentioning the only non-chemical one—ultraviolet (UV) radiation. This includes light between the wavelengths 230 to 290 nm—the so-called UV C band. These wavelengths are absorbed by DNA and RNA in bacteria and viruses. The energy of light is inversely proportional to the inverse of the wavelength, so light in the UV C range has more than twice the energy of light in the visible region. UV radiation is especially useful for destroying Cryptosporidium and Giardia.

The challenge with UV radiation is that it is absorbed or scattered by suspended matter. For it to be effective, water has to be filtered prior to UV radiation. Also, for obvious reasons, there is no residual using this chemical-free disinfection method. For these two reasons, UV radiation is used to disinfect water at the final step of wastewater treatment.

To increase ozone's efficacy of breaking down organic matter, it is sometimes combined with UV radiation. Hydrogen peroxide, H_2O_2, is also used. The tandem process is termed *advanced oxidation process* (AOP), and it is used both at the end of wastewater treatment or in drinking water treatment to oxidize hard-to-degrade organic compounds. The secret weapon of AOP is the UV-induced (*photolysis*) of ozone, or hydrogen peroxide, to create OH radicals (OH•). A radical is a neutral chemical species with one lone electron. We encountered one example in the form of the Wurster's Dye molecule in Section 3.1. Radicals are very unstable as they will go to great lengths to pair up the lone electron with another. This makes the OH• radical the strongest oxidant of all the commonly used ones. Radicals are typically found in the gas phase or organic solvents and are exceedingly rare in aqueous environments, where charged species predominate.

The predominant reaction mechanism is:

$$O_3 + H^+ + 2\,O_2^{*} \rightarrow + OH\bullet + 2\,O_2 \qquad (6.67)$$

O_2* is an oxygen molecule in an electronically excited state after absorbing UV light. It can also be viewed as a radical.

Another example of a radical in the discussion of the DPD chlorine indicator is in Section 3.1. The high oxidation potential enables it to oxidize organic compounds all the way to CO_2 and H_2O.

Advanced oxidation process is the most expensive of all disinfection/oxidation methods and is used only sparingly in both wastewater and drinking water treatment.

Table 6.3 summarizes the common disinfection alternatives. As mentioned the polarization voltages are in the same 0.2 to 0.5 V, which means that each one shows an interference to the other oxidizing agents in the table. It is unclear how one manufacturer designs one particular sensor distinct from another.

TABLE 6.3 Comparison of common disinfectants.

Disinfectant	Applications	Benefits	Drawbacks	Reduction Potential/ Polarization Voltage	Sensor Interferences
Chlorine ($HOCl$, OCl^-)	Drinking water, wastewater	Inexpensive Strong	Numerous DBPs	1.5–1.6 V 0.3 V	ClO_2, O_3, HOBr
Monochloramine (NH_2Cl)	Drinking water	No DBPs	Weak disinfectant	0.8–1.5 V 0.3 V	HOCl, O_2
Chlorine Dioxide ClO_2	Food & Beverage	Strong	Unstable. Produced on-site	1.5–1.6 V to 0.4 V	O_3, NH_2Cl
Peracetic Acid (CH_3CO_3H)	Wastewater Food & Beverage	Very strong; penetrates biofilms Harmless DBPs	Must be produced on-site; expensive	1.4–1.8 V 0.3 V	ClO_2, Br_2, HOBr
Ozone (O_3)	Pretreatment Wastewater	Strongest oxidant; kills Crypto and Legionella	No residual Corrosive	1.6–1.8 V 0.5 V	
UV light	High	No chemicals	Requires filtered water; no residual	n/a	n/a
AOP (UV + O_3)	Drinking water Wastewater	Highest oxidation potential	Expensive	n/a	n/a

8.0 REFERENCES

Analog Devices. (2013). Micropower toxic gas detector using an electrochemical sensor (Circuit Note CN-0234).

Bettenhausen, C. A. (2020). Treating wastewater with peracetic acid. *Chemistry and Engineering News, 98*(15), 16–18.

Engelhardt, T. L., & Malkov, V. B. (2013). *Chlorination, chloramination and chlorine measurement.* Hach Company LIT2129.

Harp, D. L. (2002). Current technology of chlorine analysis for water and wastewater. *Technical Information Series—Booklet No. 17*, Hach.

Hernlem, B. J., & Tsai, L. S. (2000). Titration of chlorine: Amperometric versus potentiometric. *Journal AWWA, December 2000*, 101–107.

Khan, A. A., Gaur, R. Z., Tyagi, V. K., Lew, B., Diamantis, V., Kazmi, A. A., & Mehrotra, I. (2012). Fecal coliform removal from the effl uent of UASB reactor through diffused aeration. *Desalination and Water Treatment, 39*(2012), 41–44.

Kim, S. H., Choi, J. C., Lee, J. K., & Kong, S. H. (2011). Real time detection of residual free chlorine and pH in water using a microchannel device. *Journal of Sensor Science and Technology, 20*(6), 368–374.

Kim, Y. H., & Hensley, R. (1997). Effective control of chlorination and dechlorination at wastewater treatment plants using redox potential. *Water Environment Research, 69*(5), 1008–1014.

Kodera, F., Kishioka, S., Umeda, M., & Yamada, A. (2004). Electrochemical detection of free chlorine using anodic current. *Japan Society of Applied Physics, 43*(7A), 913–914.

Koppenol, W. H. (1982). The reduction potential of the couple O_3/O_3^-. *Federation of European Biochemical Societies Letters, 140*(2), 169–172.

Kumar, L. (2013). Study of double breakpoints during chlorination of river Yamuna water, Delhi, India. *International Journal of Water Resources and Environmental Engineering, 5*(7), 370–379.

Lambrou, T. P. (2012). *A technical report for water qualitative and quantitative sensors.* [Technical report]. KIOS Research Center for Intelligent Systems and Networks, University of Cyprus.

Lee, W. H., Wahman, D. G., & Pressman, J. G. (2015). Monochloramine-sensitive amperometric microelectrode: Optimization of gold, platinum, and carbon fiber sensing materials for removal of dissolved oxygen interference. *Ionics, 21*, 2663–2674.

Moberg, L., & Karlberg, B. (2000). An improved N,N-diethyl-p-phenylenediamine (DPD) method for the determination of free chlorine based on multiple wavelength detection. *Analytica Chimica Acta.*, *407*(1–2), 127–133.

Morris, J. C. (1966). The acid ionization of HOCl from 5 to 35 °C. *Journal of Physical Chemistry*, *70*(12), 3798–3805.

Oswin, H. G., & Blurton, K. F. (1973). *Electrochemical detection cell.* U.S. Patent 3776832 A, Filed Nov. 10, 1970.

Phelan, D. M. (1996). Direct amperometric measurement versus ORP for residual control—A comparative study. Water Environment Federation Specialty Conference.

Qin, Y., Kwon, H., Howlander, M. M. R., & Deen, M. J. (2015). Microfabricated electrochemical pH and free chlorine sensors for water quality monitoring: Recent advances and research challenges. *Royal Society of Chemistry Advances*, *5*, 69086–69109.

Rajasekharan, V. V., Clark, B. N., Boonsalee, S., & Switzer, J. A. (2007). Electrochemistry of free chlorine and monochloramine and its relevance to the presence of Pb in drinking water. *Environmental Science & Technology*, *41*(12), 4252–4257.

Rao, M. V. (2002). *Acidified sodium chlorite, chemical and technical assessment.* 68th JECFA.

Reynolds, T. D., & Richards, P. A. (1996). *Unit operations and processes in Environmental Engineering.* PWS Publishing Company.

Schleicher, J. (2008). Information on the amperometric measurement of free chlorine, chlorine dioxide and ozone in water. *JUMO Book Number FAS619.*

Stanley, J. H., & Johnson, D. (1978). Amperometric membrane electrode for measurement of ozone in water. *Analytical Chemistry, 81*(13), 2144–2147.

U.S. Environmental Protection Agency. (2002). 2000 Community Water Systems.

Victorin, K., Hellstrom, K. G., & Rylander, R. (1972). Redox potential measurements for determining the disinfecting power of chlorinated water. *Journal of Hygiene, 70*(2), 313–323.

Wittbold, P., & Kim, S. M. (2021). Evaluation of commercial peroxyacetic acid products for wastewater disinfection—Considering feasibility of the on-site generation of peroxyacetic acid. *NEWEA Journal, Summer 2021*, 28–34.

Zhang, C., Brown, P. J. B., & Hu, Z. (2018). Thermodynamic properties of an emerging chemical disinfectant, peracetic acid. *Science of Total Environment*, *621*, 948–959. https://doi.org/10.1016/j.scitotenv.2017.10.195

7

Turbidity

1.0 SOLIDS IN THE WATER

When it comes to assessing water quality, the most basic question is "How much stuff is in it?" By stuff we mean solids as opposed to dissolved gases or other liquids. Solids in water can be broken into two kinds: dissolved solids and suspended solids. The total amount of dissolved solids is termed, appropriately enough, *total dissolved solids* (TDS) and the total amount of suspended solids is *total suspended solids* (TSS). Thus, *total solids* (TS) is the sum of the two, or:

$$TS = TDS + TSS \tag{7.1}$$

Measuring dissolved solids in the water is answered—mostly—by measuring the conductivity of the water. This was the topic for Chapter 4. I say "mostly" because conductivity measures dissolved salts, which consists of electrically conductive anions and cations. Though salts make up most dissolved matter, there might also be dissolved chemical compounds that do not carry a charge, especially polar organic compounds, which are poor conductors. With that caveat in mind, one uses conductivity measurements as a reasonably good surrogate for measuring dissolved solids.

The second component of solids in a liquid is suspended solids. This consists of any solid particle that ranges in size from approximately 0.5 to 100 μm. Any large particle will settle out of solution and not remain suspended for long. Any smaller particle is a dissolved solid. The division between the two is not totally arbitrary but is dependent on the characteristics of human vision. Dissolved solids do not lend a cloudy look to a water sample. Particles interfere with the clarity of water when their size is approximately the same as the wavelength of light that passes through the water to our eyes. A water sample appears clear when it contains dissolved solids that are less than 0.4 μm because visible light covers wavelengths of light from 400 to 800 nm (0.4 to 0.8 μm). Remember that light that is "visible" is visible to the human eye and not necessarily to other creatures. Bees see ultraviolet (UV) wavelengths as low as 340 nm. If the sensitivity of the human retina shifted to the UV range of 200 to 400 nm, then a solution containing particles between than 200 and 400 nm would appear cloudy.

The U.S. Enviromental Protection Agency (U.S. EPA) delineated the distinction between dissolved and suspended solids by specifying that the former consists of particles that pass through a filter with pores of approximately 0.45 μm diameter and the latter consists of particles that don't. This distinction is the impetus for the U.S. EPA test for TSS, which is as follows: Filter a sample of water through a glass microfiber filter with 0.45-μm pores. Then dry the filter at 105 °C and weigh the filter. The difference in weight represents the mass of suspended solids. The TSS value equals the mass of the filtered solids divided by the volume of water of the sample that was filtered. Units of TSS are mg/L.

So far discussion has avoided any mention of the chemical and physical makeup of the particles. On a very basic level, all particles are either living or nonliving. Nonliving particles include sand, silt, clay, colloidal particles, and organic compounds. Living material consists of microorganisms, including bacteria, fungi, algae, and protozoa. Viruses are on the order of 0.1 μm in size and do not affect the clarity of water. Table 7.1 summarizes

TABLE 7.1 Makeup of suspended solids.

Particle	Composition	Size Range (μm)	Notes
Sand	Silicate	50–2 000	Particles over 100 μm settle out
Silt	Silicate SiO_2, SiO_3	1–50	Originates from watershed runoff
Clay	Al silicate with Fe, Mg, alkali, and alkali earth cations	1–5	Forms negatively charged layers with positive ions in between
Iron precipitates	Fe_2O_3, $Fe(OH)_3$	0.0001–0.1	Nanoparticles
		1–150	Colloical; depends on polymer
Colloidal suspended solids	65% organic, 35% organic. Consists of one compound evenly dispersed in another	1–10	Precipitate out as hydroxides when treated with metal salts
Natural organic compounds	Humic acid, tannins	0.1–10	Originates from plant decay
Anthropogenic organic compounds	Carbohydrates, fats, oils, grease, proteins, phenols, volatile organic compounds, pesticides	0.5–10	Originates human activity
Viruses		0.02–0.4	
Algae		1–500	Phytoplankton, dinoflagellates
Bacteria		0.4–5	Cyanobacteria, biofilms, Legionella
Protozoans		1–1000	Ciliates, Rotifers, flagellates, amoebas, includes Giardia

the breakdown of suspended solids according to their possible constituents. The categories represented in the table are somewhat subjective, and you will find different classification schemes in other texts.

Identifying the constituents that make up a water sample is a job for a range of expensive analytical instrumentation. Turbidity measurement is an aggregate measurement much like oxydation-reduction potential (ORP) and conductivity. It can't distinguish between a carbon particle and an amoeba, but it's fast and inexpensive.

2.0 HOW LIGHT INTERACTS WITH SUSPENDED SOLIDS

2.1 Photons and Particles

Let's first understand what turbidity is and how it is measured. To do so, we need to delve into the science of light scattering. Before the quantum revolution in the early 20th century, light was considered to consist of waves of electromagnetic radiation that are characterized by a frequency and amplitude (and, just to be precise, phase). Albert Einstein and fellow German Max Planck changed our understanding of light. According to their new quantum theory, light consists of discrete packets of energy called photons. Each packet, however, has a specific amount of energy (E) that is proportional to its frequency or inversely proportional to its wavelength:

$$E = \frac{hc}{\lambda} \tag{7.2}$$

Here h is Plank's constant—6.626×10^{-27} cm^2 g/s and c is the speed of light—2.997×10^{10} cm/s.

Which picture—light as a wave or light as a collection of photons—is correct? The answer is: both. One resorts to the photon model when dealing with properties on the atomic scale and switches to the wave model when dealing with matter and light on a macroscopic scale. Both models have one parameter—the wavelength—in common so both give the same results when calculating macroscopic properties such as absorption. Both reduce to the same construct within quantum field theory. Planck's constant just might be the most important parameter in physics that you've never heard of. It ties together the world of classical physics with that of quantum mechanics.

Whether or not we use a photon, or a wave, model, the description of light interacting with a particle suspended in water the observable results are the same. There are four possible outcomes:

1. **Reflection.** Light "bounces" off a surface. The angle of incidence, as defined by the angle of the incident light with respect to a perpendicular line, equals the angle of reflection.
2. **Refraction.** Light passes from the water medium through the particle, with a differing index of refraction, and changes direction as it exits. This is the same phenomenon that allows water and lenses to bend light.
3. **Internal reflection.** Light passes from the water medium through the particle, as with refraction, but reflects off the inner surface of the particle and exits. It is a combination of refraction and reflection.

4. **Absorption.** Light is absorbed by the particle and eventually converts its energy to heat. It can be through the promotion of an electronic transition or the continuous internal refraction of light within the particle.

Figure 7.1 illustrates these five interactions.

Refraction and internal reflection are "bit players" for particles suspended in water. (In our atmosphere the internal reflection of sunlight within water droplets creates rainbows.) Absorption removes light and does not contribute to scattering though it can affect a turbidity measurement. The darker the color of a particle is the more visible light it absorbs. For example, carbon particles are very efficient light absorbers, which is why a container of powdered charcoal is very black. Absorption plays an indirect role in the most common form of turbidimeter, the nephelometer, used to measure low levels of suspended matter. It subtracts light such that the weaker scattered signal is erroneously attributed solely to scattering. Absorption plays a predominant role in turbidimeters that measure high levels of suspended solids.

The first two phenomena—reflection and refraction—manifest themselves as light that exits a sample of water at different angles than when it enters it. Together they comprise *light scattering*. The degree to which light rays at a given wavelength scatter off a suspended particle depends on the particle's dimensions as compared to the wavelength.

One caveat to keep in mind is that the above discussion of the four ways light interacts with particles is valid for single, isolated particles. When the particle density is high enough one must consider collective phenomena.

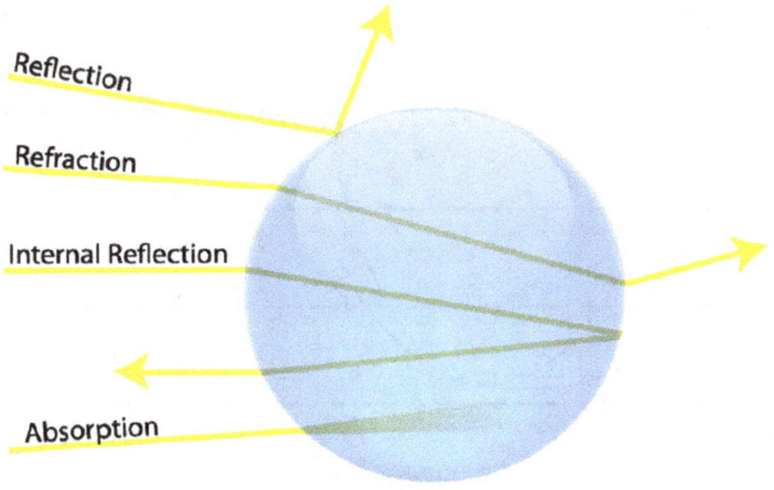

FIGURE 7.1 Possible outcomes of light impinging on a particle: reflection, refraction, internal reflection, and absorption.

2.2 (Particle) Size Matters

As Table 7.1 shows, different types of suspended matter are characterized by sizes that cover a range of nearly three orders of magnitude. The question that is most relevant to a turbidity measurement is: "What is the size distribution of particles in a sample of water?"

Figure 7.2 (Zielina, 2016) shows the particles size distribution of wastewater at the four stages of treatment: raw, after flocculation, after settling (sedimentation), and after filtration. The distribution is bimodal, with peaks around 0.5 μm and 40 μm. At each stage the number of larger particles decreases, as expected. Similar measurements of a wastewater settling pond in Tanzania showed a similar bimodal distribution with peaks at around 10 and 67 μm and a similar pattern of larger particles preferentially settling out (Izdori et al., 2018). Knowledge of the particle size distribution in samples of water is important to understand the nature of light scattering.

2.3 Light Scattering

To understand the physics of light scattering requires serious textbook material. When the particle size is on the same order of magnitude as the wavelength then *Mie Scattering* theory predicts the scattering pattern. For the purposes of this book, I just summarize the conclusions that Mie scattering theory presents. Light simply doesn't radiate or scatter out equally in

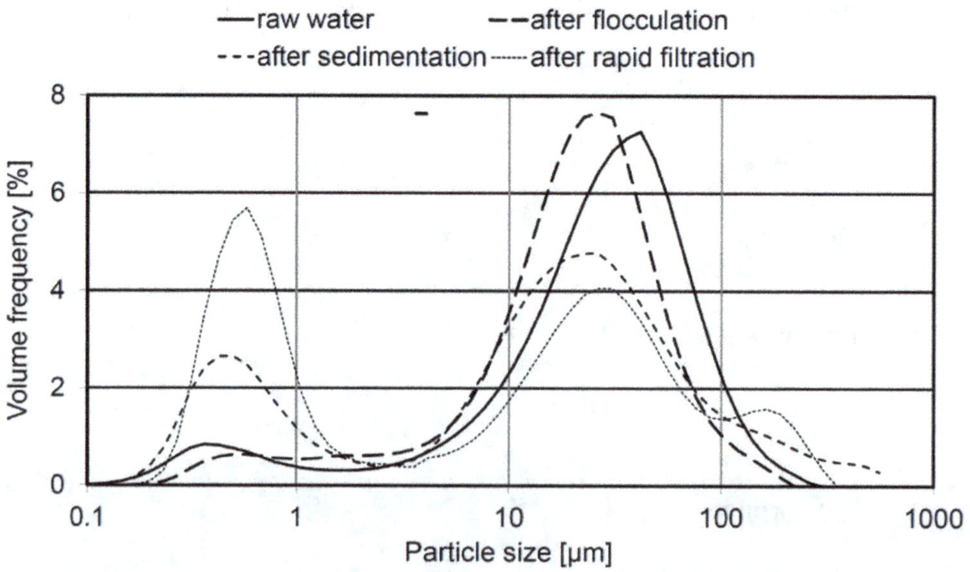

FIGURE 7.2 Particle size distribution for wastewater samples undergoing three stages of treatment from raw water to final filtration (Zielina, 2016).

all directions. Depending on their size, particles will scatter light into some directions more than others. As previously mentioned, size is relative. The key parameter is the ratio of a particle size to the wavelength of light. That ratio is embodied in the *size parameter*.

$$x = \frac{2\pi r}{\lambda} \tag{7.3}$$

Here λ is the wavelength of light, and r is the particle radius. Equation 7.3 means, for example, that red light at 600 nm will scatter off particles that have, say, radii of 1 μm, to the same degree as blue light at 400 nm scatters off 0.60-μm particles. In both cases the ratio of size to wavelength is 2/3. A size parameter of 1 translates to a particle radius that is about 1/6 of the wavelength.

The scattering of light at different size parameters is shown in Figure 7.3.

As the left side of the figure shows, particles with scattering parameters much smaller than one scatter light almost equally in all directions. The light scatters equally in forward and backward directions and with roughly half the intensity at right angles. The scattering pattern in the middle is that of particles whose size parameter is approximately 1. The light scatters preferentially in the forward direction. The scattering pattern on the right is for particles whose size parameters are much greater than one. Nearly all the scattering intensity for large particles is in the forward direction.

The changing pattern of scattering with particle size has important implications for the measurement of turbidity. Because large particles scatter most strongly in the forward direction, measuring the scattered light at a right angle will incorrectly give a lower level of scattering. As will be discussed shortly, the most common method of measuring turbidity is measuring scattering at a right angle to the incident beam. This means that measuring scattering at a right angle will give lower values for large particles versus small ones.

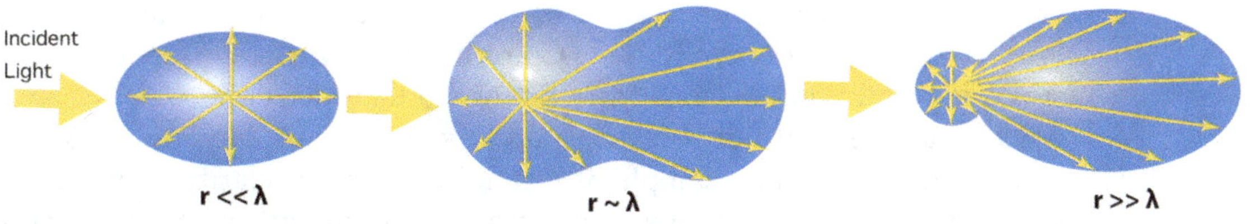

FIGURE 7.3 Scattering patterns for three cases of particle sizes. Left: The particle size is much smaller than the wavelength of incident light. Middle: Particle size is about the same size as the particle. Right: The particle size is much larger than the wavelength.

3.0 WHY TURBIDITY IS IN THE EYES OF THE BEHOLDER (OR THE METHOD)

Some properties that we ascribe to the world around us are pure constructs of the human brain. Take away the observer, and they are meaningless—that is, they cannot be defined. Turbidity is one of those properties, and it is for this reason that there are so many different instruments invented to quantify it. Every other parameter covered in this book are described by parameters that are uniquely defined, such as pH, conductivity, or dissolved oxygen. Turbidity parameters only make sense within the context of the method invented to measure it.

To understand just why turbidity is a human construct that has no unique definition consider a similar parameter that is also formulated in our brain—color. Light striking an object is composed of a range of wavelengths. The human eye is sensitive to only those wavelengths in the 400–700 nm range. What is "visible" to us may not be visible to other creatures and vice versa. If the object absorbs most of the wavelengths that are shorter than red—less than about 620 nm—and reflects most of the wavelengths greater than 620 nm, then the object looks red. A leaf looks green because it absorbs wavelengths on either side of the green range of wavelengths—less than 490 nm and greater than 580 nm—the red and blue ends of the spectrum. An object reflects what it does not absorb. Hence, a green leaf reflects wavelengths in the 490–580 nm range. The intensity of the reflected light at each wavelength is unique to that leaf. The brain makes sense of this complex spectrum of reflected light and converts it into a single parameter we call color—in this case a hue of green. Color is the brain's way of interpreting the complex makeup of wavelengths that an object reflects or emits and hits the retina of the eye. It is entirely possible that two objects illuminated by the same light source, say, the sun, appear to be the same hue, yet have two entirely different spectra.

The spectrum of 400–700 nm does not convey special physical significance. The range of wavelengths is simply the range of sensitivity of the rods and cones that make up the human retina. The brain breaks down colors into the primary colors of red, green, and blue because the cones come in three flavors—red, green, and blue. Color is not an intrinsic property of an object. It's the brain's response as a biological light detector.

One can view turbidity in the same way. It is not an intrinsic property of suspended solids in a sample of water. Turbidity, or clarity, is a quality that we humans ascribe to a sample of water (or any other liquid or gaseous medium) when objects within or behind the sample appear diffuse. One might say that, as a container of water becomes less and less clear, it looks more and more turbid. But one cannot define "clarity" any more effectively than it can define "turbidity." There are an almost unlimited number of instruments that give a quantitative measurement to this human-based parameter. Turbidity is the brain's way of making sense of the light that strikes a sample of water or other medium. Just as the brain interprets a complex spectrum of reflected light of an object with a color, the brain also interprets the light irradiating suspension of particles with a value of clarity or turbidity. The particle size distribution is to turbidity what the reflected spectrum is to color.

Another way of stating the special status of turbidity as a parameter is to acknowledge that it cannot be derived from basic units of measurement. The SI system defines seven fundamental parameters: mass, length, time, current, temperature, amount of substance, and luminous intensity. All other parameters, such as magnetic field strength, velocity, or conductivity are derived from the seven units. Turbidity, like color, cannot trace its definition solely to any of these units.

The different designs by which turbidimeters measure particle size distributions mean that the measurement of one instrument is likely not to be equal, or even proportional, to another. This means that one can calibrate two different turbidimeters using the same set of calibration standards, yet both instruments will give different turbidity readings for an abrbitrary aqueous sample. Conversely, two samples consisting of two very different compositions of suspended solids can give the same turbidity reading on those two different instruments—just as with two different objects that appear to have the same color. Another way of saying that turbidity depends on the instrument used to measure it is to call it a *method-defined parameter*. Define the instrument and the method of operation and you define the turbidity unit. It is vitally important to keep this point in mind in the following discussion of different turbidimeter designs.

4.0 DIFFERENT STRATEGIES FOR QUANTIFYING TURBIDITY

4.1 The Secchi Disk

The two oldest and simplest methods to measure turbidity both rely on the human eye as a light detector. Both are based on the simple phenomenon that scattering and absorption of a column of water increase as the column depth increases.

The first is based on the everyday observation that, as one lowers an object into a murky body of water, it eventually disappears. Pietro Secchi invented the eponymous *Secchi disk* in 1865. It is simply a 30-cm-diameter disk with a simple black-and-white pattern as in Figure 7.4. Its use is typically restricted to natural bodies of water, such as lakes, rivers, and oceans. A laboratory version is shown on the right side of the figure. The rope attached to the disk is marked off like a ruler. One lowers the disk into the water from a boat or dock. The depth at which the disk is no longer distinguishable from the surrounding water is the *Secchi depth*. At this depth the amount of light that reflects off the surface of the object equals the amount of light that backscatters from the water column above the object. This depth is inversely proportional to the turbidity, and it becomes a very simple, inexpensive means of measuring turbidity.

The determination of the Secchi depth is a very imprecise exercise. The intensity and angle of sunlight and the observer's visual acuity play a role in ascertaining when the disk "disappears." To ensure consistency, measurements should be taken in the shade between 10 a.m. and 3 p.m. The useable depth range for the standard 30-cm-diameter disk is 1 to 25 meters. Therefore, although it's not practical for water or wastewater and its accuracy is limited, the Secchi disk is still widely used today to measure turbidities of natural bodies of water.

The Secchi depth measurement is mostly an exercise in the absorption of light. The absorption by N,N-diethyl-p-phenylenediamine (DPD) after reacting with free chorine (Chapter 6, Section 3.1 is another example of an absorption measurement). A spectrophotometer can measure the tint of the DPD pink color more accurately than comparing the color of the DPD solution to a color chart. The spectrophotometer works by measuring the amount of light that shines through the sample and reaches a photodetector. At the heart of a spectrophotometer measurement is the Beer-Lambert law. It converts the transmitted light intensity to the concentration of the absorbing species:

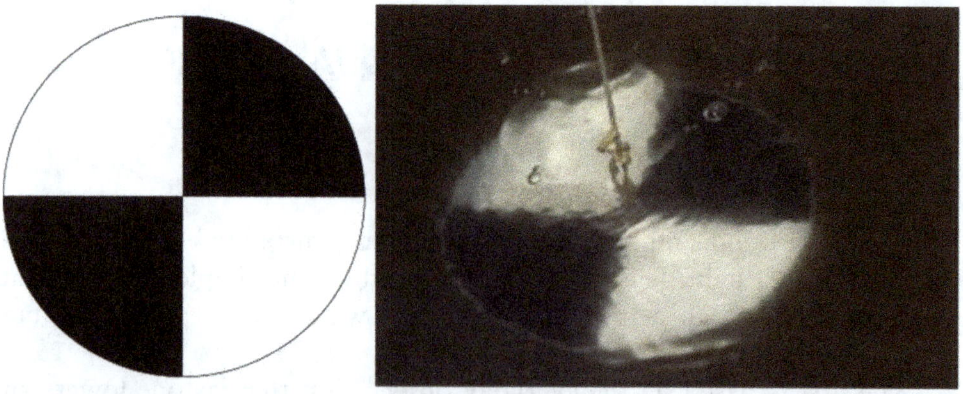

FIGURE 7.4 On the left is the Secchi disk pattern. On the right is a Secchi disk being lowered into a pond, lake, or river.

$$\frac{I(\lambda)}{I_0(\lambda)} = e^{-A(\lambda)} = e^{-k(\lambda)l} = e^{-\varepsilon(\lambda)l} \tag{7.4}$$

I/I_0 is the fraction of the incident light intensity that travels through the sample and reaches the detector. The first equality converts the unitless parameter of transmission to the unitless parameter of absorbance, A. The second equality separates the depth (l) dependence from A to create the absorption coefficient (k), which has the units of inverse length (cm^{-1}). The inverse of the absorption coefficient is called the absorption depth. When the actual depth equals the *absorption depth*, $kl = 1$ and $I/I_0 = 1/e = 0.368$—that is, the intensity of 63% of the incident light.

The third equality further separates the concentration from the absorption coefficient. The absorption coefficient is the product of c, the concentration of absorbing species (mol l^{-1} or M) and λ, the molar extinction coefficient, a parameter that is unique to the absorbing species. The larger λ is, the stronger the absorption. Its units are $M^{-1}cm^{-1}$. The insertion of λ into all three forms of the Beer-Lambert law in Equation 7.4 reminds us that absorption, whether expressed by the absorbance (A), absorbance coefficient (k) or molar extinction coefficient (λ), is a function of wavelength. Indeed, a plot of λ versus λ is the absorption spectrum. We will elaborate more on absorption spectra when we review the spectroscopy of organic matter in Chapter 9, Section 3.0. Also note that Equation 7.4 is strictly for matter that absorbs light, not for suspended particles that scatter it. Nonetheless, the same relationship holds true for the attenuation of light by a medium containing suspended particles.

The only task remaining is to convert the Secchi depth to the corresponding extinction coefficient. At a depth equal to the extinction coefficient the transmitted intensity ratio is 0.368—just as with the absorption depth defined above. In a partially arbitrary decision, the extinction coefficient equals 1.7 divided by the Secchi depth. With this definition, the transmitted light is 18% of the incident light, which is considered the threshold for detectability by the human eye. For example, if the disk disappears at 3 meters, then the extinction coefficient is $1.7/3 = 0.56\ m^{-1}$. A larger extinction coefficient means a smaller Secchi depth.

Though we have not yet defined the units of turbidity, it is useful to tie turbidity, as expressed in *Nephelometric Turbidity Units* (NTUs), to the measured Secchi depth. Myre and Shaw simultaneously measured Secchi depths and turbidity for a range of raw water samples and came up with the following empirical relationship (Myre & Shaw, 2006).

$$\text{Depth (cm)} = 244.13 \times \text{Turbidity (NTU)}^{-0.662} \tag{7.5}$$

Figure 7.5 shows excellent fit of Equation 7.5 to Myre and Shaw's actual measurements.

4.2 The Jackson Turbidimeter

The very first turbidimeter dates to 1900 when Whipple and Jackson invented the first such instrument and the first calibration standard. The principle of operation for this primitive turbidimeter is also based on the attenuation of an object at the bottom of a column of sample water. In contrast to the Secchi disk, which uses sunlight reflected off the disk, the target of the Jackson turbidimeter is a light source.

Figure 7.6 shows a schematic of the instrument invented by Whipple and Jackson. A candle illuminates the flat bottom of a glass cylinder from below. The user slowly adds the water sample while observing the illuminated liquid from above the tube. Before water is added, the candle appears as the distinct object it is. As water fills the cylinder, scattered light increasingly diffuses the image. At the point at which the scattered light intensity equals the transmitted light the visual image becomes a uniform glow, and the image of the candle disappears. The level in the cylinder at which this this occurs is the turbidity in terms of *Jackson Turbidity Units* (JTU). The

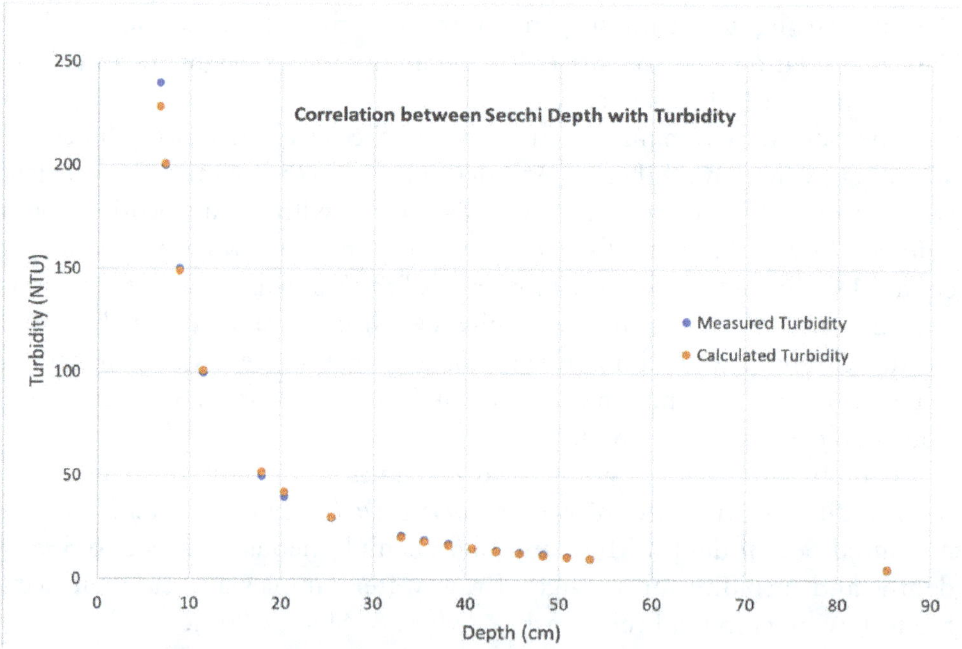

FIGURE 7.5 Correlation between the Secchi depth and turbidity (NTU) from measurements and calculated from Equation 7.5 (Data from Myre & Shaw, 2006).

FIGURE 7.6 Jackson turbidimeter. It operates just like the Secchi disk except that the light source is a candle as opposed to reflected sunlight.

JTU value is simply the inverse of the depth of the column of the sample at the point the image disappears. It is therefore analogous to the Secchi depth.

To calibrate this simple device, Whipple and Jackson made a series of calibration standards based on a weighed amount of diatomaceous earth—a very fine powder that makes long-lasting suspensions. The initial solution suspension was 1000 parts per million (ppm). By serial diluting this stock solution the pair made up a series of turbidity standards.

The use of natural compounds such as diatomaceous earth and, later, kaolin, Fuller's earth, or streambed sediments, for calibration was problematic. These mineral-based materials could be extracted and processed from different sources. No two sources are ever the same, so the same concentration of suspended solids will likely give different turbidity readings. Furthermore, determining the level of water at which the image disappears is a subjective exercise and thus prone to uncertainty. Due to constraints on the height of the glass tube, the Jackson turbidimeter is only useful for measuring turbidity levels greater than 25 JTUs—visibly cloudy water.

The Jackson turbidimeter and the Secchi disk illustrate a very important principle of optical measurements. According to the Beer-Lambert law, Equation 7.4, the attenuation of light by a sample depends on the product of the extinction depth and the thickness of sample. The latter is termed the *pathlength*. This means that a container of a water sample that is 10 cm deep and has a concentration of 1 mg/L of a particular suspension will attenuate the same amount of light as a 2-cm deep container that has a concentration of the same species of 5 mg/L. We understand this intuitively when we look at a glass of water taken from a cloudy looking swimming pool. A glass of the pool water might look crystal clear, but the deep end of the pool might look cloudy by virtue of the long path length between the bottom of the pool and the surface. The product of the path length and the concentration of the sample is the *optical depth*. It determines the measurement limit of a turbidimeter.

4.3 A Better Way to Measure Scattering

In a spectrophotometric or spectroscopic measurement, absorption of light by the analyte as a function of wavelength is the basis of the measurement. In a turbidity measurement, scattering is the basic mechanism, but absorption does play a role. Figure 7.7 shows the scattering intensity of a sample with increasing concentration. As explained above the plot could just as well be against the path length. At low concentrations of suspended solids, attenuation is nearly all by scattering. As the concentration increases, the sample absorbs a greater fraction of the incident light, thus leaving less light available for scattering. The Beer-Lambert law quantifies the absorption.

A simple extension of the Beer-Lambert law adds the attenuation of the incident light through scattering

$$\frac{I}{I_0} = e^{-(k_a + k_s)l} \tag{7.6}$$

Here k_a is the absorption coefficient, and k_s is the scattering coefficient. Both coefficients vary with the wavelength and particle size.

The shape of the scattering curve in Figure 7.7 shows scattering at 90° and is different for other angles at scattering angles. Chapter 5, Section 1 dives a little deeper on scattering angles. At the concentration level at which the incremental fraction of light absorbed equals the incremental fraction scattered the intensity no longer increases but levels off and then decreases. This point establishes the operating range of a turbidimeter that minimizes the contribution to the measured signal due to absorption. The designer of such an instrument chooses the path length of the sample cell to match the range of concentrations of suspended particles likely to be encountered.

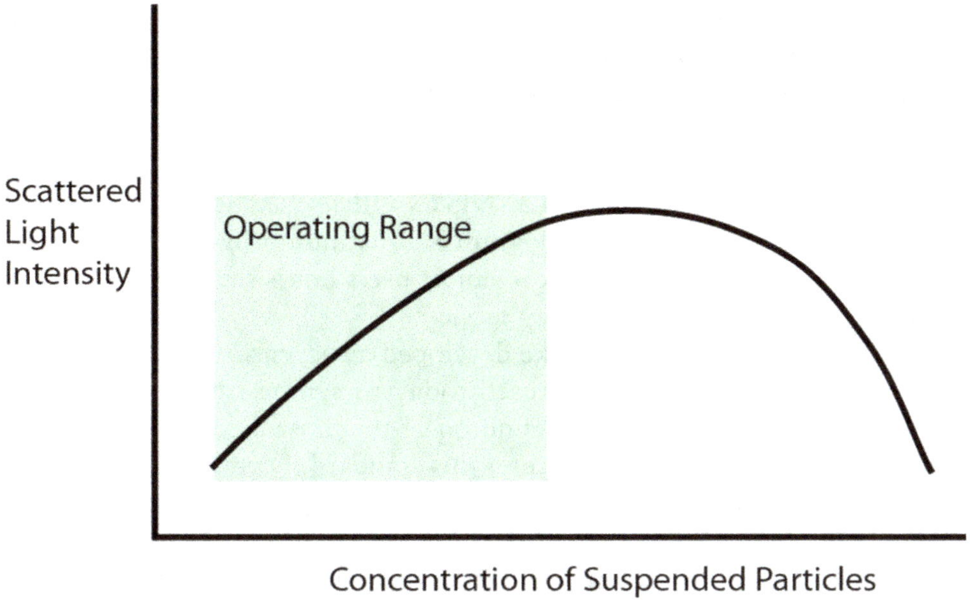

FIGURE 7.7 Conceptual plot of the scattering intensity as a function of particle density. As the particle density increases, the solids absorb a greater fraction of light, thus leaving less light available for scattering. The point at which the turbidity levels off establishes the usable operating range of the turbidimeter.

4.4 Turbidity Units and Standards

All chapters to this point in the book covered the principles of operation and calibration later. Turbidity is different. It does not exist outside a human observer. As discussed in Section 3.3, a measurement of turbidity only has meaning—that is, can be assigned a quantitative value—in the context of the instrument design and the suspended material used to calibrate it. One needs to specify *both* the design of the instrument and the choice of calibration standard to ensure that reported turbidity values can be reproduced and validated.

The very first turbidity unit of measure was based on the concentration of a solution, mg/L, of silicon dioxide, SiO_2 (1 mg/L SiO_2 is equivalent to 0.13 NTU or FTU). Natural compounds, such as silicon dioxide, kaolin (clay), and diatomaceous earth or Fuller's earth have all been used for calibration, but they can't be expected to give reproducible turbidity readings across different sources of these materials.

There are several organizations weighing in with their own definitions of turbidity standards. The two most prevalent are the U.S. EPA and European International Organization for Standardization (ISO).

Because the U.S. EPA is the main regulatory body for water and wastewater quality in this country, its definition of a *primary turbidity standard* is the one this book uses. A primary standard is one that is prepared from materials that are 100% traceable to a NIST (National Institute of Standards) standard. For instance, if a particular primary standard is a mixture of chemical compounds, then one must weigh those compounds using a scale that has been calibrated with standard weights that can trace their origin all the way to standard weights at NIST's Office Weights and Measures. (The original standard for the kilogram is a platinum-iridium bar in Paris.) This is an almost impossible task for most users but is the *raison d'être* for companies that provide *metrology* services.

In 1926 Kingsbury and Clarke developed a suspension that they called *formazin*. Their formula is still used today to synthesize most prevalent turbidity standard. In fact, the definition of the most used units of turbidity are based on a defined mixture of this standard. Formazin is a polymer suspension, formulated by the reaction formaldehyde (H_2CO) sulfate and hydrazine (N_2H_4), as shown in Figure 7.8.

Formazin is a long chain polymer, and it comes in various lengths or multiples of the building block. The polymer chains intertwine to create particles of different sizes and configurations. The particles in a Formazin suspension range in size from 0.01 to 10 μm. The left side of Figure 7.9 shows an electron microscope image of formazin particles. The individual particles havevery irregular shapes.

One of the benefits of formazin as a turbidity standard is the relative ease with which it is prepared. A mixture of 5.000 g/L of hydrazine sulfate and 50 g/L of hexamine yields a 4000 NTU solution. U.S. EPA Method 180.1 prescribes the nephelometric approach to measuring turbidity and is discussed in Section 5.1. The method includes the following recipe for making a 400 NTU formazin stock suspension:

1. Dissolve 1.0 g hydrazine sulfate [$(NH_2)_2 \bullet H_2SO_4$] in filtered water and dilute to 100 mL in a volumetric flask.
2. Dissolve 10.0 g hexamethylenetetramine [$(CH_2)_6N_4$] in filtered water and dilute to 100 mL in a volumetric flask.

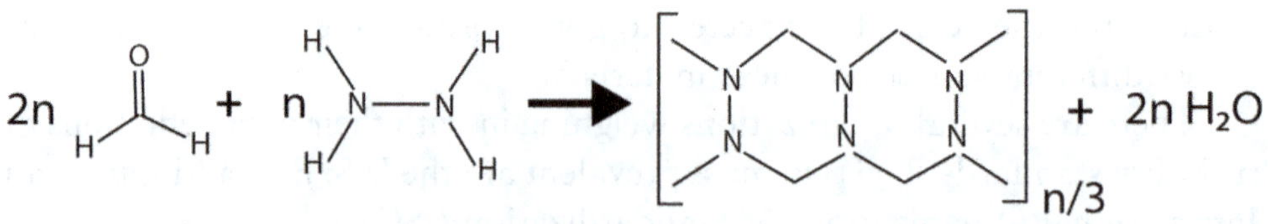

FIGURE 7.8 Synthesis of the polymer formazin from formaldehyde and hydrazine. The actual synthesis starts with hydrazine sulfate and hexamine.

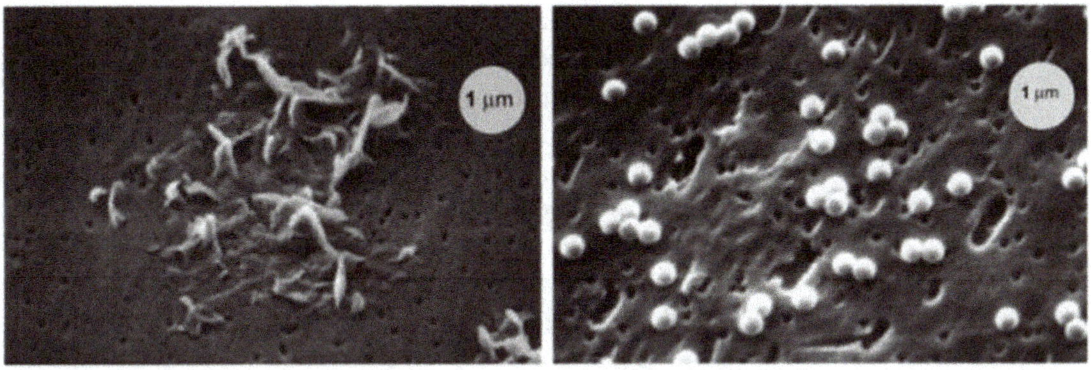

FIGURE 7.9 Scanning electron micrographs of formazin particles (left) and styrene divinylbenzene particles on right. Formazin solution was dehydrated to obtain images (Courtesy GFS Chemicals).

3. Mix 5.0 mL of hydrazine sulfate and 5.0 mL of hexamethylenetetramine solutions in a 100-mL volumetric flask and let stand 24 to 48 hours at 25±3 °C; dilute to the mark and mix.

To prepare 500 mL of 4000 NTU standard, mix 25 mL of the reagent solutions in a 500-mL flask, dilute to the mark, and mix.

A 4000 (or 400) NTU stock mixture is subsequently diluted to make whatever lower turbidity standard solution needed. For example, adding 300 mL to a 100 mL stock solution produces a 1000 NTU standard solution. Figure 7.10 shows a series of vials with increasing concentration of formazin. Most people consider a water sample whose turbidity is less than 2 NTU turbidity to be clear.

One frequently overlooked variable in the preparation and use of turbidity standards is temperature. The recipe specifies a temperature of 25±3 °C.

FIGURE 7.10 Cuvettes showing formazin samples of varying turbidity (Water Analytics).

Deviations from this standard temperature will give errors of approximately 1.2% per degree Celsius. It is also critically important to use deionized water. Even ultrapure water will scatter light in the amount of 0.02 NTU.

As Table 7.2 (Down & Lehr, 2005) shows, the more dilute the solution the shorter the shelf life (or maximum storage time). For this reason, it makes sense to have a 4000 NTU stock solution present from which to make the diluted calibration standards—typically 0.1, 1, 10, and 100 NTU.

The lasting status of formazin as a primary calibration standard is a consequence of the following desirable characteristics:

1. It is prepared from only two NIST-traceable raw materials.
2. Formazin absorbs very little light in the visible spectrum. Therefore, the decrease in transmittance when visible light is due mostly to scattering rather than absorption.
3. The formazin polymer consists of a wide range of particle shapes and sizes from 0.1 to 10 μm. Through the statistical power of large numbers of particles this means that any two suspensions are likely to have the same distribution of particle sizes. The repeatability in measured turbidity under carefully controlled conditions is approximately 1%. In practice, however, the precision of NTU readings range from 1.5% to 10% (Popocosta, 2002).

TABLE 7.2 Recommended storage times and tolerances for formazin (Down & Lehr, 2005).

Turbidity (NTU)	Shelf Life	Tolerance (NTUs)
<10	1 hour	0.1
2–20	1 day	
10–40	1 day	1
20–400	1 month	
40–100	1 month	2
100–400	1 month	5
400–1000	1 year	20
>1000	1 year	50

Formazin has some drawbacks:

1. It is a skin irritant.
2. Hydrazine sulfate is a carcinogen.
3. The suspended chemicals precipitate out.
4. Diluted formazin (<2 NTU) has a shelf life on the order of hours.

To deal with the most serious drawback of chemical lifetime, the Hach company created a modification of formazin that uses a proprietary matrix, StablCal, that stabilizes the formazin particles so that the suspension lasts for months (Sadar, 1998b). Figure 7.11 shows a comparison of the stability of formazin and StablCal as measured by the turbidity of a 2 NTU suspension. StablCal is accepted as a primary standard.

A third primary standard was designed to also counter the finite lifetime of diluted formazin standard. Styrene divinylbenzene (SDVB) co-polymer beads are suspended in a mixture with an antimicrobial chemical and a surfactant. The polymer beads are more chemically stable and less toxic than formazin. As the right side of Figure 7.9 shows, the polymeric foundation of the SDVB manifests itself in the spherical shape of the particles.

The real effect of differing turbidity standards is from their differing particle size distribution. Figure 7.2 shows the particle size distributions for

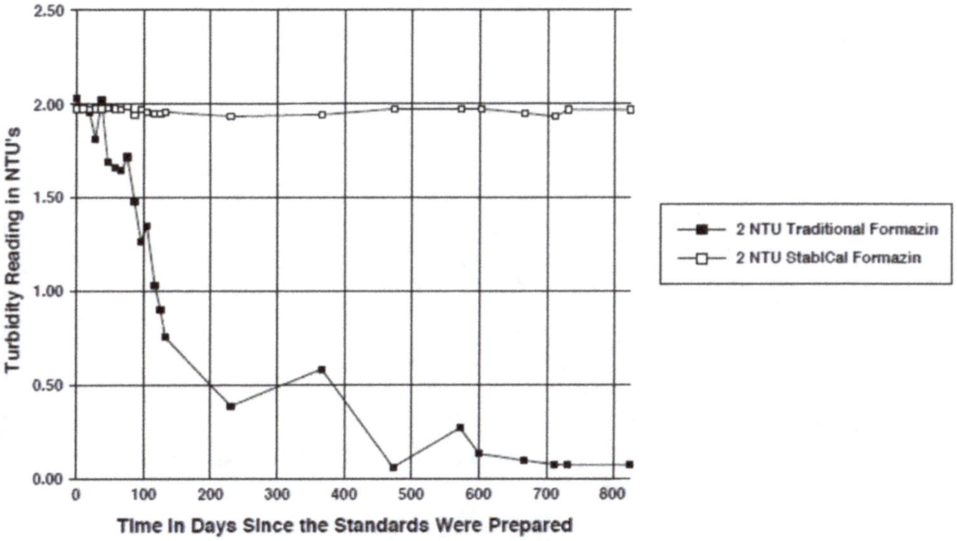

FIGURE 7.11 Stability measurement of a dilute solution (2 NTU) of formazin and StablCal (Sadar, 1998b).

water undergoing water treatment and wastewater treatment. Figure 7.12 shows the size distribution for formazin and SDVB compared to the size distribution of particles in oceans and rivers (Downing, 2008). This graph differs from the previous two in that it displays the cumulative percent composition against the particle size. One can think of this plot as a running total of the number of particles with a size diameter less than the value on the *x*-axis (or if you are mathematically inclined, the integral of the distribution function). A look back at Figure 7.3 reminds us that small particles scatter a greater fraction of light perpendicularly to the incident light than large particles. So, both SDVB and formazin, which consist mostly of particles less than 1 μm, will scatter more light at 90 degrees than river samples and—to a greater extent—oceans. That means a Nephelometric

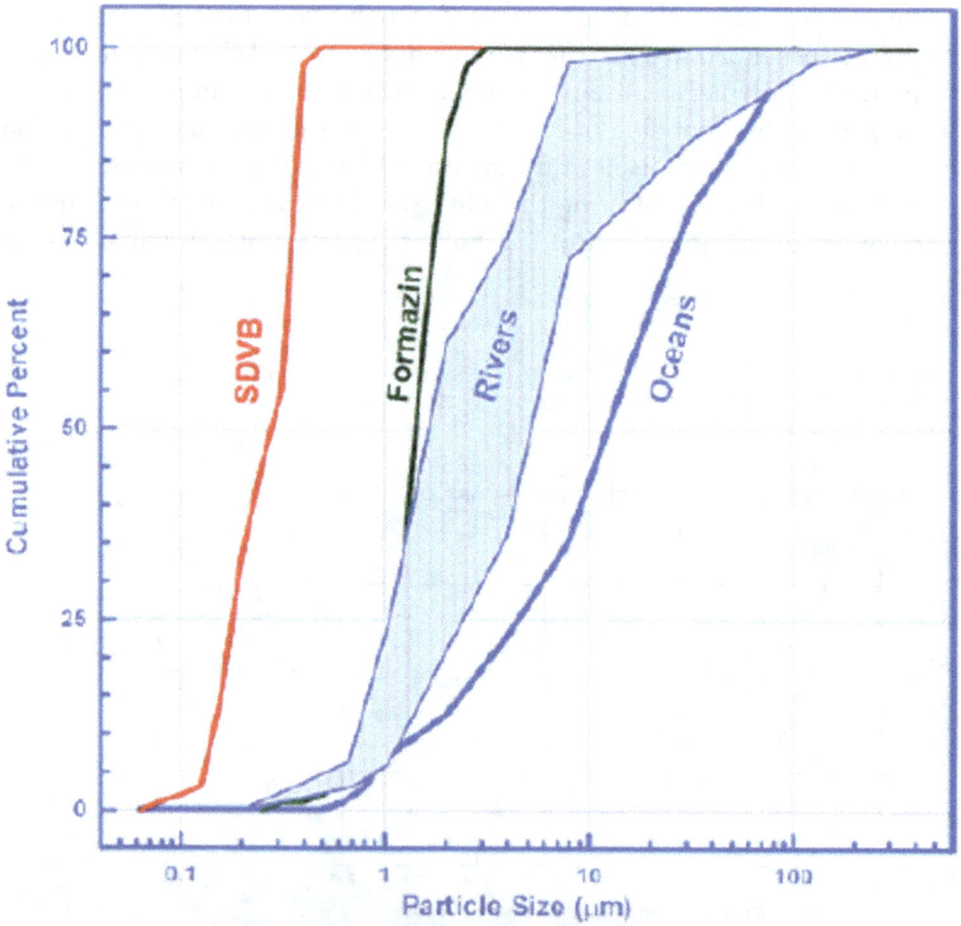

FIGURE 7.12 Particle size distribution of formazin and SDVB compared with that for oceans and rivers (Downing, 2008).

turbidimeter, which measures scattered light at 90 degrees, will read a higher turbidity value for a sample of formazin or SDVB than either river or ocean water containing the same concentration of suspended solids. Figure 7.12 (Downing, 2008) shows that formazin will yield turbidity values for raw water and wastewater that are closer to formazin than they are to SDVB. However, for the most prevalent use of turbidity measurements on finished water, the chemical stability and reproducibility of SDVB may make it an attractive alternative.

Because different turbidity standards differ in their size distributions, two standards may give the same turbidity value at one turbidity level but different turbidity values at other levels. In other words, let's say that one takes two identical turbidimeters and calibrates one at three points, say, 0.1, 1, and 10 μm, using SDVB and the other at the same three points but using formazin. If one then swaps standards and take readings at the three points, one will get different turbidity readings at one or more of the three calibration points.

For this reason, users must report turbidity readings that have either been measured on an instrument calibrated with formazin or that are calibrated using another standard (most likely SDVB) but corrected for formazin. Formazin remains the turbidity standard to which any other standard must be corrected. For instance, the concentration of SDVB for a 10 NTU calibration standard for a particular turbidimeter is adjusted so that it gives the same turbidity reading as a 10 NTU formazin standard in that machine.

To make clear that readings are reported in NTU values referenced to calibration with formazin, the ISO 7027 method, that is defined in the next section, defines the unit of turbidity as the *Formazin Nephelometric Unit*, or *FNU*. Turbidity units, like turbidity itself, are totally dependent on the design of the analyzer that measures them.

So far everything in this section applies to primary calibration. This is the calibration required for proof of compliance with U.S. EPA regulations. To verify that an instrument's calibration is still valid or for noncritical calibration, most users resort to so-called *secondary standards*. They are desirable because they don't involve the mixing of fresh formazin solutions that are necessary for accurate primary calibration. Secondary standards can be formazin solutions prepared by diluting a 4000 NTU stock solution. Some secondary standards are solid polymers that are permanently sealed in sample cells. For instance, Gelex is a polymer gel made by Hach that contains metal oxide particles. Other secondary standards are glass or plastic rods that don't required a sample cell. Before using secondary standards, one must calibrate them to a primary standard, which is usually formazin.

In the past some instrument manufacturers have called for calibration based on a carefully calibrated laboratory turbidimeter. Given the discussion in Section 3.0 on the variable readings from different turbidimeters, this practice is unadvisable. However, using a laboratory turbidimeter as a secondary calibration standard to verify calibration is very acceptable.

5.0 MODERN TURBIDIMETER DESIGN

5.1 The Standard Nephelometric Turbidimeter

As Figure 7.3 shows, the only truly reproducible way to measure scattered light is to detect it at all angles. Figure 7.13 shows the scattering intensity for angles of measurement: 0° (forward), 90° (right-angle), 30° (obtuse), and 135° (back). Note that true backscattering refers to 180° scattering. Due to the difficulty of placing the detector coincident with the light source, the term "backscatter" can refer to any source between 90° and 180°.

As first brought up in Section 4.3, none of the scattering curves are straight lines. Absorption increases with particle density, or turbidity, in accordance with the Beer-Lambert law, Equation 7.4, and attenuates the scattered light signal. For maximum signal at low turbidity values, forward scattering detection is optimal. For the greatest range of turbidity values, back scattering detection is superior but sensitivity at the low range is inferior to other angles. Clearly a turbidimeter using all four angles will

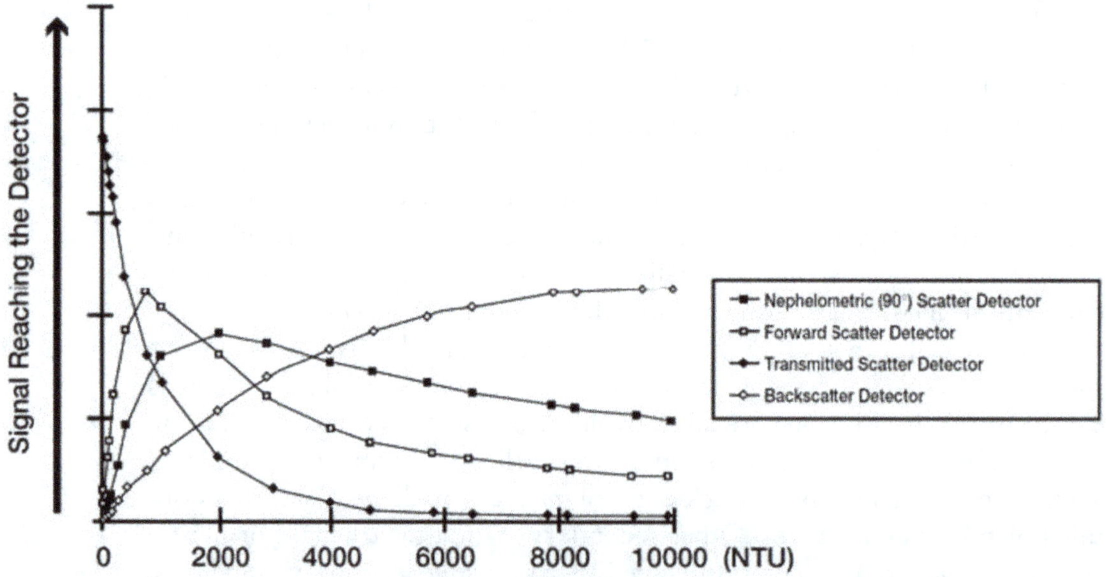

FIGURE 7.13 Scattered light intensity at the four angles is typically measured in a ratio turbidimeter with two additional detectors for forward (135°) and back (30°) scattering (Sadar, 1998a).

cover the widest range of turbidity values, but such an instrument is not a cost-effective option. If forced to design a turbidimeter using only one detector—to minimize instrument cost—a designer would be faced with choosing between measurement range (or linearity) and sensitivity. The best compromise is to measure at a right angle. The turbidity at which scattering deviates strongly from nonlinearity is around 1000 NTU—more than high enough for most applications. For this reason, most turbidimeters measure light scattered at 90°. This most common design has its own name—*nephelometric*. The term also refers to instruments that measure light scattered at 90° in aerosols in the atmosphere and in the blood stream.

To be precise in nomenclature one ought to replace the term *nephelometric* turbidimeter with a turbidimeter that satisfies the specification for turbidity measurement as prescribed by the U.S. EPA Method 180.1 and summarized in Table 7.4. This specification consists of four requirements:

1. The light source is incandescent with a color temperature between 2200 and 3000 K. As a basis of comparison, the color temperature of an incandescent bulb is 2800 K. As the color temperature increases the fraction of light output in the blue end of the spectrum increases. This light source is often referred to as *white light* though that term is ambiguous.
2. The photo detector has a spectral response that peaks between 400 and 600 nm. The detector is typically a low-cost photodiode. Some nephelometers use a filter to constrain the wavelength to the 400 to 600 nm region. For enhanced detection of small particles in a low turbidity sample by wavelengths in the blue and UV region, a turbidimeter can use a photomultiplier or vacuum photodiode. The product of the light source spectral output and photo detector spectral sensitivity determine the overall response of the instrument.
3. The detector is placed at 90° and views a solid angle of not more than 30° around the right angle.
4. The distance of the total light path including both incident light and scattered light within the sample is no more than 10 cm. This ensures that samples with turbidity values as high as 40 NTU will generate a scattered light signal that is linear with turbidity. Analogous to the Beer-Lambert law discussed in Section 4.1, the total amount of scattering is proportional to the product of the suspended solids concentration and the path length.

Since the term "turbidimeter that satisfies U.S. EPA method 180.1" is a mouthful, this book adopts the convention of using the term *nephelometric turbidimeter*. There is an almost identical specification of the nephelometric turbidimeter

The unit of turbidity when measured by a nephelometric turbidimeter that uses an incandescent light source is the NTU. It bears repeating that the design specifications of the nephelometric turbidimeter, as prescribed in U.S. EPA Method 180.1, is an integral part of the definition of NTU. The NTU unit of measurement was originally chosen such that a scattered light intensity is equal to that obtained by a 1000 mg/L solution of silicon dioxide—that is, 1 mg/L = 1 NTU. Today, a formazin mixture made according to the recipe listed in Section 4.4 and inserted into a turbidimeter that satisfies U.S. EPA Method 180.1 will, by definition, give a reading of the nominal value of a dilution of a 4000 NTU formazin solution.

Method 180.1 places a constraint on the measurement range to 40 NTU. It does allow diluting a sample with >40 NTU, so that the sample falls below this somewhat arbitrary upper limit. The NTU reading of the diluted sample is, of course, multiplied by the dilution factor to yield the actual turbidity:

$$T_{actual} = T_{diluted} \frac{Volume_{original} + Volume_{added}}{Volume_{original}} \tag{7.7}$$

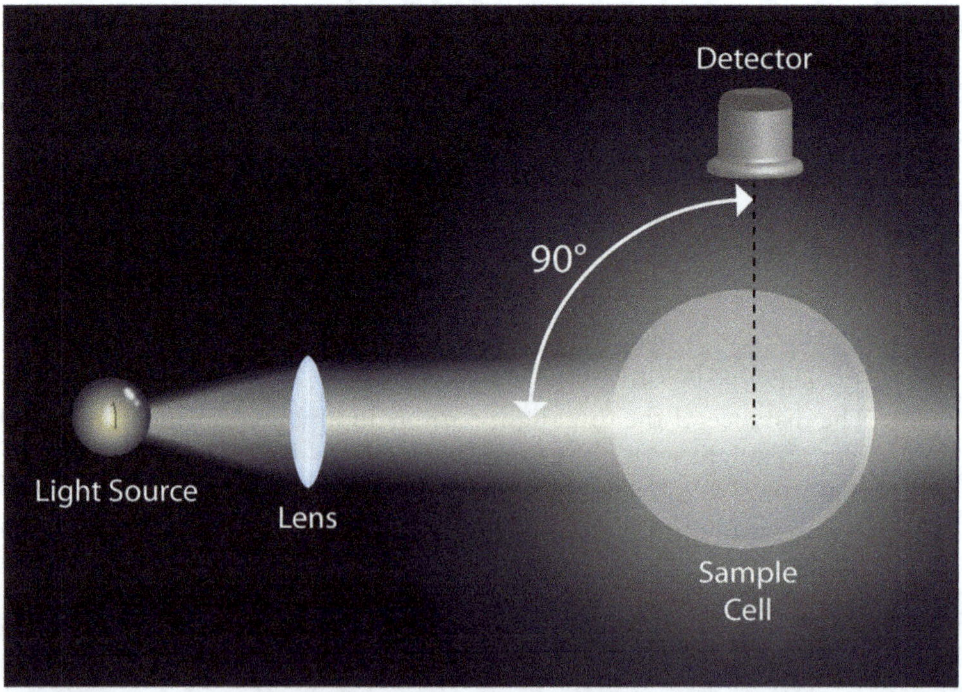

FIGURE 7.14 Schematic of turbidimeter. The detection of scattered light at a 90° angle satisfies the requirement of a nephelometric instrument (The Incandescent Source and U.S. EPA Method 180.1).

The setting of the NTU scale is tied to the problematic JTU scale in which 4000 NTU equals 1000 JTU. Strictly speaking the unit NTU may only be used for turbidity measurements made with a nephelometric turbidimeter. However, turbidity readings from a non-nephelometric analyzer may be converted to approximate NTU values using a look-up table or function that converts the raw measurement to that which would have been measured from a nephelometric turbidimeter. Just be forewarned that the conversion is highly specific to the instrument and the composition of suspended matter.

The operation of a turbidimeter is relatively simple: light hitting the photo detector creates a current, I_{90}, proportional to the light intensity. The analyzer converts the current to an NTU value, T. If the conversion factor between the two is k_{90} then another way of stating this relationship is:

$$T(NTU) = k_{90}I_{90} \tag{7.8}$$

There are two ways to configure a nephelometric turbidimeter:

1. A *side-stream* or *online* turbidimeter withdraws sample water through tubing and passes it through a sample chamber inside a light-tight housing—the *flow chanber*. The latter blocks out any ambient light and allows the instrument to record low levels of turbidimeter. Some models use a glass sample cell inside the enclosure and some use the entire interior of the chamber. The advantage of a glass sample cell is that the smaller volume means less calibration standard is needed. The savings can be considerable. The disadvantage is that any scratch, fingerprint, or condensation on the glass will scatter light and lead to an erroneously low signal.
2. A *submersible* or *in-situ* turbidimeter is based on the traditional probe body. The front end houses a window that passed the incident light into the sample and a second window at 90° to the first that collects the scattered light. It is more convenient to deploy than an online turbidimeter because no plumbing is required to install. However, it is susceptible to ambient light, so it is typically used to measure higher levels of turbidity. Many probe-based models have optional light-protected enclosures in which the turbidimeter is placed. Online turbidimeters are ubiquitous in water treatment facilities, whereas submersible sensors are common in wastewater facilities.

The left side of Figure 7.15 (Lambrou et al., 2009) shows the basic electronic design that converts the photocurrent from the photodetector to a voltage (Lambrou et al., 2009). The op-amp is a buffer, or transimpedance, amplifier that does the conversion. The voltage equals the resistance of the

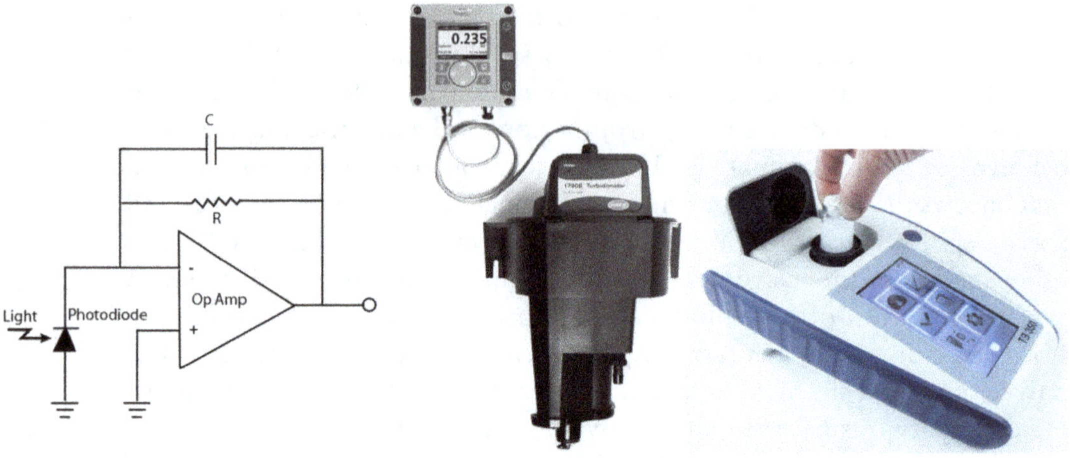

FIGURE 7.15 Left: Basic electrical design of a turbidimeter. The op-amp is a buffer (or transimpedance) amplifier that converts the current induced by the photodiode into a voltage. Right: Typical nephelometric turbidimeter. The light source, sample chamber and detector are all inside the large chamber (Middle–Courtesy Hach, Right–Courtesy Lovibond).

feedback resistor multiplied by the current. The middle figure of Figure 7.15 shows a common online turbidimeter made by the Hach Company. Not shown in this picture is the plumbing that carries water into the flow cell and out. This model uses the interior of the housing as a sample cell. The light source is at the top of the inside of the body above the water level. The detector is on the side and behind a window to the filled portion of the body. The photo on the right is of a portable turbidimeter made by Lovibond.

Table 7.3 lists the precision of reported values that are compliant with Method 180.1. Recall from Chapter 1's distinction between accuracy and precision that turbidity values that meet these reporting requirements are not necessarily accurate. Accuracy is a function of calibration.

Figure 7.13 shows that the response is approximately linear in the range of 0 to 40 NTU. Calibration requires a minimum of two points. A turbidimeter always yields a finite instrument signal even with zero analyte due to stray light and/or electronic noise. For very low turbidity values, that is, <0.1 NTU, it is necessary to use a pure water sample as a zero standard. Since the calibration curve is not a straight line, as shown in Figure 7.7, three-point calibration will give measurably more accurate results.

A turbidimeter based on U.S. EPA Method 180.1 has three drawbacks:

1. Material that absorbs just part of the visible light spectrum (400 to 700 nm) imparts a color to either the reflected or transmitted light. For instance, tannins, one of the most common suspended solids in

TABLE 7.3 U.S. EPA Method 180.1 reporting requirement for nephelometric turbidimeters.

Turbidity Range (NTU)	Record to Nearest NTU
0.0–1.0	0.05
1–10	0.1
10–40	1
40–100	5
100–400	10
400–1000	50
>1000	100

rivers, absorb light on the blue end of the spectrum so that the light that exits from a water sample containing tannins consists mostly of the remaining red end of the spectrum. Hence tannin imparts the tea color that we see in the river. The absorption of blue light narrows the measurement range.

2. As discussed in Section 2, right-angle scattering intensity is maximum when the size parameter is close to 1. For an incandescent light source, this means that the turbidimeter is mostly sensitive to particles whose size is less than 1 μm. A nephelometer that measures water with larger particles, such as wastewater, will display an artificially low value.
3. The specification of an incandescent (usually tungsten) source that emits light all across the visible spectrum is advantageous for detecting particles that encompass a range of sizes. However, it comes at a cost. The wide spectral range emitted by an incandescent means that two instruments of the same model can have light sources with different spectral output and detectors with different spectral responses. They can give different readings for the same sample. For that reason, calibration is more important for the incandescent light nephelometer than it is for an instrument using an LED or laser light source.
4. An incandescent light source loses intensity over time and must be replaced approximately annually.

5.2 The Near Infrared Alternative

These drawbacks spurred the European Union to adopt the turbidimeter specification ISO 7027. ISO Standard 7027 calls for a light source that emits at 860 nm, which is in the *near-infrared* (near-IR or NIR) section of the spectrum. The use of the term "NIR" to describe this light source can be misleading because the NIR wavelength range extends well beyond 860 nm in both directions.

The specification for an 860 nm turbidimeter is embodied in the ISO 7027 standard and is summarized in Table 7.4. The change in the light source addresses two drawbacks inherent in U.S. EPA Method 180.1:

1. The narrow wavelength of the light source translates into a less variability of different turbidimeters using different light sources. The exception is the use of an incandescent light source with a NIR filter.
2. The longer wavelength of the light source is more sensitive to the larger particles found in wastewater and natural bodies of water.

The turbidity unit of measurement for a near-IR source is the Formazin Nephelometric Unit (FNU). The definition of the FNU is tied to that of the NTU. As with the NTU and any other unit of turbidity, it is as follows: a formazin solution that gives 1 NTU in a turbidimeter satisfying U.S. EPA Method 180.1 gives 1 FNU in a turbidimeter satisfying the ISO 7027 specification. In fact, like the U.S. EPA method, ISO 7027 also restricts measurements to below 40 NTU and calls for dilution of more turbid samples.

Table 7.4 summarizes the differences in design between the turbidimeter specifications.

There is no perfect design, and the ISO 7027 has its own drawbacks. The major one is that the longer wavelength range of the NIR source is less sensitive to particles smaller than 1 μm. Though it is not affected by visible light that is absorbed by colored water, it is sensitive to particles that absorb in the NIR—mostly carbon black particles and copper sulfate (ASTM, 2011).

The main limitation for both single detector instruments is the measurement range. Absorption of the light beam by dissolved matter subtracts light that is available for scattering. As a result of absorption, and as shown in Figure 7.7, the scattering intensity increases at a rate that is less than linear—that is, "rolls over"—for turbidity values between 40 and 100 NTU. As explained above, the rollover occurs at lower turbidities for an instrument with a visible light source measuring turbidities in a colored liquid.

The three lower curves in Figure 7.17 shows the diminishing turbidity readings of three colored aqueous samples—two red wines and one beer. In this test, experimentalists added increasing concentrations of formazin

TABLE 7.4 Comparison of U.S. EPA Method 180.1 and European Standard ISO 7027.

Parameter	U.S. EPA 180.1	ISO 7027
Light Source	Tungsten, 2200–3000 K	LED, Tungsten with NIR, filter or laser
Turbidity Unit	NTU	FNU
Spectral Bandwidth	Not specified. 400–600 nm	860 ± 30 nm
Measuring Angle	90 ± 30	90 ± 2.5
Aperture Angle	Not specified	10°–20°
Path Length	10 cm	Not specified
Calibration Standard	Formazin or stabilized version (e.g., Stabil-Cal, T-Cal.)	
Measurement Range (prescribed by method)	0.01–40 NTU (linear to 40 NTU)	

to a fixed volume of the colored sample. The figure shows the actual NTU reading plotted against the NTU value of the formazin—that is, the NTU reading as if there were no scattering. If indeed there were no absorption, then the two readings would match—that is, the plot would be a 45° line. The deeper color of the red wine manifests itself in sharper turnover versus that of beer. (Clearly the beer was not Guinness.) Greater absorption manifests itself in a lower slope and, for the case of wine, a complete flattening of the plot.

Table 7.5 lists typical values for turbidities encountered in typical aqueous samples.

5.3 Sources of Error

As with all instruments, measurements can be misleading due to either instrumental errors or misinterpretation of the results. We focus this discussion on the prevalent nephelometric turbidimeter. It is a simple, precise, and cost-effective instrument. However, its configuration consisting of an incandescent light source and a single detector at 90° does lead to readings that either underestimate or overestimate the true concentration of suspended solids. It's important to understand that the NTU reading for a true nephelometric turbidimeter may be accurate. Rather, these artifacts change the interpretation of the NTU readings in terms of particle densities.

For turbidimeters that use a glass cell, stray light can result from scattering of the incident light off the surface of the sample due to condensation, scratches, fingerprints, dirt scaling, or biofouling. Some units house a

TABLE 7.5 Typical turbidity values commonly found in typical aqueous samples. (Values taken from various sources, including ISA and California State Water Control Board.)

Sample	Turbidity Range (NTU)
Pure water	0–0.01
Drinking water	0.02–0.1
Potable water	0.02–1.5
Groundwater	0.02–1
Waterbodies with sparse plant and animal life	0.02–0.1
Waterbodies with moderate plant and animal life	1–10
Waterbodies with high nutrients and planktonic life	10–50
Storm flow in rivers and streams	20–1000
Wastewater effluent	4–20
Wastewater influent	70–2000
Orange juice	300–900
Milk	>4000

desiccant or a heater to prevent condensation from forming on the cell or window. Some use an ultrasonic transducer to remove any coating from the cell. A thin coating of silicone oil minimizes scratches.

Bubbles delivered to the sample from the plumbing are a frequent cause of scattering. They can be removed by a bubble trap, which splits the flow of water into a horizontal tube going into the instrument and a vertical tube open at the top. The bubble rises in the latter and is harmlessly dispersed.

Lastly, the flow into an inline turbidimeter requires flow in a "Goldilocks" zone. A sample stream that is too high in flow can form bubbles when it encounters a pressure drop from a smaller orifice or a bend in tubing. Flow that is too low will allow larger particulates to settle on the bottom of the sample cell. Most turbidimeters have a regulator to limit the flow on the high end. For the lower limit it is up to the user to ensure that the flow is at least the minimum prescribed rate. Some low-level turbidimeters contain an algorithm that filter out the spike in turbidity caused by a bubble.

Table 7.6 summarizes the most common sources of errors and their effect on the turbidity measurement.

TABLE 7.6 Sources of error in turbidity measurements.

Interference	Consequence	Notes	Reading Error
Particle Size	The larger the particles, the more forward scattering and less right-angle scattering.	NIR light sources, as in ISO 7027 instruments, are more sensitive to larger particles.	Lower
Particle Shape	The more spherical the particle shape, the more right-angle scattering.		Higher
Of Particle Color	Absorption of part of the visible spectrum results in loss of the incident light intensity and a consequent loss in the scattered light.	Tannins, green algae, beverages	Lower
Particle Refractive Index	Scattering is proportional to the difference between the index of refraction of the water and the suspended solid.	Silica will scatter more than organic matter.	Higher
Bubbles	Entrained air bubbles cause spikes in readings.	An air purge can remove bubbles in the tubing.	Higher
Condensation on the Sample Cell	Condensation on the sample cell scatters light on the outside of the sample cell and diffuses the incident light.	Use a desiccant to remove water vapor.	Lower
Biofouling of the Inside of the Cell	Bacterial and algal colonies grow on the inside surface of the cell and scatter light.	Occasionally flush the cell and tubing with dilute hypochlorite.	Lower
Scratches and Dirt on the Sample Cell	As with condensation, scratches, fingerprints, and dirt on the sample cell scatters light before it reaches the sample.	Handle the cell with gloves. Wipe the cell with lint-free tissue.	Lower
Settling	Larger particles settle out.	Do not allow samples to sit. Shake the sample before measuring.	Lower
Readings Decrease Over Time	In a nephelometric turbidimeter, the light source has aged.	Replace the bulb.	Lower

5.4 The Ratio Turbidimeter

The choice of measuring scattered light at 90° represents the optimal compromise between range and sensitivity. But who says you can't have it all? A ratio turbidimeter consists of the usual detector at 90° plus one or more detectors at other angles.

The simplest ratio turbidimeter places a second detector at 180° to measure absorption. This absorption measurement is of particular importance because it compensates for colored samples. Turbidimeters based on the Orion Method AQ4500 (by definition, the Orion AQ4500 laboratory turbidimeter) are based on this two-detector design. This method (and model) stipulates a white light LED.

Most ratio turbidimeters contain additional detectors placed at two or more of approximately the same angles as shown in Figure 7.16. One at approximately 30° measures back scattered light and one at approximately 135° measures forward scattered light. Figure 7.13 (Liu et al., 2015) shows the scattering intensity expected for a typical sample.

The calculation of the turbidity for a four detector turbidimeter is (Liu et al., 2015):

$$T\ (NTRU) = K\frac{I_{90}}{K_0 I_0 + K_{180} I_{180} + K_{30} I_{30} + K_{135} I_{135}} \qquad (7.9)$$

The value of the individual k weighting factors and the overall conversion factor, K, are determined experimentally by fitting Equation 7.9 to a range of data and forcing the calculated turbidity to give the same turbidity

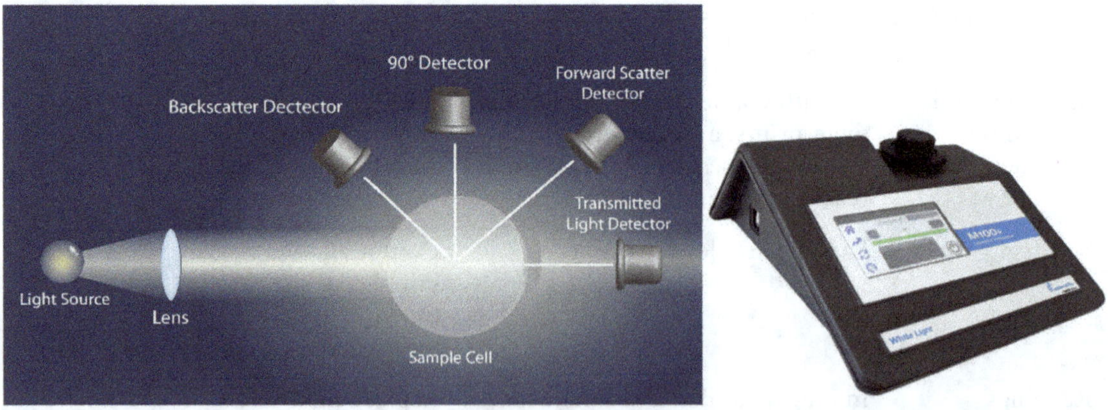

FIGURE 7.16 Left—Typical four-detector ratio turbidimeter. Right—The HF Scientific M100+ Turbidimeter is a ratio turbidimeter intended for laboratory use (Courtesy HF Scientific).

reading of a U.S. EPA Method 180.1 turbidimeter. The principles of algebra require a minimum of four turbidity samples to solve for the four *k* factors.

The ratio turbidimeter is found mostly in laboratories. The right side of Figure 7.16 shows a model made by HF Scientific. The improved resolution rendered by the additional detector can be as high as 0.0001 NTU and may not be warranted in most process control applications. Its range extends as high as the most concentrated solution of formazin that can be synthesized—approximately 7500 NTU.

The correct unit of measure for reporting a turbidity value that draws upon a ratio measurement adds the letter "R" after the first two letters of the absolute turbidity unit. Hence, an incandescent light source measurement yields NTRU values, and an 860 nm light source yields FNRU values.

Figure 7.17 shows the improvement in the operating range of the ratio turbidimeter over the single detector nephelometric turbidimeter (Sadar, 1998). The operating range is limited by the range over which the analyzer reading is roughly linear with increasing turbidity. As discussed in the previous section, all three samples are colored, but the two red wine samples absorb most strongly and limit the measurement range to below 60 NTU. When the nephelometric signal is divided by the transmission of the light source the "normalized" turbidity signal is linear to a value on the order of

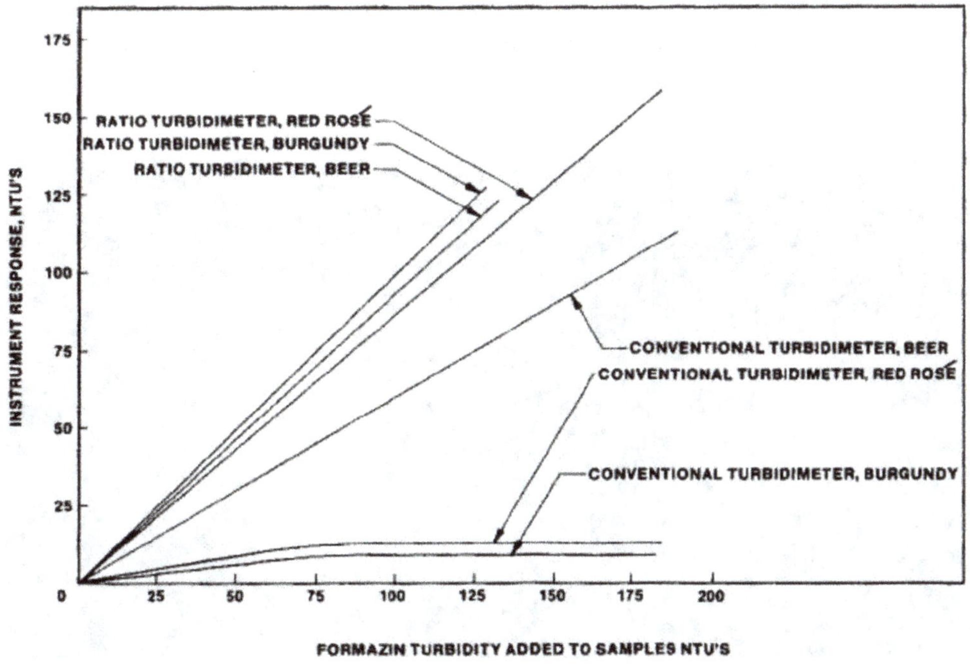

FIGURE 7.17 Response of a nephelometric turbidimeter and a ratio turbidimeter (Sadar, 1998a).

100 NTU. Multiple calibration standards enable measurements higher than the linear range up to approximately 1000 NTU. A higher range is possible with a shorter path length (<1 cm), though at the expense of sensitivity.

5.5 The Four-Beam Turbidimeter

The concept of measuring the transmitted light as well as the scattered light greatly widens the measurement range that is limited by absorption of white light by colored samples. Yet it cannot correct for diminished turbidity readings caused by variation in the output of the light source—especially an incandescent source—or of the detector. Neither can it compensate for fouling of any optical windows or sample cell (if there is one).

The four-beam turbidimeter addresses a changing light source or detector. If the fouling is uneven, then only cleaning will totally compensate for the loss of light signal. Great Lakes Instruments (now part of Hach) developed this turbidimeter in 1992, and U.S. EPA Method 180.2, also called GLI Method 2, codified it as a regulatory compliant instrument. Figure 7.18 shows how it works.

This analyzer utilizes two light sources, emitting at 860 nm, in addition to two detectors. The two detectors and two sources are located at 4 positions such that all four are 90° apart. During the first measurement phase, light source #1 turns on. Detector #2 measures the 90° scattered light, and detector #1 measures the 180° transmitted light. During the second measurement light, approximately ½ second later, source #2 turns on, and the

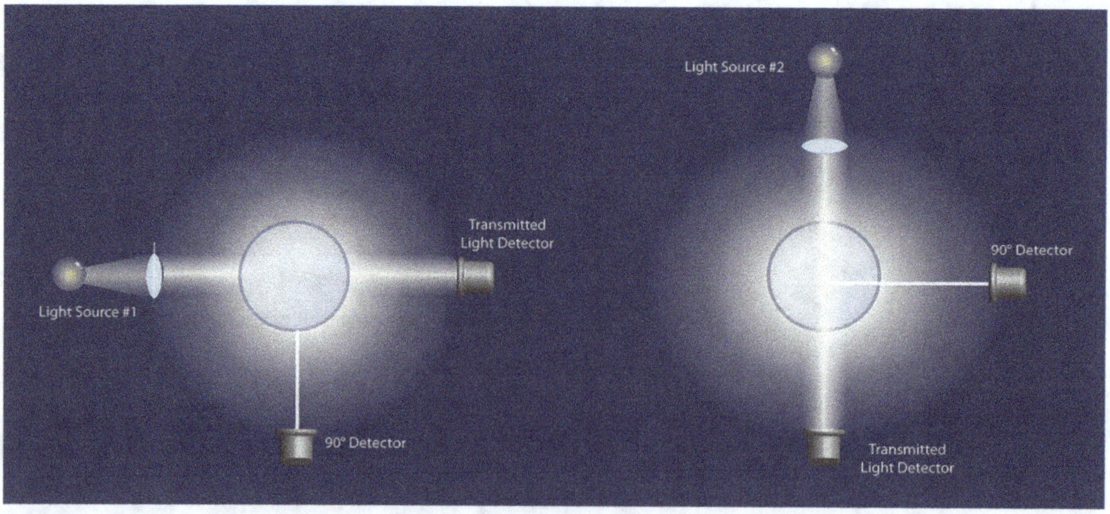

FIGURE 7.18 Schematic of a four-beam turbidimeter. The two light sources and two detectors yield two separate measurements of both scattered light and transmitted light.

detectors reverse roles: Detector #2 measures the 180° transmitted light, and detector #1 measures the 90° scattered light.

To understand how the four-beam turbidimeter compensates for source and detector changes, we write out expressions for the measured intensity on the two detectors. Let I_1 and 1_2 be the undiminished intensities of the two light sources and D_1 and D_2 be the responses of the two detectors. Let's quantify the loss of intensity of the two light sources by the coefficients $k_{1,s}$ and $k_{2,s}$ and the loss of transmission to the detectors by the coefficients $k_{1,D}$ and $k_{2,D}$. Last, let $F_{90}(C)$ be a function that quantifies the 90° scattering intensity of the medium at concentration C, and $F_{180}(C)$ be a function that quantifies the 180° transmission. The detector currents generated at the two detectors for the two 180° and two 90° measurements are

$$I_{D_1,180} = k_{1,S}\, I_1 k_{1,D} D_1 F_{180}(C) \tag{7.10}$$

$$I_{D_2,180} = k_{2,S}\, I_2 k_{2,D} D_2 F_{180}(C) \tag{7.11}$$

$$I_{D_1,90} = k_{1,S}\, I_1 k_{2,D} D_2 F_{90}(C) \tag{7.12}$$

$$I_{D_1,90} = k_{2,S}\, I_2 k_{1,D} D_1 F_{90}(C) \tag{7.13}$$

Let's now combine the four expressions into one ratio:

$$\frac{I_{D_1,90}\; I_{D2,90}}{I_{D_1,180}\; I_{D_2,180}} = \frac{F_{90}(C)}{F_{180}(C)} \tag{7.14}$$

The wisdom of constructing this ratio is immediately clear: All four k factors that quantify the loss of light intensity at the source as well as at the detector simply disappear. The ratio in Equation 7.14 is only a function of the scattering properties of the water sample. The ratio also looks very familiar. The square root of the product of the two ratios on the left side is the geometric average of them:

$$I_{90,\text{avg}} = \sqrt{I_{D_1,90} I_{D_2,90}} \text{ and } I_{180,avg} = \sqrt{I_{D_1,180} I_{D_2,180}} \tag{7.15}$$

Now the turbidity value is analogous to Equation 7.16:

$$T\ (NTRU) = K \frac{K_{90} I_{90,avg}}{K_{180} I_{180,avg}} \tag{7.16}$$

Because the light source is the same NIR one prescribed by ISO 7027, the unit of turbidity is an extension of the FNU—formazin nephelometric multibeam unit, or FNMU. Most four-beam turbidimeters report values in these more widely understood nephelometric units.

5.6 Keeping the Units Straight

We've defined several units of measurement for turbidimeter readings. This is a good place to summarize the units and add a few we have not introduced. The criteria that dictate the definition of a new turbidity unit are the wavelength of incident light, detector angle, and number of detectors used. The units we have already covered are in the first two rows. The remaining units are unlikely to appear in the world of water and wastewater turbidity measurement.

Many operators fall back on the NTU as a unit of measurement even when it is not strictly applicable. For example, a modulated four-beam turbidimeter using a near-IR light source ought to give readings in FNMUs, yet most manufacturers specify these instruments in terms of NTUs. It is important to keep in mind the lesson of Section 4.4: two readings of the same sample from two different turbidimeters but reported in NTUs are not likely to agree. To repeat the conclusion of that section, turbidity is not an intrinsic property of a sample of water—it is a measurement that depends on the design of the instrument that records it. This is a point stressed by Mike Sadar (Sadar, 2002) who proposed appending, in subscripts to the FNU unit, an abbreviation for the instrument type used for the measurement. For instance, FNU_{2B} denotes a measurement using the modulated four beam turbidimeter using a near-IR light source. The USGS and ASTM have adopted this nomenclature.

5.7 The Submersible (or In Situ) Turbidimeter

The turbidimeters that we have discussed so far draw a side stream from the process and makes measurements in a separate chamber. As we mentioned in Section 5.1, the housing minimizes outside light from interfering with measurement of low turbidity values.

For measurements that do not require high sensitivity at low turbidity levels, a probe configuration is very convenient. It's lack of immunity to stray light and its typically short light path results in a measurement range centered on higher turbidity levels. An example of a probe turbidimeter is shown in Figure 7.19. The probe has detectors at 45° and 90°. As discussed in Section 2.2 a greater fraction of light is scattered at larger angles at higher particle densities.

There is currently no U.S. EPA method that approves the use of a submersible probe.

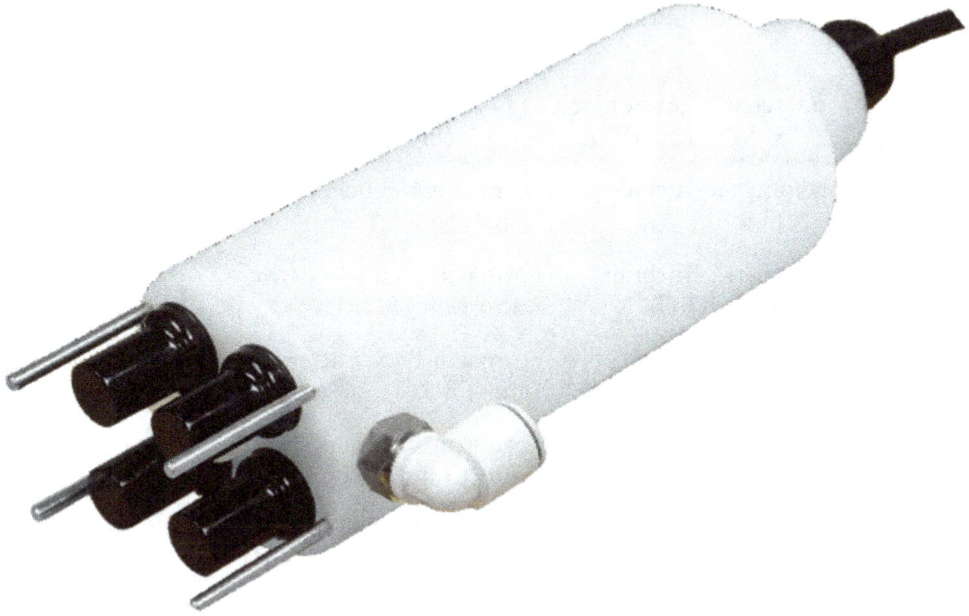

FIGURE 7.19 Example of a submersible four-beam turbidimeter. The two pairs of LED sources and detectors are clearly visible (Quadbeam Technologies).

Turbidity probes are well suited for higher turbidity levels found in wastewater and field measurements. In the more turbid water, where such a probe is likely to be found, fouling of the optical windows is more of a problem than with drinking water turbidimeters. Some probes have wiper mechanisms that clean the windows every several minutes. Others use a jet of compressed air.

5.8 Other Turbidimeter Designs and Methods

The methods described up to this point, especially U.S. EPA Method 180.1 and ISO Method 7027, provide the basis for most turbidimeters. Other methods are based on more recent novel turbidimeter designs.

It bears noting that there is an alternative method occasionally mentioned in the literature. The American Public Health Association issued Standard Methods 2130B (American Public Health Association et al., 2017). It is identical to U.S. EPA Method 180.1 with one minor difference. While it encourages use of SDVB as a secondary calibration standard, it specifies that only formazin can be used as a primary standard. U.S. EPA Method 180.1 allows other primary standards based on formazin, such as Hach's Stabil-Cal. Method 2130 B also allows turbidities as high as 1000 NTU, rather than 40 NTU prescribed by Method 180.1.

TABLE 7.7 Summary of turbidity units.

Detector Angle (Degrees)	Incandescent Light Source (U.S. EPA 180.1)	Near-IR Light Source (ISO 7027)	Typical Measurement Range
90	Nephelometric Turbidity Unit (NTU)	Formazin Nephelometric Unit (FNU)	0.02–40
90, 180 (add'l angles)	Nephelometric Turbidity Ratio Unit (NTRU)	Formazin Nephelometric Ratio Unit (FNRU)	0.01–40
30 ± 15	Backscatter Unit (BU)	Formazin Backscatter Unit (FBU)	100–10 000
30 ± 15 (add'l angles)	Backscatter Ratio Unit (BRU)	Formazin Backscatter Ratio Unit (FBRU)	10–10 000
90 + add'l sources and detectors	Nephelometric Turbidity Multibeam Unit (NTMU)	Formazin Nephelometric Multibeam Unit (FNMU)	0.01–4000
180	Attenuation Unit (AU)	Formazin Attenuation Unit (FAU)	10–10 000
90 with visible monochromatic light source	Monochromatic Nephelometric Turbidity Unit (mNTU) Sources include LEDs and lasers	n/a	0.01–1000
90	Milli-Nephelometric Turbidity Unit (mNTU)		0.001–5
90 from surface	Surface Scattering Unit (SSU)		10–10 000

Hach Method 10133 is a variation of the Method 180.2. It also specifies a detector at 90° and a path length no longer than 10 cm. As opposed to the nephelometric method, it uses a laser diode emitting between 630 and 690 nm. The laser beam enters the sample directly—that is, without the use of a sample cell. Fiber optic cables collect the scattered light and transmit it to a very sensitive photomultiplier tube (PMT). This use of a laser and PMT yield a highly sensitive turbidimeter that is ideal for levels less than 5 NTU. To accommodate the very low levels of turbidity, a new turbidity unit, the milliNTU (mNTU) was introduced, in which 1000 mNTU = 1 NTU. Turbidimeters following Method 10133 are used to monitor drinking water treatment, especially filter breakthrough.

One particularly unique design is based on a white LED light source placed above the surface of the water, thus eliminating the usual problems

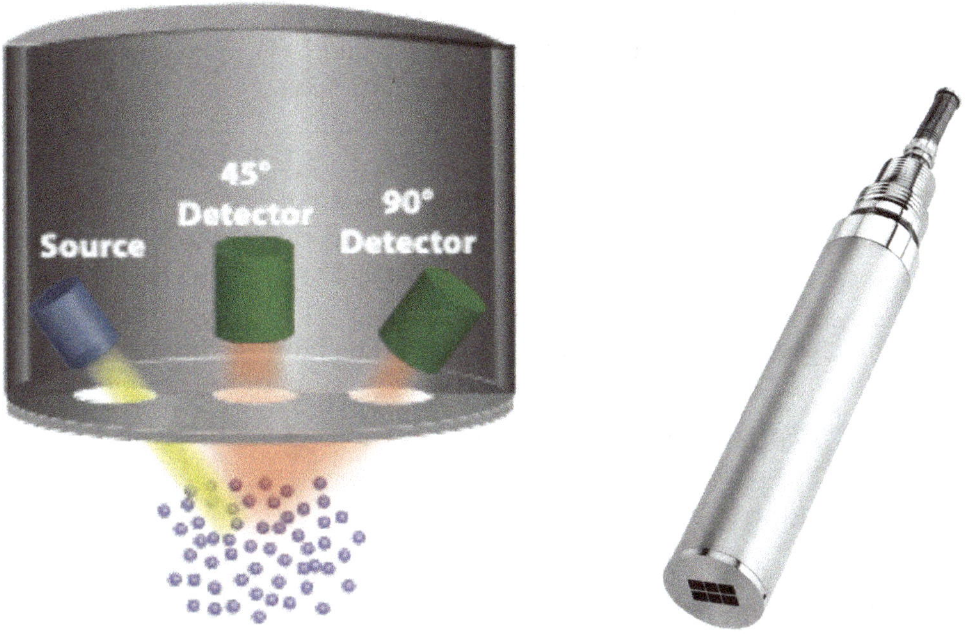

FIGURE 7.20 Left: The schematic of a turbidity probe with detectors at 45° and 90° (Courtesy Endress+Hauser). Right: The Turbimax CUS51d by Endress+Hauser has detectors at 90° and 135° (Courtesy Endress+Hauser). The 90° detector conforms to ISO 7027. These back-scatter detectors enable measurement of turbidity values as high as 4000 NTU.

of fouling, bubbles, and condensation on the sample cell. Scattering occurs just below the surface of the sample water and is collected at approximately the angle of reflection. This design by Swan was codified into an instrument specific U.S. EPA method—the Swan AMI Turbiwell Method. The measurement range of a surface scattering turbidimeter is shifted toward higher turbidity levels when compared with a conventional nephelometer. The level of precision for this instrument is approximately 0.01 NTU. Though higher than a nephelometric turbidimeter, this is rarely a problem in water treatment facilities. The maximum level of measurement is 200 NTU, which is approximately twice that of a nephelometric model.

Mitchell Method M6271 is a nephelometric method developed by Leck Mitchell in 2009 and replaces the incandescent bulb with a laser emitting at 650 ± 30 nm. The method and instrument upon which the method is based also includes a bubble trap and anti-fogging. It is similar to Hach Method 10133 discussed above. Its measurement range is up to 40 NTU. Mitchell Method M5331, a second nephelometric turbidity by Leck Mitchell, replaces the incandescent bulb with a green LED at 525 ± 15 nm (green). Otherwise, the instrument is the same as the one above, including the range up to 40 NTU.

Orion Method AQ4500 was mentioned in Section 5.4. It adds a 180° detector to the 90° detector and employs a white LED light source, which is constructed by using a yellow phosphor coating on a blue LED source. A silicon photodiode measures scattered and transmitted light. An LED source has two advantages over an incandescent source: first, its output is more stable over time. Second it can be powered instantaneously. This allows it to be pulsed so that synchronous detection can be used to reduce ambient light interference. LEDs will continue to replace incandescent sources as their spectral output is more consistent from one to another and aging of the source occurs at least an order of magnitude more slowly.

6.0 TURBIDITY IN PRACTICE

6.1 Drinking Water

The most prevalent use of turbidity measurement is to assess the treatment process in a water treatment facility. The U.S. EPA mandates that the water leaving a water treatment plant cannot exceed 1 NTU at any one time. Furthermore, 95% of samples collected during any month must have levels less than 0.3 NTU. The World Health Organization has looser standards to fit a global population. It calls for turbidity levels less than 5 NTU and preferably less than 1 NTU. There are various stages during the purification process that can call for turbidity measurements.

Raw water entering the facility may have turbidity values typical anywhere between 1 and 100 NTU—even higher if after a storm. The coagulation and flocculation stages yield floc particles, which are more likely to increase turbidity than to lower it. However, the sedimentation process that occurs in the clarifier reduces turbidity levels from the influent to between 60% and 80% (Hudson, 1981). Filtration follows clarification. On the upstream side of the filter turbidity is typically in the 20 NTU range but can briefly reach levels of 200 NTU (Davis & Lambert, 2002). Typical turbidity levels after filtration are 0.5 NTU. Membrane systems that use filtration—from microfiltration to reverse—produce filtrate with turbidity values less than 0.3 NTU. Clarification drops the turbidity to as low as 0.5 NTU before filtration. The effectiveness of filtrations is determined by measuring turbidity levels on both sides of the filter (U.S. EPA, 1999b).

Low-level turbidity levels in the filtration effluent are especially useful for detecting filter breakthrough—the process whereby passages through filtration medium or cracks in a filter membrane allow water to pass through unfiltered. The point at which breakthrough begins can provide the signal to begin backwashing or filter replacement. Most treatment facilities schedule

TABLE 7.8 Compilation of approved turbidimeter specifications (U.S. EPA).

Method	Date	Organization	Reference Title	Description
180.1	1993	U.S. EPA	Methods for the Determination of Inorganic Substances in Environmental Samples	Beam angle 90 (<30° beam angle)
ISO 7027		ISO	860 ± 30 nm	
180.2	1992	Great Lakes Instruments	GLI Method 2 Modulated Four Beam	Angles between sensors and detectors 90 ± 2.5°
10133 Rev 2	2000	Hach	Filter Track Method. Laser Nephelometry, Rev 2	
2130 B-01		Standard Methods Online		
2130 B	2005	Standard Methods	*Standard Methods for the Examination of Water and Wastewater*, 21st edition	
Mitchell M5271	2009	Leck Mitchell	Laser Nephelometry	
Mitchell M5331	2009	Leck Mitchell	LED Nephelometry	
Orion AQ4500 180.1	2009	Thermo Scientific	LED Nephelometry	
AMI Turbiwell	2009	SWAN Analytical Instruments	Continuous Measurement using a SWAN AMI Turbiwell Turbidimeter	

backwashing on a periodic basis, but the ability to delay the cleaning until it's needed can save electricity and minimize the volume of water needed to backwash. The small change in the low turbidity level of the filtered water calls for a turbidimeter with an enhanced sensitivity to low levels of turbidity. Some used for this purpose, such as the FilterTrak 660, use a red laser light source. Particle counters are also used to detect breakthrough. The laser produces a monochromatic (660 nm), intense, and highly collimated light beam that gives a high signal-to-noise ratio even for very low turbidities. This is sufficient to remove bacteria and protozoan pathogens.

6.2 Pathogen Detection

Turbidity measurements cannot detect pathogenic microbes, which include bateria and protozoa. However, microbes typically grow on surfaces of suspended solids so measuring turbidity ought to give an approximation of the concentration of pathogens. There is a correlation between pathogens and turbidity.

A corollary to this finding is that a reduction in turbidity correlates with a reduction in pathogens. Indeed, that was the case when a group (LeChevalier et al., 1991) measured turbidity and concentration of the pathogenic protozoa Cryptosporidium and Giardia. Even in developed countries these single celled organisms account for half of all waterborne illnesses. The two species responsible for most outbreaks are Cryptosporidium and Giardia. These amoebas form cysts that are on the order of 10 μm, which are large enough to be filtered out even when free floating. The cysts protect these organisms from disinfection. In fact, a chlorine concentration of 8 mg/L, at pH 7 or less, is required to kill 100% of the Giardia cysts (U.S. EPA, 1999a). The correlation between between turbidity decrease and Cryptosporidium and Giardia cyst removal in raw water was marked by scatter, since the cysts were just a small fraction of the particulate matter, but the correlation was still clear. However, only raw water that contains these pathogens showed the correlation. For raw water that does not contain the pathogens in substantial numbers, there was little correlation between their removal and the turbidity of the treated water.

A study carried out at several treatment facilities (Patnia et al., 1997) on the West Coast showed that bringing turbidity levels to less than 0.1 NTU was capable of a log-5 (i.e., a factor of 10 000) reduction in these parasites. This is still a factor of 10 higher than the log-6 removal mandated by the Enhanced Surface Water Treatment Rule, so disinfection is still needed for the final tenfold reduction. However, the correlation between turbidity removal and pathogen reduction was not linear—lowering turbidity from

0.3 to 0.1 NTU produced a large log 1 (factor of 10) reduction in Cryptosporidium and Giardia.

A review of five studies on the correlation between turbidity in treatment plant effluent and gastrointestinal illness (Mann et al., 2007) concluded that two of the five did indeed show a positive correlation. At this time one can only say that high turbidity levels in a drinking water facility can serve as a warning for an increase in microbial levels.

6.3 Groundwater

The measurement of groundwater turbidity is the single most important parameter for gauging contamination. Most turbidity values for raw groundwater are less than 5 NTU (U.S. Geological Society, 2018), which explains the attraction of obtaining drinking water from groundwater. The suspended solids that are prevalent in groundwater are inorganic (i.e., mineral), as opposed to organic and biological contaminants found in surface water. Groundwater that travels through layers of silt and clay are likely to have turbidity levels higher than 5 NTU and require filtration. More importantly, elevated turbidity readings indicate groundwater contamination by organic compounds that are not being filtered by the subsurface.

During the development of a new well or remediation of an existing one sequential turbidity readings track the purging of sediment and contaminants in the well bore. One study (Tease & Dowling, 2012) showed a correlation between turbidity values and contamination of groundwater by gasoline for a well but only for a narrow range of purging flow at a purge rate of approximately 100 mL/min.

6.4 Surface Water

Excessive amounts of suspended matters have a greater effect on the water quality of rivers, streams, and oceans than any other parameter. Erosion caused by precipitation results in high levels of suspended solids. Agricultural runoff contains high levels of phosphorous and nitrogen, which spawn algal blooms. To quantify the amount of matter carried by surface water, one needs to measure the concentration of suspended solids. TSS values vary with time. If one knows the flow (volume per unit time) as a function of time, then the total load of suspended solids load over a time period, *T*:

$$SSL = \sum_{0}^{T} TSS(t)\, Q(t) \Delta t \tag{7.17}$$

$Q(t)$ is the flow and is in units of volume per unit time, such as m^3/sec. Measurements are made every Δt seconds. The summation sign means that measurements, made every Δt seconds, are added up until a time, *T*, is reached. The unit of SSL is weight—for example, kilograms or pounds. The process of moving suspended matter from land to a body of water is termed *sediment loading*.

TSS levels not only vary in time, but also in space. To make TSS measurements, as the next section describes, would be an impossible task. Turbidity measurements are easy to make and can be converted to TSS data. Figure 7.21 shows measurements of TSS, suspended sediment concentration (SSC), flow, and turbidity in the Little Arkansas River in Kansas (Fink, 2005). Turbidity readings were made by several YSI sondes deployed at points in the river representative of high and low flow levels. All TSS measurements were done on grab samples. The turbidity readings track the TSS readings very well.

Because of the turbidity levels are commensurate with large amounts of suspended matter, the 90° nephelometer is not adequate to measure the full range. Instruments that measure absorption only or backscatter are common. Many of these turbidimeters do not adhere to a U.S. EPA or ISO method, but that requirement is not as crucial for environmental monitoring as it is for municipal water treatment. Ultimately TSS values are the ones that count.

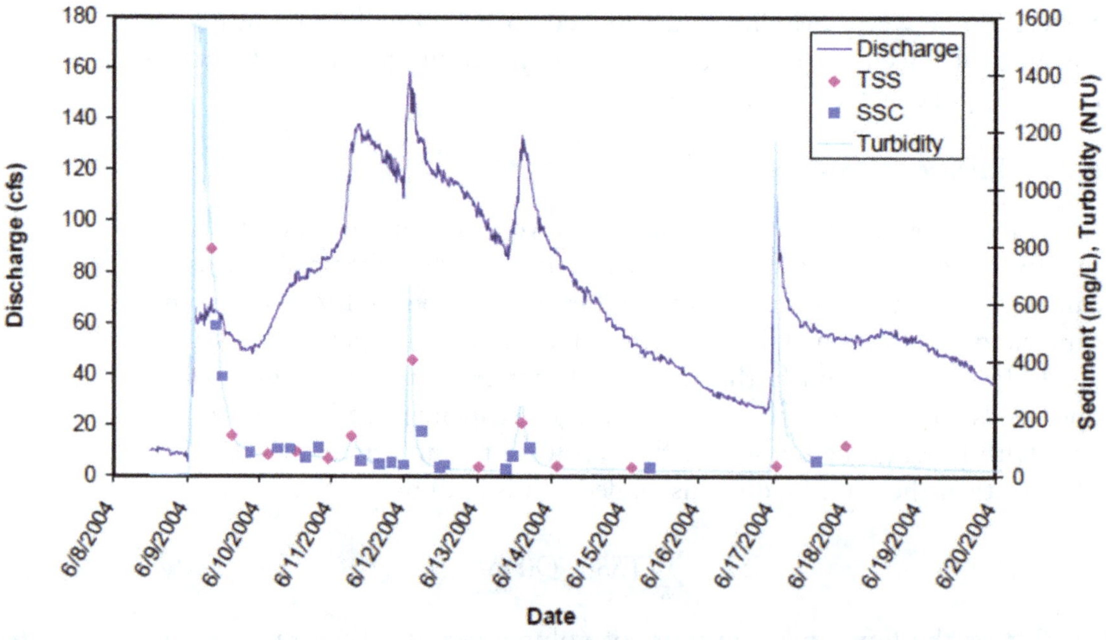

FIGURE 7.21 Turbidity, TSS, SSL, and flow for the Little Arkansas River in Kansas (Fink, 2005).

7.0 TSS MEASUREMENTS AND TURBIDITY

The measurement of suspended solids started the discussion of turbidity in the beginning of the chapter. Let's now review that measurement process. The steps of this simple but time-consuming method as prescribed by the U.S. EPA Method 160.2 or ASTM D5903 (ASTM, 2018). The range of TSS values that this method measures is 4 to 20 000 mg/L. The procedure is as follows:

1. Dry a glass fiber filter in an oven at 103 °C in an aluminum foil dish for a period long enough that its weight remains constant. The filter should pass any solids smaller than 0.45 μm. Weight is measured to within 0.1 mg.
2. Filter a known volume of water.
3. Place the filter containing the trapped solids in the oven and dry it at 103 °C in an aluminum foil dish until the weight remains constant.
4. Subtract the initial weight from the final weight and divide the resulting suspended solids weight by the volume of sample. The units are reported as mg/L.

The entire process can take from 6 to 24 hours. This stands in stark contrast to a turbidity measurement, which takes seconds. Given that the procedure for measuring TSS values is so time-consuming, it only makes sense to measure turbidity values and transform them to TSS values. To do so requires knowledge of the correlation between the two. It would be very convenient if they were proportional to each other. Indeed, over a limited range (<1000 NTU), they are, but the proportionality constant differs from one aqueous sample to the next. The proportionality can be expressed as a simple linear expression:

$$TSS = a\,T + b \tag{7.18}$$

One expects the intercept, *b*, to be zero within the experimental uncertainty. However, changing conditions during the collection of samples sometimes leads to a non-zero intercept. The proportionality constant, *a*, is the conversion factor that converts TSS units to turbidity units. To derive the value for *a* one typically takes samples of the water at a different location. The constant is specific for not just a specific location but also a specific condition and specific instrument.

A simple example is colored water. It will absorb part of the white light of an U.S. EPA 180.1–compliant nephelometer and diminish the scattered light measurement. A more subtle example is the measurement of turbidity and TSS values in natural waters. Take the example of measurements in several streams in Singapore. The value of *a* increased after a rain event by an average of 80% compared to just prior to the event (Daphne et al., 2011). The wet weather washed fine mineral and organic material into the river. As explained in Section 2, the smaller particles scatter more visible light and scatter more of it in the 90° direction where it is picked up by the detector of a nephelometric turbidimeter.

The proportionality constant in field measurements changes not only over time but over different geography. Figure 7.22 shows TSS versus turbidity measurements taken over a 2-week period in Baird Creek, Wisconsin, and at two locations along the creek (Fink, 2005). Although the correlation between TSS and turbidity as displayed is excellent, the disparity between the slopes corresponding to the two sites differ by a factor of 2.

For wastewater treatment, streams, rivers, and precipitation runoff the range of turbidity values can extend beyond 10 000 NTU. For this broad range the relationship between turbidity and TSS concentration is nonlinear. The extension of the linear relationship, for turbidities >1000 NTU is the power law formula

$$T = aTSS^b + c \tag{7.19}$$

Note that the relationsip in Equation 7.19 shows T in terms of TSS, which is the reverse of the linear Equation 7.18. Typical values of b are close to 1, indicating a relationship that is close to linear. Figure 7.23 shows turbidity versus TSS relationships for samples of water mixed with soil, silt and clay (Holliday et al., 2003). Exponential coefficients (b) varied from

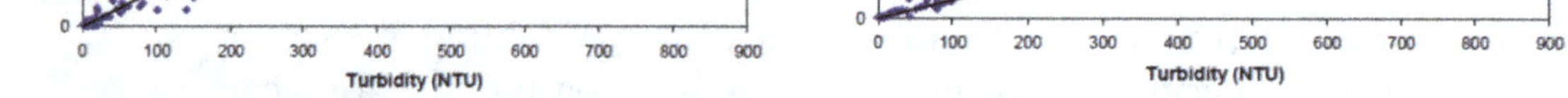

FIGURE 7.22 Turbidity and TSS measurements taken at two sites in Baird Creek, Wisconsin. The correlation between the two is excellent, but the proportionality constants differ by a factor of 2 (Fink, 2005).

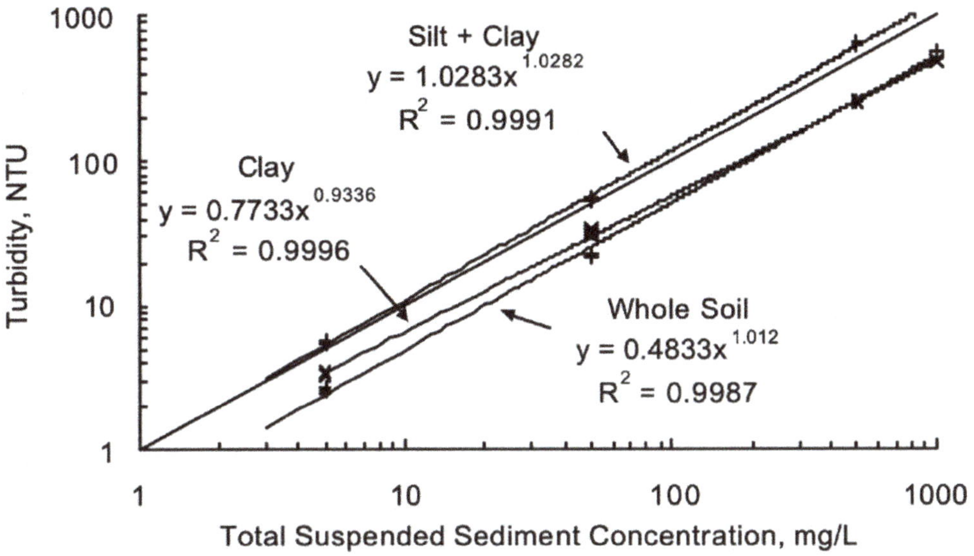

FIGURE 7.23 Power law relationship between turbidity and TSS values for suspensions of soil, clay, and silt (Holliday et al., 2003).

0.94 and 1.01. Similar studies of soils suspended in water taken at six sites of mining runoff gave larger b values that varied from 1.31 to 1.45 (Perkins, 2013).

The bottom line is that turbidity measurements can be used as surrogates to time-consuming TSS measurements, but the relationship needs to be determined experimentally for each location, condition, and instrument (Holliday et al., 2003).

A discussion of measuring TSS would not be complete without mentioning that there is a second surrogate technique—acoustic Doppler metering (Wood, 2014). The Doppler technique is well known in water instrumentation for measuring flow but is not widely known for measuring suspended solids. A Doppler sensor works via a transducer that sends an acoustic pulse in the ultrasound frequency range—typically 600 KHz to 1 MHz—through a fluid containing particles or bubbles. As a wave strikes a solid particle, part of it is reflected (*backscatters*) and is picked up by a receiver that is usually immediately adjacent to the transducer. If the particle is stationary, then the wave returns at the same frequency. If the particle is moving away from the transducer (or, to be exact, has a component of its velocity that is opposite the direction of flow), then the frequency of the return wave decreases. If the particle is moving toward the transducer/receiver, then the frequency of the wave shifts toward a higher value. The shift in frequency is proportional to the magnitude of the flow. Hence the Doppler flow analyzer measures flow by measuring the frequency shift. In addition to the change in frequency the strength of the return pulse is proportional to the concentration of particles and, hence, the total suspended solids. Doppler measurements of suspended solids are done in rivers and streams. Two (or more) transducers are aimed at opposing angles with respect to a line perpendicular to the direction of flow. The meter calculates TSS values by averaging the backscatter intensities of both beams.

8.0 REFERENCES

American Public Health Association, American Water Works Association, & Water Environment Federation. (2017). *Standard methods for the examination of water and wastewater* (23rd ed.) American Public Health Association.

ASTM. (2011). *Standard guide for the use of various turbidimeter technologies for measurement of turbidity in water.* ASTM Standard D7726-11.

ASTM. (2018). *Standard test methods for filterable matter (total dissolved solids) and nonfilterable matter (total suspended solids) in water.* ASTM D5907-18.

Daphne, L. H. X., Utomo, H. D., & Kenneth, L. Z. H. (2011). Correlation between turbidity and total suspended solids in Singapore rivers. *Journal of Water Sustainability, 1*(3), 313–322.

Davis, J., & Lambert, R. (2002). *Engineering in emergencies: A practical guide for relief workers* (2nd ed.). Warwickshire, UK: RedR/ITDG Publishing.

Down, R. D., & Lehr, J. H. (Eds.). (2005). *Environmental instrumentation and analysis handbook*. John Wiley & Sons.

Downing, J. (2008). Comparison of suspended solids concentration and turbidity. Application Note Code 2Q-AA. Campbell Scientific.

Fink, J. C. (2005). The effects of urbanization on Baird Creek, Green Bay, Wisconsin. [Thesis (M.S.)]. University of Wisconsin-Green Bay, 167.

Garcia, A., Perez, M. A., Grillo, G. J., & Tejerina, J. (2005). New design of low-cost four-beam turbidimeter by using optical fibers. *Instrumentation and Measurement Technology Conference*. Ottawa, Canada.

Hannouche, A., Chebbo, G., Ruban, G., Tassin, B., Lemaire, B. J., & Joannis, C. (2011). Relationship between turbidity and total suspended solids concentration within a combined sewer system. *Water Science and Technology, 64*(12), 2445–2452.

Holliday, C. P., Rasmussen, T. C. & Miller, W. P. (2003). Establishing the relationship between turbidity and total suspended sediment concentration,

Proceedings of the 2003 Georgia Water Resources Conference, University of Georgia.

Hudson, H. E. (1981). *Water clarification processes, practical design and evaluation.* Van Nostrand Reinhold Environmental Engineering Series.

Imran, H., Farooq, S., & Qaisar, S. (2008). Chlorination and water quality monitoring within a public drinking water supply in Rawalpindi Cantt (Westridge and Tench) area, Pakistan. *Environmental Monitoring and Assesment*, *158*(1-4), 393-403.

Izdori, F., Semiao, A. J. C., & Perona, P. (2018). Empirical characterization of particle size distribution spatial dynamics for helminth eggs detection in waste stabilization ponds. WSP, *Water*, *10*(2), 138–144.

Lambrou, T., Anastasiou, C. C., & Panayiotou, C. G. (2009). A nephelometric turbidity system for monitoring residential drinking water quality. *Sensor Applications, Experimentation, and Logistics—First International Conference*, SENSAPPEAL 2009, Athens.

LeChevallier, M. W., Norton, W. D., & Lee, R.G. (1991). Giardia and Cryptosporidium spp. in filtered drinking water. *Applied and Environmental Microbiology*, *57*(9), 2617–2621.

Liu, H., Yang, P., Song, H., Guo, Y., Zhan, S., Huang, H., & Che, Y. (2015). Generalized weighted ratio method for accurate turbidity measurement over a wide range. *Optics Express, 23*(25), 32703–32717.

Mann, A. G., Tam, C. C., Higgins, C. D., & Rodrigues, L. C. (2007). The association between drinking water turbidity and gastrointestinal illness: A systematic review. *BMC Public Health*. https://pubmed.ncbi.nlm.nih.gov/17888154/.

Myre, E., & Shaw, R. S. (2006). *The turbidity tube: Simple and accurate measurement of turbidity in the field* [Master's thesis]. Michigan Technological University.

Patnia, N. L., Jacangelo, J. G., Cummings, L., Wilczak, A., Riley, K., Oppenheimer, J., & Watson, M. (1997). Optimization of filtration for cyst removal, Project #703. Water Research Foundation.

Perkins, R. L. (2013). *Turbidity monitoring on Minnesota construction sites* [Master's thesis]. University of Minnesota.

Popacosta, K. (2002). Turbidity calibration standards evaluated from a different perspective. *Turbidity and Other Sediment Surrogates Workshop*. Reno, NV.

Sadar, M. (1998a). Turbidity science. *Technical information series*, Booklet No. 11. Hach.

Sadar, M. (1998b). Turbidity standards. *Technical information series,* Booklet No. 12. Hach.

Sadar, M. (2002). Turbidity instrumentation—An overview of today's available technology. *Turbidity and Other Sediment Surrogates Workshop.* USGS.

Tease, B., & Dowling, T. (2012). The influence of groundwater sample flow rate on suspended solids, geochemical and contaminant concentrations and its effect on trend analysis. *Proceedings of the Annual International Conference on Soils, Sediments, Water and Energy, 17*(15), 200–220.

U.S. Environmental Protection Agency. (1999a). Giardia: Drinking water health advisory (EPA-822-R-99-008).

U.S. Environmental Protection Agency. (1999b). Turbidity through the treatment process. *EPA guidance manual,* Chapter 10, 10–25.

U.S. Geological Society. (2018). *USGS national field manual for the collection of water-quality data.* https://www.usgs.gov/mission-areas/water-resources/science/national-field-manual-collection-water-quality-data-nfm?qt-science_center_objects=0#qt-science_center_objects

Wood, M. S. (2014). Estimating suspended sediment in rivers using acoustic dopplermeters (USGS Fact Sheet 2014-3038). http://pubs.usgs.gov/fs/2014/3038

Zielina, M. (2016). Monitoring of the process in water treatment plant. *Environment Protection Engineering,* 42(3), 167–177.

8

Advanced Electrochemical Sensors—ISEs and Voltammetry

This book has covered most of the instrumentation one is likely to find in an industrial or municipal facility that uses or treats water. Except for the optical dissolved oxygen sensor and turbidimeter, electrochemistry has been the foundation of all of them. Electrochemical measurement modalities were broken down into two major categories—potentiometric sensors (pH and oxydation-reduction potential [ORP]) and amperometric (dissolved oxygen and free chlorine—and one niche category—conductometric (conductivity). The optical dissolved sensor and turbidimeter are both based on optical sensing: the former on quenching of ruthenium fluorescence by oxygen and the latter on scattering by suspended particles.

For the sake of completeness, we survey two groups of sensor technologies that are less commonly used but that play an important role in larger facilities or laboratories. I include them because I am convinced that they are likely to transition from the laboratory to the treatment facility or manufacturing floor. This chapter won't go into the same level of depth as in previous chapters, but it will go just deep enough that the reader can pick up one of these instruments and have enough familiarity to know where to start.

This chapter dives into two advanced versions of electrochemical sensors. *Ion selective electrodes* (ISEs) are variants of the potentiometric sensor, of which the pH and ORP probes are the most common examples. Voltametric sensors are generalized versions of the amperometric sensor, of which the dissolved oxygen and chlorine sensor are the most common examples.

1.0 ION SELECTIVE ELECTRODES

1.1 Back to the Nernst Equation

The first group of instruments ought to be familiar. They are potentiometric, and the output they produce is modeled by the Nernst equation. The examples covered in the first part of the book were the pH and ORP probes. The ORP sensor uses a process electrode that is nothing more than a noble metal catalytic surface, upon which oxidation and reduction reactions occur. It has minimal selectivity toward any of the thousands of possible redox species. The pH probe is the first example of an ISE. The glass process electrode is constructed to be particularly sensitive to the concentration of one specific ion—the H^+ ion.

To understand the ISE, take a second look at the Nernst equation that was first presented in Chapter 2, Section 2.0. Consider the following generalized redox half-reaction:

$$aA + bB + ne^- \leftrightarrow cC + dD \tag{8.1}$$

Recall that it is standard practice to present all reduction and oxidation reactions as reduction reactions. A reaction that is an oxidation with respect to the standard hydrogen electrode (SHE) will be written as a reduction but characterized by a negative potential. Remember that, by virtue of Equation 2.16, a positive potential means a negative free energy change, which further means that the reaction proceeds spontaneously. Most of the redox half-reactions presented in the book consider only one species being reduced or oxidized but that Equation 8.1 is a generalization that considers the possibility of additional species.

The Nernst equation yields the potential of an oxidation or reduction reaction as a function of the concentration of the redox active species:

$$E = E^0 - \frac{RT}{nF}\ln Q;\; Q = \frac{[C]^c[D]^d}{[A]^a[B]^b} \tag{8.2}$$

E^0 is the standard potential, and its value is referenced to the SHE. It is positive for reduction reactions and negative for oxidation reactions. It is important to understand that the Nernst equation does not give the value of E^0. It must be determined experimentally. It does give the actual potential given the standard potential and concentrations of all chemical species involved in the redox reaction.

For the ORP sensor any number of redox reactions occur at the Pt electrode. E is the sum of constituent Nernst equations for each reaction, with each one characterized by its own standard potential and concentration of participating species.

For the pH probe, there is no actual redox reaction but a pseudo one involving charge transfer. The ISE is a thin piece of glass constructed of a silicon dioxide framework in which either Na^+ or Li^+ ions replace some Si^{4+} ions. In the very thin layers—the so-called gel layers—on both surfaces of the glass H^+ ions replace Na^+ (or Li^+) ions.

$$H^+(aq) + Si-O \leftrightarrow Si-O-H^+ \tag{8.3}$$

The conditioning of the electrode is complete when there is a stable population of Si-O-H^+ complexes within the gel layer. Changes in pH change the ratio between the changing H^+ ions in solution, and relatively constant Si-O-H^+ ions in the gel layer generate a Nernst potential that varies with the sample pH. Equation 3.11 is the Nernst equation for this charge transfer reaction.

$$E = E^0 - \frac{RT}{nF}\ln\frac{[SiO^-H^+][Li^+]}{[H^+][SiO^-Li^+]} \tag{8.4}$$

The pH sensitive glass caps a glass ampoule containing a saturated KCl solution and an Ag-AgCl wire. The inside of the pH glass faces a constant pH 7 ($[H^+] = 10^{-7}$ M) while the outside surface faces a sample of varying pH. The Nernst potential on the outside surface is subtracted from the Nernst potential on this inside surface, thus cancelling E^0 and giving us the familiar linear relationship between the probe potential and pH (Equation 3.21). E^0 is zero, and Q is nothing more than the H^+ ion activity (or, approximately, the concentration).

We can generalize the pH probe's principle of operation to all ISEs. The charge transfer reaction that takes place in each one is a conversion between the target ion, X^+, in the sample and the target ion complexed with some chemical species, A, in the membrane:

$$X^+ + A \leftrightarrow XA^+ \tag{8.5}$$

Note that we could have just as easily made X an anion. Also, A can be an anion, A^-, which would render XA a neutral species. The corresponding Nernst equation is:

$$E = E^0 - \frac{RT}{nF}\ln\frac{[XA^+]}{[X^+][A]} \tag{8.6}$$

The process electrode of the ISE is, like the pH electrode, a glass (or plastic) ampoule filled with KCl, capped with the ISE membrane, and containing an Ag-AgCl wire. Whereas the pH electrode has a fixed, neutral (pH 7) solution inside the electrode, the ISE has a fixed concentration of the analyte ion in the KCl solution. In a conditioned probe $[XA^+]$ is relatively constant so Equation 8.6 collapses to:

$$E = E^0 - \frac{RT}{nF}\ln\frac{[XA^+]}{[X^+]} - \frac{RT}{nF}\ln\frac{1}{[X^+]} = \text{const} + \frac{RT}{nF}\ln[X^+] \tag{8.7}$$

The overall potential of the process electrode is the difference between the Nernst potentials, given by Equation 8.7, for the outside and inside surfaces. If the concentration of the analyte, X^+, in the sample is equal to that of X^+ inside the electrode, then the overall potential is zero. Well, not exactly zero. We need to add the small asymmetry potential across the width of the membrane (E_2), diffusion potential of the liquid junction (E_6), the difference between the potential of the Ag-AgCl electrode in the process electrode (E_5), and the potential of the Ag-AgCl electrode in the reference electrode (E_4). (Refer to Table 3.1 for a summary of the six potentials that contribute to the output of the pH probe.)

The factor of RT/nF in the second part gives the now familiar 59.16 mV/pH unit (at 25 °C) that every seasoned pH probe user knows. Remember that n in Equation 8.1 is the magnitude of the charge that scales the 59.16 mV value. Because the H^+ ion has a single positive charge $n = +1$, the slope of the pH response curve, as given by the Nernst equation, remains as 59.16 mV/pH. For an ion with a multiple charge, one divides 59.16 by the charge multiplicity (n) to obtain the change in sensor voltage as a function of the log of the concentration. So, for instance, a probe that measures Ca^{2+} changes its voltage by 59.16/2 mV, or 29.58 mV for every tenfold change in concentration.

The major characteristic of potentiometric sensors that one needs to remember is that, ideally, they produce zero current. The insertion of a very high impedance voltmeter or galvanometer between the process and reference electrode ensures a condition of zero current—the opposing half-reactions are equal and opposite. In practice, the impedance of the voltmeter is finite and some current, on the order of nanoamps, exists.

Although the Nernst equation is the basis for all potentiometric ISE measurements, the measurement from a pH probe is fundamentally different from all other ISEs. The pH value one seeks is the logarithm of the concentration of H^+ ions. For ISEs one simply wants to know the concentration. This shift has serious ramifications for the accuracy of the measurement. An error of 5 mV in a pH measurement leads to a pH error of less than 0.1 pH units. But an error of 5 mV in the measurement of a monovalent ion (e.g., Na^+) will result in a 8% error in the concentration. For a divalent ion measurement (e.g., Ca^{2+}), the 5 mV error results in an 16% error.

1.2 Ionic Sites—The Neglected Costars of Potentiometry

There is one point that is glossed over in nearly every explanation of ISEs, including the previous section. It was first brought up in the chapter on pH, and I repeat it here. It might seem obvious that the concentration of the analyte ion adsorbed into the membrane is proportional to the concentration of the ion in the sample. In fact, many texts paint this picture of the pH probe or ISE. But this interpretation is wrong: the potential that arises at the interface of the membrane and sample is given by the Nernst equation, restated in Equation 8.6. According to the famous equation the potential is proportional to the ratio of the redox reaction's product concentration(s) to reactant concentration(s). As discussed in the pH chapter, there is no chemistry, in the traditional sense, at the membrane of an ISE. The reactant is the ion in solution, and the product is the ion complexed with some molecular entity in the membrane. In the previous section, X^+ is the ion, and A is the ionophore. For a pH probe A is the Si-O framework of silica.

The potential at the ISE surface varies with the logarithm of the ratio of analyte ions in solution (sample or reference) to the ratio of analyte ions bound to ionophores, $[XA^+]/[X^+]$. If the concentration of bound K^+ ions in the membrane increases in proportion to that of K^+ ion in the sample, then the ratio stays constant—even though the absolute concentration of K^+ in solution varies. The Nernst equation therefore predicts that the ISE will *not* respond to changes of ion concentration in the sample. Its output will be constant.

The only way an ISE can possibly work is for the ionophores in the membrane to selectively adsorb the target ions but keep their concentration in the membrane layer constant. In other words, the membrane must selectively adsorb the ion of interest, but it must build up a concentration of the ion to a fixed value and maintain it at that value.

The requirement is similar to that of a buffer. A pH buffer, for example, minimizes the pH of a solution when acid or base is added. For the pH buffer the mechanism is a reaction that can go in either direction, as generalized in Equation 3.33. For an ISE, the mechanism varies depending on the type of ISE—crystalline, glass (pH), or polymeric.

For an ISE the mechanism that minimizes the change of ion concentration in the membrane is the introduction of a second ingredient in the membrane that has been called various terms, including *ion excluders* and *ionic sites*. The latter is used mostly today. The term *site* encompasses chemical compounds that are added to the membrane as well as parts of a membrane structure, as in the silicate network of a pH probe. Figure 8.1 illustrates ionic sites that accompany ionophores for measuring potassium ions.

Ionic sites carry the opposite charge of the ion of interest. For instance, the ionic site for the potassium ion, K^+, can be one of the two negatively charged organic compounds shown in Figure 8.8. These large anions are initially paired with cations—typically Na^+ or K^+—to maintain charge neutrality. During the conditioning period K^+ ions displace any Na^+ or other cations by virtue of the potassium ion's affinity toward ionophores. In short, ionophores attract the ion of interest while ionic sites fix their number. Notice, in Figure 8.1, that there is an overabundance of ionic sites compared to ionophores. This is necessary to maintain a (more-or-less) fixed concentration of K^+ ions in the membrane and, thus, generate a Nernstian potential when exposed to K^+ ions in solution. If the concentration of ionic sites were to exceed that of ionophores then the membrane would continue to adsorb cations, including K^+, which would allow the latter's concentration to fluctuate and diminish the Nernst response. An excellent review article by Bühlmann and Chen (2012) discusses the oft-neglected role of ionic sites.

1.3 It's the Activity, Not the Concentration (Again)

There is one source of error in ISEs discussed in the pH chapter that bears repeating. The Nernst equation is really a function of redox species' *activities*—not their *concentrations*. Only when a species is infinitely dilute does its activity equal its concentration. The activity for each species is always less than its concentration. To extract concentration information, one must know the proportionality between the activity and concentration, which is given by the activity coefficient. Equation 3.33 first recast the

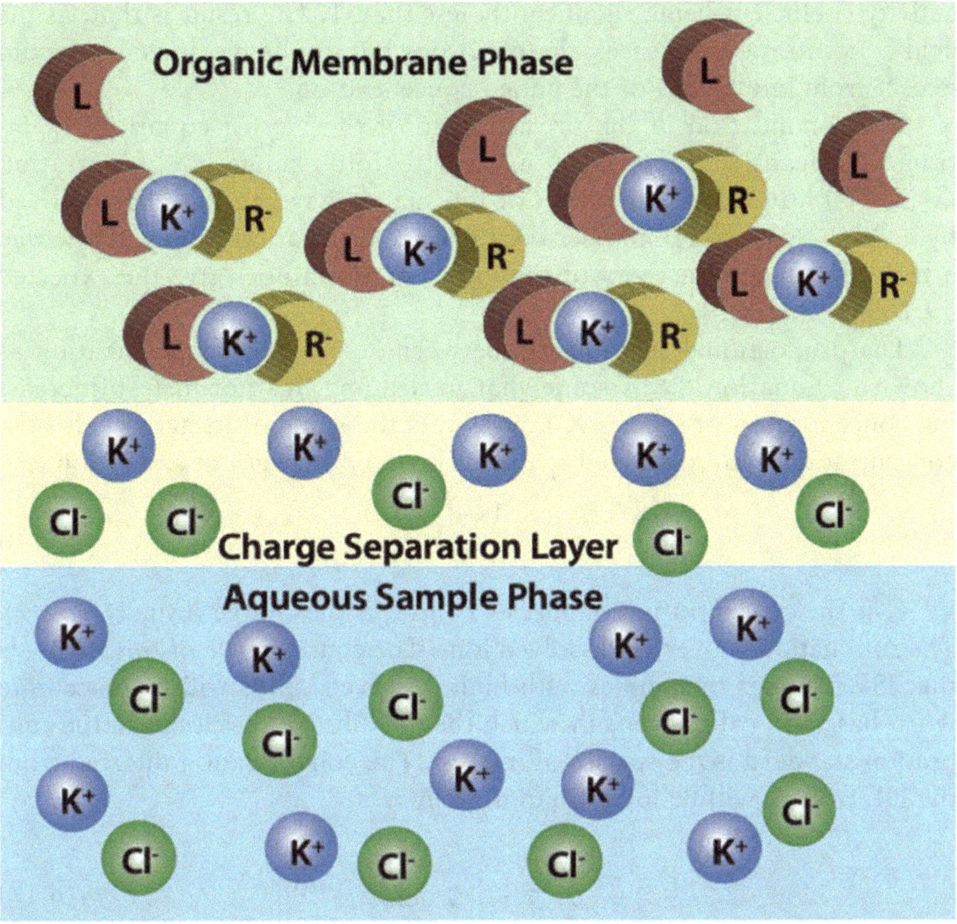

FIGURE 8.1 Schematic of an ISE membrane ion. The analyte is K^+ ion, balanced in the sample by Cl^- anions. L represents the ionophore and R^- represents the ionic sites. The latter's negative charge balances the positive charge of the target K^+ ions (Derived from Bühlmann & Chen, 2012).

Nernst equation in the more exact formulation in which a_x is the activity of species x, and Y_x is the activity coefficient that relates the concentration to the activity.

$$E = E^0 - \frac{RT}{nF} ln \frac{a_c^{\ n_c} a_d^{\ n_d} \ldots}{a_a^{\ n_a} a_b^{\ b} \ldots} \ ; a_X = \gamma_X[X] \qquad (8.8)$$

The activity coefficient takes on a value that approaches 1 as the solution of X becomes increasingly dilute and approaches 0 as it becomes saturated. The activity of a redox active species manifests itself in pH readings at pH values less than 2 in which the high H^+ concentration results in the activity coefficient being significantly less than 1. The result is that, as the acid concentration increases, the pH drops but not as much as one would expect from knowledge of the added concentration.

The ramification of the precise form of the Nernst equation, Equation 8.8, means that the 59.16 mV per tenfold change in concentration (at 25 °C) no longer holds. It's not unusual to have an ISE characterized by a slope closer to 50 mV per decade of concentration change. However, a tenfold change in activity of the ion in question does give the expected 59.16 mV change.

The proportional relationship between activity and concentration as shown in Equation 8.8 is somewhat misleading. It's not just an increasing concentration of species X that lowers the value of its activity. It's the concentration of *all* dissolved species. The *ionic strength* of a solution is:

$$I_X = \frac{1}{2}\sum_i C_i z_i^{\,2} \qquad (8.9)$$

C_i is the concentration of species i and z_i is the charge (usually 1 or 2). The summation is over all dissolved ions. The consequence of this reality is that ISE readings in samples with high dissolved solids will be inaccurate when based on calibrations in which the only dissolved species in the calibration standard is the species of interest. The concentration inferred from the ISE voltage will be lower than it really is.

$$E = E^0 - \frac{RT}{nF}\ln\left[a_i + \sum_j K_{ij} a_j^{n/m}\right] \qquad (8.10)$$

The analyte of interest is denoted by i, interfering ions are denoted by j, n is the charge of the analyte i, and m is the charge of interfering analyte j. $K_{i,j}$ is the selectivity coefficient and serves as a weighting factor for species (i) contributing to the potential.

One can visualize Equation 8.10 as a set of charge transfer reactions all occurring simultaneously at the surface of the working electrode. Each reaction contributes to the exchange current, as explained in Chapter 2, Section 4. Typical values for the current exchange density of an analyte, such as H^+ or K^+, are 10^{-3} A cm^{-2}. The actual value is dependent on concentration. The adsorption of K^+ ions into a valinomycin membrane is characterized by an exchange current density of 2.1×10^{-3} A cm^{-2}. The current densities for interfering ions, including other alkali metals, range from 5×10^{-4} to 1.2×10^{-6} A cm^{-2}. However, for rubidium (Rb^+), the exchange current density is 4.7×10^{-3} A cm^{-2}. Fortunately, this ion is present in very small concentrations.

Equation 8.10 explains why ISEs typically produce a large voltage offset even when the concentration of the analyte in the process equals its concentration of the electrode's interior.

For the sake of simplicity, suppose that there is only one interfering ion. Let's suppose that there is no analyte and only one interfering ion, *j*.

$$E_j = E_j^0 - \frac{RT}{nF} ln K_{j,i}(a_i)^{Z_j/Z_i} = E_j^0 - \frac{RT}{nF} ln K_{j,i} - \frac{Z_j RT}{Z_i nF} ln a_i \qquad (8.11)$$

The only term that varies with the activity of the interfereing ion, *j*, is the third term. If the the concentration of the interfering ion is constant, then Equation 8.11 gives a constant potential. Add this constant to the potential of the charge transfer reaction of the analyte in solution being adsorbed into the membrane. The result is a voltage offset.

The ramification of Equation 8.11 is that one should calibrate the ISE in standards that contain dissolved ion concentrations similar to those found in the actual sample. This is the *raison d'être* behind *ionic strength adjustment* (ISA) solution. The composition of an ISA solution can vary, but the critical component is a salt—usually KCl in concentrations ranging from 0.1 to 4 M. For the measurement of ammonium, the presence of the interfering K^+ ion would lead to an artificially high reading. So KCl is out as an ISA for the measurement of ammonium. Typically, lithium hydroxide (LiOH) takes its place. Added to it is less than 10% of a chelating agent (usually EDTA) to complex interfering cations. The ISA solution for a K^+ ion is NaCl.

The use of an ISA is critically important with the fluoride ISE. For this ISE more than a simple salt is needed. That's because of the following interferences to either the fluoride concentration or the selectivity of the ISE:

1. Polyvalent cations, such as Fe^{3+} and Al^{3+}, combine with F^+ ions to form complexes. A chelating agent, cyclohexylenedinitrilotetraacetate (CDTA), is added to complex with these cations and therefore prevent them from tying up the fluoride ions.
2. Hydroxide ions (OH^-) interfere with F^- ions on the ISE, so it is necessary to maintain an acidic pH. On the other side of neutral, hydrogen ions (H^+) react with fluoride ions to create bifluoride ions (HF^{2-}). The Goldilocks's zone for pH is between 5.0 and 5.5. To maintain this range acetic acid (CH_3CHOOH) is added to the ISA solution as a buffer.

The salt used is NaCl, and the resulting mixture is called *total ionic strength adjustment buffer* (TISAB). Figure 8.2 (Noh & Coetzee, 2007)

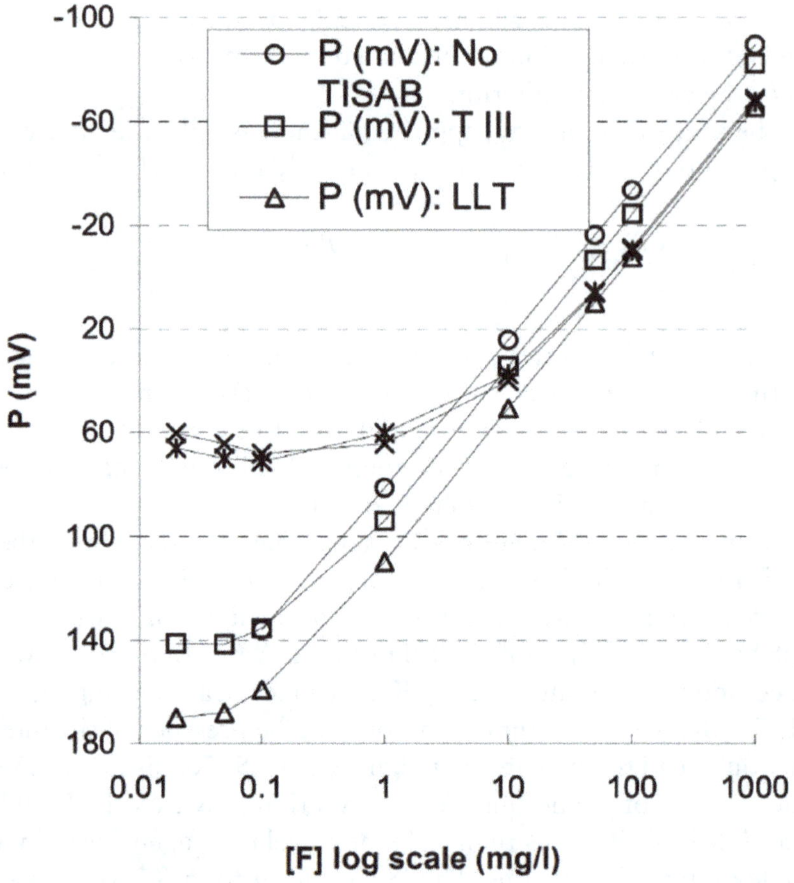

FIGURE 8.2 ISE performance with and without TISAB. The solid line shows the probe output for a solution of pure NaF. The top pair of curves is the sensor response in the absence of TISAB. The middle pair represents the addition of a conventional TISAB solution, and the bottom represents a customized TISAB solution (Noh & Coetzee, 2007).

shows the expanded linear range with the addition of TISAB solution. The top, curved pair of line displays the fluorine ISE output for the case in which there is no ISA present in the water. The middle pair shows the improved linearity with the addition of the TISAB as just described. The lower line shows the probe output with a modified TISAB solution that has been adjusted to miminize interference from iron and aluminum ions. The lower limit of the measurement range improves from about 10 mg/L all the way down to 0.1 mg/L.

1.4 Selectivity—The Lower Limit

Ion selective electrodes output a voltage according to the Nernst equation for the ion of interest. Implicit in the formalism of the modified Nernst

Equation (Equation 8.10) is that potentiometric ISEs, as well as any other sensor, display a cross-sensitivity to unwanted ions.

Let's go back to our prototypical pH electrode. This exchange of protons constitutes the exchange current. Other ions exchange across the gel layer as well, especially Na^+ ions. The ratio of the current exchange of H^+ ions to that of Na^+ ions is a measure of the selectivity of the glass membrane toward H^+ ions. The sodium error, discussed in Chapter 3, Section 7.3, is only noticeable for pH levels greater than 11. On the opposite end of the pH scale, acid error, is caused by the diminished activity of H^+ ions, but this is due to the distinction between activity and concentration. Because pH measurement is logarithmic, even for the extreme pH values where acid or alkaline errors do manifest themselves, the error in pH units is a lot less than the error of the underlying H^+ concentration.

The same cannot be said for most other ISE electrodes, which display a non-negligible sensitivity to interfering ions and which can't minimize the contribution of these ions simply because their contribution is minimized by being included in a logarithm. Table 8.2 lists most ISEs commercially available and the species that interfere with the target analyte. The interference values aren't listed but, when they are, they are expressed as a ratio. For instance, an interference of 10^{-3} means that the addition of the interfering species at 1000 times the concentration of that of the analyte will produce the same probe voltage. In most cases, the interference can be quantified. One either keeps the concentration of the analyte fixed, adds an increasing amount of the interfering ion and measures the response, or keeps the concentration of the interfering ion and increases the concentration of the analyte ion. Typically, there are fewer than three major interfering ions that make an appreciable contribution to the total potential. For a NH_4^+ probe, K^+ is the major interfering ion of concern.

There is one other source of an offset that is minor but can become critical for the measurement of very low (nanomolar or less) concentrations. That is the diffusion of the analyte ion through a thin polymer membrane. Because the current in a potentiometric sensor is ideally zero, the diffusion of the analyte ion from the process into the electrode must be balanced by the diffusion of an ion of the same charge from the interior of the electrode into the process (Bakker & Pretsch, 2005).

1.5 Donnan Failure—The Upper Limit

Section 1.2 explained how ionic sites maintain a fixed concentration of analyte ions in the membrane so that the analyte ion concentration in the process can generate a Nernstian response. That Nernstian response will occur as long as there are more ionophores than ionic sites, and as long as there are enough ionic sites to balance the oppositely charged analyte ions that immerse themselves in the ion selective membrane.

If the concentration of analyte ions exceeds the capacity of the ionic sites, then the ions will adsorb into the membrane and pull in oppositely charged ions to maintain charge balance. For instance, suppose we wish to measure K^+ in a KCl solution. The ionic site constituents will be anions. When the latter are 100% paired with K^+ ions, additional K+ cations are attracted to the ionophore along with Cl^- anions. The result is a flattening and downward trend of probe voltage with increasing K^+ concentration. The end of the Nernstian response with increasing analyte ion is the Donnan Failure and is illustrated in Figure 8.3 (Bühlmann et al., 1998). Curves *a* through *e* represent ratios of ionic sites to ionophores of 0.05, 2, 50, 100, and 500. Ironically, the larger the binding energy between the ionophore and analyte, the lower the analyte concentration at which Donnan failure occurs.

1.6 Construction of an Ion Selective Electrode

Potentiometric sensors consist of a reference electrode and a measurement, or process, electrode. In our investigation of pH and ORP sensors we restricted our discussion on the reference electrode to the standard hydrogen electrode and the more practical Ag-AgCl electrode.

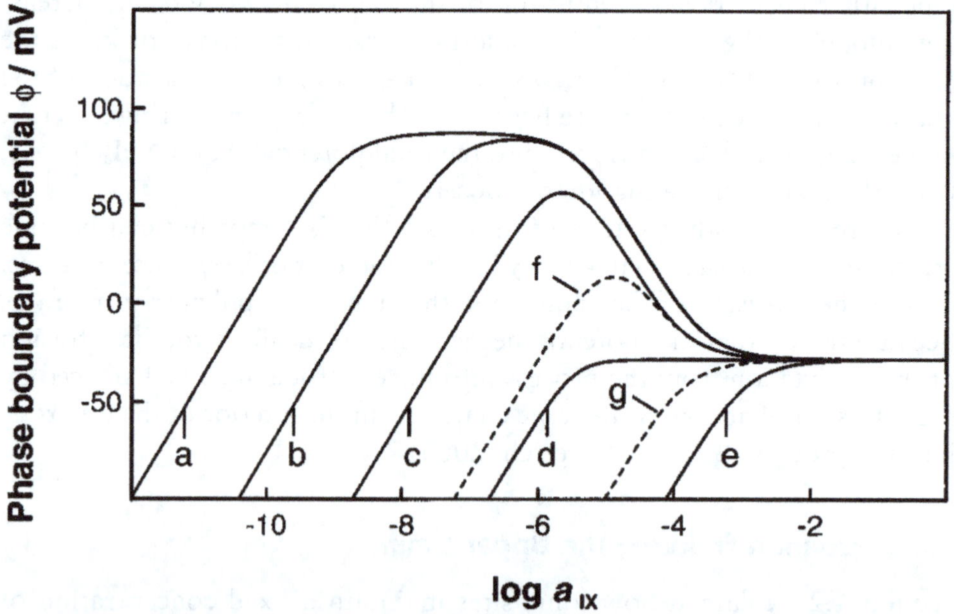

FIGURE 8.3 Nernst response of ISE limited by ionic sites. The rollover of probe voltage is the result of Donnan failure. Curves *a* through *e* represent ratios of ionic sites to ionophores of 0.05, 2, 50, 100, and 500 (Bühlmann et al., 1998).

$$AgCl(s) + e^- \leftrightarrow Ag(s) + Cl^-;\ E^0 = 0.199V(\text{Sat'd KCl}) \qquad (8.12)$$

To construct an Ag-AgCl electrode one simply immerses a silver wire coated with an AgCl film in a solution of saturated KCl.

The half-reactions of the reference electrodes proceed with small charge transfer resistances, as explained in Chapter 5, Section 2.0. In simple terms, this means that they proceed with ease so that the actual working potentials are very close to the so-called standard half-cell potentials. Alternatively, one can say that the reference reactions are *reversible*. This is a necessary characteristic of any reference half-reaction because it must be able to respond to a reduction or oxidation half-reaction at the process electrode. Two other necessary criteria for a good reference electrode, that both the calomel and Ag-AgCl electrodes satisfy, are that they are *stable* and *reproducible*.

Now let's look at the process electrode that measures the desired chemical species of an ISE. The champion of selective electrodes, in terms of measurement range, is the pH probe. Its range of 14 pH units is a whopping 10^{14}:1. That's equivalent to a single ruler that can measure distances from 10 000 km—the distance across the United States—to 0.1 µm—the size of particles too small to even scatter visible light. The basis of the pH probe's enormous measurement range is the large number of spaces in the SiO_2 framework that comprises the pH probe glass membrane in which Na^+ (or Li^+) atoms are situated and which can be replaced by H^+ ions. These spaces are the ionic binding sites, and their concentration determines the range of analyte—H^+ ion—that the electrode can cover. For the pH electrode that concentration of binding sites is on the order of 10 M. That means a concentration of H^+ ions between 10 M and 10^{-14} M will produce a measurable change in potential as determined by the Nernst equation. A typical ion exchange membrane consists of approximately 10^{-2} M binding sites, so the measurement ranges for other ISE's are more limited.

On the side facing the sample, the concentration of H^+ ions in water at pH 7 will create a potential on the outer glass surface that balances the potential on the inner surface. Hence the total potential of the glass membrane is ideally zero. Likewise, the ISE electrode generates a zero potential across the membrane as long as the concentration—or to be more exact, the activity—of the analyte on the inside surface the membrane matches the activity of the analyte ion in solution. That zero will shift by several mV depending on the potentials of both Ag-AgCl electrodes, any asymmetry potential and the junction potential. Figure 8.4 (Benoit et al., 2013) shows the calibration curve for an ISE selective to silver ions (Ag^+). Note that the zero crossing—the part of the curves at which the probe output is 0 mV—occurs at an Ag^+ concentrations of approximately 10^{-8} M and that the measure range is from 10^{-7} to 10^{-2} M.

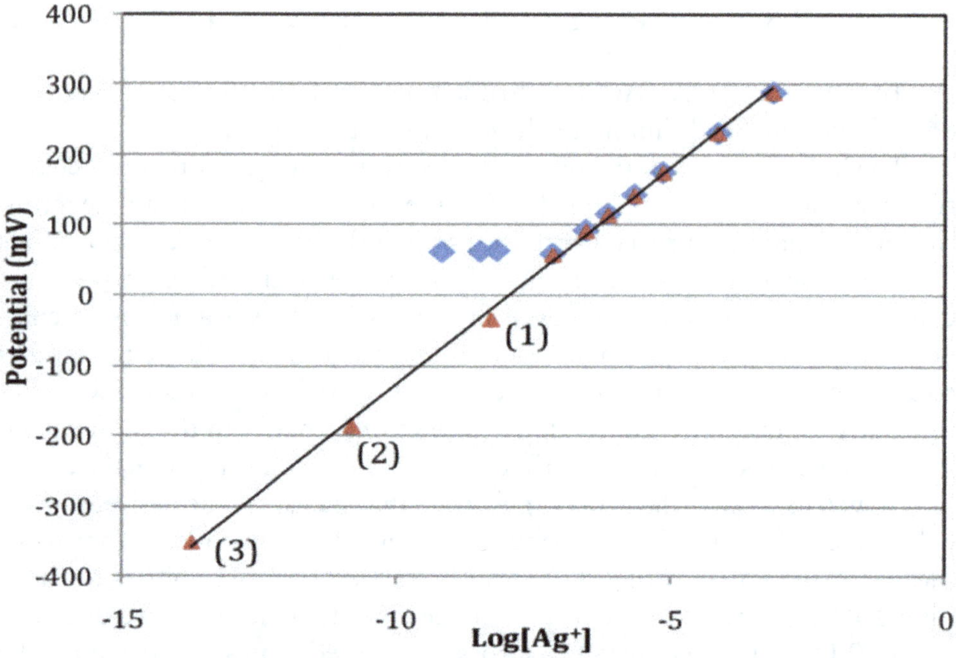

FIGURE 8.4 ISE calibration curve for an Ag^+ ISE. The Nernstian response is from 10^{-7} to 10^{-2} Molar (Benoit et al., 2013).

Recall that the potential of the Ag-AgCl electrode in the reference electrode feeds into one side of the voltmeter while the potential of the Ag-AgCl electrode in the process electrode feeds the other side. Both contain the same Ag-AgCl wire immersed in a KCl solution saturated with AgCl. The difference (assuming we neglect the potentials resulting from minute differences between the two electrodes) is that the KCl electrolyte in the process electrode has a fixed concentration of the analyte ion so that the potential generated at the surface of the ISE membrane changes with changes in the sample analyte concentration.

Figure 8.5 shows the schematic of a generalized ISE.

Because most ISE electrodes have a shorter lifetime than a pH electrode, especially those based on polymer membranes, they are commonly available as a separate electrode to be paired with an external reference electrode. This saves cost over the combination probe, which must be tossed when the ISE membrane fails.

Ion selective electrodes can be broken down into three types according to the nature of the component that provides the selectivity: glass, solid state (or crystalline), and polymer. The following sections describe each of the three by focusing on one or two examples.

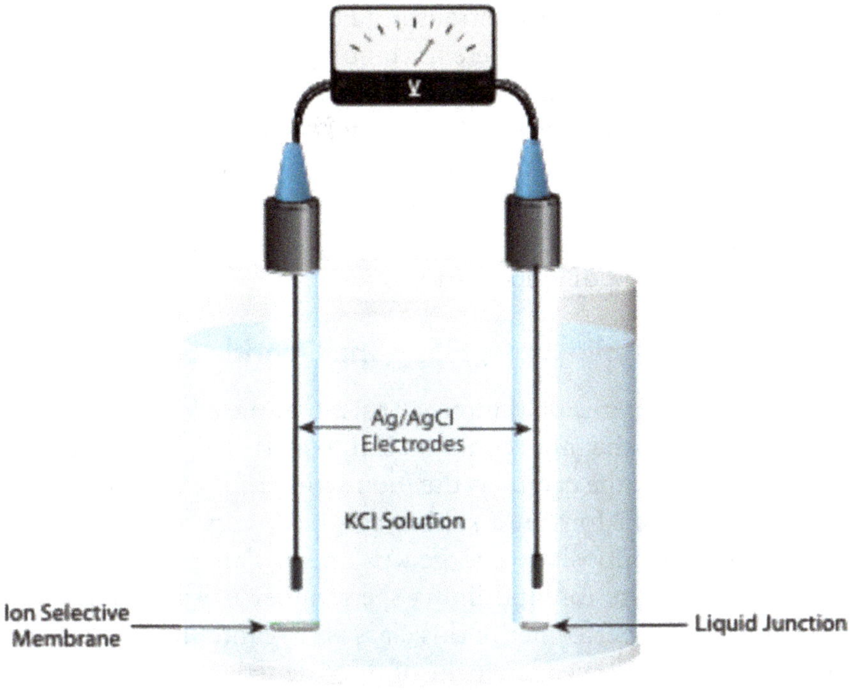

FIGURE 8.5 Generalized schematic of an ion selective electrode. Both the process and reference electrode contain an internal Ag-AgCl electrode. The voltage that the voltmeter in the analyzer measures is due to the additional potential created by the analyte diffusion into the ion selective membrane.

1.7 Glass ISEs

The most common glass-based ISE is the famous pH probe. No other ISE is capable of measuring an analyte over as high a range (10^{14} to 1) and with as high selectivity over interfering ions (10^{11} to 1 for Na^+ ions). But there are others.

In general, glass substrates exchange monovalent (+1) cations. Silica glass—SiO_2—can be made selective toward H^+, Na^+, and Ag^+. Chalcogenide glass—that is, glass that contains sulfur—is selective toward the divalent (+2) ions lead and cadmium. The same principles just discussed are at work for all glass-based sensors: the positively charged cations insert themselves into the molecular framework where they are stabilized through their attraction to the negatively charged oxygen atoms.

Most formulations of glass predispose them to be selective to H^+ ions, thus making the pH electrode a natural fit for glass electrodes. Corning 015 glass is a typical one used for pH probes and has a composition of 22% Na_2O, 6% CaO, and 72% SiO_2. The Na^+ ions in the glass matrix are displaced by H^+ ions—hence the basis for its role as a pH sensor. When the probe is immersed in NaOH solutions at pH levels higher than 11, the high concentration of Na^+ ions compete for binding sites in the glass matrix and give rise to alkaline error, described in Chapter 3, Section 7.3.

The weaker affinity of the Si-O framework for Na^+ ions than for H^+ ions can be altered to favor the Na^+ ions by changing the composition of the glass. For instance, adding Al^{2+} ions in the form of Al_2O_3 makes the glass selective for Na^+. Table 8.1 summarizes the composition of glasses tailored toward the alkali metal cations.

1.8 Crystalline (Solid State) ISEs—The Fluoride Sensor

Crystalline, or solid-state, membranes consist of ionic crystals that are insoluble in water (unlike the most common salt, NaCl). The exchange of one of the ions that make up the crystal is the foundation of the Nernstian response of this solid-state ISE. Ions in a salt crystal have very strong electrostatic attractions to their oppositely charged partners. This attraction limits the mobility of the analyte ion and limits the number of solid-state potentiometric ISEs. The star performer in this class is the lanthanum fluoride, LaF_3, ISE that is used for the measurement of fluoride, F^-, ions.

The exchange of fluoride ions between the sample and crystal membrane is:

$$LaF_3 \leftrightarrow LaF_2^{\ 1} + F^- \tag{8.13}$$

To increase the very low mobility of F^- ions through the LaF_3 crystal, the latter is doped with europium fluoride (EuF_2). This salt has one less fluoride ion than its host, and the resulting vacancy in the crystal lattice increases the mobility of fluoride ions.

TABLE 8.1 Summary of glass based ISEs.

	Glass Composition				Selectivity Coefficient
Analyte	Na_2O	Li_2O	Al_2O_3	SiO_2	10^{11}
H^+	22		6	72	
Li^+		15	25	60	
Na^+	11		18	71	10^3
K^+	27		5	68	

The only significant interfering ion is OH^-. For this reason, the pH of a solution being measured, or a calibration standard, must be maintained on the acidic side of neutral—between 5 and 7. Figure 8.2 displays the linear measurement range of the fluoride ISE, which is 10^{-6} to 1 M, or 0.018 to 1800 parts per million (ppm).

The range of possible crystalline membranes is limited by the mobility of ions in most crystals. Most ions exhibit poor mobility and therefore lack the conductivity required of an ISE. On the other hand, a highly soluble salt would not last long in solution. To overcome this double whammy of a limitation some crystallne ISE's use a mixture of two salts: One contains the ion of interest while the other is a salt of very low solubility but with the same anion (if the ion of interest is a cation) or cation (if it's anion). The so-called *heterogeneous membrane* is either a pressed pellet of the two salts or a poymer in which the crystals of the two salts are embedded. The low solubility of the companion salt constrains the solubility of the ion of interest and the result is a sensor that responds to that ion. Examples include the metal ions such as Cu^{2+}, Pb^{2+}, and Cd^{2+}, which are present as the sulfide salt (e.g., PbS) and mixed with the the nearly insoluble silver sulfide (AgS) salt. For anions such as the halogen ions the cation in common is typically Ag^{2+}.

1.9 Polymer Membrane ISEs—The Potassium and Ammonia Sensors

Polymer membranes (aka *ion exchange membranes*) solve the problem of low conductivity that limit the use of crystalline and glass ISEs. They are constructed of a polymer matrix—typically polyvinyl chloride (PVC)—that incorporates an ion selective species. PVC is by far the most common ion

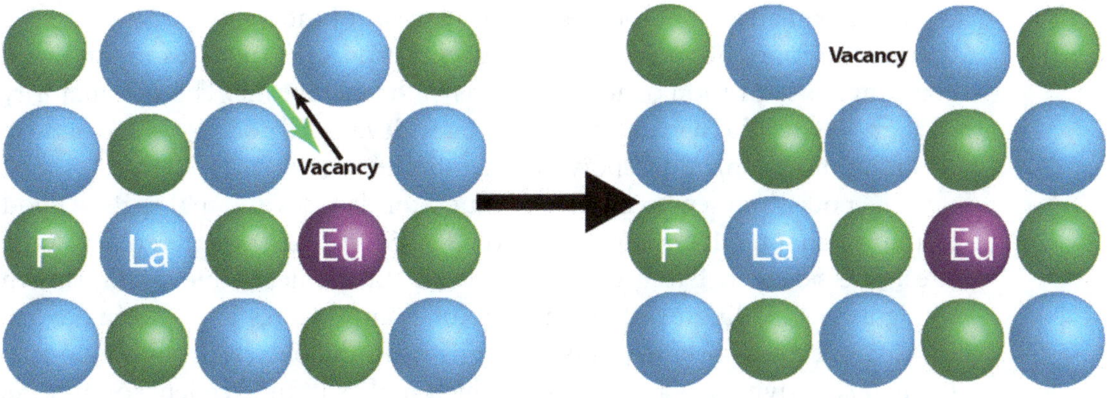

FIGURE 8.6 LaF_3 crystal doped with EuF_2. The EuF_2 molecule has one less F atom and creates an opening in the lattice just right for a F^- ion.

selective substrate. It is normally a solid plastic. By adding a plasticizer in a ratio to PVC of approximately two-to-one, the substrate becomes liquid-like. Ions are thus able to diffuse into the membrane. Other membrane materials include silicone rubber, polyurethanes, acrylates, and perfluoropolymers. These materials are mostly nonpolar and hydrophobic. There are three reasons for this material choice:

1. The membrane will dissolve very slowly in water.
2. A membrane constructed of polar compounds would attract other, potentially interfering ions, and would therefore lessen the selectivity of the ISE.
3. The microscopic porosity of polymers enables fast transit of the analyte ion in and out of the membrane, giving response times in fractions of a second.

The "secret sauce" for the ISE is the *ionophore*. There are two requirements this chemical species must meet. One is obvious, and one is not:

1. The ionophore must be selective to the analyte ion. Typically, the selectivity is a minimum of two orders of magnitude—that is greater than 100:1.
2. The energy barrier for the ion to migrate from the polar water environment into the nonpolar, hydrophobic membrane environment must not be too great. This is a common problem for small ions, such as Li^+, that are so strongly bound to water (*hydrophilic*) that the energy barrier to break its ion-dipole bonds is so high that they cannot leave their aqueous environment.

Approximate 1% of the membrane consists of the ion selective species—that is, the ionophore. A smaller concentration of ionic sites maintains a constant concentration of adsorbed ions in the membrane. The vast majority of ISEs—as well as virtually all new ones developed today—are based on this ionophore embedded polymer membrane.

The potassium sensor has been around for decades and makes for a good example. Chapter 1, Section 2 used the K^+ ISE to illustrate the role of ionic sites. The sensor is based on a large, cyclic molecule, valinomycin, shown in Figure 8.8. Valinomycin is an antibiotic, and it does its assassins' work on bacteria by adsorbing potassium ions from the bacteria's mitochondria. Its high selectivity toward K^+ ions, which makes it an excellent ionophore. Ions that interfere with valinomycin are, in order of decreasing effect, Rb^+, Cs^+, $NH_4{}^+$, Na^+, Ca^{2+}, Mg^{2+}, and Li^+. Examination of Figure 8.7 shows

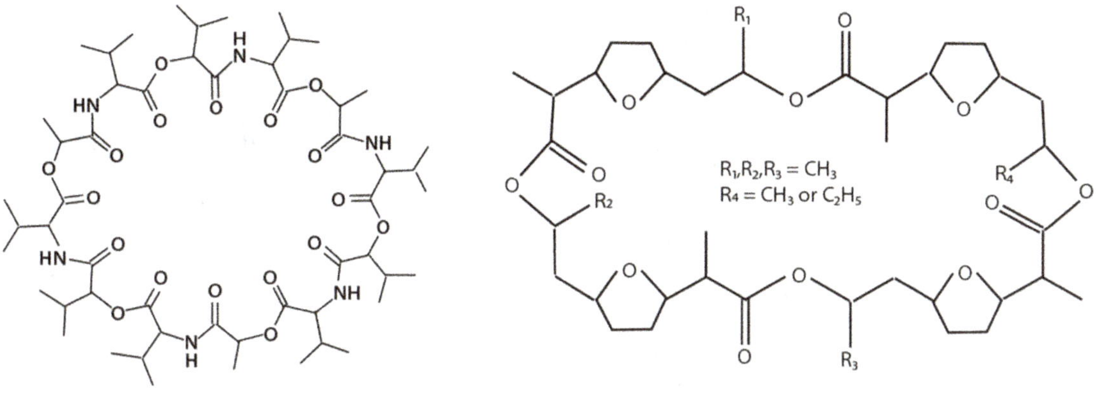

FIGURE 8.7 Potassium ion ISEs. Left: Valinomycin is a complex biological molecule that is highly selective to K^+ ions. Right: Nonactin/Monactin. For nonactin, all R groups are methyl (CH_3). Formonactin R_4 is ethyl (C_2H_5).

why this ionophore complexes so well with cations. The negatively charged (nucleophilic) O atoms all point into the center of the ring and attract positively charged species. The exterior of the ring consists of mostly C-H bonds, which are nonpolar and immiscible in water. They therefore must be embedded in a nonpolar medium like PVC. As K^+ ions diffuse into the polymeric membrane loaded with valinomycin they build up in concentration until their concentration matches the concentration of ionic sites. At this point the concentration of ions in the membrane becomes constant, and the ISE can no longer respond to changes in sample concentration. The left side of Figure 8.7 shows the polymer membrane in which the valinomycin is dispersed.

Valinomycin was one of the first two ISEs based on antibiotics to be discovered. The other is a family of cyclic compounds collectively called *nactins*. Members of this family differ through the four attached alkyl groups. The first two nactins are nonactin and monactin. These two are highly selective toward ammonium but also respond to K^+ and the same interfering ions listed for valinomycin. Like valinomycin its inward facing O atoms attract cations while its mostly nonpolar C-H groups allow it to be embedded in a hydrophobic medium.

The higher selectivity of nonactin toward ammonium ions than potassium ions make it the preferred ionophore for NH_4^+. However, because it does complex with K^+ ions, many ammonium sensors have a second sensor containing a valinomycin membrane to make a separate K^+ measurement and subtract it from the NH_4 measurement.

The second ingredient that is necessary to maintain a constant concentration of K^+ ions in the membrane is the ionic site compound. Figure 8.8

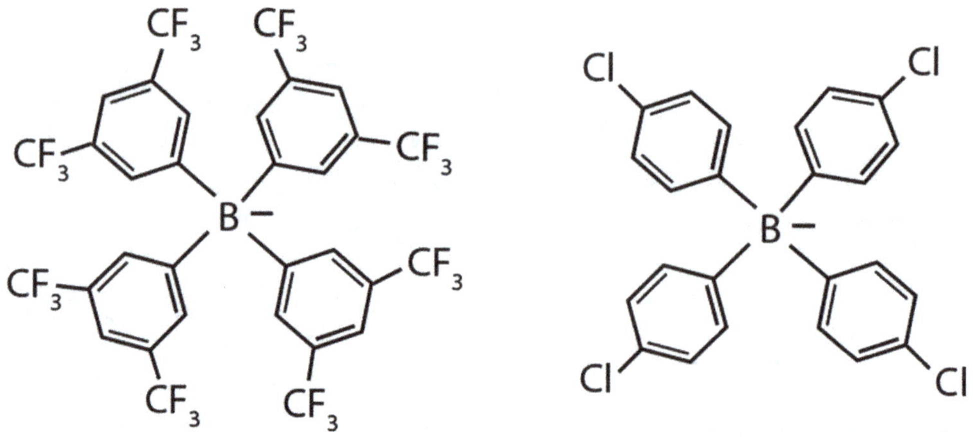

FIGURE 8.8 Tetraphenyl borate anions as ionic site anions for the K1 ISE.

shows two such tetraphenyl borate anions that are commonly used. These anions originate as salts with Na^+ or K^+ cations. As the membrane is conditioned with an inner electrolyte containing K^+ (and Cl^-) ions and the conditioning sample, also containing K^+, the ionophores attract K^+ ions and the borate anions, as ionic sites, fix their concentration.

Potassium and ammonium ions are both cations and both bind to their ionophore through ion-dipole interactions. The dipole for both ionophores are the oxygen atoms that are electrophilic, which results in them carrying a partial negative charge.

Ionophores for anions are harder to construct. The most common mode of attraction to an anion is the hydrogen bond. When an H atom bonds to an electrophilic atom, like O or N atoms, its one and only electron is drawn toward the latter, leaving it with a partial positive charge. That positive charge attracts anions. However, if the pH of the sample is much greater than 7, then the prevalent OH^- anions pluck off H^+ atoms to make H_2O molecules—that is, the hydroxyl anion *deprotonates* the ionophore molecule containing the H atom, thereby changing it from an anion attracting molecule to a cation attracting one. Thus, ionophores relying on hydrogen bonding, such as those for sulfate, only work in acidic samples.

A second mode of attracting anions is the use of metal cations in organic compounds—metal organic ionophores. Examples include Mn^{2+}, Co^{3+}, Ga^{3+}, Mo^{5+}, Ru^{2+}, Rh^{3+}, Pd^{2+}, In^{3+}, Sm^{4+}, Hf^{3+}, Hg^{2+}, and U^{6+}, In these ionophores OH^- anions also compete with the anion for positive charge. One example is the ionophore Mn(II) porphyrin (a molecule of "fused" benzene rings) for the measurement of CO_3^{2-} anions.

For all polymer based ISEs the optimum ratio of ionic sites to ionophores is a function of the binding energy of the ionophore to the analyte versus that for interfering ions. This is a complicated subject that we won't delve into further.

One last category of ISE is a modification of conventional membranes to enable the measurement of neutral species—especially a biological compound. In most cases this involves adding a second layer that contains an enzyme or antibody selective for the analyte. Many biological reactions are redox reactions, and most redox reactions involve the generation or consumption of H^+ ions. That means that the analyte sensitive layer produces a change in pH, and a conventional pH probe can read that change. For instance, a sensor for glucose incorporates the glucose oxidase enzyme. The enzyme accelerates the oxidation by oxygen of glucose to gluconate (Caras et al., 1985).

$$\text{Glucose} + O_2 \xrightarrow{\text{Glucose Oxidase}} \text{Gluconate} + H^+ + H_2O_2 \qquad (8.14)$$

The concomitant release of H^+ ions allows a layer of this enzyme to graft onto a pH probe. Thus, a drop in pH can correlate to the concentration of glucose. Sveringhaus electrodes measure dissolved CO_2 by adsorbing it into a bicarbonate buffer grafted onto a pH probe and measuring the change in pH. The use of antibodies and enzymes to measure neutral biological compounds of medical concern is a large subject that is undergoing rapid development.

Table 8.2 summarizes the most common ISEs. The third column lists interfering ions and discussion on this major limitation is discussed in Section 1.4. The values for the limit of detection (LOD) for one analyte vary as much as three orders of magnitude from one source to another. For instance, I have read LOD's for the polymer based Ca^{2+} ISE between 5×10^{-7} M and 3×10^{-9} M^{-10}. The lower LOD are mostly due to improvements in the membrane composition and/or differences in the calibration method.

1.10 Calibration

Calibrating any ISE can be done in the same manner as with any other sensor. One simply measures the response of the sensor at two or more known concentrations of analyte and draws a curve through the points. For the range of concentrations over which the curve is a straight line one can calculate the slope and offset of the straight line and use this number to convert any measured voltage to a concentration value.

As mentioned, no sensor in existence can match the enormous range of the pH probe—1 to 10^{14}. The pH probe is capable of this amazing feat because of the abnormally high concentration of binding sites for H^+ ions.

TABLE 8.2 Ions detected by ISEs (Sources: Hanna Instruments; Kamata et al., 1988; Malon et al., 2003; Mettler Toledo; Qin et al., 2000; Sokalski et al., 1999).

Species	Membrane	Limit of Detection (Molar)	Major Interferences	Reference Solution
Hydrogen (H^+)	Silica Glass	10^{-14} to 10^1	Na^+	4M KCl
Lithiu (Li^+)	Polymer	10^{-5} to 1	K^+, Na+, Ca^{2+}, Mg^{2+}	3M KCl
Potassium (K^+)	Valinomycin	10^{-6} to 1	$TRIS^-$, NH_4^+, Cs^+, H^+	0.1 M NaCl
Sodium (Na^+)	Silica Glass	10^{-6} to 1	H^+ (pH>pNa+3), Li^+, K^+	0.1 M $CaCl_2$
Silver (Ag^+)	Solid State	10^{-7} to 1	Hg^{2+}, proteins	1 M KNO_3
Lead (Pb^{2+})	Chalcogenide Glass	10^{-6} to 0.1	Hg^{2+}, Ag^+, Cu^{2+}	1 M KNO^3 (Calomel reference)
Cadmium (Cd^{2+})	Mixed pellet	10^{-7} to 10^{-1}	Fe^{2+}, Pb^{2+}, Hg^{2+}, Zn^{2+}, Ag^+, Cu^{2+}	1 M KNO_3
Calcium (Ca^{2+})	Polymer	3×10^{-6} to 1	Pb^{2+}, Fe^{2+}, Zn^{2+}, Cu^{2+}, Cu^{2+}	1 M KNO_3
Copper (Cu^{2+})	Solid State	10^{-6} to 0.1	Pb^{2+}, Ca^{2+}, Ni^{2+}, Co^2+, Mg^{2+}	1 M KNO_3
Ammonium (NH_4^+)[1]	Polymer	10^{-5} to 1	K^+	0.1 M NH_4Cl
Fluoride (F^-)	LaF_3 crystal	10^{-6} to 1	OH^-	10% KNO_3
Chloride (Cl^-)	AgCl crystal	10^{-4} to 1	Hg^{2+}, Cl^-, I^-, S^{2-}, CN^-, $S_2O_3^{2-}$	1 M KNO_3
Bromide (Br^-)	Solid State	5×10^{-6} to 1	I^-, CN^-, NH_4^+, Cl^-	1 M KNO_3
Iodide (I^-)	AgI crystal	10^{-6} to 1	Hg^{2+}, S^{2-}, CN, $S_2O_3^{2-}$	1 M KNO_3
Sulfide (S^{-2})	Ag_2S crystal	10^{-5} to 1	Hg^{2+}, proteins	1 M KNO_3

(continued)

TABLE 8.2 (*Continued*)

Species	Membrane	Limit of Detection (Molar)	Major Interferences	Reference Solution
Cyanide (CN^-)	AgI crystal	10^{-6} to 10^{-2}	Cl^-, Br^-, I^- S^{2-}	1 M KNO_3
Thiocyanate (SCN^-)	AgSCN crystal	10^{-5} to 1	Cl^-, I^-, Br^-, CN^-, S^{2-}, $S_2O_3{}^{2-}$	$NaNO_3$
Nitrate ($NO_3{}^{2-}$)	Polymer	10^{-5} to 1	Cl^-, Br^-, $NO_2{}^-$, $SO_4{}^{2-}$, $Cl0_4{}^-$	0.04 M $(NH_4)_2SO_4$

All of the ISEs in common use today have linear measurement ranges more in the 10^6:1 range so calibration that extends beyond this range will yield a curved line. The linear range is only between 10^{-3} to 1 M (35 ppm to 35 000 ppm).

Note that the probe output is approximately zero at 10^{-2} M and that the output at zero chloride is $^-100$ mV. The sensor output is zero when the concentration of the analyte inside the electrode is equal to that outside (assuming that the ionic strength is also equal on both sides of the membrane). As explained in Section 1.4, the lower limit is determined by interfering ions and the diffusion of the analyte ion *through* the membrane. The upper limit is determined by the number of ionophores and ionic sites—the latter always being less than the former.

The previous section highlighted the reality that the ISE potential, as determined by the Nernst equation is dependent on the ion activity rather than ion concentration. The result is that the voltage output of a probe in a sample filled with competing ions will yield a lower measured concentration unless calibration is done via the addition of ion strength adjustment solution.

1.11 Measurement—Direct or Incremental

If one takes care to calibrate an ISE using ISA solution and, possibly, pH adjustment or complexing agents, then calibration will yield accurate concentrations. If the ionic strength of your calibration standards matches the ionic strength of your sample, then you can be assured that your calibration will give accurate measurements. But if the ionic strengths are not equal, either because one did not add ISA to your calibration standard, or the sample composition changes, then fear not. One can use a measurement technique based on taking *incremental* measurements as follows.

The most common of these techniques is *standard addition*. It works by measuring the change in probe output upon addition of a known volume of a standard calibration solution. The procedure is as follows:

1. Start with a known volume of sample and measure the probe output.
2. Add a small volume of calibration standard (V_{std}) that will approximately double the sample's analyte concentration (V_{samp}). This means that the calibration standard will have a much higher concentration (C_{std}) solution of the analyte than that of the sample (C_{samp}). For instance, if the sample volume is 100 mL, and your estimate for the concentration is on the order of 10 ppm, you might add 10 mL of standard containing 100 ppm of the analyte.
3. Measure the new probe output and subtract the initial probe output to get the change in probe voltage (ΔV).
4. Using the slope of the Nernst equation at the sample and temperature, calculate the actual concentration according to:

$$C_{sample} = \frac{\dfrac{V_{std}C_{Std}}{V_{samp}}}{\left[1+\dfrac{V_{std}}{V_{samp}}\right]10^{\frac{\Delta V}{slope}}-1} \tag{8.15}$$

The slope, at 25 °C, is −59.16 for +1 cations (e.g., K^+), −59.16 for −1 anions (e.g., Cl^-), −29.58 for +2 cations (e.g., Ca^{2+}), and −29.58 for −2 anions (e.g., S^{2-}).

An alternate method, *sample addition*, is to reverse the method—that is, start with a known volume of standard calibration solution and add a smaller known volume of sample. To double the concentration of the standard solution one should choose its concentration such that the guesstimated sample concentration is much higher. For instance, if calibration standard volume is 100 mL, and your estimate of the sample concentration is on the order of 10 ppm, you would first prepare a calibration standard of only 1 ppm of analyte and add 10 mL of sample.

In this case the calculated sample concentration is given by:

$$C_{sample} = \frac{C_{std}\left[1+\dfrac{V_{samp}}{V_{std}}\right]10^{\frac{\Delta V}{slope}}-1}{\dfrac{V_{samp}}{V_{std}}} \tag{8.16}$$

As alternates to both these methods, one can sometimes subtract the analyte by adding a known solution of a chemical that is known to completely react with the analyte.

The corresponding method to standard addition is *standard subtraction* and the corresponding method to sample addition is sample subtraction. The two equations written above work for these two variations by just making a change of signs.

Standard or sample addition is an invaluable measurement method when the sample is either variable in composition or would otherwise require frequent calibration with varying additions of ISA solution.

2.0 THE pH ISFET

The modern pH probe is a most impressive analyzer. If the Guinness Book of Records had an entry for sensor covering the widest range, it would win with its measurement range of 14 orders of magnitude. It is stable, requires little maintenance and is inexpensive. But it has one problem. The glass membrane that comprises the sensing element has a very high impedance. This requires a voltmeter whose impedance is at least as high (approximately 1 GΩ). To minimize the impedance, pH electrode manufacturers make the glass membrane as thin as possible—a fraction of a millimeter. The glass membrane is the pH probes Achilles heel. It's easy to break, and if it does so in a process involving food, beverage or pharmaceutical manufacturing, the result can be catastrophic. The presence of a piece of glass in these processes means the entire production line has to be shut down, and its contents potentially discarded.

The search for a pH probe with a solid-state process electrode led to the development of the *ion selective field effect transistor*, or ISFET. To understand how it works, let's first understand the principles of operation of the field effect transistor, or FET. Figure 8.9 is a schematic of the device.

The FET is a variation of a transistor, in which a voltage signal to be measured induces a sizeable current between two electrodes that is proportional to the signal.

The ISFET pictured in Figure 8.9 is a *p*-channel FET. It starts with a chip of pure, crystalline silicon (Si). The chip is doped with an element that has one *less* electron than its silicon host. Boron (B) fits the bill as it is sits in the column to the left of Si in the periodic table. Having one less electron than the surrounding Si atoms means that each B atom donates the opposite of a negative charge—a *hole*. We say this material is *p-type* and denote the resulting doped silicon as p-Si. The two end sections of the chip are then doped with an element that contains one *more* electron than silicon.

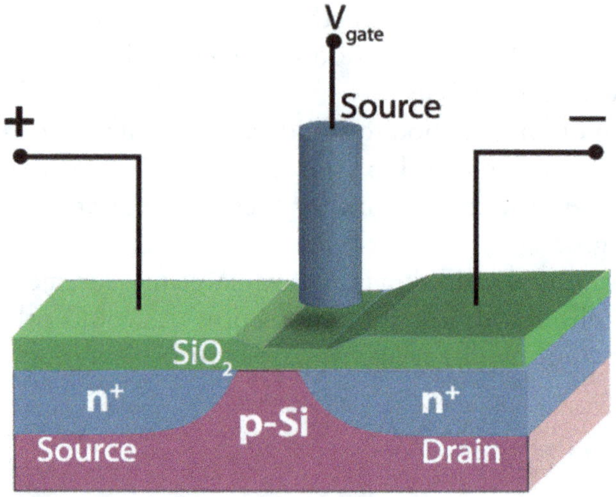

FIGURE 8.9 Construction of an ISFET pH probe.

Phosphorous (P), arsenic (As), and antimony (Sb) are all in the column to the right of silicon, so they work nicely. The extra electron renders the ends of the doped silicon *n-type*. Note that the Si substrate does not pick up a negative charge at the ends or positive charges in the body. The doping element simply injects neutral atoms that have one more or one less electron than the silicon atoms surrounding them. The electrons or holes have no "home" in the silicon structure, so they are both free to wander around.

When a voltage is applied between the *n*-type ends, current flows from the positive end, the *source*, to the negative end, the *drain*. A potential applied to the *p*-type mid-section, the *gate*, modulates the current. By changing the electric field between the two ends it changes the conductivity. A positive potential attracts electrons from the positively doped silicon substrate and increases the current of electrons. The larger the positive potential, the larger the induced current. A negative potential attracts positive charges—holes—and restricts the flow of current.

This design converts a small, high-impedance voltage into a large, easily measured current. This device is called a field effect transistor because the charge on the gate electrode induces an electric field in the *p*-type silicon substrate. The electric field that the gate electrode imposes on the Si substrate changes its conductivity.

The potential applied to the gate of an FET comes from the external signal it is designed to amplify. Because the pH value is a measure of positively charged H^+ ions, we only need to make the gate sensitive to H+ ions to turn the FET into a pH probe. As discussed in Chapter 3, Section 3.2, one can consider the mechanism of pH sensing by glass to be a charge transfer reaction between a H^+ ion and a Si-O group.

$$SiO^- + H^+ \rightarrow Si - OH \qquad (8.17)$$

In addition to binding to silicon oxide, H^+ ions also bind to crystalline materials such as silicon nitride (Si_3N_4), aluminum oxide (Al_2O_3), and tantalum oxide (Ta_2O_5). These three materials exhibit superior stability and faster response and are frequently deposited as a thin layer (via chemical vapor deposition) on top of the SiO_2 gate material. Any one of these three materials are used to construct a gate for a pH sensitive FET.

Establishing a potential for the gate is only half the battle. That potential must be referenced to ground, and that ground must be the reference electrode. This is the same reference electrode of a glass-based probe, and it establishes a continuous electrical pathway to the ion selective gate electrode via the same liquid junction. Figure 8.10 is a schematic of an actual pH ISFET.

The Nernst equation still predicts the behavior of the ISFET, but the voltage change at the gate electrode is not the parameter the analyzer measures, as with the conventional pH probe. As explained above, the signal is the change in current between the source and drain in response to the H^+ charge concentration on the gate. The actual measurement occurs through a feedback circuit. The ISFET circuit imposes an external voltage between the reference electrode and the source to oppose the voltage created by the adsorption of H^+ ions on the gate. By doing so the circuit maintains a constant current between the source and drain. The magnitude of this compensating voltage is proportional to the pH of the solution.

The same ion selective membranes used for conventional potentiometric sensors can be used for ISFETs. These include metal ions, K^+, Na^+, Ag^+, Pb^{2+}, and Cd^{2+}, and anion NO_3^-. The membranes are typically constructed of plasticized PVC. The potentials of these membranes tend to drift so some manufacturers add an Ag-AgCl layer between the gate electrode and the ion selective membrane.

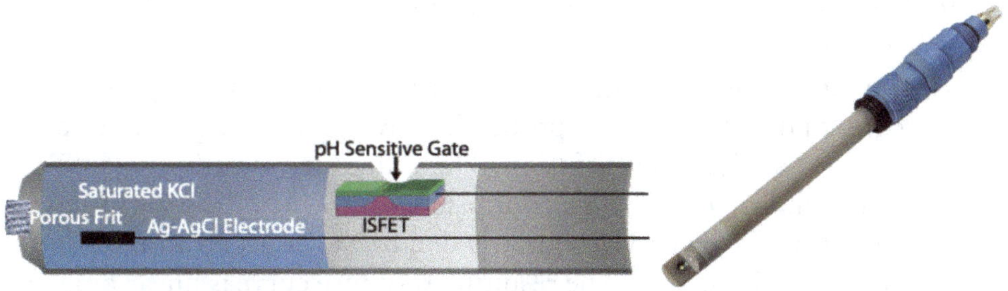

FIGURE 8.10 Left: Schematic of actual pH ISFET. The reference electrode and liquid junction are the same as in a glass-based pH probe. Right: pH ISFET (Courtesy Endress+Hauser).

2.1 Limitations

If one can find materials that show a preference toward a particular ion, one can make an ion sensitive electrode based on potentiometry.

The ISFET electrode has several advantages over the conventional glass pH probe:

1. The pH ISFET is virtually unbreakable so that it can be used in food, beverage, and pharmaceutical applications.
2. Its solid-state construction also enables it to be cleaned more aggressively without fear of breakage.
3. There is no gel layer to be formed and maintained, so it does not need to be kept hydrated.
4. It has the fast response of an FET device.

On the flip side the ISFET has several disadvantages when compared to its glass cousin:

1. The ISFET is more expensive.
2. The signal is a current. Therefore, the ISFET probe cannot be connected to a high-impedance voltmeter used for a glass probe.
3. Oxidants such as chlorine will permanently damage the gate electrode.
4. The selective elements in most ISEs have shorter lifetimes than pH sensors. This is especially true for sensors based on organic compounds. The hydrophobic ionophores slowly leach out of the polymer (usually PVC) membrane.
5. All FETs behave, to some degree, like a photoconductor—that is, light striking the gate surface generates electrons. This means that measurements must be shielded from strong light sources.
6. The ISFET output drifts to a larger degree than its conventional ISE cousin.
7. The measurement range of ISEs is several orders of magnitude narrower than that of the pH probe. The lower limit is due to interfering ions and analyte ion diffusion. The minimum sensitivity of ISEs is greater than 10^{-9} M. This contrasts with the 10^{-14} M sensitivity of the pH probe. The upper limit is due to the concentration of ionophores and ionic sites. The resulting response curve is linear only in a range on the order of 10^{-7} to 1 M. One can measure concentrations extending beyond the linear range but that requires multiple calibration points to capture the nonlinear response. The next section shows

how voltammetry has lowered the sensitivity to ions by about two orders of magnitude.

8. The signal is highly temperature dependent. Higher temperatures, like light, increase the injection of electrons into the *p*-type silicon substrate.

For these reasons ISFETs (especially pH ones) have gained acceptance principally in the food, beverage, and pharmaceutical markets.

3.0 VOLTAMMETRY

3.1 Amperometry With a Third Dimension

The amperometric sensors discussed in Chapters 5 and 6 measure only one species (O_2 and HOCl). The choice of applied voltage and membrane material determine that species. The electrochemical oxygen sensor discussed in Chapter 5 was designed so that oxygen reacts at the cathode to form OH^- and is the predominant reaction. The reduction reaction creates a stream of electrons, and the magnitude of the resulting current is proportional to the dissolved oxygen concentration. Any other reaction at the cathode constitutes interference to the dissolved oxygen measurement. The potential that maximizes the reduction of oxygen is approximately 0.7 V with respect to the Ag-AgCl reference. Likewise, the reduction of HOCl at the cathode to Cl^- and OH^- provides a current whose magnitude is proportional to the concentration of HOCl. The potential that maximizes the reduction of HOCl is approximately 0.3 V. The potential applied to the cathode in any amperometric sensor tracks the standard state reduction potentials given in Table 2.1. Typically, the potential is slightly larger because an overpotential, as defined in Chapter 5, Section 2, is required to overcome the capacitance at the electrode from the accumulation of ions on either side of the electrode surface.

The different potentials at which different molecules and ions reduce or oxidize is the foundation of an amperometric probe that can measure a range of chemical species. Figure 8.11 shows that the basis of this new sensor is nothing more than the 3-electrode potentiostatic sensor we discussed in Chapter 6, Section 4.3 for measuring free chlorine. The reference electrode is an Ag-AgCl electrode in KCl that maintains the working electrode at a precisely defined potential. The working electrode is the one at which either reduction or oxidation of a species of interest occurs. Splitting the anode into a counter electrode and reference electrode ensures that no current flows to or from the reference electrode, thus maintaining its voltage as set by the

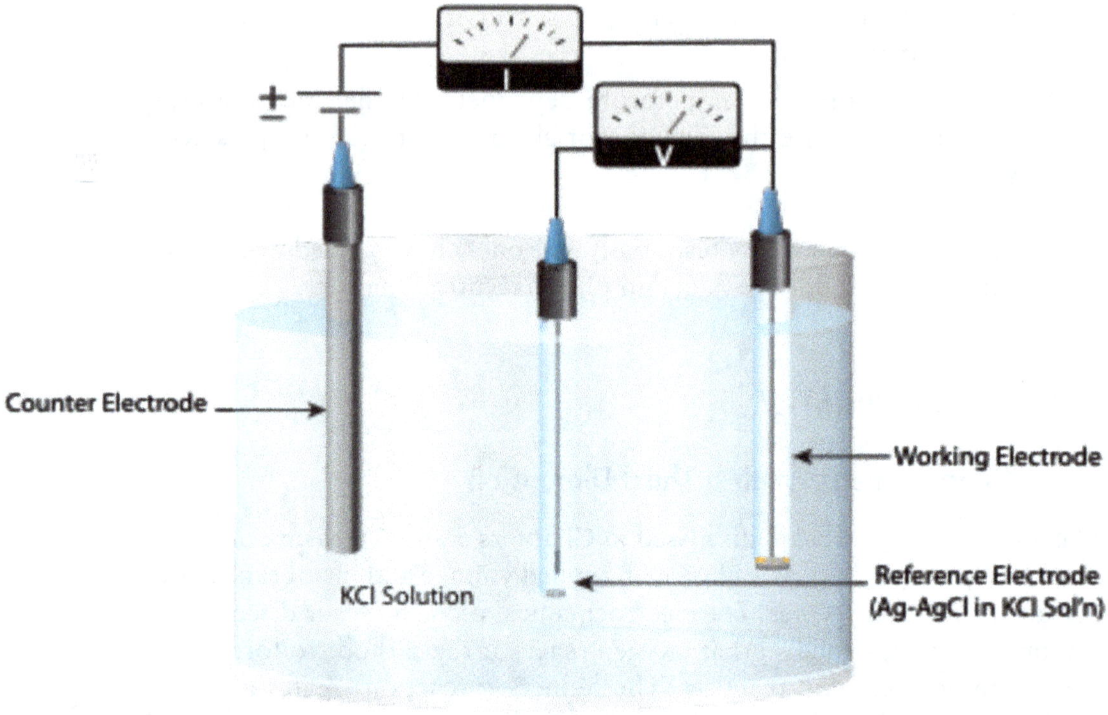

FIGURE 8.11 Schematic of voltametric analyzer. The reference electrode maintains the working electrode at the precise voltage. Current flows between the counter electrode and the working electrode.

Nernst equations. The counter electrode provides the source of electrons for a reducing reaction or the sink for an oxidizing one.

Chapter 5, Section 3.1 discussed the mass transport of O_2 molecules to the membrane where they reduce on the cathode to OH^- ions at the cathode. If the flow of water across the membrane replenishes dissolved oxygen molecules inside the membrane, the current of the dissolved oxygen probe is dependent only on the diffusion of molecules through the membrane. In the absence of a membrane the role of mass transport of oxygen molecules is more complex.

Let's review the three modalities of mass transport to bring dissolved species from the bulk of the solution to the electrode surface: diffusion, convection, and migration. *Diffusion* is the movement of ions toward the electrode due to a concentration gradient. To define diffusion, we must first define *flux*. This is the rate at which material flows across an imaginary surface. Think of water flowing through a full pipe and counting the number of water molecules as they cross an imaginary plane in the pipe that is perpendicular to the flow. One can also imagine ions drifting toward and striking the surface of an electrode. Units of flux are number of particles (or moles) per unit of area per unit of time. For purposes of illustration, we frame our discussion around the following generalized redox reaction, in which the oxidant, *Ox*, is reduced to *Re*:

$$Ox + ne^- \leftrightarrow Re \tag{8.18}$$

Every *Ox* ion that strikes the electrode surface reduces to *Re*. If *Ox* ions are being reduced faster than new ones can reach the surface, then the surface concentration of *Ox* species is zero. The concentration increases from zero, at the surface of the electrode, and levels off at the concentration of *Ox* ions equal to that of the bulk solution. The shading in Figure 8.12 represents the concentration of *Ox* ions. The change in that concentration represented by the shaded area is called the *gradient*. It is the slope of the graph of concentration plotted against the distance from the working electrode. When you empty a packet of sugar in your coffee the concentration of sugar crystals at the bottom of the cup is greater than in the rest of the coffee, and the sugar gradually dissolves and mixes with the bulk of the coffee. The concentration gradient decreases until it is zero, and the sugar is completely dissolved throughout the cup of coffee. Similarly, the concentration gradient

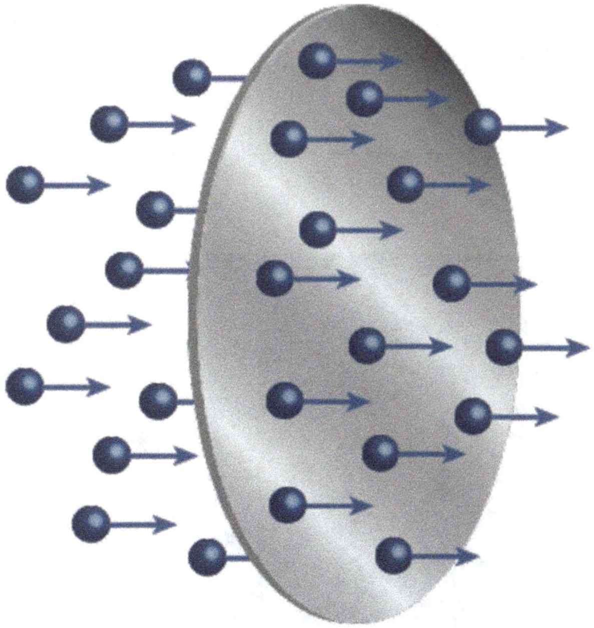

FIGURE 8.12 Illustration of flux—the amount of material that crosses an imaginary plane that is perpendicular to the flow. If the plane is the surface of the electrode, then the flux is proportional to the number of molecules (or atoms, ions, or particles) that strike the surface.

likewise drives the diffusion of *Ox* ions (or molecules) toward the electrode surface where they reduce to *Re* ions (or molecules). The distance from the electrode surface to the point in the bulk where the concentration levels off demarcates the *diffusion layer*. (The surface of the electrode is part of the electrical double layer that forms on every electrode. The double layer forms a capacitor as described in Chapter 5, Section 2.)

Figure 8.13 shows two graphical interpretations of how diffusion creates a concentration gradient. On the left is a plot of the concentration, at one particular time, as a function of the distance from the electrode in which the concentration is proportional to the blue color. One can easily make out the boundaries of the diffusion layer. On the right is a family of representations, in a time sequence, of the concentration as a function of distance from the electrode surface. At time zero (at the top of the figure), when the voltage is initially applied to the electrode, the *Ox* species just starts to reduce. The *Ox* concentration gradient at this initial time is very sharp, as shown in the top right of Figure 8.13. As diffusion brings more *Ox* ions to the electrode the gradient becomes more and more shallow. The diffusion layer is proportional to the gradient, i.e. faster diffusion means a wider diffusion layer.

In our discussion of the amperometric dissolved oxygen probe, Equation 5.29 expressed the probe output current as a function of the dissolved oxygen concentration. That expression emanated from one of the most basic equations in all of physics—Fick's first law of diffusion:

$$J = -D\frac{\partial C}{\partial x} \approx -D\frac{\Delta C}{\Delta x} = -D\frac{(C_{bulk} - C_{surface})}{W} \tag{8.19}$$

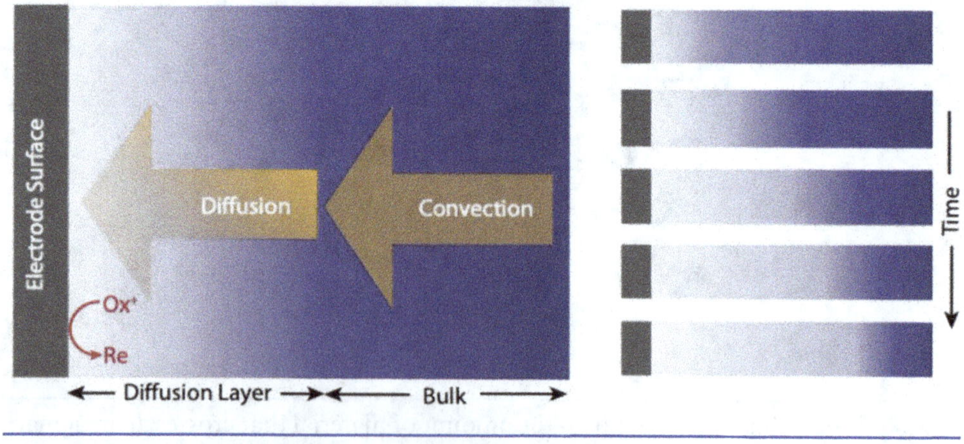

FIGURE 8.13 Diffusion layer on the surface of the working electrode. On the left is a graphical depiction of concentration and on the right is the concentration gradient at several times as diffusion occurs.

The derivative, $\frac{\partial C}{\partial x}$, is the concentration gradient, which changes with distance from the electrode. The approximation in the middle sets the change in concentration to the overall change in concentration between that of the surface of the electrode and that at a distance from the surface at which the bulk concentration is constant. That distance, Δx, becomes, in the right-hand side, the width of the diffusion layer, W. For a membrane covered sensor, W is simply the thickness of the membrane. For this setup, in which the membrane is absent, it is an approximation of the diffusion layer. As Figure 8.13 shows, the concentration gradient is not constant across the diffusion layer but is actually curved.

Diffusion is a passive phenomenon. The *Ox* ions (or atoms or molecules) move from the bulk of the liquid to the electrode surface with no help from the outside environment. The driving force is the concentration gradient. If the reaction rate of *Ox* ions reducing to *Re* ions is fast enough, the concentration of *Ox* ions will be zero at the surface. From an earlier discussion on electrode surfaces in Chapter 5, Section 2, we say that the electrode is (mostly) nonpolarized—that is, there is little electrical resistance to electron transfer in the redox reaction occurring at the working electrode.

Convection, on the other hand, is the motion of *Ox* ions with some mechanical assistance. Just as stirring the sugar in a coffee cup speeds up the mixing of the sugar, stirring will drive *Ox* ions from the bulk of the sample to the surface. In an electrochemical cell convection usually comes about through a mechanical stirrer.

A third mode of mass transport is *migration*. This is simply the movement of ions via the attraction or repulsion between the analyte ion and the electrode. Obviously, neutral molecules are not affected. If the solution in which the analyte is dissolved is ionic or polar (like water) and the analyte is in low concentration, as it almost always is, then the solvent will screen the analyte charges and negate their migration. Migration can therefore be neglected in the modeling of amperometry or voltammetry.

The diffusion layer is typically very thin—less than 10 μm. Convection has little effect on the concentration gradient within the diffusion layer. This is shown on the left side of Figure 8.13. However, it does affect the concentration at the outer surface of the diffusion layer. Sufficient convection will maintain a constant concentration of the analyte species being oxidized or reduced. This steady-state concentration gradient ensures a steady-state redox current.

So far everything in this arrangement of electrodes is the same as it is for the 3-electrode, amperometric free chlorine sensor. Notice, however, that there is no membrane surrounding the three electrodes to restrict the redox reaction to the chemical species of interest. A membrane may or may not be part of a voltametric analyzer.

The key feature that converts an amperometric sensor to a voltametric analyzer is the varying voltage. As the voltage between the working and reference electrodes changes the analyzer records the current. As the analyzer swings the potential from zero to increasingly positive values, the latter eventually approaches the potential at which an analyte reduces. The result is a sharp increase in current. As the applied voltage swings from zero to increasingly negative values the latter approaches the potential at which an analyte oxidizes. The magnitude of the current for either an oxidation or reduction reaction is proportional to the rate at which the analyte undergoes its redox reaction.

3.2 Linear Sweep Voltammetry

Let's now see how what happens to the redox current when we *sweep* the voltage of the working electrode over a predetermined range. This is the basis of a voltametric measurement.

The current read by the analyzer is proportional to the flux of Ox ions striking the electrode surface where they are reduced. The units of flux are moles/cm^2-sec. To convert flux to current we multiply it by Faraday's constant (the charge per mole of electrons) the area of the electrode (A) and the number of electrons exchanged per mole (n). The current into the electrode is therefore:

$$i \approx \frac{nFAD}{W}(C_{\text{bulk}} - C_{\text{surface}}) \tag{8.20}$$

As this approximate equation shows, the current varies proportionally to the change in concentration of the analyte from the bulk of the sample to the electrode surface. It also varies inversely with the width of the diffusion layer. The ratio of the former to the latter is the *concentration gradient*.

For the case of the ideal, nonpolarized working electrode C_{surface} is close to zero. The Ox analyte is consumed faster than mass transport can resupply it, and Equation 8.20 simplifies to the following:

$$i_l \approx \frac{nFAD}{W}C_{\text{bulk}} \tag{8.21}$$

This is analogous to Equation 5.29, which expressed the current of a dissolved oxygen sensor as a function of dissolved oxygen concentration. Equation 8.21 is a simple equation from which one can determine the concentration of the analyte, C_{bulk}, from the current, providing one knows the width of the diffusion layer. W is unique to the redox species being considered.

There are about a dozen unique strategies for varying the potential, and each has its own advantages and drawbacks. Let's start with the most basic strategy—*linear sweep voltammetry*, or LSV. In this simplest of experimental strategies, the analyzer simply records the current as it sweeps the voltage at a constant (i.e., linear) rate on the working electrode from a starting potential to a final one. The resulting plot of the current against the voltage is the *voltammogram*.

The voltammogram has two limiting cases:

1. With sufficient convection—that is, stirring—C_{bulk} remains constant, and the current, I_l, is also constant. The plot on the left in Figure 8.14 illustrates this case. A typical voltametric analyzer rotates the working electrode to ensure adequate convection.
2. In the absence of convection, i_l, increases to a peak value and drops. The plot on the right in Figure 8.14 illustrates this case.

The gray line near the bottom of the two plots is the *residual current*. It must be subtracted from the total current to derive the current that emanates from the actual redox reaction—the *Faradaic current*. The residual current is composed of two sources:

1. The current that arises from redox reactions other than the one for the analyte.
2. The *charging current* forms the electrical double layer at the surface of the electrode. It typically increases as the driving voltage increases and gives rise to an upward sloping baseline.

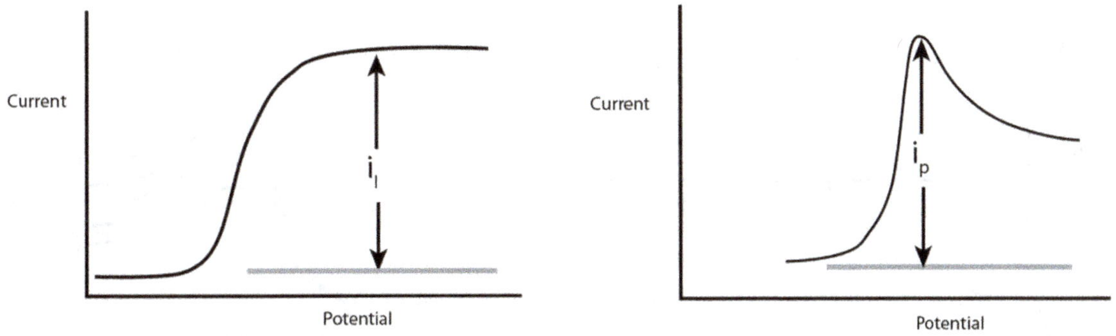

FIGURE 8.14 The voltammogram on the left results from a constant concentration gradient created by convention and is characterized by a limiting current (i_l). The voltammogram on the right results from a changing diffusion layer in the absence of convection and is characterized by a peak current (i_p).

Now let's see how the sweep rate affects the voltammogram. Figure 8.15 shows a family of voltammograms for different rates. Note that, as the voltage scan rate increases, the current magnitude also increases. At first glance this seems counterintuitive as slowing down a measurement generally increases the signal or the signal-to-noise. This is because, as the applied voltage changes more slowly, the diffusion layer grows and, as predicted by Equation 8.21, the current decreases. Note also that the position of peak voltage, E_p^0, does not change with the change in voltage. This is true if the redox rate is faster than the scan rate. This is certainly valid when the working electrode is a close approximation of a nonpolarized one. Also recall that, for a nonpolarized electrode, the Nernst equation is valid. However, do not confuse the voltage at which peak current occurs (per mole of Ox), E_p^0, with the standard redox potential contained in the Nernst equation. It is always higher and is close to the overpotential chosen for the driving voltage of an amperometric sensor. The next section shows how to calculate E^0 from a voltammogram.

3.3 Extracting the Standard Potential and Concentration From the Voltammogram

The goal of most voltametric measurements is to determine the redox species concentration. For the case of a peak current and a linear sweep of the voltage, the *Randles-Sevcik equation* relates the peak (i_p) current to the concentration:

$$i_p(\mu A) = 4.464 \times 10^5 nFAC \frac{nFDv}{RT}^{1/2} = 0.269 n^{3/2} A(Dv)^{1/2} C(at\ 25\ ^\circ C) \qquad (8.22)$$

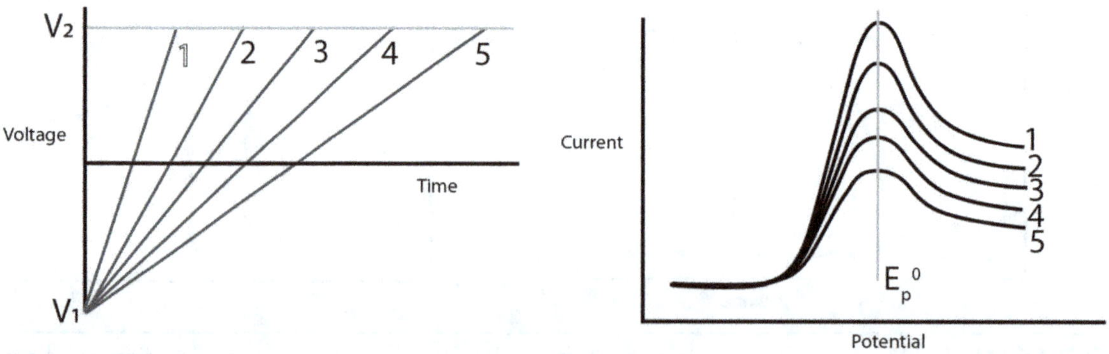

FIGURE 8.15 Principle of linear sweep voltammetry. On the left is a series of linear sweeps in which the potential on the working electrode increases at 5 different scan rates. On the right is the resulting voltammograms. E_p is the over-potential.

A is the working electrode's surface area, D is the diffusion constant, and v is the rate at which the voltage on the working electrode increases.

Though Equation 8.22 yields the concentration of the target analyte, in practice, it's not easy to use. The diffusion constant must be known, and there may be no accurate way to separate out the residual current from the peak current. A more practical approach is to simply measure the current over a range of known concentrations and construct a calibration curve. The inset in Figure 8.16 shows such a calibration curve for the oxidation of Fe^{+2} to Fe^{+3} (Robles et al., 2014). The residual current shows up clearly in the calibration line as a large offset, which can then be subtracted from the total current to determine the concentration.

As an alternative use of the Randles-Sevcik equation, if one knows the diffusion coefficient and the concentration, one can calculate the diffusion coefficient and the diffusion layer thickness of the species.

The other piece of information we may want to extract from voltammetry is the redox standard potential. In most cases one starts a voltametric exercise with the knowledge of the redox couple and its standard potential. But there might be a case when there is a mixture of redox active species, and one needs to determine E_0 in order to identify each one. Alternatively, the

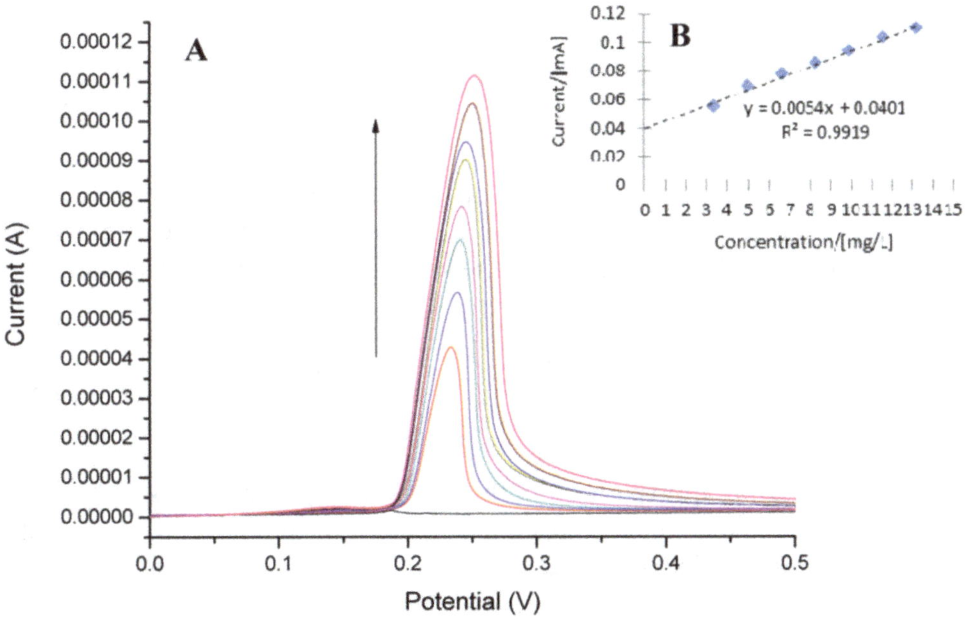

FIGURE 8.16 Calibration of voltametric analyzer for the oxidation of mercury. The inset graph is the plot of the peak current against known concentrations. This provides a calibration curve. Note that the residual current has been subtracted on the polarogram's y-axis (Robles et al., 2014).

reaction being studied might be new and does not have a known standard potential. As mentioned in the last paragraph, E_p^0 is the value of the overpotential used to drive an amperometric sensor, but it is always of a higher magnitude than the standard state reduction potential.

We can derive the standard state potential from a linear sweep voltammogram provided that the following assumptions are valid:

- Reaction at the electrode surface is faster than the rate of diffusion or the voltage sweep rate. This means that the Nernst equation is valid and can be used to calculate the ratio of concentrations of the redox couple—that is, [*Re*]/[Ox].
- The diffusion layer is constant through convection or a rotating disk electrode. This means that there is a limiting current as shown on the left side of Figure 8.14.
- The bulk solution initially contains only *Ox* species—that is, [*Re*] = 0.
- The solution at the surface of the electrode contains only *Re* species, i.e., [Ox] = 0.

From the Nernst equation and Equation 8.21, one gets an equation that relates the voltage, for which the current is ½ the limiting current (I_l), to the standard state potential (Harvey, 2019):

$$E_{1/2} = E_o - \frac{RT}{nF} ln\left[\frac{K_O}{K_R}\right] = E_o + \frac{0.05916}{n} log\left[\frac{K_R}{K_O}\right] (\text{at } 25\ °\text{C}) \quad (8.23)$$

The potential at which I = $\frac{1}{2}i_l$ is $E_{1/2}$ and called the *half-wave potential.* K_0 is the collection of parameters in Equation 8.21 that accompany the concentration of Ox, or C_{bulk}. K_R is the analogous factor for the concentration of reducing species:

$$K_O = \frac{nFAD}{\delta_O};\ K_R = \frac{nFAD}{\delta_R} \quad (8.24)$$

To a very good approximation, the diffusion layer thickness is the same for both *Ox* and *Re*, which means that δ_O equals δ_R and K_O equals K_R. With this approximation Equation 8.23 collapses to the eminently satisfying:

$$E_{1/2} = E^0 \quad (8.25)$$

To determine the standard redox potential, E^0, simply requires us to measure the peak or limiting current, divide it by 2, and find the potential on the voltammogram that gives the half-current value. This treatment has involved introducing several approximations. Section 3.8 shows how a cyclic voltammogram yields a more accurate determination of E^0.

3.4 A Real Example

This section shows a real linear sweep voltammogram. Take the case of the Fe^{+2} and Fe^{+3} redox couple that has appeared in prior discussions as the quintessential redox couple. Only this time we view the couple as a complex with an organic molecule. *Ferrocene* is $Fe(C_5H_5)_2$ with the iron atom in the +2 state. It oxidizes to the *ferricenium* cation $Fe(C_5H_5)_2{}^+$.

$$Fe(C_5H_5)_2 \rightarrow Fe(C_5H_5)_2{}^- + e^- \quad (8.26)$$

3.5 The Voltametric Advantage

The real power of voltammetry is its ability to measure a range of analytes. Figure 8.18 is a conceptual drawing of a voltammogram for two analytes subject to a constant diffusion layer—as in the left side of Figure 8.14. The right side of Figure 8.18 is a plot of the change in current as in the right side of Figure 8.14 but for two analytes. This upwardly ascending staircase of peak currents, one for each of the two redox species, can be extended to multiple analytes. As an example, see Figure 8.20.

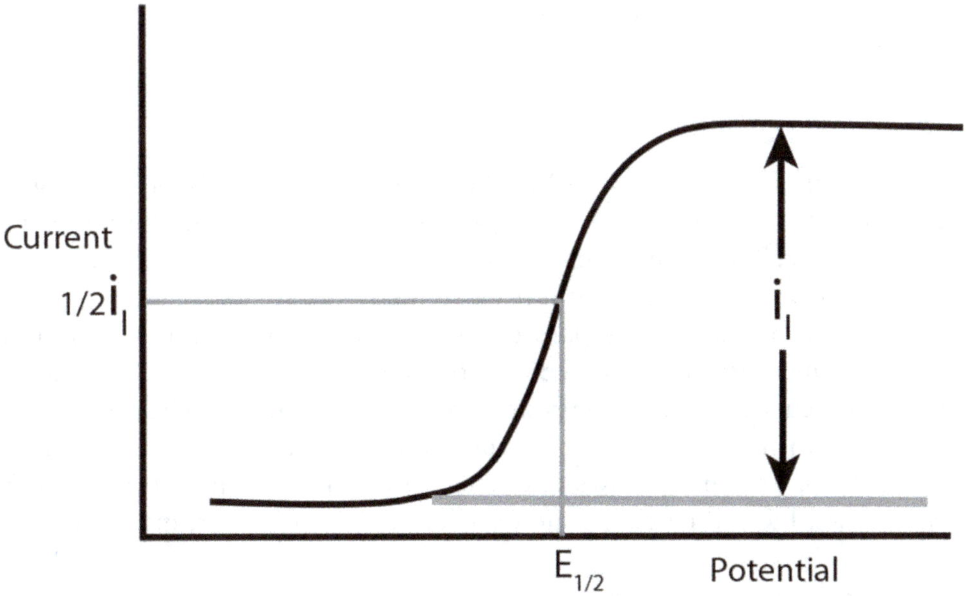

FIGURE 8.17 Determination of the half-wave potential from a linear sweep voltammogram.

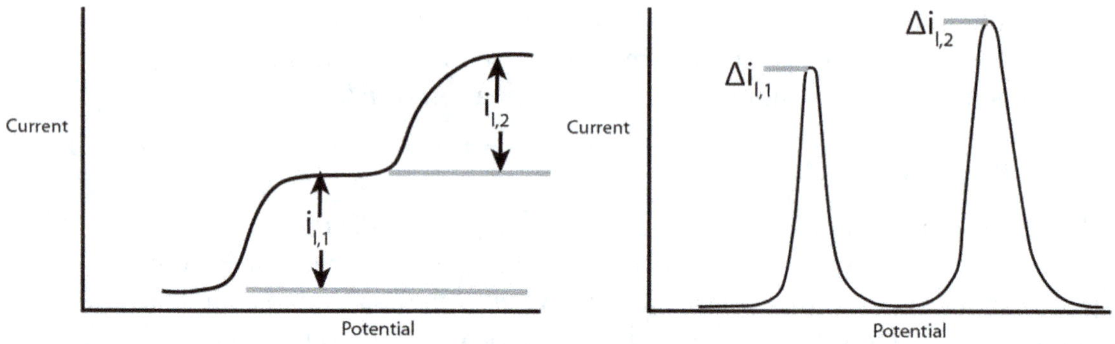

FIGURE 8.18 Idealized voltammogram. On the left is a plot of current from (or to) the working electrode as a function of the applied voltage. On the right is the change in current, which is equivalent to a derivative plot. These voltammograms show two species being analyzed in one linear sweep.

The right-hand plot in Figure 8.18 takes the same data from the left-hand figure and plots the *change* in current instead of the absolute magnitude of the current. In mathematical terms, this is a plot of the first derivative of the current $\frac{\partial i}{\partial E}$. The integral of the curve, $\int \frac{\partial i}{\partial E} dE$, is the current. The red line in the left-hand plot represents the residual current. The derivative of a constant is zero, so the right-hand plot has a zero offset. The figure, as shown, represents a level maximum current as described in the left side of Figure 8.14. For a voltammogram that reaches a peak value and then falls, as in the right side of Figure 8.14, the derivative plot would show a dip below the zero axis.

3.6 Polarography

The discussion of the Clark cell in the discussion of the dissolved oxygen probe noted that the term *polarographic* is often used as an alternative to *amperometric*. The term is actually reserved for a voltametric analyzer in which the working electrode is a drop of mercury.

Polarography is not a separate voltametric technique but a variation in which a mercury electrode replaces the inert metal working electrode. A pool of mercury feeds a capillary opening in glass or metal cylinder. A drop of mercury forms on the bottom of the cylimder, and redox reactions occur on the drop. The drop grows and eventually falls, at which time a fresh drop grows. Liquid mercury is an excellent choice as a working electrode:

1. The limitation of negative potentials from the reduction of H^+ to H_2 is lifted due to the high overpotential characteristic of a mercury electrode.

2. Metal analytes form an amalgam with mercury, which enhances the signal.
3. A fresh surface on which redox reactions occur is constantly being generated. This minimizes contamination of the electrode surface and yields very reproducible currents.
4. The mercury drop stirs the sample during its fall and thus eliminates the need for a stirring mechanism.

On the downside:

1. The positive voltage range, where oxidations occur, is limited.
2. The mercury source must be pure. When exposed to air, mercury readily oxidizes.
3. Mercury is an expensive and highly toxic metal that is becoming increasingly restricted in all applications.

Figure 8.19 shows the setup for a dropping mercury drop electrode (MDE) and an actual voltammogram, or *polarogram*, for Cd^{+2}.

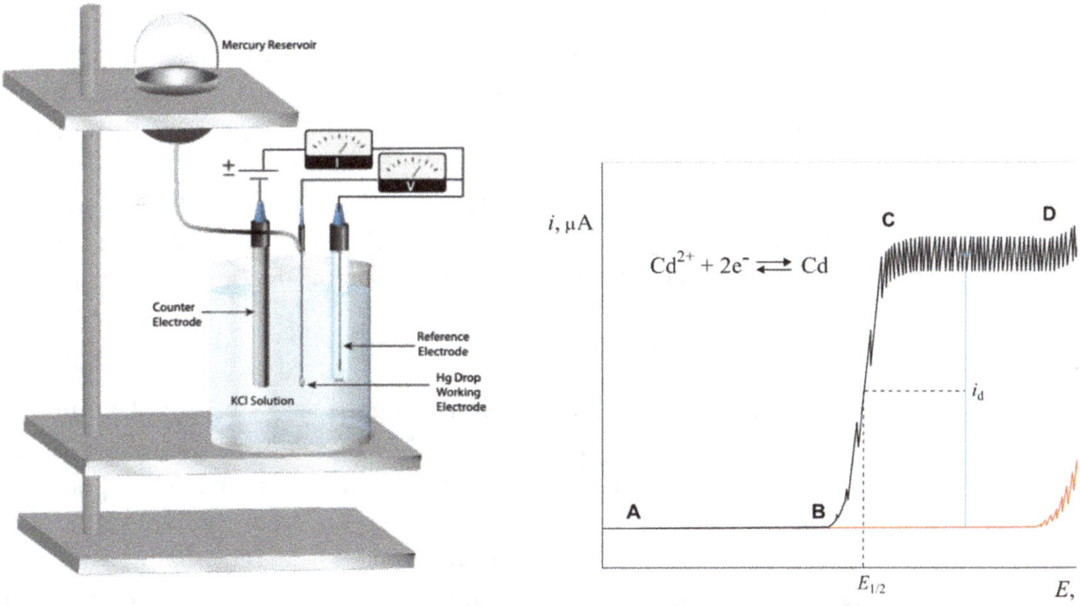

FIGURE 8.19 Left: Setup for the hanging mercury drop electrode. Right: Polarogram for the measurement of Cd^{+2} in 1 M HCl. The bottom trace is of 1 M HCl to give a baseline signal (E. Generalic, https://glossary.periodni.com/glossary.php?en=polarogram).

The voltammogram for this linear sweep method shown in Figure 8.19 differs from the voltammogram discussed in Chapter 3, Section 3 in that the peak current has very rapid oscillations. This is a result of the dynamics of the mercury drop formation. As the drop grows the surface area expands and the reaction rate increases. When the drop falls the signal plummets, only to increase again as a new drop grows. A variation of the dropping mercury drop electrode is the hanging drop mercury electrode (HDME) where the Hg drop remains stable for the course of one sweep. This eliminates the oscillations but at the cost of the mercury surface becoming increasingly contaminated.

The determination of E^0 from the half-wave potential is done just as it is for linear sweep voltammetry, as shown on the right of Figure 8.19. The calculation of concentration, however, requires a different mathematical treatment—the *Ilkovic equation*:

$$i_p = 706nCD^{1/2}m^{2/3}t^{1/6} = K_pC \tag{8.27}$$

In this equation, n, C, and D are the usual electron number, concentration (moles/l) and diffusion coefficient (cm^2/sec), m is the rate of mercury flow (mg/sec) and t is the time (s) between drops (i.e., the duty cycle). The current i_p is in µA. All parameters, except C, are constants and can be collected into one constant, K_p.

Though this example shows only a linear sweep voltammogram the HDME can be used in any of the other variants discussed. As just mentioned, the real power of the HDME is its ability to measure the concentrations of a wide range of analytes. Figure 8.20 shows a polarogram with four metal ions: Cu^{+2}, Pb^{+2}, Cd^{+2}, and Zn^{+2}. The right-hand plot is a differential plot just like the right side of Figure 8.18. In practice one would use one of the variations of voltage generation in the HDME that are discussed in the next section to extract accurate standard state potentials and concentrations.

3.7 Variations on a Pulse

For measurement of ions in the ppm range linear sweep voltammetry is simple, fast, and relatively inexpensive. The precision is limited by the rate of diffusion. For measurement of concentrations in the parts per billion (ppb) range, variations of the simple strategy are required.

Pulsing the working electrode minimizes the redox time and, thus, increase in the diffusion layer. Making the measurement in a small fraction of the pulse at the end of the pulse minimizes the charging current. There are several variations for pulsed voltammetry but two are the most popular. Figure 8.21 show the voltage sequence and the resulting current for both.

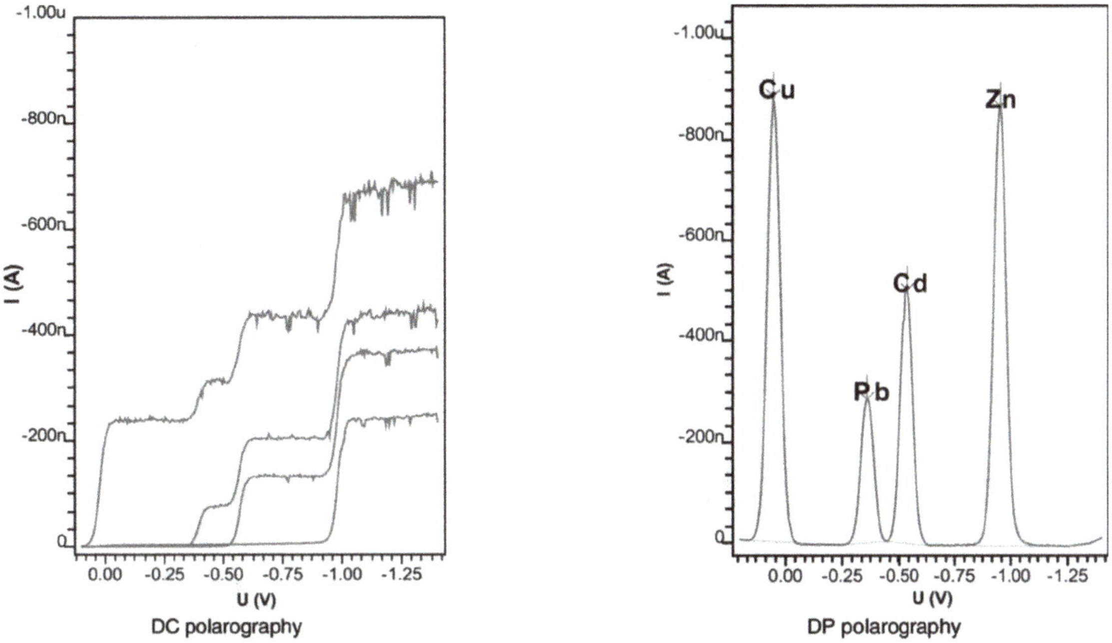

FIGURE 8.20 Left: Polarograms of a solution containing the three metal ions of copper, lead, cadmium, and zinc. Right: A derivative plot of the left-hand plot (Courtesy Metrohm AG).

They two pulsed techniques are:

1. *Normal Pulse Polarography* (NPP). The linear sweep is replaced by a series of equally spaced pulses of increasing potential. Between each pulse the voltage returns to its baseline value. The actual measurement duration occurs in a small fraction of time at the end of the pulse width. As the left side of Figure 8.21 shows the pulse duration, t_p, the actual measurement, t_m, duration the duration between pulses, τ, and the increase in pulse voltage, ΔE_p, are all constant. Values for τ are on the order of seconds (as it is the time for the Hg drop to develop), t_p is on the order of 100 msec and t_m is on the order of 20 msec. If one draws a line through the tops of the pulses the straight line of a linear sweep emerges. The little black rectangle on the right side of each pulse is the measurement time. As stated above, this delay in measurement allows the charging current to decay faster than the decrease in the Faradaic current.
2. *Differential Pulse Polarography* (DPP). As in NPP the analyzer outputs a sequence of pulses of increasing voltage. Unlike NPP the voltage between pulses does not drop to the baseline value. Instead, the voltage drops by a set amount (ΔE_b). The analyzer takes a current

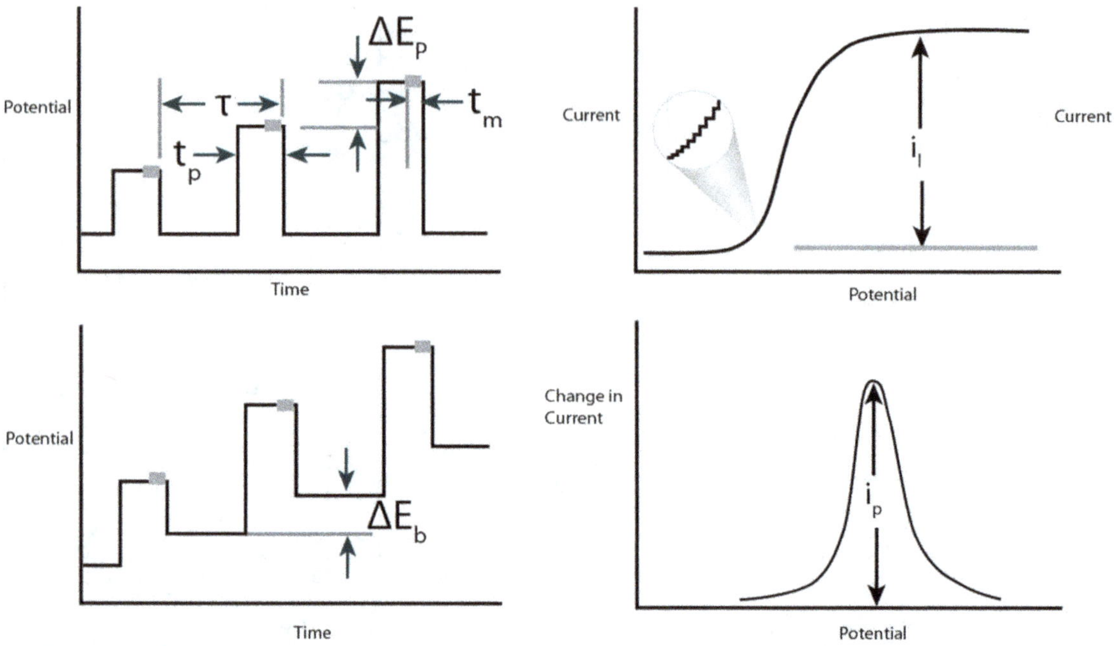

FIGURE 8.21 Top: Schematic of normal pulsed polarography. On the left is the sequence of pulses against a constant background potential. The small gray rectangle is the measurement time during the last few milliseconds of each pulse. On the right is the resulting current versus function. The inset shows that the current is a sequence of current jumps that correspond to the potential pulses. Bottom: Potential and current plots for differential pulse polargraphy. The difference is the increasing background.

measurement at this lower voltage and, after a time t_p, again at the higher voltage. The difference between the two measured currents, ΔI, corresponds to the difference between the two successive voltages, ΔE_p. The plot of ΔI versus ΔE_p is thus an approximation of the derivative of the NPP curve. As with NPP, the current measurement is made only over the last few milliseconds of each pulse. Differential plots make it easier to determine the voltage for the peak or limiting current.

The combined advantages of the improved signal-to-noise and resolution of pulsed and differential polarographic techniques are dramatically displayed in Figure 8.22 in which the chlordiazepoxide (Lubrium) is shown as a linear sweep polarogram and a differential pulse polarogram (Hackman et al., 1974).

Another example is shown in Figure 8.23, which shows a sequence of DPP plots of silver ion (Ag^+) at varying known concentrations. A plot of the peak current versus the ammonia concentration shows a high degree of precision and a zero offset. Contrast this with a similar calibration plot

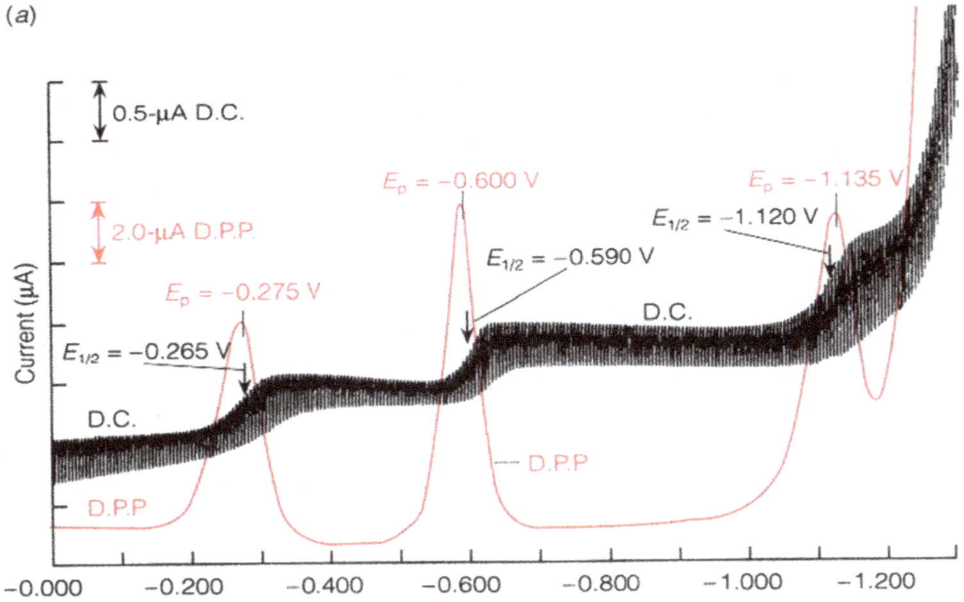

FIGURE 8.22 A linear sweep polarogram contrasted with a differential pulse polarogram. The dramatic enhancement in resolution is apparent from the latter. The analyte is chlordiazepoxide (Lubrium) (Brooks & Hackman, 1975).

obtained using linear sweep voltammetry in Figure 8.16. Because this figure plots the differential voltametric signal the plot goes through zero. For measurement of species in the ppb range, differential pulse polarography is the most used technique.

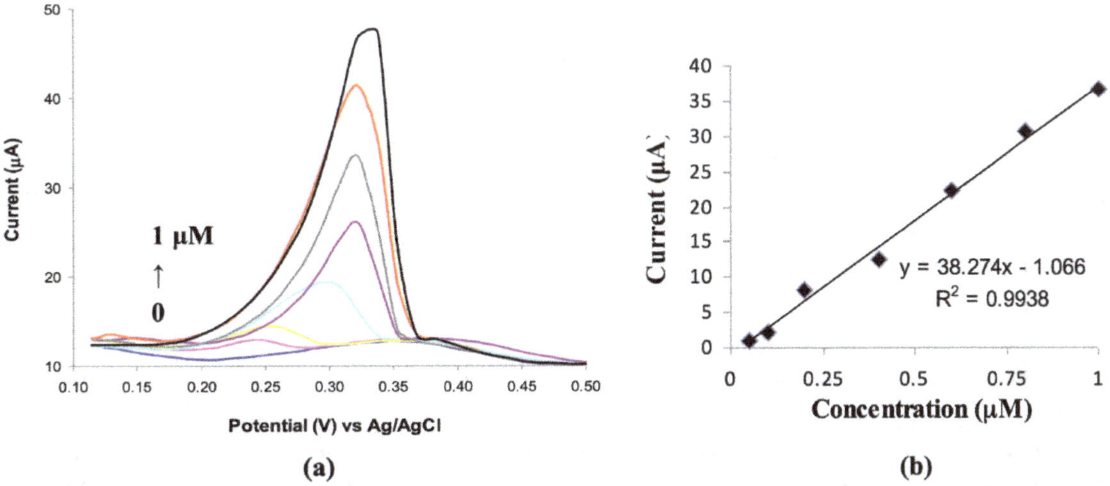

FIGURE 8.23 Calibration of polarographic sensor for measurement of silver (Ag^+). The peak current is measured for each known concentration of Ag^+ and then plotted on the right-hand side (Radulescu et al., 2010).

3.8 Cyclic Voltammetry

This is the most common form of voltammetry that does not involve the hanging drop mercury electrode. It is just a linear sweep in one potential direction followed by a sweep in the backward direction. Why the "return trip"? If the redox reaction is reversible then the reverse voltammogram will have the same shape as the forward voltammogram. The voltage ramp is simply a linear voltage sweep in one direction followed by a mirror image sweep in the opposite direction.

Figure 8.24 shows the voltage profile and resulting *cyclic voltammogram*.

In linear sweep voltammetry, there is one peak, and one can use the Randles-Sevcik equation (8.22) to determine the concentration from the size of the peak. Recall that one must separate out the residual current to get an accurate peak current value. The residual current shows up as a sloping baseline, which manifests itself as the current at the end of the sweep (the *asymptotic limit*). In a cyclic voltammogram, there are two sweeps, and each one has its own sloping background. Figure 8.24 shows the sloping baselines for each sweep as determined by the extrapolation of the diminishing currents at the end of both sweeps. If the redox reaction is truly reversible, then both peak currents—$I_{p,c}$ and $I_{p,a}$—are equal, and one can average them to get a more accurate concentration determination.

Equation 8.28 shows the current for the reduction of ferricenium back to ferrocene:

$$Fe(C_5H_5)_2^- + e^- \rightarrow Fe(C_5H_5)_2 \quad (8.28)$$

As the voltage decreases, the current resulting from the oxidation of ferrocene continues, albeit at an increasingly lower value. At the same

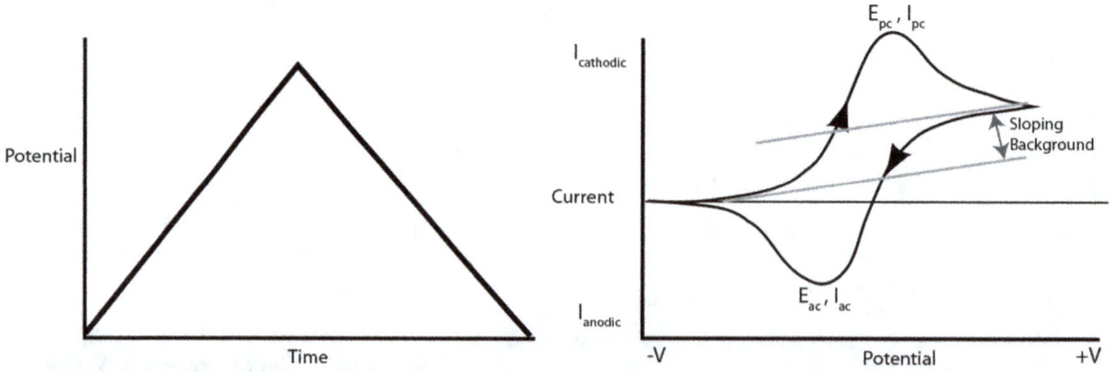

FIGURE 8.24 On the left is the voltage sweep for cyclic voltammetry. On the right is a typical voltammogram. The peak current for both forward and backward sweeps should be of the same magnitude with respect to the limiting currents on both ends. I_{pc} is the peak cathodic current at E_{pc} and I_{pa} is the peak anodic current at E_{pa}.

time, the positive current from the reduction of ferricenium increases. Remember from the discussion in Chapter 5, Section 2 and Equations 5.11 and 5.12, that there is never just one cathodic current or one anodic current. Both currents occur at the same time and on same surface of the electrode. The ratio between the two is a function of the applied potential. At negative potentials the cathodic current is greater than the anodic current and vice versa.

The second piece of information we can derive from the cyclic voltammogram is the standard redox potential. Some mathematical trickery with the Nernst equation leads to the following equation.

$$E_{midpoint} = \frac{1}{2}(E_{p,anode} + E_{p,cathode}) = E_0 + \frac{RT}{nF}\ln\left(\sqrt{\frac{D_R}{D_O}}\right) \quad (8.29)$$

The difference between the peaks of the forward and reverse scan is:

$$\Delta E = [E_{p,anode} - E_{p,cathode}] = \frac{59.16}{n} mV \text{ (at 25 °C)} \quad (8.30)$$

This gives us the familiar 59.16 mV. It's not a very useful measurement, but it does give us a good sanity check of the voltametric scan.

So far, we have dealt with redox reactions that are reversible. The anodic branch and the cathodic branch are symmetric. If the reaction is not reversible the voltammogram will certainly show a lack of symmetry, like the one in Figure 8.25. We use the forward scan only to study the reduction reaction. The reverse scan describes one or more oxidation reaction for a different set of redox partners.

3.9 Stripping Voltammetry

The most sensitive of all voltametric techniques is based on the concept of *stripping*. The strategy is to concentrate the analyte onto the electrode just prior to applying the usual voltage sweep. The pre-concentration step boosts the concentration of the analyte on the surface of the electrode by orders of magnitude. *Anodic stripping voltammetry* (ASV) is used to measure metal cations, while *cathodic stripping voltammetry* (CSV) is used to measure anions.

To illustrate this technique, consider a metal ion, say, copper, in an anodic stripping voltammetry. If one wants to measure the concentration copper ion, Cu^{+2}, in solution, but the redox measurement we measure is the oxidation of the neutral copper atom:

$$Cu \rightarrow Cu^{+2} + 2e^- \quad (8.31)$$

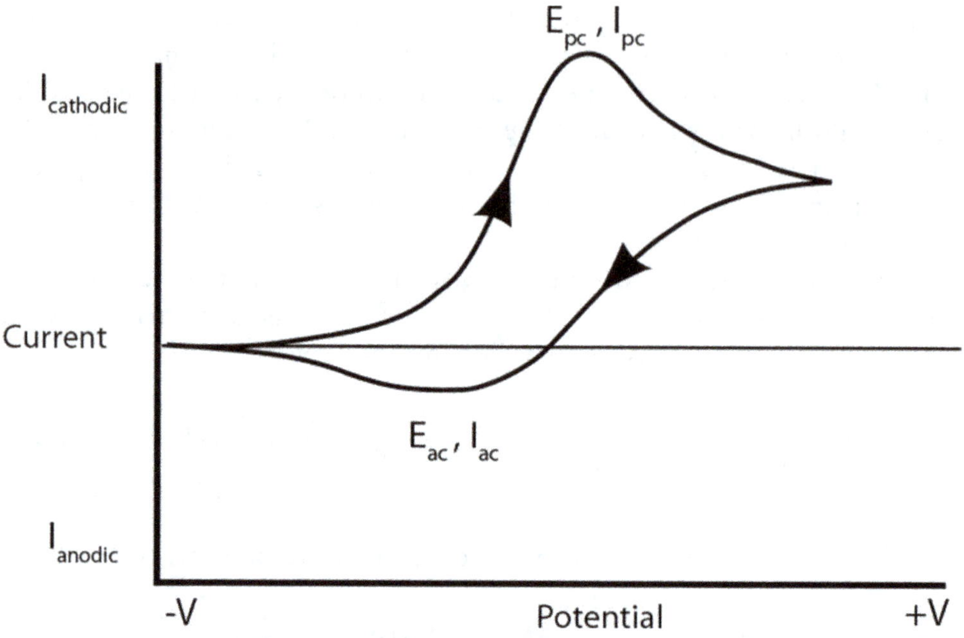

FIGURE 8.25 Cyclic voltammogram for a species that undergoes irreversible reduction. The reduction of the cation produces a species that is resistant to oxidation. The forward scan is still useful for extracting concentration information.

The setup for ASV is very similar to polarography except that, in place of a hanging mercury drop, the working electrode usually consists of a smooth electrode (usually glassy carbon) covered with a thin mercury film. (Some ASV analyzers use a hanging drop electrode and a solid platinum or gold electrode with a polymer membrane. However, the enrichment factor for the mercury film electrode is much higher than that for the hanging drop.) There are three major steps involved in a measurement. Figure 8.26 illustrates the sequence, and the following description uses Cu^{2+} as the example analyte.

1. In the pre-concentration step, the polarity of the working electrode is opposite of that for the usual linear sweep. For Cu^{2+} this means applying a negative potential. Cu^{2+} ions in solution are reduced to neutral copper. The reduction is the opposite of Equation 8.30. The copper forms a mercury amalgam.

$$Hg + Cu^{+2} + 2e^{-} \rightarrow Cu(Hg) \qquad (8.32)$$

During the deposition phase a stirring mechanism insures a steady buildup of copper. This happens over a period of minutes.

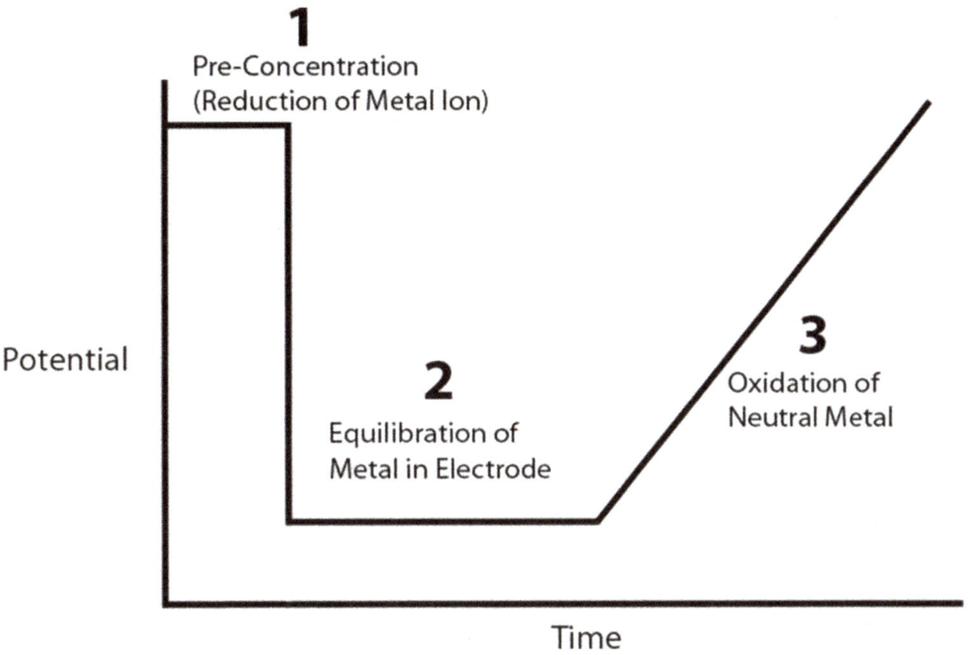

FIGURE 8.26 Sequence of voltage changes in an anodic stripping (of a metal ion) voltametric measurement.

2. While the polarity of the electrode is still negative the stirring ceases to allow the concentration of analyte in the amalgam to reach an equilibrium concentration. This phase lasts about 30 seconds.
5. In the measurement phase the potentiostat applies an increasingly positive potential to oxidize the Hg atoms in the amalgam:

$$Cu(Hg) \rightarrow Cu^{+2} + 2e^{-} \quad (8.33)$$

Figure 8.27 shows the actual voltage profile and resulting current for the measurement of AS^{+3} using ASV (Rodriguez et al., 2012). The peak current yields the concentration using the Randles-Sevcik equation, but the resulting concentration value depends on the thickness of the mercury film, the pre-concentration time, and the rest time in addition to the usual electrode area, diffusion constant and sweep rate. The only way to convert the current to concentration is to carry out a calibration with known concentrations and a reproducible set of operating parameters.

Figure 8.28 shows an anodic stripping voltammogram for the three metal species of zinc, cadmium, and lead (Willemse et al., 2011). This setup adds a square wave voltage to the stripping step and a rotating disk electrode that

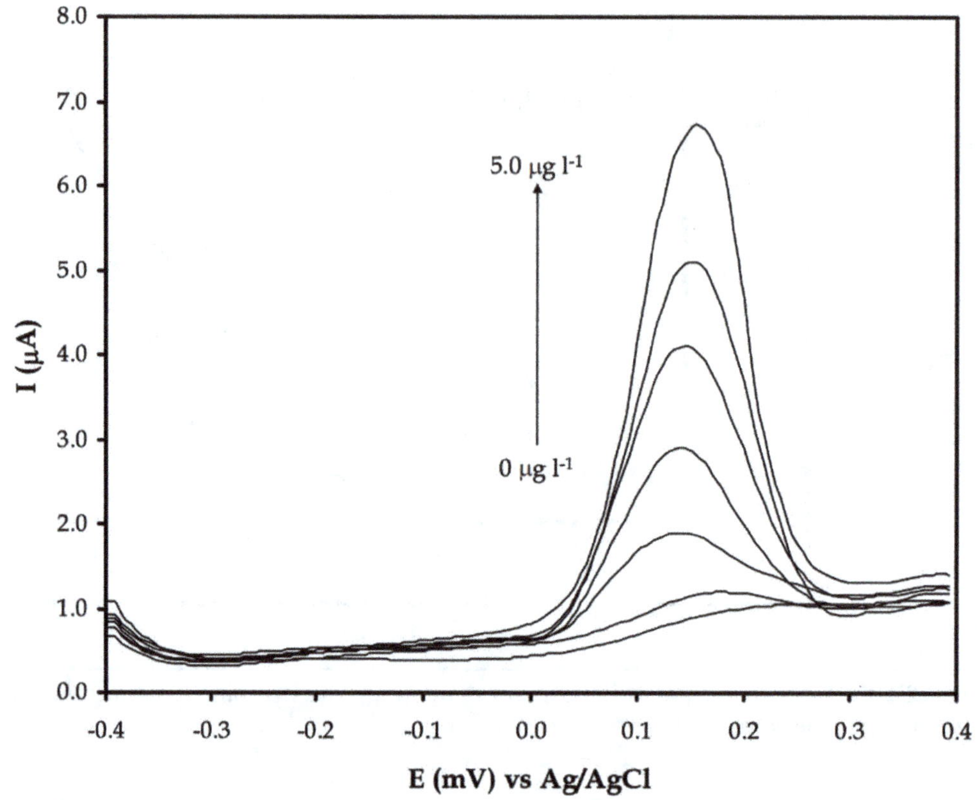

FIGURE 8.27 ASV for the measurement of Cu^{2+}. On the bottom is the voltage profile for the pre-concentration phase and stripping phase. On the top is the electrode current (Rodriguez et al., 2012).

we cover shortly. A plot of current versus concentration in the upper right corner shows that the limit of detection is less than 0.1 μg/L or 100 ppb.

Anodic stripping voltammetry can be used to measure most metal ions, including bismuth, cadmium, copper, gallium, indium, lead, thallium, tin, and zinc. Cathodic stripping voltammetry is used to measure anions such as bromide, chloride, iodide, mercaptans, sulfide, and thiocyanate.

3.10 Rotating Disc Electrode (Hydrodynamic Voltammetry)

The dropping mercury electrode created its own stirring mechanism. Section 3.2 discussed the benefit of convection to maintain a constant diffusion layer thickness. For analyzers that don't use the HMDE the most convenient way to increase convection in a manner that is simple, adjustable, and reproducible is to replace the stationery electrode with a rotating disk electrode (RDE).

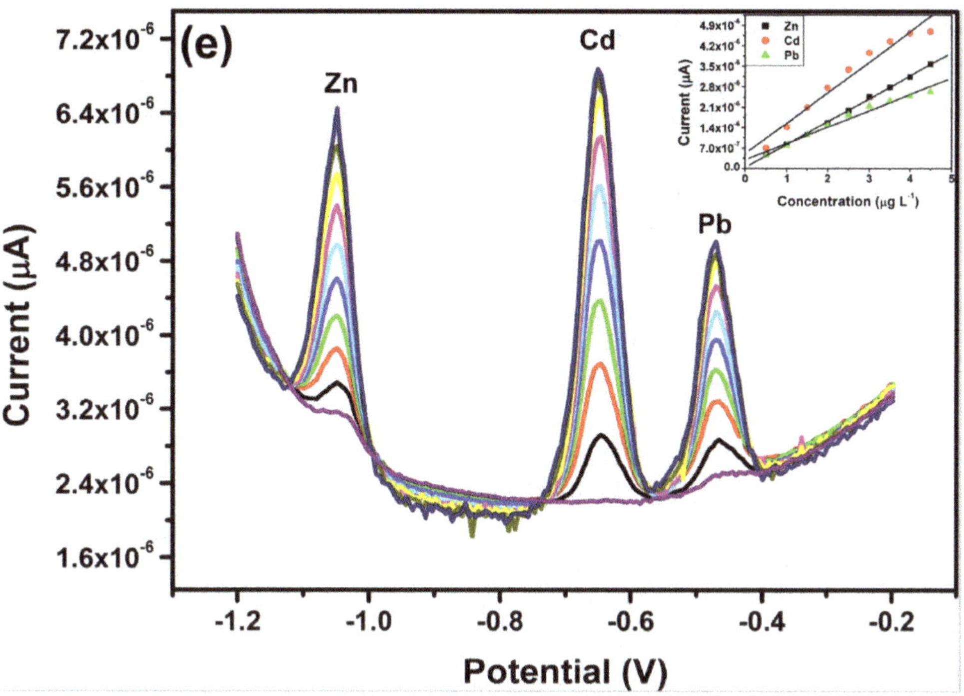

FIGURE 8.28 ASV of three metals (Zn, Cd, and Pb). The inset figure is a plot of current versus potential and shows lower limits of detection (< 100 ppb) (Willemse et al., 2011).

The rotating disk does something very useful to the sample solution that a simple stirring mechanism does not. It creates *laminar* flow along the surface of the disk—that is, the flow across the disk consists of layers of constant flow in a direction outward from the center.

The Levich equation gives an approximation for the limiting current in terms of the solution viscosity, ν (cm^3s^{-1}), the angular frequency, ω (s^{-1}), and the usual electrode area, A, and diffusion constant, D.

$$i \approx 0.620\ n\ F\ A\ D^{2/3}\ \nu^{-1/6}\ \omega^{1/2}\ C \tag{8.34}$$

The technique using an RDE in a voltametric setup is sometimes called *hydrodynamic voltammetry*. It can be applied to a mercury electrode and a solid one. When applied to a mercury electrode, the rotating disk provides an alternative for method to the hanging drop for maintaining a constant diffusion layer. There are other methods that transport the solution past the electrodes to set up a constant diffusion layer, but we do not go into them here.

3.11 Ion Transfer Stripping Voltammetry

Until this point, a necessary condition for an ion to be detected is that it undergoes oxidation or reduction. However, there is still help for those ionic species that don't. In *Ion Transfer Stripping Voltammetry* (ITSV), the working electrode (usually gold, platinum, or glassy carbon) is capped by or two membrane layers. Figure 8.29 shows the sensor design used by Kim and Amemiya (2008). The external layer contains an ionophore that selectively adsorbs the analyte ion. The inner layer contains a second polymer that is redox active. To show how it all works we'll look at a sensor tailored toward the measurement of perchlorate, ClO_4^-.

1. In the pre-concentration phase the potentiostat applies the positive voltage needed for the PVC (polyvinyl chloride) membrane to preferentially adsorb the perchlorate anion. The buildup of negative charge from the adsorbed ClO_4^- expels anions in the PVC (labeled Y^-) into the POT (poly-3-octylthiophene) layer. The POT molecules oxidize to supply electrons to the potentiostat. In effect, the Y^- anions carry the charge from the ClO_4^- anions, and the readily oxidized POT molecules serve as proxies to the ClO_4^- anions.
2. In the stripping and measurement phase, the potentiostat voltage turns negative, and the flow of electrons reduces the POT^+ ions back to neutral POT molecules. The Y^- anions return to the PVC membrane and expel the ClO_4^- back into solution. The net result is that the perchlorate anions are recycled and preserved.

The electrode in this example is a rotating disk electrode. The stripping sequence is combined into a cyclic voltammogram.

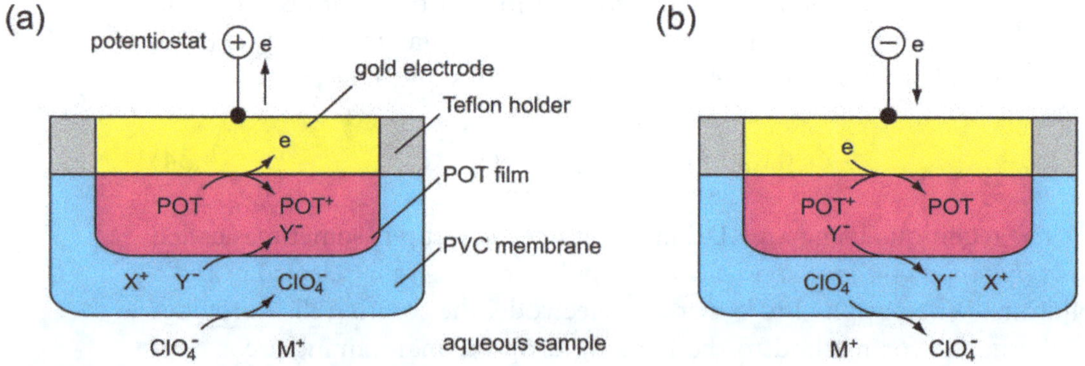

FIGURE 8.29 Ion transfer stripping voltammetry of perchlorate. On the left the pre-concentration phase, and on the right, is the stripping phase. The POT molecules undergo oxidation and reduction, respectively, as surrogates for the ClO_4^- ions (Kim & Amemiya, 2008).

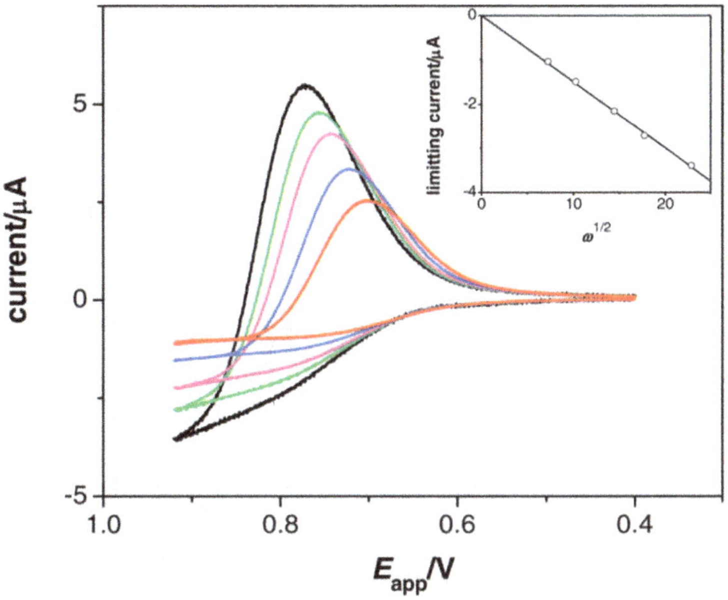

FIGURE 8.30 Cyclic voltammetry combined with ion transfer stripping voltammetry for perchlorate ions. The family for curves is for varying speeds of the rotating disk electrode. The plot in the corner is the limiting current vs. the square root of the angular velocity (Kim & Amemiya, 2008).

3.12 Summary and Comparison

The most significant advantage that voltammetry has over competing analytical techniques, such as ion chromatography, is its relative simplicity and cost. In contrast to the other electrochemical techniques this book has covered, there are many options in the configuration of the voltametric analyzer. Figure 8.31 shows an example of a system by Metrohm.

Table 8.3 summarizes the voltametric and polarographic techniques. As a rule, the higher the sensitivity the lower the accuracy. For stripping voltammetry, the measurement error is about 25% and is the result of interferences, signal noise and variations in the charging current and deposition time.

Let's just review some general considerations.

1. In the ppm range linear sweep, or cyclic, voltammetry is adequate. Differential pulse polarography is the most used technique for measurement in the ppb range. For measurement in the sub-ppb range, stripping voltammetry is the clear choice.
2. Both the accuracy and precision of most voltametric techniques is in the range of a few percent for concentrations in the ppm range. The accuracy is dependent largely on the baseline current, which is due to

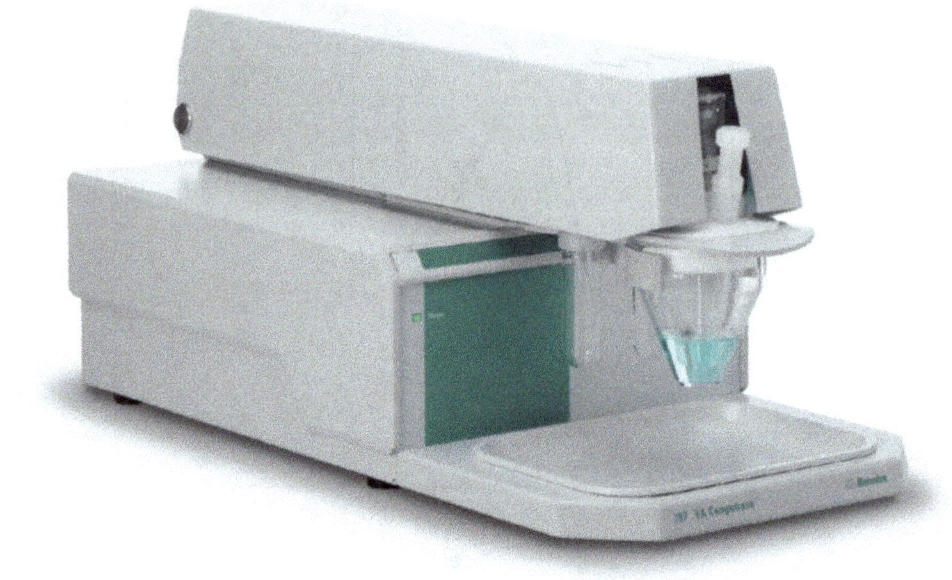

FIGURE 8.31 The Metrohm 797 Computrace is a voltametric analyzer that can be configured and programmed to work in most of the modes of operation discussed (Courtesy Metrohm AG).

redox reactions of interfering species and the charging current. For sub-ppb analyte concentrations by stripping voltammetry, the accuracy can be as low as 25%. The reduced accuracy is due—in addition to interferences and charging current—variations in the experimental parameters, such as deposition time.

3. The voltage range of a voltammogram or polarogram determines the resolution. For linear sweep voltammetry the minimum is approximately ±0.3 V. For differential pulse it is about ±0.05 V

TABLE 8.3 Review of voltametric and polarographic techniques.

Linear Sweep Polarography	5×10^{-3} M
Gated direct current Polarography	1×10^{-5} M
Normal Pulse Polarography	10^{-7} to 10^{-8} M
Differential Pulse Polarography	10^{-8} M
Square Wave Polarography	10^{-8} M
Hydrodynamic Voltammetry (RDE)	10^{-9} M
Stripping Voltammetry	10^{-10} to 10^{-11} M

4. To measure metal ion concentrations, the pH is lowered to between 1 and 2. This ensures that metal solids are converted into metal ions. Sulfuric acid is most commonly used.
5. To exclude dissolved oxygen from being reduced it is purged by passing nitrogen gas through the sample.
6. The working electrode is typically platinum, gold, glassy carbon, or any of these capped with a mercury film. The counter electrode is either a platinum wire or a glassy carbon rod. The reference electrode is most often Ag-AgCl but can be one of the alternative electrodes used for potentiometric measurements: calomel, normal hydrogen electrode, or Ag-$AgNO_3$.
7. The solution is chosen so that the analyte is soluble. If the solvent is an organic solvent, like ethanol, acetonitrile, diethyl ether, dimethyl sulfoxide, or dichloromethane, then a salt or polar compound is added to create an electrolyte solution that minimizes migration of the analyte to the electrode. Whatever that added chemical is, it must not oxidize or reduce at a potential that is within the operating range of the instrument. Common salts are those that include K^+, Na^+, Cl, ClO_3^-, $Cl0_4^-$, and SO_4^{2-}.
8. The scan rate has a strong effect on the signal strength. It seems counterintuitive that a slower scan rate produces a smaller current and was explained in Section 3.2.
9. Whenever possible, a calibration using a sequence of known standards is preferable. When not convenient one can use the method of standard addition, as explained in Section 1.11.

4.0 REFERENCES

Bakker, E., & Pretsch, E. (2005). Potentiometric sensors for trace-level analysis. *Trends in Analytical Chemistry, 24*(3), 199–207.

Benoit, R., Wilkinson, K. J., & Sauve, S. (2013). Partitioning of silver and chemical speciation offree Ag in soils amended with nanoparticles. *Chemistry Central Journal, 7(*71), 1–7.

Brooks, M. A., & Hackman, M. R. (1975, October). Trace level determination of 1,4-benzodiazepines in blood by differential pulse polarography. *Analytical Chemistry*, *47*(12), 2059–2062.

Bühlmann, P., Amemiya, S., Yajima, S., & Umezawa, Y. (1998). Co-ion interference for ion-selective electrodes based on charged and neutral ionophores: A comparison. *Analytical Chemistry*, *70*(20), 4291–4303.

Bühlmann, P., & Chen, L. D. (2012). Ion-selective electrodes with ionophore-doped sensing membranes. In P. A. Gale & J. W. Steed (Eds.), *Supramolecular chemistry: From molecules to nanomaterials* (pp.). John Wiley & Sons, Ltd. https://doi.org/10.1002/9780470661345.smc097

Caras, S. D., Petelenz, D., & Janata, J. (1985). pH-Based enzyme potentiometric sensors. Part 2. Glucose field effect transistor. *Analytical Chemistry, 57*(9), 1920–1923.

Harris, D. C. (2010). *Quantitative chemical analysis*. W. H. Freeman & Company.

Harvey, D. (2019). http://chemwiki.ucdavis.edu/Core/Analytical_Chemistry/Analytical_Chemistry_2.0/11_Electrochemical_Methods/11D%3A_Voltammetric_Methods

Kamata, S., Bhale, A., Fukunaga, Y., & Murata, H. (1988). Copper (II) selective electrode using thiuram disulfide neutral carriers. *Analytical Chemistry, 60*(22), 2464–2467.

Kim, Y., & Amemiya, S. (2008). Stripping analysis of nanomolar perchlorate in drinking water with a voltametric ion-selective electrode based on thin-layer membrane. *Analytical Chemistry, 80*(15), 6056–6065.

Lai, C. Z., Joyer, M. M., Fierke, M. A., Petkovich, N. D., Stein, A., & Bühlmann, P. (2009). Subnanomolar detection limit application of ion-selective electrodes with three-dimensionally ordered macroporous (3DOM) carbon solid contacts. *Journal of Solid State Chemistry, 13(*1), 123–128.

Malon, A., Radu, A., Qin, W., Qin, Y., Ceresa, A., Maj-Zurawska, M., Bakker, E., & Pretsch, E. (2003). Improving the detection limit of anion-selective electrodes: An iodide-selective membrane with a nanomolar detection limit. *Analytical Chemistry, 75*(1), 3865–3871.

Noh, J-H., & Coetzee, P. P. (2007). Evaluation of the potentiometric determination of trace fluoride in natural and drinking water with a fluoride ISE. *Water South Africa, 33*(4), 519–529.

Qin, W., Zwicki, T., & Pretsch, E. (2000). Improved detection limits and unbiased selectivity coefficients obtained by using ion-exchange resins in the inner reference solution of ion-selective polymeric membrane electrodes. *Analytical Chemistry, 72*(4), 3236–3240.

Radulescu, M., Chira, A., Radulescu, M., Bucuur, B., Bucur, M. P., & Radu, G. L. (2010). Determination of silver(I) by differential pulse voltammetry using a glassy carbon electrode modified with synthesized N-(2-Aminoethyl)-4,4'-Bipyridine. *Sensors, 10*, 11340–11351.

Robles, I., Godínez, L. A., Manríquez, J., Rodríguez, F., Rodríguez, A., & Bustos, E. (2014). Electrochemical detection of mercury removal from polluted bentonite and quartz using different removing agents. In M.C. Hernandez-Soriano (Ed.), *Environmental risk assessment of soil contamination*, 1235–1238.

Rodriguez, J. A., Barrado, E., Vega, M., Castrillejo, Y., & Lima, J. L. F. C. (2012). Sequential injection Anodic Stripping Voltammetry at Tubular Gold Electrodes for Inorganic Arsenic Speciation. *InTech*, March, 2012, 203–218.

Sokalski, T., Ceresa, A., Fibbioli, M., Zwickl, T., Bakker, E., & Pretsch, E. (1999). Lowering the detection limit of solvent polymeric ion-selective membrane electrodes. 2. Influence of composition of sample and internal electrolyte solution. *Analytical Chemistry, 71*(6), 1210–1214.

Willemse, C. M., Tihomelang, K., Jahed, N., Baker, P. G., & Iwuoha, E. I. (2011). Metallo-Graphene nanocomposite electrocatalytic platform for the determination of toxic metals. *Sensors, 11*(4), 3970–3987.

9

Organic Matter

TOC (total organic carbon), BOD, COD, POC, NPOC, DOM, and NOM. Aside from consisting of three letters (with one exception), and with "O" being in the middle, these initials have one thing in common: they measure the amount of organic matter in a sample of water. All organic compounds consist of at least one carbon atom bonded to at least one hydrogen atom. Most consist of multiple C atoms, and most of them also bond to H, O, and/or N atoms. The combinations of these elements give rise to millions of actual organic compounds. So, it's not surprising that trying to give a quantitative measure of organic matter concentration is like trying to quantify the concentration of animal species, from bacteria to elephants, in one cubic mile. Is it any wonder that trying to assign a single number to the concentration of organic compounds is a daunting task? Each parameter listed above is the result of a unique method for making this measurement.

Organic matter is central to both water treatment and wastewater treatment. In water treatment, knowing its concentration is necessary to adjust the dosage of coagulants to remove it. Insufficient removal of organic compounds (>10 mg/L) leaves behind residual compounds that will react with free chlorine and form disinfection by-products (DBPs). To minimize the formation of DBPs, U.S. Environmental Protection Agency (U.S. EPA) has phased in its Stage 2 DBP Rule, created in 2012. The rule mandates the monitoring of total organic carbon (TOC) for about 75 000 larger water treatment facilities (U.S. EPA, 2014). Any locations throughout these facilities' distribution system where DBP concentrations exceed the maximum contaminant level (80 µg/L for total trihalomethanes) must monitor TOC. To remove DPBs, facilities add metal salt coagulants and filter through granulated activated carbon or filter through ultra- or nano-filtration.

In wastewater treatment, knowing organic matter content is necessary throughout the process. The major component of wastewater is organic matter and optimizing the rate of aeration needed to break it down is the largest cost of running a treatment facility. Having real-time measurements of organic matter is necessary for continuous optimization of aeration. As in water treatment facilities, chlorination of the filtered water of a wastewater treatment facility is necessary to destroy microbes. As with a facility's, monitoring for organic matter ensures that DBPs will not exceed the U.S. EPA or state standard. Lastly, the effluent must meet maximum discharge limits on organic matter, typically expressed as biological chemical demand (BOD) (see Chapter 5, Section 6.2). Though time-intensive BOD measurements may still be required, real-time monitoring for process control is only possible with a surrogate measurement, such as TOC or dissolved organic carbon (DOC).

Although TOC, by its definition, is a measure of only carbon, the natural organic matter (NOM) from which TOC is derived also contains other elements. On average TOC compounds are approximately 50% carbon, 35% oxygen, 5% hydrogen, 3% nitrogen, and low levels of phosphorus, sulfur, and metals (Westerhoff, 2006). Of these other elements, nitrogen is especially important as it forms other DBPs with chloramine (e.g., nitrosamines), as mentioned in Chapter 6, Section 7.

In industrial water applications the monitoring of organic matter to minimize it is important in cooling towers, boilers, filtration systems, disinfection and, of course, the effluent of industrial wastewater treatment systems.

This final chapter presents the two groups of instrumentation that measure organic matter. TOC analyzers take the approach of counting every single carbon atom in a sample of water. It does this by combusting all organic compounds down to carbon dioxide. It provides an unequivocal, bottom-line result, but it is slow and expensive. Spectral measurements measure absorption at wavelengths that are typically associated with organic compounds. The analyzers are inexpensive and fast, but their results are open to interpretation and therefore carry some uncertainty.

1.0 FROM BOD TO COD

Chapter 5, Section 6.2 discussed the measurement of BOD as a proxy measurement of the amount of organic matter present in water. BOD measures the amount of organic matter consumed by bacteria through the latter's uptake of dissolved oxygen. The organic matter is considered to be *biodegradable*. It is a very indirect matter of measuring organic matter and requires 5 days to complete a measurement. Why 5 days? It's a remnant from Victorian times and is the time it takes (contaminated) water in the Thames River to flow from London to the sea.

Organic compounds that are not oxidized by bacteria are not measured in a BOD test. COD addresses this shortfall by measuring the amount of organic matter that can be chemically oxidized by a strong oxidant—usually potassium dichromate, $K_2Cr_2O_7$. The reduction is from hexavalent chromium, Cr^{6+}, to trivalent chromium Cr^{3+}. The oxidation converts organic carbon to CO_2. The redox reaction for a generalized organic compound, $C_nH_aO_bN_c$ is:

$$C_nH_aO_bN_c + dCr_2O_7^{2-} + (8d + c)H^+ \rightarrow nCO_2 + \frac{a + 8d - 3c}{2}H_2O + cNH_4 + 2dCr^{3+} \tag{9.1}$$

Essentially, dichromate gives up its oxygen atoms to the carbon atoms in the organic molecules to create CO_2. The amount of dichromate required to oxidize all organic matter will differ from one compound to another. It is therefore impossible to assign an organic matter concentration in units of mg/L or ppm for a mixture of compounds. Instead, the units of COD are the same as those for BOD—mg O_2/liter. Because oxygen is not the oxidant, the actual units are *mg O_2 equivalent/liter*.

U.S. EPA Method 410.4 prescribes the process (O'Dell, 1993). In the *digestion* step an excess of K_2CrO_7 and sulfuric acid are added at 150 °C for 2 hours to a sample until all organic carbon has been oxidized. The sulfuric acid is added to bring the pH below 2 to ensure that there is sufficient acidity to make Equation 9.1 go forward. Silver can be added to catalyze the oxidation. See the method for details of the procedure.

After the 2 hours for oxidation to complete has elapsed, the mixture is left to cool to between 20 °C and 25 °C. There are two methods to determine the amount of dichromate consumed—back titration and colorimetric.

The colorimetric measurement is based on the different wavelengths that the two Cr oxidation states absorb. The Cr^{3+} oxidizing agent, $Cr_2O_7^{2-}$, is yellow and absorbs at 420 nm, whereas Cr^{3+} is blue and absorbs between 600 and 620 nm. A spectrophotometer measures absorbance at 600 nm. To assign a COD value to the absorbance value a calibration curve is necessary.

Constructing a calibration curve for single analytes is straightforward. One prepares several dilutions of the analyte and measures the absorbance for each standard. Because there is no single analyte for COD, a surrogate must be used—potassium hydrogen phthalate (KHP). The calibration typically encompasses the range of 0 to 500 mg O_2/L. At 600 nm this range of KHP COD values corresponds to absorbances of 0 to 0.2 cm^{-1}. If the absorbance range of the actual sample is higher than 0.2 cm^{-1} then a dilution to bring it into this range is necessary.

The titration method is analogous to the back titration of free chlorine discussed in Chapter 6. A reducing agent—usually ferrous ammonium sulfate, $Fe(NH_4)_2(SO_4)_2 \bullet 6H_2O$ or FAS—is added to reduce the excess dichromate. This is the same reducing agent used in the back titration of Wurster dye to DPD discussed in Chapter 6, Section 3.1. To visually identify when all excess dichromate has been reduced to Cr^{3+} *ferroin* is added. Ferroin is a complex of phenyl groups and ferrous (Fe^{2+}) iron and is bright red. At the point at which the FAS has reduced all excess dichromate the ferroin oxidizes and turns from red to blue. Subtracting the excess FAS concentration from the initial concentration of FAS gives the amount of FAS that reacted with the dichromate, and the latter gives the amount of COD. As Equation 9.1 shows, the number of moles of dichromate per mole of organic compound is *d*. Because the organic content in the sample will undoubtedly consist of a wide range of compounds, one can only estimate the average number of C atoms in this range. The unknown method carries with it a level of uncertainty that varies with the distribution of organic compounds.

The BOD measurement reflects the oxidation of organic compounds by oxygen, through the intermediary of bacteria. The COD measurement reflects the oxidation of organic compound by the stronger oxidizing agent, permanganate, and at high temperature. Therefore, the COD value of a sample of water is generally higher than the corresponding BOD one. Dissolved and suspended compounds may also be present that can be oxidized and will lead to a spuriously large COD measurement. To suppress the oxidation of Cl^- ions to Cl_2 mercury is added. Mercury, in the form of $HgSO_4$, reacts readily with chloride to create the precipitate $HgCl_2$.

2.0 TOC—BURN AND MEASURE

Organic matter comes in a nearly infinite range of compounds from simple 5-atom methane (CH_4) to biological compounds containing thousands of atoms. The spectroscopic techniques the remaining chapter discusses provide inexpensive and fast methods of estimating the total amount of organic compounds in solution. But there is one bottom line that is based on one simple fact: all organic compounds are built around carbon atoms. If we can measure the total number of carbon atoms, we can get the next best thing to measuring every conceivable organic molecule.

Counting carbon atoms may seem like an insurmountable problem, but it is doable. Another basic fact about carbon compounds is that they all oxidize to one carbon-based molecule—CO_2. All chemical reactions go "downhill in energy," and, for all organic compounds, carbon dioxide lies at the bottom of the hill.

The TOC analyzer gives a measure of carbon atoms by oxidizing all organic compounds in the water to CO_2 and "counting" CO_2 molecules. The carbon dioxide that forms in water can be purged from the sample such that all of it converts to CO_2 gas. Carbon dioxide gas is very easy to measure—usually through its absorption of infrared light. The absorption of infrared light by carbon dioxide is the topic for the most pressing ecological discussion of our era. It's the basis of the greenhouse effect and the alleged source of global warming.

It is important to point out one set of carbon-based compounds that is not included in TOC. There are some carbon-based compounds of carbonate (CO_3^{2-}) and bicarbonate (HCO_3^-). Together they comprise *total inorganic carbon* (TIC). The TOC analyzer, however, measures all carbon, or *total carbon* (TC). The units of TC, TOC, and TIC are mg/L. The total carbon is obviously the sum of organic and inorganic carbon:

$$TC = TOC + TIC \tag{9.2}$$

All TOC analyzers have some means of separating out TIC from TC to obtain TOC. In most cases there is no need to measure the TIC fraction, so the user uses the following two-step so-called direct method for measuring TOC. One first adds an acid, typically phosphoric acid, to bring the pH below 3, and raises the temperature to approximately 250 °C. The low pH converts carbonate and bicarbonate to CO_2—just as Equation 3.6 predicts. A carrier gas sweeps away the CO_2 due to these two species, and then the real business of breaking down the leftover TOC starts.

To determine the actual TIC present in a sample, one follows the *differential* method, which involves two samples. First one decomposes all carbon compounds using one of the methods described below and determines the TC. Then one adds acid to a second sample and, in the absence of decomposition, measures the CO_2 resulting from CO_3^{2-} and HCO_3^-.

The breakdown of total carbon to total organic and total inorganic carbon doesn't end there. TOC compounds can be further classified as being purgeable (POC) or non-purgeable (NPOC). Purgeable organic carbon compounds are those that have high vapor pressure—that is, they have low boiling points and vaporize rapidly at room temperature. These include *aromatic* compounds—based on benzene—or small chain *aliphatic* compounds—based on chains of carbon atoms attached to each other through single or double bonds. The name *aromatic* is most fitting since compounds that vaporize easily are those that we smell. All are based on the six-carbon ringed benzene molecule.

The carrier gas that carries away CO_2 also carries away the POC compounds before they have a chance to be oxidized to CO_2. Therefore the direct method of measuring TOC actually measures NPOC. If there are purgeable compounds in solution, then one should use the differential method. The direct method works well for surface, ground, or drinking water because POCs are minor constituents in clean water.

Figure 9.1 shows the classification hierarchy of carbon compounds. As the figure shows, NPOC further breaks down into suspended and dissolved matter. The same is true for TIC. The breakdown of NPOC is relevant to the workings of a TOC analyzer only insofar that particulates do not oxidize as readily as dissolved compounds. This may be a problem for persulfate oxidation. *Dissolved organic carbon* (DOC) is of particular importance. To measure it one filters the sample through a 0.45-µm filter. As described in Chapter 7, Section 1 that is the somewhat arbitrary distinction between dissolved and suspended solids.

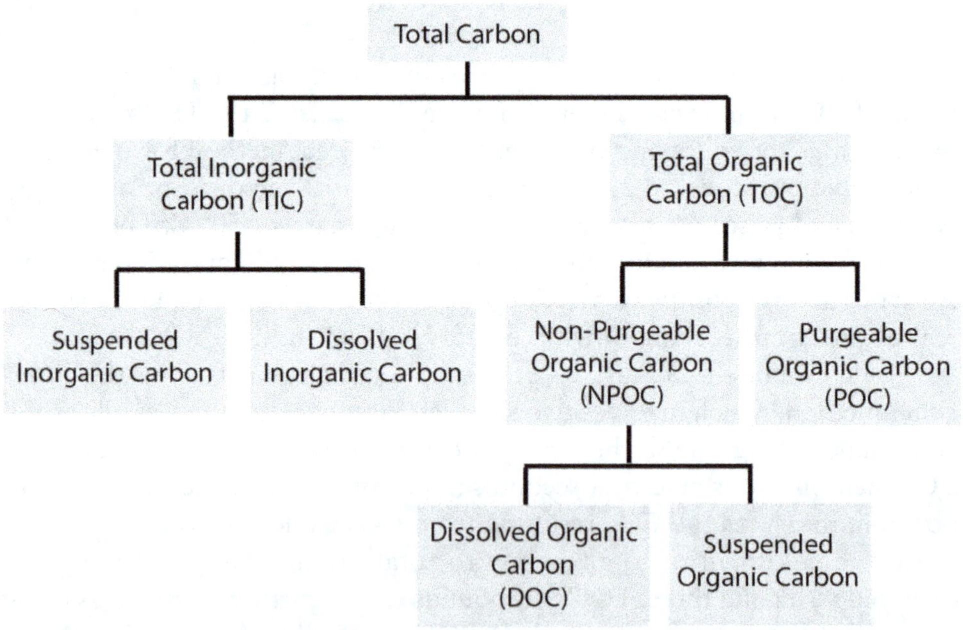

FIGURE 9.1 Classification of carbon containing compounds.

2.1 Making Carbon Dioxide From Carbon Compounds

There are several methods for breaking down TOC compounds and oxidizing the C atoms to carbon dioxide: combustion and chemical oxidation.

The simplest method for oxidizing organic compounds is simply heating the sample solution to several hundred degrees Celsius in the presence of a catalyst and oxygen. The thermal decomposition breaks apart all organic matter the same way the combustion engine in your car reduces nearly all compounds that make up gasoline or diesel to CO_2 and H_2O. The high temperature drives away all water and an inert gas, usually nitrogen, carries away the leftover CO_2 to be detected and measured.

One of the most ubiquitous oxidizing agents in the world is one that is a virtual nonplayer in water. That is the free radical, OH•. (The dot denotes the single, unpaired electron.) The cleansing of the atmosphere of organic pollutants is largely the work of OH• radicals. Chapter 6, Section 3.1 introduced the concept of free radicals in a discussion on the DPD molecule. Though radicals contain at least one unpaired electron they are, nonetheless, electrically neutral. Radicals are driven toward pairing that electron with another chemical specie and are thus highly reactive. The DPD molecule introduced in the chapter on free chorine is a rare example of a stable free radical.

To achieve the difficult task of creating OH radicals in water one starts with sodium persulfate, $Na_2S_2O_8$, and dissolves it in water to form the persulfate anion:

$$Na_2S_2O_8 \rightarrow 2\ Na^+ + S_2O_8^{2-} \tag{9.3}$$

Next one heats it to produce either the OH radical or hydrogen peroxide, H_2O_2.

$$S_2O_8^{2-} + 2\ H_2O \rightarrow 2\ SO_4^{2-} + H^+ + 2\ OH\bullet \tag{9.4}$$

$$S_2O_8^{2-} + 2\ H_2O \rightarrow 2\ SO_4^{2-} + H^+ + H_2O_2 \tag{9.5}$$

Any H_2O_2 that forms oxidizes to OH radicals:

$$2\ H_2O_2 \rightarrow 2\ OH\bullet + O_2 \tag{9.6}$$

This method is the fastest of the three methods, but it is not as effective in oxidizing suspended organic particles or compounds that are not easily oxidized.

2.1.1 UV Oxidation

A third method of breaking down carbon containing compounds is ultraviolet (UV) absorption in the presence of a catalyst. The technique owes its workings through the arrangement of electronic energy levels. Most have excited levels at high enough energies that UV light is required to reach them. Upon absorbing UV light to reach excited electronic states, most organic molecules break apart. With enough UV excitation, all organic compounds can be reduced to single carbon atoms, which can then be oxidized to form CO_2. Analyzers using UV radiation combine this method with a catalyst or the persulfate oxidant to ensure complete breakdown. It is mostly used for samples with low organic content.

2.1.2 Supercritical Water Oxidation

Supercritical water oxidation (SCWO) is a technique that might just as well be called "oxidation on steroids." The supercritical point is a temperature and pressure at which all three phases of a chemical compound exist—gas, liquid, and solid. For water the supercritical point occurs at 375 °C (770 °F) and 320 psi. Remember that, at 1 atmosphere of pressure (14.7 psi) liquid water can't rise above 100 °C. GE's Sievers InnovOx analyzer introduces the oxidant peroxydisulfate ($S_2O_8^{2-}$) to oxidize the organic matter.

2.2 Detection of CO_2

The final determination of the TOC content is to measure the carbon dioxide that forms from the decomposition of all organic compounds and the oxidation of C atoms to CO_2. There are two methods for doing this: non-dispersed infrared absorption and conductivity measurement.

The most common method for measuring gaseous CO_2 concentration is to measure its absorption of infrared light. Figure 9.2 shows the absorption spectrum of carbon dioxide. The units on the horizontal axis are in cm^{-1} and vary inversely with wavelength units. They work well in the infrared region. The main absorption feature at 2400 cm^{-1} is equivalent to 4.2 µm, which is within the near infrared wavelength range.

To measure the CO_2 concentration, the analyzer directs a near infrared light beam through the sample and measures the intensity of the near-infrared (NIR) light after being absorbed by the CO_2 gas. Because the NIR source is not dispersed by a grating or prism, as it is in a spectrometer, it is usually called *non-dispersed infrared*, or NDIR.

The signal from NIR absorption is in real time. To convert it to a TOC concentration the analyzer must integrate over the signal over time. To correct for changes in flow of the carbon dioxide two mass flow sensors at

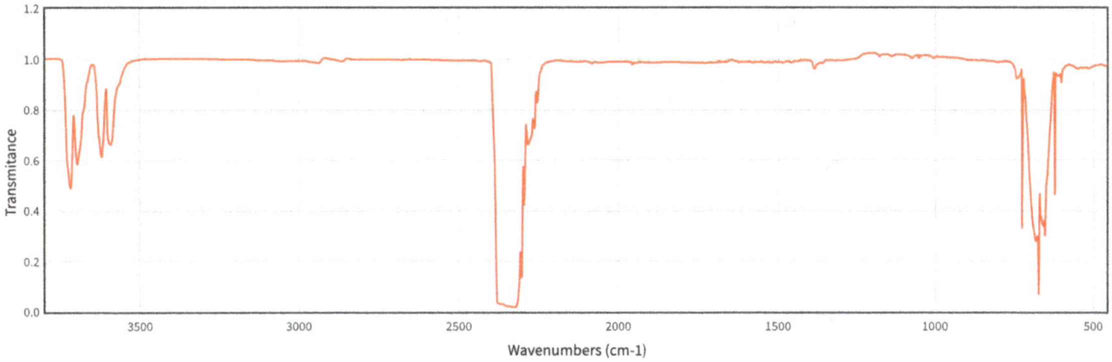

FIGURE 9.2 Absorption spectrum of CO_2. The horizontal x-axis is in units of cm^{-1}. The main absorption peak is at 2400 cm^{-1}, which is equivalent to 0.42 μm in the near infrared (https://webbook.nist.gov/cgi/cbook.cgi?ID=C124389&Units=SI&Type=IR-SPEC&Index=1#IR-SPEC).

the front and the back of the chamber measure the pressure difference and convert that difference to a flow value.

NDIR detection is commonly used for measuring NPOC. This is because the detector signal follows the time evolution of the CO_2. The analyzer can "watch" the passage of CO_2 generated from the dissolved gas and then switch on the oxidation step after the nascent CO_2 has passed. By integrating the signal only over the actual TOC based CO_2 the analyzer makes a background-free measurement.

Conductivity measurement of TOC is simple, inexpensive, and works well at low ranges of TOC levels but is subject to interferences at higher levels. For it to work with minimal interference CO_2 must be the predominant oxidation species. The gas dissolves in the sample and forms a mixture of carbonic acid, bicarbonate (HCO_3), and carbonate (CO_3^{2-}). In the absence of existing ions, the conductivity of the solution is proportional to the concentration of the HCO_3^- and CO_3^{2-} ions. A simple conductivity sensor becomes the detector. Because dissolved salts are certain to be found in the water a measurement before oxidation must be made and subtracted from the measurement following oxidation. The TOC concentration then becomes proportional to the differential conductivity measurement. To reduce the interference from existing ions, some analyzers insert a membrane that allows only CO_2 to pass through into a pure water sample.

The third detection scheme is a variation of NIR absorption. Gases that absorb light gain energy that they release non-radiatively, as described in Chapter 5, Section 4.1 discussion of fluorescence. That non-radiative release is in the form of molecular motion, which, on a macroscopic level, is nothing more than heat. In the Static Pressure Concentration—Non-Dispersive Infrared Radiation (SPC-NDIR) detector, CO_2 gas flows into a chamber and an NDIR beam at 4.3 μm is focused into the chamber. Unlike the NDIR detector there is no exit port, so the pressure of CO_2 gas builds up. As it does the CO_2 absorbs the NIR light and heats up. A thermopile—a very sensitive temperature element—measures the temperature increase. After the chamber is evacuated a second beam at 3.9 μm—a wavelength that is not absorbed by CO_2—is focused by the same set of optics. The thermopile records a lower temperature. The difference in temperature is proportional to the amount of CO_2.

2.3 Calibration

The final operation of the TOC analyzer is to convert the absorption of NIR light by CO_2 into the TOC concentration in the sample. The strength of the absorption depends on the rate of decomposition, the flow-rate of CO_2 gas entrained in the carrier gas, and the fraction of gas that the IR beam irradiates. Though one can, in principle, measure all these parameters, it would be difficult and subject to error.

It's a lot easier to measure the analyzer response to a series of calibration standards. U.S. EPA Method 415.3 (Potter & Wimsatt, 2009) specifies calibration using potassium hydrogen phthalate, $C_8H_5O_4K$, or KHP, as the organic carbon surrogate. The method calls for four specified dilutions plus a blank for a total of 5 calibration points. Figure 9.3 shows the results of calibrating the OI Analytical 9210p analyzer.

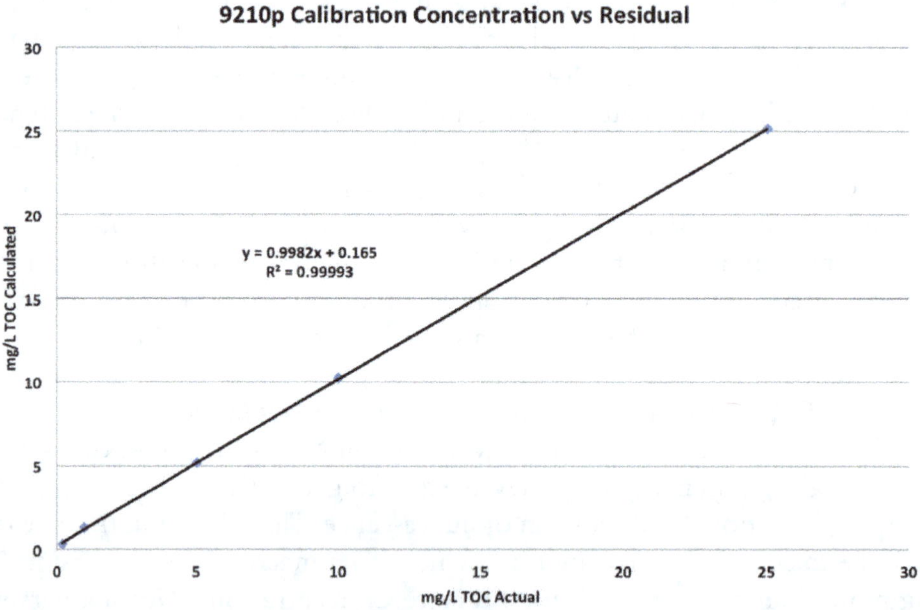

FIGURE 9.3 Calibration of the OI Analytical 9210p analyzer using KHP and the 5 KHP solutions specified by U.S. EPA Method 415.3. This analyzer uses persulfate oxidation (OI Analytical).

TOC measurement by breakdown of organic matter to carbon dioxide remains the gold standard of measuring organic matter. There are two common issues that affect the accuracy of a TOC measurement. The first is overloading the sample chamber such that less than 100% of the organic matter oxidizes to CO_2. The result is a TOC measurement that is low and deposition of organic residue in the combustion chamber and detector. It will also supply organic matter to subsequent measurements and cause an artificially high TOC value. This artifact is called *carryover*.

Table 9.1 summarizes TOC methods according to the conversion of organic compounds to carbon dioxide (HIS Engineering 360).

2.4 Surrogate Measurements With UV Absorption

A TOC analyzer is a very sophisticated analyzer with a price tag to match. It is also not, by nature, a continuous measurement, though it can be automated for process control. Fortunately, there is a surrogate measurement that can yield a good estimate of TOC or its cousin—NOM (natural organic matter).

TABLE 9.1 Summary of methods for measuring CO_2 in TOC analyzers (IHS Engineering 360).

Applications	TOC Method	Comments
High purity water	UV or UV/ persulfate	Samples with less than 50 ppb TOC. UV or UV/persulfate provides the most trouble-free operation and the ease of detecting very low-level TOC.
Water for injection	UV or UV/ persulfate	Samples with less than 500 ppb TOC. Combustion technology can measure in this range, but UV or UV/persulfate is still the easier method.
Drinking or source water	UV/persulfate or combustion	For meeting precision and accuracy requirements, UV/persulfate is the better technology. Combustion should be used if capturing the particulated organic carbon as part of the TOC value is important.
Salty water	Combustion	Combustion technology is not affected by chlorides present in seawater and other salty matrixes.
Industrial waste effluent	Combustion	Combustion technology was designed to handle particulates and chlorides, and tough to oxidize compounds typical of industrial waste.

Most organic compounds have strong absorption in the UV. To be more specific aromatic organic compounds absorb strongly in the UV. Because particulates scatter more light than they absorb, the TOC that can be measured via UV absorption is more appropriately termed DOC—dissolved organic carbon. Fortunately, DOC is the last of the 3-letter acronyms we'll meet that give some measure of the amount of organic matter. It joins BOD, COD, NOM, and TOC. To ensure that suspended particles don't influence the absorption measurement, one filters the sample through a 0.45 μm filter.

Mercury lamps have been around for about a century and output light from excited Hg atoms mostly at 254 nm. Dissolved organic carbon conveniently, especially aromatic ones, absorb UV light at this wavelength. Let's be clear: it isn't the case that all aromatic organic compounds absorb light at the magical wavelength of 254 nm. Most absorb over a range of wavelengths, and 254 nm falls within that range. UV sterilizers have used mercury lamps to disinfect everything from water to lab equipment.

UV absorption at 254 nm by aromatic DOC is a very inexpensive way to give an estimate of TOC. Although it measures the aromatic content of DOC and may be influenced by the presence of other UV absorbing species, for many applications, it's a good surrogate for TOC measurement. Furthermore, aromatic compounds are more reactive than aliphatic compounds (consisting of chains of C atoms), so the amount of aromatic carbon represents the amount of organic matter that is likely to react with free chlorine and form DBPs.

The absorption of 254 nm has become a parameter on its own—*UV Absorbance* (UVA). To derive units of absorption that describe its magnitude, we pick up the discussion in Chapter 7, Section 7.4.

We define the absorbance (A) as the attenuation of light as quantified in Equation 7.4, which I repeat here:

$$\frac{I(\lambda)}{I_0(\lambda)} = e^{-A(\lambda)} = e^{-k(\lambda)l} = e^{-\varepsilon(\lambda)cl} \tag{9.7}$$

Recall that the wavelength (λ) in parentheses reminds us that the absorbance, absorption coefficient (k), and molar extinction coefficient (ε) are all functions of the wavelength—in this case 254 nm. The absorption coefficient at 254 nm is also called the *spectral absorption coefficient* (SAC) or UV_{254}. As the second equality in Equation 9.7 states, UVA (or SAC or k_{254}) is measured by dividing the fractional absorbance (A) by the path length (l) of the cell:

$$k(254\ nm) = \frac{A}{l} = UVA = SAC \tag{9.8}$$

To measure the absorbance, A, requires a blank and simple math. We can't measure absorbance directly, but we can measure transmission—the amount of light that passed through a sample cell and is measured by a photodetector (I/I_0). Because the sample cell and, to a lesser extent, water itself, absorb and scatter light, we make a measurement of the transmission of light through a sample cell containing pure water. This gives us the transmitted light intensity in the absence of sample—I_0. We next switch this blank cell with an actual cell containing the sample and measure the new transmitted intensity value—I. The ratio of the sample to the ratio gives us the transmission as a fraction, I/I_0. The light intensity measurement units don't matter because they cancel out in this fraction. At 254 nm, the fractional transmission is called the *Ultraviolet Transmission*, or UVT.

The Beer-Lambert law allows us to convert UVT to A. We take the logarithm of Equation 9.7 to give A in terms of I/I_0.

$$A(254\ nm) = -\ln\left[\frac{I(254\ nm)}{I_0(254\ nm)}\right] = -\ln(UVT) = k_{254}l = c\varepsilon_{254\ nm}l \quad (9.9)$$

For some applications, mainly UV disinfection, UVT measurement is the critical parameter. It is important that the UVT value exceeds a minimum threshold to ensure that the UV radiation is strong enough throughout the entire sample volume to do its job of killing pathogens.

The last two equalities in Equation 9.9 is a reminder of the relationship between the transmission (UVT), absorption coefficient (k_{254}), and molar extinction coefficient ε_{254}. Don't worry about the negative sign in front of ln(UVT). The logarithm of a fraction less than one is always negative so that negative sign will cancel the one in the equation. It should be obvious why we separate the absorption (A) from the absorption coefficient (k). A is unitless, and k has the units of cm^{-1}. The latter is divorced from the depth of the absorption medium. Let's say we have two sample cells: One is 1 cm deep and the other is 100 cm deep but has a concentration that is 1% of the 1 cm cell. Both will have the same A value but their k values will be different by a factor of 100. The exponential relationship between transmittance and absorbance means that, for a pathlength of 1 cm, the transmission of light decreases by $1/e$, or 1/2.718.

To check the accuracy of the spectrophotometer, one prepares a KHP (potassium phthalate) *spectrophotometer check solution* at a measured concentration that gives roughly the same absorbance as the sample.

The measurement of UVA is not just a more convenient alternative to TOC methods. In some cases, it can be a more reproducible measurement. Take, for example, the measurement of total trihalomethanes potential

(TTHMFP) in surface water to be used for drinking water. TTHMFP is determined by chlorinating a sample volume of water and measuring the concentration of THMs. The two plots in Figure 9.4 are of measurements taken from six Rocky Mountain watersheds: three rivers, a reservoir, a canal, and a tunnel. A total of six different techniques were used—TOC, UVA, and four fluorescence-based measurements—to compare with TTHMFP measurements. The two plots reproduced in Figure 9.4 show the plots for TOC versus TTHMFP and UVA versus TTHMFP (Beggs et al., 2013).

A comparison of the two shows that UVA correlates more strongly with TTHMFP than TOC. The basis for this strong agreement between UVA and organic matter in these surface water measurements are due to the predominance of aromatic compounds. Aromatic compounds dominate the chemical precursors to DBPs.

Of the aromatic compounds found in raw surface water sources for drinking water, the largest class is that of humic compounds. These are large molecules consisting of a large number or aromatic rings as illustrated in Figure 9.5 showing one humic compound (Stevenson, 1982).

Note that this large aromatic molecule shows an extended pattern of alternating carbon-carbon double bonds and single bonds. The consequence of this pattern is that the electrons that comprise the bonds become *delocalized*—that is, they spread themselves over the network of bonds. By doing so they lower their energy. These so-called *conjugated* electrons absorb lower energy light in the visible rather than the UV range. The same molecular structure of the compound carotene is what gives carrots their orange color—they absorb blue wavelengths. Humic compounds are similarly colored and give surface waters their characteristic reddish-brown color.

Measuring the color—that is, the absorption of visible wavelengths of light—is another way of estimating the concentration of aromatic organic

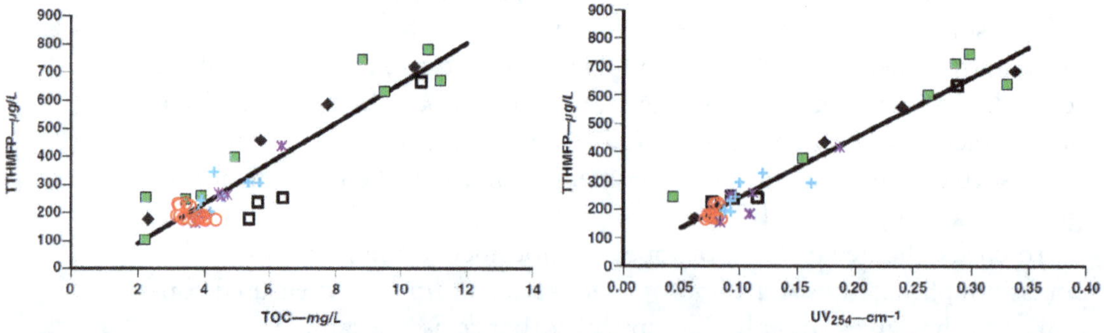

FIGURE 9.4 Correlation between TTHMFP and TOC and between TTHMFP and UVA. The plot of UVA (UV_{254}) shows a better correlation than the plot of TOC (Beggs et al., 2013).

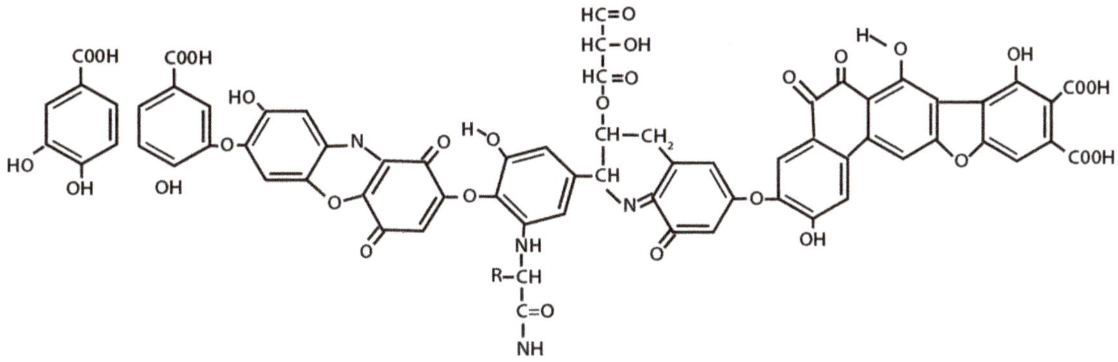

FIGURE 9.5 Molecular structure of humic acid. The multiple aromatic rings (denoted by double lines) imbue a high degree of aromaticity. The network of alternating double and single C-C bonds results in lower electronic energy levels and absorption of visible wavelengths (Structure based on Stevenson, 1982).

compounds. As explained in Chapter 7, Section 3, color is a very subjective parameter. Nonetheless the *platinum-cobalt color scale* was developed to quantify the intensity of color. The full range of this scale consists of 1000 platinum-cobalt units (PCUs). This full-scale measurement is the intensity of a 500-ppm platinum cobalt solution, which is yellow. Clear lake water is less than 10 PCU. Lakes with good visibility have color between 20 and 50 PCU. Water that is visibly colored can have color as high as 500 PCUs.

Figure 9.6 shows the reduction of color, TOC, COD, and BOD during treatment of pulp and paper effluent (Sonkar et al., 2019). The correlation between color and the four organic parameters is clear.

A similar study color of raw water from the Hillsborough River in Tampa, Florida (Kornegay et al., 2000) showed a very strong correlation between TOC and color and demonstrated that measuring color can indeed be a surrogate measurement to TOC when humic and large aromatic compounds predominate.

Combining TOC with UVA measurements provides the basis for a third measurement of organic matter. It is the basis of U.S. EPA Method 415.3 (Potter & Wimsatt, 2009). As with pure TOC measurements, this method calls for first filtering the water sample through a 0.45-µm filter. To estimate the fraction of organic matter that is composed of reactive, aromatic compounds, one divides UVA by DOC to determine the fraction of organic compounds that is aromatic. That ratio is called *Specific UV Absorbance*, or SUVA.

$$SUVA = UVA / DOC \tag{9.10}$$

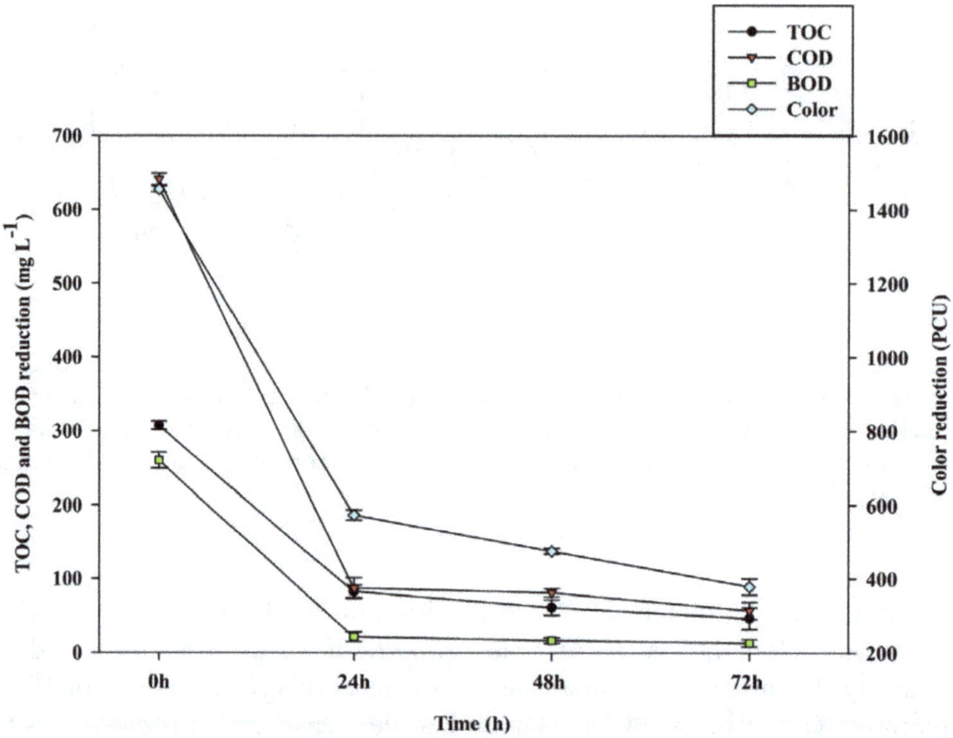

FIGURE 9.6 Decrease of TOC, BOD, and COD and color during treatment of pulp and paper waste (Sonkar et al., 2019).

As expressed in Equation 9.10, DOC is in units of mg/L, and UVA is in units of cm^{-1}. However, to make the units consistent, centimeters are converted to meters. This requires multiplying the ratio by 100:

$$SUVA\left(\frac{L}{mg-m}\right)=\frac{UVA(cm^{-1})}{DOC\left(\frac{mg}{L}\right)}\times 100\left(\frac{cm}{m}\right) \tag{9.11}$$

The generation of the parameter SUVA from the ratio of UVA to DOC explains why U.S. EPA Method 415.3 encompasses both techniques from one water sample. Figure 9.7 shows the measurements of UVA and TOC for raw water used by the Harwood's Mill water treatment facility in Newport News, Virginia. If one accepts the approximation that TOC is composed mostly of DOC—that is, the amount of suspended organic matter entering the facility is dissolved—then dividing the UVA values by TOC values gives SUVA values. If Figure 9.4 had plotted TOC versus UVA, instead of the other way around, one could simply take the slope of the line through the data points as the average SUVA. The dotted line is the slope of the 1/SUVA line for small values of UVA up to point A and under the assumption that the line should pass through zero. The presence of a finite amount of TOC accompanied by zero UV is due either to the experimental uncertainty of the measurements or carbon compounds that absorb at wavelengths below 254 nm.

Figure 9.7 shows the excellent correlation between the percentage of aromatic compounds in organic carbon versus SUVA in samples of surface water collected from 16 states (Weishaar et al., 2003). The percentage of aromaticity was determined by ^{13}C nuclear magnetic resonance.

Though SUVA measurements are an indicator of the degree of aromaticity in dissolved organic compounds, they are influenced by other parameters such as pH, nitrate, and iron (Weishar et al., 2003).

Even when water is not dominated by aromatic compounds UVA measurements serve as surrogate measurements for TOC. It is instantaneous and inexpensive and can be used as a real-time monitor of organic matter.

The main drawback of estimating DOC from UV measurements is the interference of non-carbon compounds that also absorb in the UV region. These include the commonly found ions of iron, nitrite, nitrate, and bromide (Eaton et al., 1995).

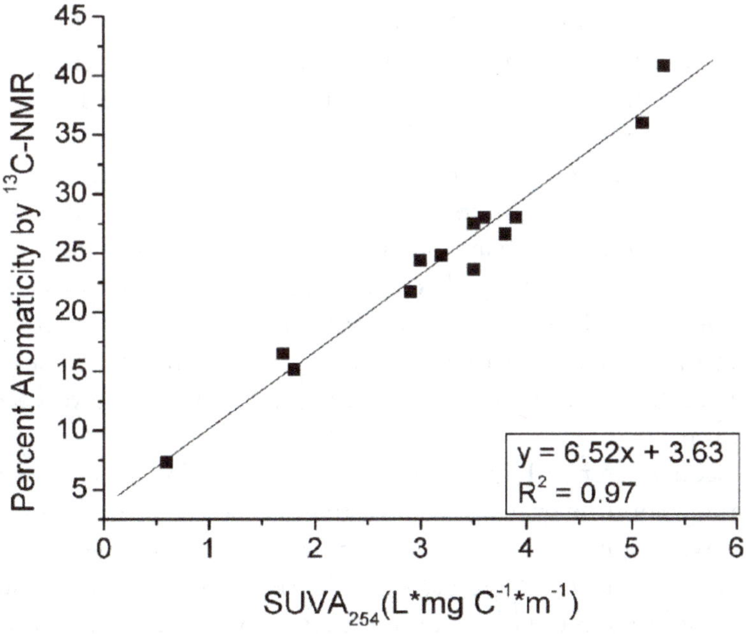

FIGURE 9.7 Correlation between TOC and SUVA (specific UV absorbance). This data is from samples of stream rivers in 16 states. The aromatic concent was determined ferom ^{13}NMR measurements (Weishaar et al., 2003).

3.0 FROM 254 NM TO BEYOND—SPECTROSCOPY

3.1 The Spectrometer

So far this discussion of light absorption to measure DOC has been limited to measurement to one specific wavelength—254 nm. The convenience of using an inexpensive light source makes a compelling case for monitoring water at 254 nm as a surrogate for TOC measurements—at least for continuous monitoring needs. To be sure, one still needs to calibrate a U_{254} sensor with known DOC concentrations obtained from a TOC analyzer.

To take the next logical step in measuring DOC through spectral absorption, one might record the absorption of water over a range of wavelengths. After all, the absorption signature of a chemical is spread over a range of wavelengths—not just one. The pattern of absorption and emission is unique for every chemical species and can be used to identify the identity and concentration of that species. The absorption of light by wavelength by an element, compound, or ionic species is an *absorption spectrum.*

Before proceeding further, it is important to distinguish between the spectral absorbance of chemical species of interest and the spectral absorbance of reagents that react with and correlate with the species. The colorimetric methods for measuring chlorine determine free chlorine concentrations by reacting the free chlorine species with a dye—usually DPD. The reaction creates a magenta compound that absorbs light at 530 nm. One can assess the intensity of the magenta color either by eye or with a spectrophotometer. Measuring the direct absorption of the species itself, instead of the colored reagent, has the potential of revealing much more information.

Recording an absorption spectrum requires a *spectrometer.* Figure 9.8 shows how it works. A *dispersive element* separates light according to its wavelength. Some use a prism but most use a *grating*—a piece of glass or metal that is etched with very closely spaced grooves. The third component in a spectrometer is the *detector*—typically either a CCD or CMOS array. In older spectrometers, the diffraction grating rotates, and as it does, it brings one wavelength into view at a time. Thus, it requires time to capture just one spectrum.

The spectrometer in Figure 9.8 records *absorption spectra* using single detector and a rotating grating. The width of the aperture defines the narrow range of wavelengths collected at each sample point. It defines the resolution of the spectrometer and is typically on the order of 0.1 nm. If one were to replace the light source with a sample that emits light (as a result of being heated or absorbing light itself), the modified setup would record an emission spectrum. *Optical emission spectrometry* turns dissolved analytes into gaseous analytes by evaporating the dissolved chemicals into a chamber that

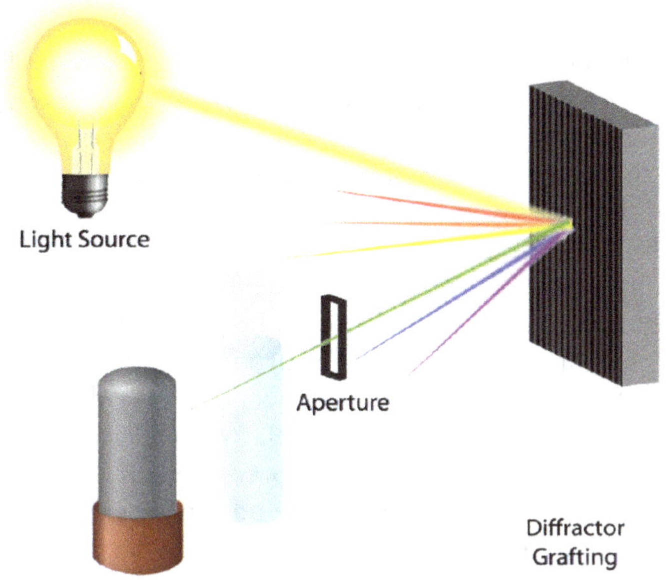

FIGURE 9.8 Schematic of a spectrometer. The diffraction grating disperses the light, and the aperture selects a narrow band of wavelengths. Most field spectrometers dispense with the aperture and image the entire spectrum on a detector array.

is heated up to a point where the chemical species emit light. Another example of emission is the fluorescence of the ruthenium complex that underlies the basis of operation of the optical dissolved oxygen probe discussed in Chapter 5, Section 4.2. The complex absorbs blue light at around 455 nm and fluoresces at the longer 613 nm. Oxygen quenches the fluorescence and the amount of the quenching correlates to the concentration of dissolved oxygen. For the purposes of DOC measurement, the spectrometer operates in absorption mode in which a UV light source passes through the aqueous sample the transmitted light is recorded as a function of wavelength.

Although grating spectrometers offer the highest possible wavelength resolution, most spectrometers relinquish price, complexity, and resolution for cost, compactness, and speed. They record the entire spectrum in one exposure by replacing the single element detector with an array of detectors (aka *pixels*). Instead of the width of the slit determining the resolution the width of each detector does. The silicon photodiode is the most used detector as it is responsive to the visible and UV range—from 1800 to 1200 nm. There are several variations of the Si photodiode that mirror the choices available for imaging sensors. This includes mainly CCD and CMOS-based arrays.

3.2 Spectroscopy of Water

Absorption and emission spectra are common diagnostic tools in the analysis of gaseous species but not so much for aqueous ones. Interactions between compounds and ions result in broad, featureless peaks and valleys, as opposed to the sharp features of the spectra of gases. Figure 9.9 shows a spectrum of the atmosphere. The sun is the light source. The features that almost look like noise represent the absorption by gaseous species such as water vapor and carbon dioxide. These many lines are characteristic of gaseous spectra. Figure 9.10 shows the absorption spectra of a sample of water from the red to the infrared. The absorption features are much more amorphous, making it more difficult to identify individual compounds.

The transmission of light at a particular wavelength, λ, is related to the sample's absorption by the form of the Beer-Lambert law in Equation 9.9.

A range of wavelengths centered around λ passes through the aperture and the sample but, for the sake of simplicity, one can treat the light being absorbed as a single wavelength in the center of the range. A detector behind the aperture measures the light intensity, $I(\lambda)$. $I_0(\lambda)$ is the intensity of light that passes through the sample cell in the absence of a sample. This measurement, often called a *reference measurement*, compensates for changes in source light intensity and transmission changes through the sample cell caused by fogging, smudges, biofilm accumulation, and so on. Remember that absorption (or transmission) is a function of the wavelength. Plotting

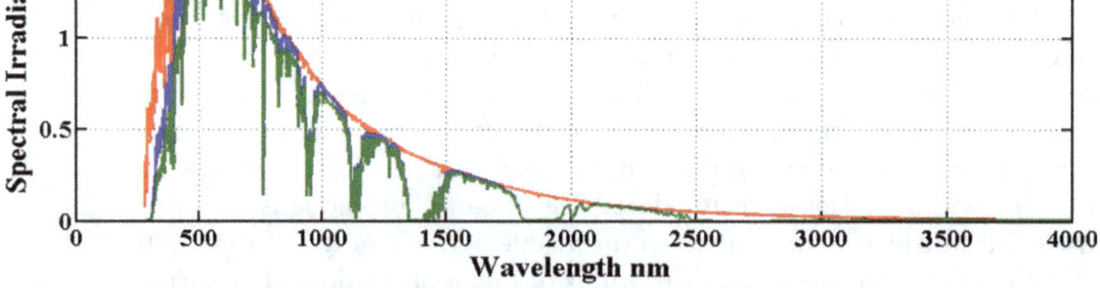

FIGURE 9.9 Absorption spectrum of Earth's atmosphere from the near UV to the near IR. The sharp features reveal the existence of chemical species (Mirghani, 2021).

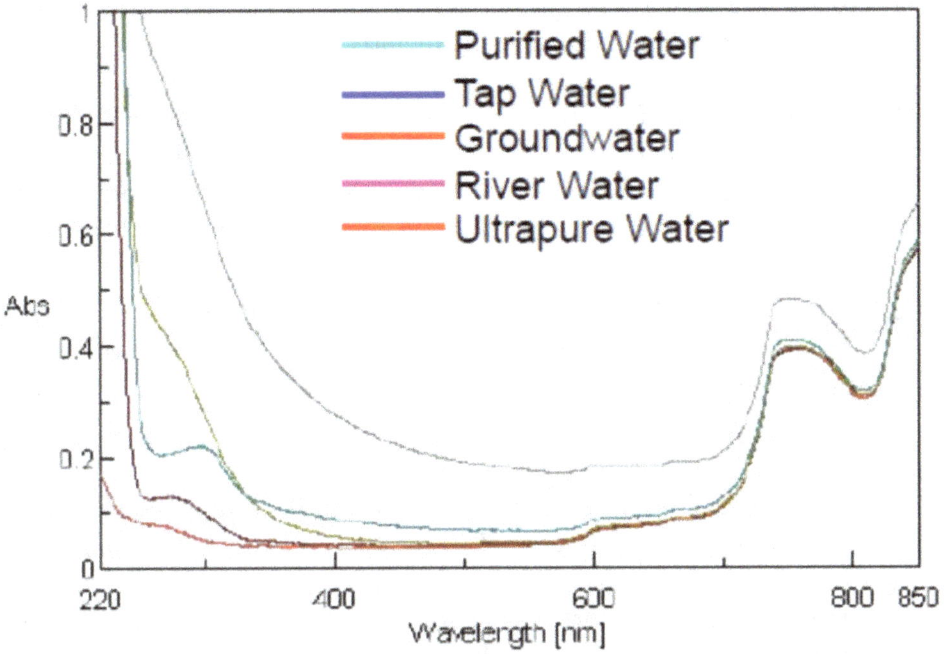

FIGURE 9.10 UV-visible absorption spectrum of water. The difference between the pure water spectrum and others is due to dissolved compounds (Courtesy Jasco).

any of the absorption parameters (A, k, or ε) against the wavelength gives an absorption spectrum.

Spectral features in the visible to UV regions of an absorption arise from the stimulation of electrons from one electronic energy level to a higher one. Water absorbs very weakly in the visible region. After all it's colorless. Because of the separation of electronic energy levels, water absorbs light deep in the UV as well as the infrared.

Figure 9.10 shows an absorption spectrum of water from ultrapure to river water and in the ultraviolet and visible wavelength range. The presence of dissolved substances in the water is most apparent in the wavelength range between 200 and 400 nm. Not many species absorb in the visible range, 400 to 650 nm. If they did, they would appear colored. This leaves the region between 200 and 400 nm as the most optimal for measuring dissolved compounds, including organic matter.

Due partly to the interactions between the highly polar water molecules and the electrons of the analyte molecule or ion, the absorption bands are very broad and featureless. This means that absorption bands are more likely to look like distortions of water's background spectrum instead of clearly differentiated bands. Moreover, bands belonging to different species but close together in wavelength can merge together and be difficult to untangle.

There are three special categories of chemicals of interest to water quality that can be determined via a spectrometer:

1. Organic compounds. The impetus for this section is to extend the measurement of DOC at a single wavelength, 254 nm, to a range of wavelengths encompassing this wavelength. Organic compounds, especially those with a high degree of aromaticity, absorb between 180 and 350 nm.
2. Nitrite-nitrate. Nitrate has a strong absorption band centered at 210 nm and a weaker one centered at 280 nm. Nitrite also absorbs at the former wavelength and is difficult to separate out from nitrate. Some sensor manufacturers refer to the absorption band at 210 nm as belonging to nitrate plus nitrite.
3. Color. As discussed in Chapter 2, Section 4 color is considered a highly subjective parameter. However, it can be quantified using the Platinum-Cobalt scale. ASTM standard D1209-00, also known as the American Public Health Association color, specifies color by stipulating absorbances four wavelengths as shown in Table 9.2 (ASTM International, 2019).

   By comparing the actual absorption to dilutions of the stock solution, one can assign a color to a sample.
4. Turbidity. The dependence of the scattering is not critically dependent on wavelength. The incandescent lights source found in most U.S. turbidimeters cover the entire visible light region while the IR light source found in most European turbidimeters covers a 30-nm span of wavelengths centered around 860 nm. Turbidity is a measure of the light scattered by suspended particles. However, one can view it as an additional loss mechanism that, when added to absorption, accounts for the transmission of light through the sample. By subtracting absorption from the transmitted light spectrum, the analyzer can determine the light loss from scattering and, hence, the turbidity.

TABLE 9.2 Absorbances for Platinum-Cobalt 500-unit stock solution (ASTM International, 2019).

Wavelength (nm)	Absorbance (m^{-1})
430	0.110 to 0.120
455	0.130 to 0.145
480	0.105 to 0.120
510	0.055 to 0.065

3.3 The Spectral Analyzer

Figure 9.11 shows the schematic of a probe-based spectral analyzer. The design is analogous to a probe-based turbidimeter. The entire optical train, including the UV light source, collection optics, sample compartment, dispersive element, and detector array are contained in the probe. The sample compartment is not a container but simply a notch cut into the probe into which windows on each opposing face pass a collimated beam.

An alternate design is based on a side stream measurement. This design is analogous to a surface mounted turbidimeter. Tubing withdraws the process and plumbs it into a quartz sample cell. The design does require a delivery system but relaxes the constraints of fitting a complex optical train into the constrained volume of a probe.

3.4 Chemometrics to the Rescue

Figure 9.12 shows the characteristic absorption features for DOC, nitrate, color, and turbidity. The figure uses the parameter COD as a surrogate for DOC. This is a consequence of calibration done using a COD laboratory measurement as described below. One could just as easily use another surrogate for DOC, such as BOD or TOC, if the independent calibration technique measures that parameter.

To understand why the calibration methods discussed in earlier sections don't work look at this figure. The absorption spectrum of these chemical species superimposed on the absorption spectrum of water in this UV-visible wavelength range is a very broad line. The strong absorption of

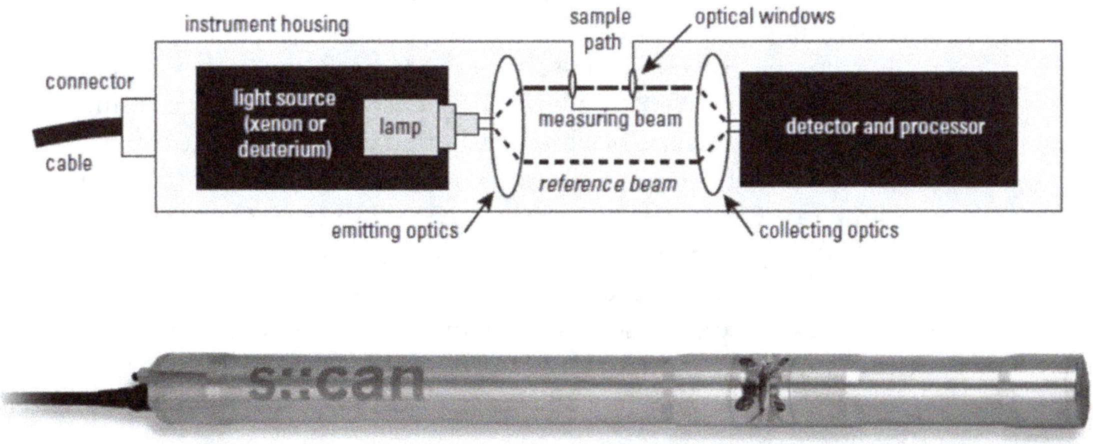

FIGURE 9.11 Schematic of a spectral analyzer in which the light source and detector are both contained in the probe. The sample cell is actually a notch in the probe through which the UV source passes. The actual probe is below. Not shown is the controller (Courtesy S::CAN).

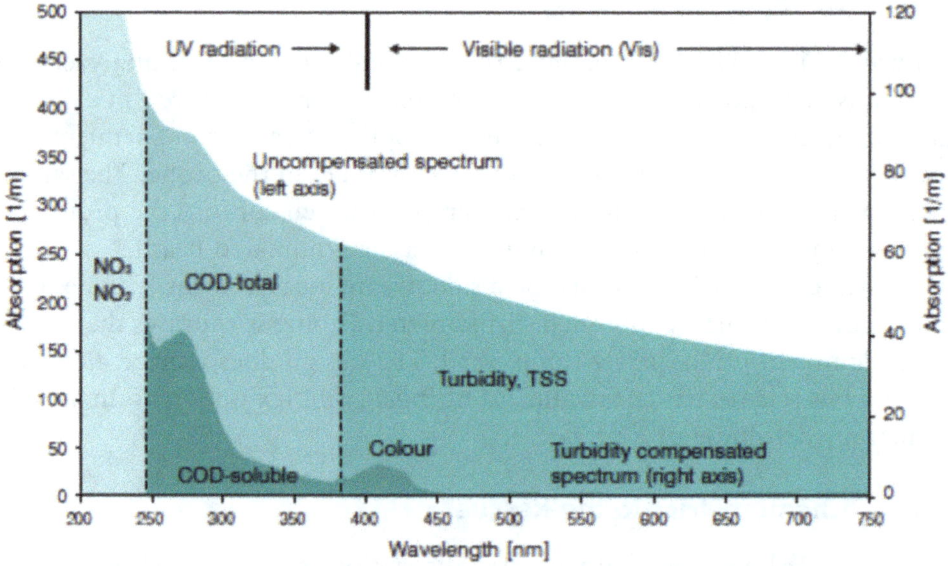

FIGURE 9.12 UV-visible spectrum of water showing the wavelength ranges and absorption features of the major analytes: Nitrate, COD, and turbidity (Courtesy S::CAN).

nitrate overlaps the DOC absorption between 200 and 230 nm. Bromide at 65 mg/L swamps both DOC and nitrate features between 200 and 220 nm.

Traditional spectral analysis assumes that a spectrum of several constituents is the sum of the spectra of the individual consituents, is virtually worthless for such featureless spectra as these two. Furthermore, the Beer-Lambert law, Equation 7.4, increasingly breaks down when species concentrations increase. Until the late nineties, this kind of spectrum would have been impossible to analyze.

Fortunately, spectral analysis methods based on statistics have made possible the extraction of species concentrations from very amorphous spectra. There are several techniques but all come under the umbrella term *chemometrics*, also termed *multivariate analysis*. The most used methods are: Principal Components Analysis (PCA), Principal Components Regression (PCR), and Partial Least Squares Regression (PLS).

The mathematics needed to explain them are well beyond the scope of this book. A comparison of some of these methods can be found in the review article by Lepot et al. (2016). The concept behind PCA is simple to understand from a mathematical standpoint though harder to interpret. The procedure is as follows:

One takes a series of spectra from samples differing in their concentrations of species. The spectra can be thought of as functions with the independent variables being the sample number and the wavelength of light and the dependent variable being the absorption at every wavelength. Then one teases out, mathematically, a new set of independent variables that are independent of each other and that can be used to reconstruct the original spectra. One then comes up with a large number of new independent variable but, ideally, one wants the minimum number that will fit the original spectra with an error (*residual error*) that is manageably small. We would expect the number of these new variables to be equal to the number of unique constitutents in the water sample. Indeed, this is approximately true. The caveat is that DOC is not one species but a group of organic compounds that can vary widely from sample to sample. The same is true for other parameters, such total suspended solids (TSS) or turbidity or color.

Calibration for every unique sampling site enables one to define the constituents, or parameters, of DOC (COD), TSS (turbidity), and color and to assign an identity and concentration value for every constituent—our original goal. To do so one simply takes a series of samples, differing in concentration or magnitude, and independently measures the concentration/magnitude via certified laboratory methods. From this set of independent measurements, called a *training set*, one can then construct the chemometric model with which one can measure unknown samples from the same site.

To increase the precision of spectral analysis some researchers have extended the absorption technique to measure the fluorescence of organic compounds over a range of excitation wavelengths. This creates a three-dimensional spectrum versus the two-dimensional spectrum discussed above and can identify a greater number of organic compounds (Murphy et al., 2014). The complexity of this technique and the associated equipment is likely to relegate this technique to the laboratory for the foreseeable future. However, in an era in which data analytics and machine learning are a part of our vernacular, it's clear that spectral analysis will grow in importance for measuring organic carbon in every type of environment from surface water to wastewater.

4.0 REFERENCES

ASTM International. (2019). Standard test method for color of clear liquids (platinum-cobalt scale). *Book of standards volume*: 06.04 D1209-05.

Beggs, K. M. H., Billica, J. A., Korak, J. A., Rosario-Ortiz, F. L., McKnight, D. M., & Summers, R. S. (2013). Spectral evaluation of watershed DOM and DBP precursors. *Journal of AWWA, 105*(4), E173–E188.

Eaton, A. D., Clesceri, L. S., & Greenberg, A. E. (1995). Standard method 5910B: Ultraviolet absorption method. *Standard methods for the examination of water and wastewater* (19th ed.). American Public Health Association.

Kornegay, B. H., Kornegay, K. J., & Torres, E. (2000). *Natural organic matter in drinking water: Recommendations to water utilities*. AWWA Research Foundation.

Lepot, M., Torres A., Hofer T., Caradot N., Gruber G., Aubin J. B., & Bertrand-Krajewski J. L. (2016). Calibration of UV/Vis spectrophotometers: A review and comparison of different methods to estimate TSS and total and dissolved COD concentrations in sewers, WWTPs and rivers. *Water Research, 101*, 519–534.

Mirghani, M. (2021). Spectral irradiance (https://www.mathworks.com/matlabcentral/fileexchange/52938-spectral-irradiance), MATLAB Central File Exchange. Retrieved August 17, 2021.

Murphy, K., Bro, R., & Stedman, C.A. (2014). *Chemometric analysis of organic matter fluorescence,* Chapter 10, 339–375.

O'Dell, J. W. (Ed.). (1993). *Method 410.4, Revision 2.0: The determination of chemical oxygen demand by semi-automated colorimetry*. U.S. EPA.

Pellerin, B. A., Bergamaschi, B. A., Downing, B. D., Saraceno, J. F., Garret, J. D., & Olsen, L. D. (2013). Optical techniques for the determination of nitrate in environmental waters: Guidelines for instrument selection, operation, deployment, maintenance, quality assurance, and data reporting, water quality, collection of water data by direct measurement (Chapter 5, Book 1, Section D of *Collection of water data by direct measurement*).

Potter, B. B., & Wimsatt, J. C. (Eds.). (2009). *U.S. EPA Method 415.3, Rev 1.2—Determination of total organic carbon and specific UV absorbance at 254 nm in source water and drinking water.* U.S. Environmental Protection Agency.

Rieger, L., Langergraber, G., Thomann, M., Fleischmann, N., & Siegrist, H. (2004). Spectral in-situ analysis of NO2, NO3, COD, DOC and TSS in the effluent of a WWTP. *Water Science and Technology, 50*(11), 143–152.

Sonkar, M., Kumar, M., Dutt, D., & Kumar, V. (2019). Treatment of pulp and paper mill effluent by a novel bacterium Bacillus sp. IITRDVM-5 through a sequential batch process. *Bioctalysis and Agricultural Biotechnology, 20*, 1–10.

Stevenson, F. J. (1982). *Humus chemistry; genesis, composition, reactions.* New York: Wiley-Interscience.

U.S. Environmental Protection Agency. (2014). Stage 1 and stage 2 disinfectants and disinfection byproducts rules.

Weishaar, J. L., Aiken, G. R., Bergamaschi, B. A., Fram, M. S., Fujii, R., & Mopper, K. (2003). Evaluation of specific ultraviolet absorbance as an indicator of the chemical composition and reactivity of dissolved organic carbon. *Environmental Science Technology, 37*(20), 4702–4708.

Westerhoff, P. (2006). Chemistry and treatment of disinfection byproducts in drinking water. *Southwest Hydrology*, Nov–Dec, 20–23.

Index

www.ingramcontent.com/pod-product-compliance
Lightning Source LLC
La Vergne TN
LVHW060556240526
839643LV00003B/24